SHADOWS OF THE MIND

Roger Penrose is the Rouse Ball Professor of Mathematics at the University of Oxford. He has received a number of prizes and awards, including the 1988 Wolf Prize for physics which he shared with Stephen Hawking for their joint contribution to our understanding of the universe.

BY ROGER PENROSE

The Emperor's New Mind:
Concerning Computers, Minds,
And The Laws Of Physics

Shadows Of The Mind:
A Search For The Missing Science
Of Consciousness

Roger Penrose

SHADOWS OF THE MIND

A Search for the Missing Science of Consciousness

Published by Vintage 1995

2 4 6 8 10 9 7 5 3 1

First published in Great Britain by Oxford University Press, 1994

Vintage
Random House, 20 Vauxhall Bridge Road, London SW1V 2SA

Random House Australia (Pty) Limited
20 Alfred Street, Milsons Point, Sydney
New South Wales 2061, Australia

Random House New Zealand Limited
18 Poland Road, Glenfield,
Auckland 10, New Zealand

Random House South Africa (Pty) Limited
PO Box 337, Bergvlei, South Africa

Random House UK Limited Reg. No. 954009

A CIP catalogue record for this book is available from the British Library

ISBN 0 09 958211 2

Papers used by Random House UK Ltd are natural, recyclable products made from wood grown in sustainable forests. The manufacturing processes conform to the environmental regulations of the country of origin

Printed and bound in Great Britain by
Cox & Wyman, Reading, Berkshire

PREFACE

This book may be regarded as, in some sense, a sequel to *The Emperor's New Mind* (here abbreviated ENM). Indeed, I shall continue the theme that ENM initiated, but what I have to say here can be read entirely independently of that book. Part of the motivation for writing again on this subject arose originally from a need for detailed replies to a number of queries and criticisms that various people have raised in connection with arguments put forward in ENM. However, I shall be presenting a case here that stands completely on its own and which explores some new ideas going well beyond those of ENM. One of the central themes of ENM had been my contention that by use of our consciousness we are enabled to perform actions that lie beyond any kind of computational activity. However, in ENM this idea was presented as a somewhat tentative hypothesis, and there was a certain vagueness as to what types of procedure might be encompassed under the heading of 'computational activity'. The present volume provides what I believe to be a much more powerful and rigorous case for this general conclusion, and it applies to any kind of computational process whatever. Moreover, a far more plausible suggestion than could be provided in ENM is presented here, for a mechanism in brain function whereby a non-computational physical action might indeed underlie our consciously controlled behaviour.

My case has two distinct strands to it. One of these is essentially negative, in that I argue strongly against the commonly held viewpoint that our conscious mentality—in all of its various manifestations—could, in principle, be fully understood in terms of computational models. The other strand to my reasoning is positive, in the sense that it represents a genuine search for a means, within the constraints of the hard facts of science, whereby a scientifically describable brain might be able to make use of subtle and largely unknown physical principles in order to perform the needed non-computational actions.

In accordance with this dichotomy, the arguments in this book are presented in two separate parts. Part I provides a thorough and detailed discussion, strongly supporting my thesis that consciousness, in its particular manifestation in the human quality of 'understanding', is doing something that mere computation cannot. I make clear that the term 'computation' includes both 'top-down' systems, which act according to specific well-understood algorithmic procedures, and 'bottom-up' systems, which are more loosely programmed in ways that allow them to learn by experience. Central to the arguments of Part I, is the famous theorem of Gödel, and a very thorough examination of the relevant implications of Gödel's theorem is provided. This greatly extends earlier arguments provided by Gödel himself, by Nagel and Newman, and by Lucas; and all the various objections that I am aware of are answered in detail. In relation to this, some thorough arguments are given against bottom-up systems (as well as top-down ones) being capable of ever achieving genuine intelligence. The conclusions are that conscious thinking must indeed involve ingredients that cannot be even *simulated* adequately by mere computation; still less could computation, of itself alone, evoke any conscious feelings or intentions. Accordingly, the mind must indeed be something that cannot be described in any kind of computational terms.

In Part II, the arguments turn to physics and to biology. The line of reasoning, though containing portions which are decidedly more tentative than the rigorous discussion of Part I, represents a genuine attempt to understand how such non-computational action might arise within scientifically comprehensible physical laws. The basic principles of quantum mechanics are introduced afresh—it being not required that the reader have any prior knowledge of quantum theory. The puzzles, paradoxes, and mysteries of that subject are analysed in some depth, using a number of new examples which graphically illustrate the important roles of non-locality and counterfactuality, and of deep issues raised by the phenomenon of quantum entanglement. I shall argue strongly for a need for a fundamental change, at a certain clearly specified level, in our present quantum-mechanical world-view. (These ideas relate closely to recent work by Ghirardi, Diósi, and others.) There are significant differences between the ideas that I shall argue for here and those promoted in ENM.

I am suggesting that a physical non-computability—needed for an explanation of the non-computability in our conscious actions—enters at this level. Accordingly, I require that the level at which this physical non-computability is significant must have importance in brain action. It is here where my present proposals differ most substantially from those of ENM. I argue that whereas neuron signals may well behave as classically determinate events, the synaptic connections between neurons are controlled at a deeper level, where it is to be expected that there is important physical activity at the quantum–classical borderline. The specific proposals I am making require that there be large-scale quantum-coherent behaviour (in accordance with proposals that have been put forward by Fröhlich) occurring within the microtubules in the cytoskeletons of neurons. The suggestion is that this quantum activity ought to be non-computationally linked to a computation-like action that has been argued by Hameroff and his colleagues to be taking place along microtubules.

The arguments that I am presenting point to several places where our present-day pictures fall profoundly short of providing us with a scientific understanding of human mentality. Nevertheless, this does not mean that the phenomenon of consciousness must remain outside scientific explanation. I argue strongly, as I did in ENM, that there should indeed be a scientific path to the understanding of mental phenomena, and that this path should start with a deeper appreciation of the nature of physical reality itself. I feel that it is important that any dedicated reader, wishing to comprehend how such a strange phenomenon as the mind can be understood in terms of a material physical world, should gain some significant appreciation of how strange indeed are the rules that must *actually* govern that 'material' of our physical world.

Understanding is, after all, what science is all about—and science is a great deal more than mere mindless computation.

Oxford R.P.
April 1994

ACKNOWLEDGEMENTS

For their assistance to me in the writing of this book, there are many people to whom I am greatly indebted—too many to thank individually, even if I could recall all their names. However, I am especially grateful to Guido Bacciagaluppi and Jeremy Butterfield for their criticisms of portions of an early draft, in which they unearthed an important error in the reasoning, as it had stood at the time, in what is now part of Chapter 3. I am grateful also to Abhay Ashtekar, Mary Bell, Bryan Birch, Geoff Brooker, David Chalmers, Francis Crick, David Deutsch, Soloman Feferman, Robin Gandy, Susan Greenfield, Andrew Hodges, Dipankar Home, Ezio Insinna, Dan Isaacson, Roger James, Richard Jozsa, John Lucas, Bill McColl, Claus Moser, Graeme Michison, Ted Newman, Oliver Penrose, Jonathan Penrose, Stanley Rosen, Ray Sachs, Graeme Segal, Aaron Sloman, Lee Smolin, Ray Streater, Valerie Willoughby, Anton Zeilinger, and especially Artur Ekert for various pieces of information and assistance. There have been innumerable correspondents and people offering verbal comments concerning my previous book *The Emperor's New Mind*. I thank them here—even though most of them are still awaiting replies to their letters! Without my having the benefit of their differing viewpoints concerning my earlier book, it is unlikely that I should have embarked upon the daunting task of writing another.

I am grateful to the organizers of the Messenger Lectures at Cornell University (which I gave under the same title as the final section of this book), the Gifford Lectures at the University of St Andrews, the Forder Lectures in New Zealand, the Gregynogg Lectures at the University of Aberystwyth, and a distinguished lecture series at the Five Colleges, Amherst, Massachusetts, in addition to innumerable 'one-off' talks in different parts of the world. These gave me the opportunity to air my views and to obtain valuable reactions from the audiences. I thank the Isaac Newton Institute in Cambridge, and also Syracuse University and Penn State University for their hospitality in awarding me, respectively, a Visiting Distinguished Professorship in Mathematics and Physics, and the Francis R. Pentz and Helen M. Pentz

Distinguished Professorship of Physics and Mathematics. I thank also the National Science Foundation for support under contracts PHY 86-12424 and PHY 43-96246.

Finally, there are three people who deserve special mention. Angus MacIntyre's selfless assistance and support in checking over my arguments concerning mathematical logic in Chapters 2 and 3, and in providing me with many needed references, has been enormously helpful. I offer him my very particular warm thanks. Stuart Hameroff taught me about the cytoskeleton and its microtubules—structures that, two years ago, I had not even known existed! I thank him greatly for this invaluable information, and for his assistance with checking the major part of Chapter 7. And I shall always remain indebted to him for opening my eyes to the wonders of a new world. He, like the others I thank, is of course in no way responsible for any of the errors that undoubtedly remain in this book. Most of all, my beloved Vanessa is owed particular thanks on several counts: for explaining to me why parts of the book needed rewriting; for soul-saving assistance with the references; and for her love, her patience, and her deep understanding, especially when I continually underestimate the time that my writing takes me! Oh yes, and I thank her also for—unbeknown to her—partially supplying me with a model, in my *imagination*, for the Jessica of my little story. It is my regret that I did not actually know her at that age!

Figure acknowledgements

The publishers either have sought or are grateful to the following for permission to reproduce illustration material.

Fig. 1.1 A. Nieman/Science Photo Library.
Fig. 4.12 from J. C. Mather *et al.* (1990), *Astrophys. J.*, **354**, L37.
Fig. 5.7 from A. Aspect and P. Grangier (1986), in *Quantum concepts in space and time* (ed. R. Penrose and C. J. Isham), pp. 1–27, Oxford University Press.
Fig. 5.8 from the Ashmolean Museum, Oxford.
Fig. 7.2 from R. Wichterman (1986), *The biology of paramecium*, 2nd edn., Plenum Press, New York.
Fig. 7.6 Eric Grave/Science Photo Library.
Fig. 7.7 from H. Weyl (1943), *Symmetry*, © 1952 Princeton University Press.
Fig. 7.10 from N. Hirokawa (1991), in *The neuronal cytoskeleton* (ed. R. D. Burgoyne), pp. 5–74, Wiley-Liss, New York.

CONTENTS

NOTES TO THE READER

Certain portions of this book differ very greatly from others, with regard to the degree of technicality that is involved. The most technical parts of the book are Appendices A and C, but there would be no significant loss for most readers if they simply ignore all the appendices. The same may be said of the more technical parts of Chapter 2, and certainly of Chapter 3. These are mainly for those readers who need to be persuaded of the strength of the case I am making against any purely computational model of human understanding. The more persuadable (or hurried) reader may, on the other hand, prefer a relatively painless route to the essentials of this case. This route is obtained by merely reading the fantasy dialogue of §3.23, preferably preceded by Chapter 1, and by §2.1–§2.5 and §3.1.

Some of the more serious mathematics encountered in this book arises in relation to the discussions of quantum mechanics. This occurs especially with regard to the Hilbert space descriptions of §5.12–§5.18 and, most particularly, the discussions in §6.4–§6.6 centred around the density matrix—important for the appreciation of why we shall eventually need an *improved* theory of quantum mechanics! My general advice to non-mathematical readers (or even to mathematical ones, for that matter), upon their encountering a daunting-looking mathematical expression, is simply to pass it by—as soon as it becomes clear that no further examination of it will easily conjure up any further understanding. It is true that the subtleties of quantum mechanics cannot be fully appreciated without some acquaintance with its elegant but mysterious mathematical underpinnings; yet something of the flavour of that subject will nevertheless come through even when its mathematics is totally ignored.

In addition, I must offer an apology to the reader on a quite different matter. I can well appreciate that she or he might take exception, were I to refer to her or him in a way that appears to make presumptions as to her or his sex—and I shall certainly not do so! Yet, in the kind of discussion that will frequently be encountered in this book, it may be necessary to refer to some *abstract* person such as an 'observer' or a 'physicist'. It is clear that there is no implication as to the sex of such an individual, but the English language does not have a neutral gender third-person singular pronoun. Repeated use of phrases like 'he or she' are certainly awkward. Moreover the modern tendency to use 'they', 'them', or 'their' as singular pronouns is grammatically offensive; nor can I see any merit grammatically, stylistically, or humanistically in alternating between 'she' and 'he' when referring to impersonal or metaphoric individuals.

Accordingly, I have adopted the policy, in this book, of generally using the pronouns 'he', 'him', and 'his' when referring to an abstract person. This is to carry *no* implication as to his sex. He is not to be thought of as male; nor is he to be taken as female. There may, however, be some suggestion that he is to be sentient, so referring to him as an 'it' seems inappropriate. I trust that no female reader will feel offended that my (abstract) three-eyed colleague on α-Centauri is referred to 'he' in §5.3, §5.18, and §7.12, nor that this pronoun is used also for the entirely impersonal individuals of §1.15, §4.4, §6.5, §6.6, and §7.10. On the other hand, I trust that no male reader will be offended by the fact that I use the pronoun 'her' both for the clever spider of §7.7 and for the devoted and sensitive elephant of §8.6 (for the clear-cut reason that, in this case, they are both known to be *actually* female), nor that the intricately behaving paramecium of §7.4 is also referred to as 'she' (being taken as 'female' for the inadequate reason that she is directly capable of reproducing her kind)—nor, even, that Mother Nature is herself a 'she', in §7.7.

As a final comment, I should point out that the page references to *The Emperor's New Mind* (ENM) always pertain to the original hardback version. The pagination of the US (Penguin) paperback is basically the same, but the non-US (Vintage) paperback numbering differs, being closely approximated by the formula

$$\frac{22}{17} \times n,$$

where n is the hardback page-number given here.

PROLOGUE

Jessica always felt slightly nervous when she entered this part of the cave. 'Daddy? Suppose that great boulder fell down from where it's wedged between those other rocks. Wouldn't it block our way out, and we'd never, ever, ever get home again?'

'It would, but it won't', replied her father a little distractedly, and somewhat unnecessarily brusquely, as he seemed more interested in how his various plant samples were accustoming themselves to the dank and dark conditions in this, the most remote corner of the cave.

'But, how do you know it won't, Daddy?' Jessica persisted.

'That boulder's probably been there for many thousands of years. It's not going to come down just when we're here.'

Jessica was not at all happy with this. 'Surely, if it's going to fall down sometime, then the longer it's been there, the more likely it's going to fall down now?'

Jessica's father stopped prodding at his plants and looked at Jessica, with a faint smile on his face. 'No, it's not like that at all.' His smile became more noticeable, but now more inward. 'Actually, you could even say that the longer that it's been there, the *less* likely that it's going to fall down when we're here.' No further explanation was evidently forthcoming, and he turned his attentions back to his plants.

Jessica hated her daddy when he was in these moods—no she didn't; she always loved her daddy, more than anything or anybody, but she still wished he didn't have moods like that. She knew they were something to do with him being a scientist, but she still didn't understand. She even hoped that she might someday be a scientist herself, but if she did, she'd make sure she never had moods like that.

At least she'd stopped worrying that the boulder might actually fall down and block the cave. She could see that her daddy wasn't afraid that it might, and her daddy's confidence made her feel confident too. She didn't understand her daddy's explanation, but she knew that her daddy was always right about that kind of thing—or at least he *almost* always was. There was that argument about the clocks in New Zealand, when mummy said one thing but daddy insisted that the opposite thing was true. Then three hours later daddy came down from his study and said that he was sorry, and he was wrong, and she'd been right all along! That *was* funny! 'I bet that mummy could have been a scientist too if she'd wanted to be', she thought to herself, 'and she wouldn't have had strange moods like daddy has'.

Jessica was more careful to put her next question at a moment when her daddy had just finished what he had just been doing and hadn't quite started what he was just going to do. 'Daddy? I know that the boulder isn't going to fall down. But let's just imagine that it did, and we're trapped here for the rest of our lives. Would the whole cave get very dark? Would we be able to breathe?'

'What an unpleasant thought!' answered Jessica's father. Then he looked carefully at the shape and size of the boulder, and at the opening in the cave. 'Mmmm', he said, 'yes, I think that the boulder would fill the entrance-hole very tightly. There would certainly be some space for air to get in and out, so we wouldn't suffocate. As for light, well I think that there would be a small roundish crack at the top that would let some light in, but it would be very dark—much darker than it is now. But I'm pretty sure that we could see all right, once we'd got used to it. It wouldn't be very nice, I'm afraid! But I can tell you one thing: if I had to live here the rest of my life with anyone, then I'd rather it was with my wonderful Jessica than with anyone else in the whole wide world—and with mummy too, of course.'

Jessica remembered why she loved her daddy so much! 'I want mummy in here too, in my next question, because I'm going to suppose that the boulder fell down before I was even born and you and mummy had me here in the cave, and I grew up with you here . . . and we could keep alive by eating all your funny plants.'

Her father looked at her a little oddly, but said nothing.

'Then I would never have known any life at all *except* this here in the cave. How could I know what the real world outside was like? Could I know that there are trees in it, and birds, and rabbits and other things? Of course you could *tell* me these things, because you'd known them yourselves before you got trapped, but how would *I* know—I mean how would I really know *myself*, rather than just having to believe what you said?'

Her father stopped and thought for a few minutes. Then he said: 'Well, I suppose that every now and again, on a sunny day, a bird might fly exactly in line between the sun and the crack, and then we could see its shadow on the cave wall behind. Of course its shape would be distorted somewhat, on the rather irregular-shaped wall, but we could learn how to correct for that. If the crack was small and round enough, then the bird might cast quite a clearly defined shadow, but if not, then we might have to make other kinds of corrections also. Then, if the same bird were to fly across many times, we might begin to get quite a good picture of what it actually looks like, and how it flies, and so on, just from its shadow. Again, when the sun was low in the sky, there might happen to be a tree, suitably positioned between the sun and our crack, with its leaves waving, so that we could begin to get a picture of this tree, too, from its shadow. And perhaps from time to time a rabbit might jump up into the way of our crack, so we could begin to picture it from its shadow too.'

'That's interesting', said Jessica. She paused for a few moments, and then said: 'Do you think that it would be possible for us to make a real scientific discovery, while stuck down here in the cave? Imagine that we had made a big discovery about the outside world, and then we were in here having one of your big conferences, trying to persuade everyone else that we were right—of course all the other people at the conference (and you as well) would have to have been brought up in the cave too, otherwise it's cheating. But it's all right for them to be brought up in the cave too, because you've got lots and lots of funny plants and we could *all* live off them!'

This time, Jessica's father visibly winced, but still he said nothing. He looked pensive for several minutes. Then: 'Yes, I think it would be possible. But, you see, the hardest thing would be to try to persuade them that any outside world existed at all. All that they would know about would be the shadows, and how they moved about and changed from time to time. To them, the complicated wiggling shadows and things on the cave wall would be all that there was to the world. So, part of our task would be to convince people that there actually *is* an outside world that our theory refers to. In fact, these two things would go together. Having a good theory of the outside world would be an important part of making people accept that it was really there!'

'OK Daddy, what's our theory?'

'Not so fast . . . just a minute . . . here it is: the earth goes round the sun!'

'*That*'s not a very new theory.'

'No—it's actually nearly twenty-three centuries old—nearly as old as the length of time that boulder's been wedged there near the entrance! But in our imagination we've all spent our entire lives in the cave, and the people wouldn't have ever heard of such an idea before. We'd have to convince them that there was really such a *thing* as the sun—and even as the earth, for that matter. The idea is that the simple elegance of our theory in explaining all sorts of fine details of the movement of the light and shadows would eventually persuade most of the people at the conference that not only is there actually a very bright thing out there—that we're calling the "sun"—but that the earth is in continual motion around it, spinning on its axis all the time.'

'Would it be very difficult to persuade them?'

'It certainly would! In fact we'd have to do two quite different kinds of thing. First, we'd need to show how our simple theory explains in a very accurate way an awful lot of very detailed data concerning how the bright spot, with its shadows, moves about on the cave wall. Now, some people might be persuaded by this, but others would point out that there's a much more "common-sense" picture in which the sun moves around the earth. In detail, that picture would be more complicated than the one that we are putting forward. But these people would prefer to stick to their complicated one—reasonably enough—because they simply couldn't accept the possibility that the cave was moving around at about a hundred thousand kilometres per hour, as our theory would require.'

'Gosh, is it *really* doing that?'

'Yes, that sort of thing. So for the second part of our argument, we'd have to change tack completely and do something that a lot of people at the conference would think was completely irrelevant. We'd be rolling balls down tracks and swinging pendulums and that sort of thing—just to show that the physical laws that govern the behaviour of things in the cave would be unaffected if the whole contents of the cave were moving in any direction you like, at any speed you like. This would show them that they wouldn't actually feel anything if the cave moves around at an enormous speed. That's one of the important things that Galileo had to show—you remember about him from that book I got you?'

'Of course I do! Oh dear, the whole thing sounds awfully complicated. I bet that lots of people at our conference will go off to sleep, just as I've seen them do at real conferences when you're giving a lecture.'

Jessica's father reddened just noticeably. 'I expect you're right! Yes, but I'm afraid that this is what science is often like: lots and lots of details, most of which can seem very boring and sometimes almost completely irrelevant to the picture that you're trying to get across, even if that final picture might have a striking simplicity to it, as with our idea that the earth spins as it goes round a thing called the sun. Some people might not feel that they need to bother with all the boring details because they find the idea plausible enough anyway. But the real sceptics would want to check through everything, looking for possible loopholes.'

'Thank you daddy! I always like it when you talk to me about things like this, when you sometimes get all red and excited. But can we go back now? It's getting dark and I'm tired and hungry—and a bit cold.'

'Come on then.' Jessica's father put his jacket over her shoulders, collected up his things and put his arm around her, to guide her out of the now darkening cave entrance. As they made their exit, Jessica looked up at the boulder again.

'You know, I think I agree with you daddy. That boulder's going to stay up there *more* than another twenty-three centuries!'

Part I: Why We Need New Physics to Understand the Mind

The Non-Computability of Conscious Thought

1

Consciousness and computation

1.1 Mind and science

What is the ultimate scope of science? Is it just the *material* attributes of our universe that are amenable to its methods, whereas our *mental* existence must forever lie outside its compass? Or might we someday come to a proper scientific understanding of the shadowy mystery of minds? Is the phenomenon of human consciousness something that is beyond the scope of scientific enquiry, or may the power of scientific method one day resolve the problem of the very existence of our conscious selves?

There are those who would believe that we may actually be close to a scientific understanding of consciousness; that this phenomenon holds *no* mystery, and even that all the essential ingredients may already be in place. They argue that it is merely a matter of the extreme complication and organizational sophistication of our brains that at present limits our understanding of human mentality—a complication and sophistication certainly not to be underestimated, but where there are no matters of principle that might take us beyond our present-day scientific pictures. On the other end of the scale are those who would maintain that the matters of the mind and spirit—and the very mystery of human consciousness—are things that we cannot ever hope to address adequately by the cold and calculational procedures of an unfeeling science.

In this book, I shall attempt to address the question of consciousness from a scientific standpoint. But I shall strongly contend—by *use* of scientific argument—that an essential ingredient is missing from our present-day scientific picture. This missing ingredient would be needed in order that the central issues of human mentality could ever be accommodated within a coherent scientific world-view. I shall maintain that this ingredient is itself something that is *not* beyond science—although, no doubt, it is an appropriately expanded scientific world-view that we shall need. In Part II of this book, I shall try to guide the reader in a very specific direction aimed at such an extension of our present-day picture of the physical universe. It is a direction that involves an important change in the most basic of our physical laws, and I shall be fairly specific about what the nature of this change must be and of how it might apply to the biology of our brains. Even with our limited present understanding of the nature of this missing ingredient, we can begin to point to where it must be making its mark, and how it should be providing one vital contribution to whatever it is that underlies our conscious feelings and actions.

Though, of necessity, some of the arguments I shall give are not altogether simple, I have tried to make my case as clearly as I can, using only elementary notions where possible. In places, some mathematical technicalities are introduced, but only when they are necessary, or are otherwise helpful for improving the clarity of the discussion. Whilst I have learnt not to expect everyone to be persuaded by the kinds of argument that I shall be presenting, I would suggest, nevertheless, that these arguments do deserve careful and dispassionate consideration; for they provide a case which should not be ignored.

A scientific world-view which does not profoundly come to terms with the problem of conscious minds can have no serious pretensions of completeness. Consciousness is part of our universe, so any physical theory which makes no proper place for it falls fundamentally short of providing a genuine description of the world. I would maintain that there is yet no physical, biological, or computational theory that comes very close to explaining our consciousness and consequent intelligence; but that should not deter us from striving to search for one. It is with such aspirations in mind that the arguments of this book are presented. Perhaps someday the fully appropriate collection of ideas will be brought about. If so, our philosophical outlook can hardly be other than profoundly altered. Yet, all scientific knowledge is a two-edged sword. What we actually *do* with our scientific knowledge is another matter. Let us try to see where our views of science and the mind may be taking us.

1.2 Can robots save this troubled world?

As we open our newspapers or watch our television screens, we seem to be continually assaulted by the fruits of Mankind's stupidity. Countries, or parts of countries, are set against one another in confrontations that may, from time to time, flare into hideous warfare. Excessive religious fervour, or nationalism, or separate ethnic interests, or mere linguistic or cultural differences, or the self-seeking interests of particular demagogues, may result in continuing unrest and violence, sometimes boiling over to outbursts of unspeakable atrocity. Oppressively authoritarian regimes still subjugate their peoples, keeping them in check by the use of death squads and torture. Yet, those who are oppressed, and who might seem to have a common purpose, are often locked in conflict with one another, and when given a freedom that they may have been long denied, may seem to choose to use that freedom in horribly self-destructive ways. Even in those fortunate countries where there is prosperity, peace, and democratic freedom, resources and manpower are squandered in apparently senseless ways. Is this not a clear indication of the general stupidity of Man? Though we believe ourselves to represent the pinnacle of intelligence in the animal kingdom, this intelligence seems sadly inadequate to handle many of the problems that our own society continues to confront us with.

Yet, the positive achievements of our intelligence cannot be denied. Among these achievements are our impressive science and technology. Indeed, whilst it must be admitted that some of the fruits of this technology are of distinctly questionable long-term (or short-term) value, as is borne witness by numerous environmental problems and a genuine fear of a technology-induced global catastrophe, it is this same technology that has given us our modern society, with its comforts, its considerable freedoms from fear, disease, and need, and with its vast opportunities for intellectual and aesthetic expansion, and for mind-broadening global communication. If this technology has opened up so many potentialities and, in a sense, increased the scope and the power of our individual physical selves, can we not expect much more in the future?

Our senses have been vastly extended by our technology, both ancient and modern. Our sight has been aided and enormously increased in power by spectacles, mirrors, telescopes, microscopes of all kinds, and by video-cameras, television, and the like. Our hearing has been aided, originally by ear-trumpets, but now by tiny electronic devices, and greatly extended by telephones, radio communication, and satellites. We have bicycles, trains, motor cars, ships, and aeroplanes to aid and transcend our natural forms of locomotion. Our memories are helped by printed books, films—and by the huge storage capacities of *electronic computers*. Our calculational tasks, whether simple and routine, or of a massive or sophisticated kind, are also vastly extended by the capabilities of modern computers. Thus, not only does our technology provide us with an enormous expansion of the scope of our *physical* selves, but it also expands our *mental* capabilities by greatly improving upon our abilities to perform many routine tasks. What about mental tasks that are not routine—tasks that require genuine *intelligence*? It is natural to ask whether these also will be aided by our computer-driven technology.

There is little doubt in my own mind that there is indeed, implicit in our (frequently computer-driven) technological society, at least one direction with an enormous potential for enhancing intelligence. I refer, here, to the educational possibilities of our society, which could gain great benefit from diferent aspects of technology—but only if it is used with sensitivity and understanding. Technology provides the potential, by use of well-produced books, film, television, and interactive computer-controlled systems of various kinds. These, and other developments, provide many opportunities for expanding our minds—or else for deadening them. The human mind is capable of vastly more than it is often given the chance to achieve. Sadly, these opportunities are all too frequently squandered, and the minds of neither young nor old are provided the openings that they undoubtedly deserve.

But many readers will ask: is there not a rather different possibility for the vast expansion of a mental capability, namely that alien electronic 'intelligence' which is just beginning to emerge from the extraordinary advances in computer technology? Indeed, we already frequently turn to computers for intellectual assistance. There are many circumstances in which unaided human intelligence is far from adequate to assess the probable consequences of alternative actions. Such consequences may lie considerably beyond the scope of human computational powers; thus it is to be expected that the computers of the future will greatly expand this role, where hard computational fact provides an invaluable aid to human intelligence.

Yet may not computers eventually achieve very much more than just this? Many experts claim that computers offer us, at least in principle, the potential for an *artificial* intelligence that will ultimately exceed our own.[1] When computationally controlled robots reach the level of 'human equivalence', then it will not take long, they argue, before they race enormously beyond our own puny level. Only *then*, these experts would claim, shall we have an authority with sufficient intelligence, wisdom, and understanding to be able to resolve the troubles of this world that humanity has created.

How long will it be before this happy state of affairs is to come about? There is no clear consensus among these experts. Some would measure the timescale in terms of many centuries, whilst others claim that this human equivalence is just decades away.[2] The latter would point to the very rapid 'exponential' growth of computer power and base their estimates upon comparisons between the speed and accuracy of transistors, and the relative slowness and sloppy action of neurons. Indeed, electronic circuits are already over a million times faster than the firing of the neurons in the brain (the rate being some 10^9/s for transistors and only about 10^3/s for neurons*), and they have an immense precision in timing and accuracy of action that is in no way shared by neurons. Moreover, there is a great deal of randomness in the brain's 'wiring' that, it would seem, could be vastly improved upon by the deliberate and precise organization of electronic printed circuits.

There are some areas where the neuron structure of the brain does provide a numerical advantage over present-day computers, although these advantages might be relatively short lived. It is argued that in total neuron number (some hundreds of thousands of millions), human brains are at the moment ahead of computers with respect to their transistor counts. Moreover, on the average, there are a good deal more *connections* between different neurons than there

*The Intel Pentium chip has over three million transistors on a 'slice of silicon' about the size of a thumbnail, each capable of performing 113 million full instructions per second.

are connections between transistors in a computer. In particular, Purkinje cells in the cerebellum can have up to 80 000 synaptic endings (junctions between neurons), whereas for a computer, the corresponding number is only about three or four at most. (I shall have some comments to make about the cerebellum later; cf. §1.14, §8.6.) Moreover, most of the transistors of today's computers are concerned just with memory and not directly with computational action, whereas it might be the case that with the brain such computational action could be more widespread.

These temporary advantages for the brain could easily be overcome in the future, particularly when massively 'parallel' computational systems become more developed. It is to a computer's advantage that different units can be combined together to form larger and larger ones, so the total number of transistors could, in principle, be increased almost without limit. In addition, there are technological revolutions waiting in the wings, such as the replacing of the wires and transistors of our present computers by appropriate optical (laser) devices, perhaps achieving, thereby, enormous increases in speed, power, and miniaturization. More fundamentally, our brains would appear to be *stuck* with the numbers that we have at present, and we have many further constraints, such as having to grow from a single cell. Computers, on the other hand, can be deliberately constructed so as to achieve all that is eventually needed. Though I shall later be pointing to some important factors that are not yet being taken into account by these considerations (most particularly, a significant level of activity that underlies that of neurons), an impressive-looking case can indeed be made that on any issue of merely computing power, if computers do not have the advantage over brains already, then they *will* certainly have it before too long.

Thus, if we are to believe the strongest of the claims of the most outspoken of the proponents of artificial intelligence, and accept that computers and computer-guided robots will eventually—and even perhaps before too long—exceed all human capabilities, then the computers will be able to do immeasurably more than merely assist *our* intelligences. They will actually have immense intelligences of their own. We could *then* turn to these superior intelligences for advice and authority in all matters of concern—and the humanity-induced troubles of the world could at last be resolved!

But there appears to be another logical consequence of these potential developments that may well strike us as genuinely alarming. Would not these computers eventually make human beings themselves superfluous? If the computer-guided robots turn out to be our superiors in every respect, then will they not find that they can run the world better without the need of us at all? Humanity itself will then have become obsolete. Perhaps, if we are lucky, they might keep us as pets, as Edward Fredkin once said; or if we are clever, we might be able to transfer the 'patterns of information' that are 'ourselves' into robot form, as Hans Moravec (1988) has insisted; or perhaps we will *not* be that lucky and will just *not* be that clever . . .

1.3 The $\mathscr{A}$, $\mathscr{B}$, $\mathscr{C}$, $\mathscr{D}$, of computation and conscious thinking

But are the relevant issues merely those of computing power, or of speed, accuracy, or memory, or perhaps of the detailed way in which things happen to be 'wired up'? Might we, on the other hand, be doing something with our brains that cannot be described in computational terms at all? How do our feelings of conscious awareness—of happiness, pain, love, aesthetic sensibility, will, understanding, etc.—fit into such a computational picture? Will the computers of the future actually have *minds*? Does the presence of a conscious mind actually influence behaviour in any way? Does it make sense to talk about such things in scientific terms at all; or is science in no way competent to address issues that relate to the human consciousness?

It seems to me that there are at least four different viewpoints[3]—or extremes of viewpoint—that one may reasonably hold on the matter:

$\mathscr{A}$. All thinking is computation; in particular, feelings of conscious awareness are evoked merely by the carrying out of appropriate computations.

$\mathscr{B}$. Awareness is a feature of the brain's physical action; and whereas any physical action can be simulated computationally, computational simulation cannot by itself evoke awareness.

$\mathscr{C}$. Appropriate physical action of the brain evokes awareness, but this physical action cannot even be properly simulated computationally.

$\mathscr{D}$. Awareness cannot be explained by physical, computational, or any other scientific terms.

The point of view expressed in $\mathscr{D}$, which negates the physicalist position altogether and regards the mind as something that is entirely inexplicable in scientific terms, is the viewpoint of the mystic; and at least some ingredient of $\mathscr{D}$ seems to be involved in the acceptance of religious doctrine. My own position is that questions of mind, though they lie very uncomfortably with present-day scientific understanding, should not be regarded as being forever outside the realms of science. If science is yet incapable of saying much that is of significance concerning matters of the mind, then eventually science must enlarge its scope so as to accommodate such matters, and perhaps even modify its very procedures. Whereas I reject mysticism in its negation of scientific criteria for the furtherance of knowledge, I believe that within an expanded science and mathematics there will be found sufficient mystery ultimately to accommodate even the mystery of mind. I shall expand on some of these ideas later on in this book, but for the moment it will be sufficient to say that I am rejecting $\mathscr{D}$; and I am attempting to move forward along the path that science has set out for us. If you are a reader who believes strongly that $\mathscr{D}$, in some form, must be right, I ask that you bear with me and see how far we can get along the scientific road—and try to perceive where I believe that this road must ultimately be taking us.

Let us consider what seems to be the other extreme: the viewpoint $\mathscr{A}$. Those who adhere to the standpoint that is often referred to as *strong AI* (strong Artificial Intelligence) or sometimes *hard* AI, or *functionalism*,[4] would come under this heading—although some people might use the term 'functionalism' in a way that could include certain versions of $\mathscr{C}$ also. $\mathscr{A}$ is regarded by some as the only viewpoint that an entirely scientific attitude allows. Others would take $\mathscr{A}$ to be an absurdity that is barely worth serious attention. There are undoubtedly many different versions of viewpoint $\mathscr{A}$. (See Sloman (1992) for a long list of alternative computational viewpoints.) Some of these might differ with regard to what kind of thing would be counted as a 'computation' or as 'carrying out' a computation. Indeed, there are also adherents of $\mathscr{A}$ who would deny that they are 'strong AI supporters' at all, because they claim to take a different view as to the interpretation of the term 'computation' from that of conventional AI (cf. Edelman 1992). I shall address these issues a little more fully in §1.4. For the moment it will be sufficient to take this term simply to mean the kind of thing that ordinary general-purpose computers are capable of doing. Other proponents of $\mathscr{A}$ might differ as to how they interpret the meaning of the words 'awareness' or 'consciousness'. Some would not even allow that there *is* such a phenomenon as 'conscious awareness' at all, whereas others would accept the existence of this phenomenon, but regard it as just some kind of 'emergent property' (cf. also §4.3 and §4.4) that comes along whenever a sufficient degree of complication (or sophistication, or self-reference, or whatever) is involved in the computation that is being performed. I shall indicate my own interpretation of the terms 'consciousness' and 'awareness' in §1.12. Just for now, any differences in possible interpretation will not be greatly important for our considerations.

The strong-AI viewpoint $\mathscr{A}$ is what my arguments in ENM were most specifically directed against. The length of that book alone should make it clear that, while I do not myself believe that $\mathscr{A}$ is correct, I *do* regard it as a serious possibility that is worthy of considerable attention. $\mathscr{A}$ is an implication of a highly operational attitude to science, where, also, the physical world is taken to operate entirely computationally. In one extreme of this view, the universe itself is taken to be, in effect, a gigantic computer;[5] and appropriate subcomputations that this computer performs will evoke the feelings of 'awareness' that constitute our conscious minds.

I suppose that this viewpoint—that physical systems are to be regarded as merely computational entities—stems partly from the powerful and increasing role that computational simulations play in modern twentieth-century science, and also partly from a belief that physical objects are themselves merely 'patterns of information', in some sense, that are subject to computational mathematical laws. Most of the material of our bodies and brains, after all, is being continuously replaced, and it is just its *pattern* that persists. Moreover, matter itself seems to have merely a transient existence since it can be converted from one form into another. Even the *mass* of a material body, which provides a precise physical measure of the quantity of matter that the body contains, can in appropriate circumstances be converted into pure energy (according to Einstein's famous $E=mc^2$)—so even material substance seems to be able to convert itself into something with a mere theoretical mathematical actuality. Furthermore, quantum theory seems to tell us that material particles are merely 'waves' of information. (We shall examine these issues more thoroughly in Part II.) Thus, matter itself is nebulous and transient; and it is not at all unreasonable to suppose that the persistence of 'self' might have more to do with the preservation of *patterns* than of actual material particles.

Even if we do not think that it is appropriate to regard the universe as simply being a computer, we may feel ourselves operationally driven to viewpoint $\mathscr{A}$. Suppose that we have a robot that is controlled by a computer and which responds to questioning exactly as a human would. We ask it how it feels, and find that it answers in a way that is entirely consistent with its actually possessing feelings. It tells us that it is aware, that it is happy or sad, that it can perceive the colour red, and that it worries about questions of 'mind' and 'self'. It may even give expression to a puzzlement about whether or not it should accept that *other* beings (especially human beings) are to be regarded as possessing a consciousness similar to the one that it claims to feel itself. Why should we disbelieve *its* claims to be aware, to wonder, to be joyful, or to feel pain, when it might seem that we have as little to go on with respect to other human beings whom we *do* accept as being conscious? The operational argument does, it seems to me, have some considerable force, even if it is not entirely conclusive. If all the *external* manifestations of a conscious brain, including responses to continual questioning, can indeed be completely imitated by a system entirely under computational control, then there would indeed be a plausible case for accepting that its *internal* manifestations—consciousness itself—should be also considered to be present in association with such a simulation.

The acceptance of this kind of argument, which basically is what is referred to as a *Turing test*,[6] is in essence what distinguishes $\mathscr{A}$ from $\mathscr{B}$. According to $\mathscr{A}$, any computer-controlled robot which, after sustained questioning, convincingly behaves *as though* it possesses consciousness, must be considered *actually* to be conscious—whereas according to $\mathscr{B}$, a robot could perfectly well behave exactly as a conscious person might behave without itself actually possessing any of this mental quality. Both $\mathscr{A}$ and $\mathscr{B}$ would allow that a computer-controlled robot could convincingly *behave* as a conscious person does, but viewpoint $\mathscr{C}$, on the other hand, would not even admit that a fully effective simulation of a conscious person could ever be achieved merely by a computer-controlled robot. Thus, according to $\mathscr{C}$, the robot's actual lack of consciousness ought ultimately to reveal itself, after a sufficiently long interrogation. Indeed, $\mathscr{C}$ is much more of an *operational* viewpoint than is $\mathscr{B}$—and it is more like $\mathscr{A}$ than $\mathscr{B}$ in this particular respect.

What about $\mathscr{B}$ then? I think that it is perhaps the viewpoint that many would regard as 'scientific common sense'. It is sometimes referred to as *weak* (or *soft*) AI. Like $\mathscr{A}$, it affirms a view that all the physical objects of this world must behave according to a science that, in principle, allows that they can be computationally simulated. On the other hand, it strongly denies the operational claim that a thing that behaves externally as a conscious being must necessarily be conscious itself. As the philosopher John Searle has stressed,[7] a computational simulation of a physical process is a very different thing from the actual process itself. (A computer simulation of a hurricane, for example, is certainly no hurricane!) On view $\mathscr{B}$, the presence or absence of consciousness would depend very much upon what actual physical object is 'doing the thinking', and upon what particular physical actions that object is performing. It would be a secondary matter to consider the particular computations that might happen to be involved in these actions. Thus, the action of a biological brain might evoke consciousness, whilst its accurate electronic simulation might well not. It is not necessary, in viewpoint $\mathscr{B}$, for this distinction to be between biology and physics. But the actual *material* constitution of the object in question (say, a brain), and not just its computational action, is regarded as all-important.

The viewpoint $\mathscr{C}$ is the one which I believe myself to be closest to the truth. It is more of an operational viewpoint than $\mathscr{B}$ since it asserts that there are external manifestations of conscious objects (say, brains) that differ from the external manifestations of a computer: the outward effects of consciousness cannot be properly simulated computationally. I shall be giving my reasons for this belief in due course. Since $\mathscr{C}$, like $\mathscr{B}$, goes along with the physicalist standpoint that minds arise as manifestations of the behaviour of certain physical objects (brains—although not necessarily only brains), it follows that an implication of $\mathscr{C}$ is that *not* all physical action can be properly simulated computationally.

Does present-day physics allow for the possibility of an action that is in principle impossible to simulate on a computer? The answer is not completely clear to me, if we are asking for a mathematically rigorous statement. Rather less is known than one would like, in the way of precise mathematical theorems, on this issue.[8] However, my own strong opinion is that such non-computational action would have to be found in an area of physics that lies *outside* the presently known physical laws. Later on in this book, I shall reiterate some of the powerful reasons, coming from within physics itself, for believing that a new understanding is indeed needed, in an area that lies intermediate between the 'small-scale' level, where quantum laws hold sway, and the 'everyday' level of classical physics. However, it is not by any means universally accepted, among present-day physicists, that such a new physical theory is required.

Thus there are at least two very different standpoints that could come under the heading of $\mathscr{C}$. Some $\mathscr{C}$-believers would contend that our present physical understanding is perfectly adequate, and that we should look to subtle types of behaviour within conventional theory that might be able to take us outside the scope of what can be achieved entirely computationally (e.g. as we shall examine later: chaotic behaviour (§1.7), subtleties of continuous as opposed to discrete action (§1.8), quantum randomness). On the other hand, there are those who would argue that the physics of today really offers us no reasonable scope for non-computability of the type needed. Later in this book I shall give what I believe are powerful reasons for adopting $\mathscr{C}$ according to this stronger standpoint—which requires some fundamentally new physics to be involved.

Some people have tried to contend that this really places me in camp $\mathscr{D}$, since I am arguing that we must look beyond the reaches of known science if we are ever to find any kind of explanation of the phenomenon of consciousness. But there is an essential difference between this strong version of $\mathscr{C}$ and the viewpoint $\mathscr{D}$—particularly with regard to the issue of *methodology*. According to $\mathscr{C}$, the problem of conscious awareness is indeed a scientific one, even if the appropriate science may not yet be at hand. I strongly support this viewpoint; I believe that it must indeed be by the methods of science—albeit appropriately extended in ways that we can perhaps only barely glimpse at present—that we must seek our answers. That is the key difference between $\mathscr{C}$ and $\mathscr{D}$, whatever similarities there may seem to be in the corresponding opinions as to what *present-day* science is capable of achieving.

The viewpoints $\mathscr{A}$, $\mathscr{B}$, $\mathscr{C}$, $\mathscr{D}$, as defined above, are intended to represent extremes, or polarities, of possible stances that one might choose to take. I can accept that some people may feel that their own viewpoints do not fit clearly into any of these categories, but perhaps lie somewhere between them, or cut across some of them. There are certainly many possible gradations of belief between $\mathscr{A}$ and $\mathscr{B}$, for example (see Sloman 1992). There is even a view, not uncommonly expressed, that might best be regarded as a combination of $\mathscr{A}$ and $\mathscr{D}$ (or perhaps $\mathscr{B}$ and $\mathscr{D}$)—a possibility that will actually feature significantly in our later deliberations. According to this view, the brain's action is indeed that of a computer, but it is a computer of such wonderful complexity that its imitation is beyond the wit of man and science, being necessarily a divine creation of God—the 'best programmer in the business'![9]

1.4 Physicalism vs. mentalism

I should make a brief remark about the use of the words 'physicalist' and 'mentalist' that are often used to describe opposing standpoints in connection with the issues addressed by $\mathscr{A}$, $\mathscr{B}$, $\mathscr{C}$, and $\mathscr{D}$. Since $\mathscr{D}$ represents a total denial of physicalism, believers in $\mathscr{D}$ would certainly have to be counted as mentalists. However, it is not at all clear to me where the line between physicalism and mentalism is to be drawn in relation to the other three viewpoints $\mathscr{A}$, $\mathscr{B}$, and $\mathscr{C}$. I think that holders of viewpoint $\mathscr{A}$ would normally be thought of as physicalists, and I am sure that the vast majority of *them* would say so. However, there is something of a paradox lurking here. According to $\mathscr{A}$, the *material* construction of a thinking device is regarded as irrelevant. It is simply the computation that it performs that determines all its mental attributes. Computations themselves are pieces of abstract mathematics, divorced from any association with particular material bodies. Thus, according to $\mathscr{A}$, mental attributes are themselves things with no particular association with physical objects, so the term 'physicalist' might seem a little inappropriate. Viewpoints $\mathscr{B}$ and $\mathscr{C}$, on the other hand, demand that the actual physical constitution of an object must indeed be playing a vital role in determining whether or not there is genuine mentality present in association with it. Accordingly, it might well be argued that these, rather than $\mathscr{A}$, represent the possible physicalist standpoints. However, it seems that such terminology would be at variance with some common usage, the term 'mentalist' being often regarded as more appropriate for $\mathscr{B}$ and $\mathscr{C}$, since here mental qualities are regarded as being 'real things' and not just as 'epiphenomena' that might arise incidentally when (certain types of) computations are performed. In view of such confusions, I shall tend to avoid the use of the terms 'physicalist' and 'mentalist' in the discussions that follow, and refer, instead, to the specific viewpoints $\mathscr{A}$, $\mathscr{B}$, $\mathscr{C}$, and $\mathscr{D}$ as defined above.

1.5 Computation: top-down and bottom-up procedures

I have not been at all explicit, thus far, about what I am taking the term 'computation' to mean, in the definitions of $\mathscr{A}$, $\mathscr{B}$, $\mathscr{C}$, and $\mathscr{D}$, of §1.3. What *is* a computation? In short, one can simply understand that term to denote the activity of an ordinary general-purpose computer. To be more precise, we must take this in a suitably idealized sense: a *computation* is the action of a *Turing machine*.

But what is a Turing machine? It is, indeed, a mathematically idealized computer (the theoretical forerunner of the modern general-purpose computer)—idealized so that it never makes any mistakes and can run on for as long as is necessary, and so that it has an unlimited storage space. I shall be a little more explicit about how Turing machines may be precisely specified in §2.1 and Appendix A (p. 117). (For a much more thorough introduction, the interested reader is referred to the descriptions given in ENM, Chapter 2, or else to Kleene (1952) or Davis (1978), for example.)

The term 'algorithm' is frequently used to describe the action of a Turing machine. I am taking 'algorithm' to be completely synonymous with 'computation' here. This needs a little clarification, because some people take a more restrictive point of view with regard to the term 'algorithm' than I am proposing to do here, taking it in the sense of what I shall refer more specifically to as a 'top-down algorithm'. Let us try to understand what the terms 'top-down' and its antithesis 'bottom-up' are to mean in the context of computation.

A computational procedure is said to have a *top-down* organization if it has been constructed according to some well-defined and clearly understood fixed computational procedure (which may include some preassigned store of knowledge), where this procedure specifically provides a clear-cut solution to some problem at hand. (Euclid's algorithm for finding the highest common factor* of two natural numbers, as described in ENM, p. 31, is a simple example of a top-down algorithm.) This is to be contrasted with a *bottom-up* organization, where such clearly defined rules of operation and knowledge store are not specified in advance, but instead there is a procedure laid down for the way that the system is to 'learn' and to improve its performance according to its 'experience'. Thus, with a bottom-up system, these rules of operation are subject to continual modification. One must allow that the system is to be run many times, performing its actions upon a continuing input of data. On each run, an assessment is made—perhaps by the system itself—and it modifies its operations, in the light of this assessment, with a view to improving this quality of output. For example, the input data for the system might be a number of photographs of human faces, appropriately digitized, and the system's task is to decide which photographs represent the same individuals and which do not. After each run, the system's performance is compared with the correct answers. Its rules of operation are then modified in such a way as to lead to a probable improvement in its performance on the next run.

The details of how this improvement is to be arranged, in any particular bottom-up system, are not important for us here. There are many different possible schemes. Among the best-known systems of bottom-up type are the so-called *artificial neural networks* (sometimes referred to, a little misleadingly, simply as 'neural networks' or 'neural nets') which are computer learning programs—or else specifically constructed electronic devices—that are based upon certain ideas about how the organization of a system of connections of neurons in the brain is thought to be improved as that system gains its experience. (The question of how the brain's system of neuron interconnections *actually* modifies itself will be an important one for us later; cf. §7.4 and §7.7.) It is clearly also possible to have computer systems that combine elements of both top-down and bottom-up organizations.

The important thing for our purposes here is that both top-down and bottom-up computational procedures are things that can be put on a general-purpose computer and are therefore both to be included under the heading of

*To American readers: this is the 'greatest common divisor', or 'GCD'.

what I am referring to as *computational* and *algorithmic*. Thus, with the bottom-up (or combined) systems, the *way* in which the system modifies its procedures is itself provided by something entirely computational which is specified ahead of time. This is the reason that the entire system can indeed be implemented on an ordinary computer. The essential *difference* between a bottom-up (or combined) system and a top-down one lies in the fact that with bottom-up systems, the computational procedure must contain a 'memory' of its previous performance ('experience'), so that this memory can be incorporated into its subsequent computational actions. For the moment, the details of this are not particularly important, but some further discussion will be given in §3.11.

According to the aspirations of *artificial intelligence* (abbreviated 'AI'), one strives to imitate intelligent behaviour, at whatever level, by some kind of computational means. Here, both top-down and bottom-up organizations have been frequently used. Originally it was the top-down systems that seemed most promising,[10] but now bottom-up systems of the artificial neural network type have become particularly popular. It would seem that it is in some kind of *combination* of top-down and bottom-up organization that we must expect to find the most successful AI systems. There are different kinds of advantage to be gained from each. The successes with top-down organization tend to be in areas in which the data and the operational rules are clearly delineated and of a very well-defined computational kind, such as with certain specific mathematical problems or chess-playing computer systems or, say, with medical diagnosis in which sets of rules are given for diagnosing different diseases, based on accepted medical procedures. The bottom-up organization tends to be useful when the criteria for decisions are not very precise or are ill-understood, such as with the recognition of faces or sounds, or perhaps in prospecting for mineral deposits, where the improvement of performance by experience is the basic behavioural criterion. In many such cases, there could actually be elements of *both* top-down and bottom-up organization (such as with a chess computer that learns from its experiences, or where some clear-cut theoretical geological understanding is incorporated in a computational device aiding the search for mineral deposits).

I think it would be fair to say that only with certain instances of top-down (or primarily top-down) organization have computers exhibited a significant superiority over humans. The most obvious example is in straightforward numerical calculation, where computers would now win hands down—and also in 'computational' games, such as chess or draughts (checkers), and where there may be only a very few human players able to beat the best machines (more about this in §1.15 and §8.2). With bottom-up (artificial neural network) organization, the computers can, in a few limited instances, reach about the level of ordinary well-trained humans.

Another distinction between different kinds of computer systems is that which differentiates the *serial* from the *parallel* architecture. A serial machine is one that does its computations one after the other, in a step-by-step action, whereas a parallel one does many independent computations simultaneously, the results of these different computations being brought together only when appropriately many of them have been completed. Again, theories about how the brain might operate have been instrumental in the development of certain parallel systems. It should be emphasized, however, that there is not really a distinction of *principle* between serial and parallel machines. It is always possible to simulate parallel action serially, even though there are some kinds of problem (but by no means all) for which a parallel action can solve the problem more efficiently, in terms of computing time, etc., than a serial one can. Since I shall be concerned here mainly with matters of principle, the distinctions between parallel and serial computation will not be of much relevance to us.

1.6 Does viewpoint $\mathscr{C}$ violate the Church–Turing thesis?

Recall that according to viewpoint $\mathscr{C}$ the conscious brain is supposed to act in a way that is beyond computational simulation, whether top-down, or bottom-up, or whatever. Some people, in expressing their doubts about $\mathscr{C}$, might partially base these doubts on an assertion that $\mathscr{C}$ would contradict the (generally believed) so-called *Church thesis* (or Church–Turing thesis). What is Church's thesis? In its original form, put forward by the American logician Alonzo Church in 1936, it asserted that anything that could reasonably be called a 'purely mechanical' mathematical process—i.e. anything *algorithmic*—could be achieved within a particular scheme discovered by Church himself, called the *lambda calculus* (λ-calculus)[11] (a scheme of very considerable elegance and economy of concept; see ENM pp. 66–70 for a brief introductory account). Shortly afterwards, in 1936/7, the British mathematician Alan Turing found his own much more persuasive way of describing algorithmic processes, in terms of the action of theoretical 'computing machines' that we now call *Turing machines*. The Polish-born American logician Emil Post (1936) also developed a somewhat similar scheme to Turing's a little afterwards. Church's calculus was soon shown, by Church and by Turing separately, to be equivalent to Turing's (and hence also Post's) concept of a Turing machine. Moreover, modern general-purpose computers arose, to a considerable extent, out of Turing's very conceptions. As mentioned above, a Turing machine is, in fact, completely equivalent, in its action, to that of a modern computer—with the specific idealization that the computer must in principle have access to an unlimited storage capacity. Thus, the original Church's thesis is now seen simply to assert that mathematical algorithms are precisely the things that can be carried out by an idealized modern computer—which, with the *definition* of the word 'algorithm' that is now usually adopted, becomes mere tautology. There is certainly no contradiction with $\mathscr{C}$ involved in accepting this form of Church's thesis.*

It is, however, probable that Turing himself had something further in mind: that the computational capabilities of any *physical* device must (in idealization) be equivalent to the action of a Turing machine. Such an assertion would go well beyond what Church seems originally to have intended. Turing's own motivations for the development of the concept of 'Turing machine' were based on his ideas of what a human calculator might be able to achieve in principle (see Hodges 1983). It seems likely that he viewed physical action in general—which would include the action of a human brain—to be always reducible to some kind of Turing-machine action. Perhaps one should call this (physical) assertion 'Turing's thesis', in order to distinguish it from the original (purely mathematical) assertion of 'Church's thesis', which is in no way contradicted by $\mathscr{C}$. This, indeed, is the terminology that I shall adopt in this book. Accordingly, it is this *Turing's thesis*, not Church's thesis, that would be contradicted by viewpoint $\mathscr{C}$.

1.7 Chaos

There has been a great deal of interest, in recent years, in the mathematical phenomenon that goes under the name of 'chaos', where physical systems seem to be able to behave in wild and unpredictable ways (Fig. 1.1). Does the phenomenon of chaos provide the needed non-computable physical basis for a viewpoint of the nature of $\mathscr{C}$?

Chaotic systems are dynamically evolving physical systems, or mathematical simulations of such physical systems, or just mathematical models studied for their own sake, in which the future behaviour of the system depends extremely critically upon the precise initial state of the system. Although ordinary chaotic systems are completely deterministic and computational, they can, *in practice*, behave as though they are not deterministic at all. This is because the accuracy according to which the initial state needs to be known, for a deterministic prediction of its future behaviour, can be totally beyond anything that is conceivably measurable.

An example that is often quoted in this connection is the detailed long-range prediction of the weather. The laws governing the motion of air molecules, and also of the other physical quantities that might be relevant to computing the

*It sometimes arises, in some mathematical discussion, that a procedure is encountered that is 'obviously' algorithmic in nature, even though it may not be at all immediate how to formulate that procedure in the form of a Turing-machine or lambda-calculus operation. In such cases, one may assert that such an operation must indeed exist 'by Church's thesis'. See Cutland (1980), for example. There is nothing wrong with proceeding in this way, and there is certainly no contradiction with $\mathscr{C}$ involved. In fact, this kind of use of Church's thesis pervades much of the discussion of Chapter 3.

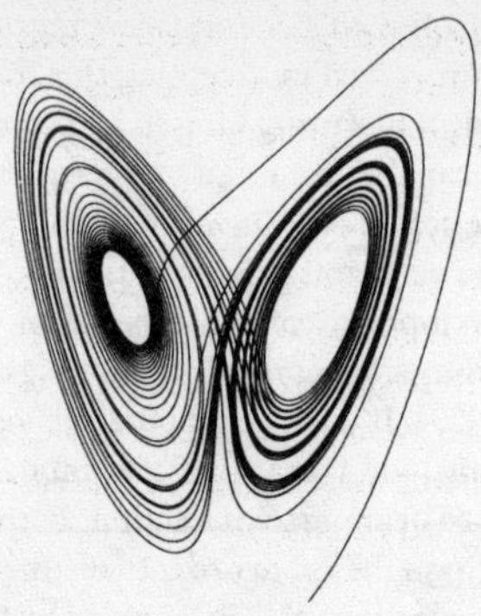

Fig. 1.1. The Lorentz attractor—an early example of a chaotic system. By following the lines, one moves from the left-hand lobe to the right-hand one and back again in a seemingly random fashion, and which lobe one finds oneself in at a given moment depends critically on the starting point. Yet the curve is defined by a simple mathematical (differential) equation.

weather, are all perfectly well known. However, the weather patterns that may actually emerge, after only a few days, depend so subtly on the precise initial conditions that there is no possibility of measuring these conditions accurately enough for reliable prediction. Of course the number of parameters that would have to enter into such a computation would be enormous, so it is perhaps not surprising that prediction, in this case, might prove to be virtually impossible in practice.

On the other hand, such so-called chaotic behaviour can occur also with very simple systems, such as those consisting of only a small number of particles. Imagine, for example, that you are asked to pocket the fifth snooker ball E in a crooked* and very well-spaced-out chain A, B, C, D, E, by hitting A with the cue so that A hits B, causing B to hit C, then C to hit D, and finally D to hit E into the pocket. The accuracy needed for this is, in general, far in excess of the abilities of any expert snooker player. If there were 20 in the chain, then even if the balls were perfectly elastic exact spheres, the task of potting the final ball would be far beyond the most accurate machinery of modern technology. In effect, the behaviour of the later balls in the chain would be random, despite the fact that the Newtonian laws governing the behaviour of the balls is mathematically completely deterministic and in principle effectively computable. No computation could predict the *actual* behaviour of the later balls in the chain, simply because there would be no way of determining enough accuracy for the actual

*In an earlier draft of this book I had not included the word 'crooked' here. If the balls are all precisely arranged in a straight line, then the feat turns out to be quite easy, as I learned to my surprise when testing things myself. There is a fortuitous stability that occurs with precise alignment, but this does not hold in the general case.

initial position and velocity of the cue or of the positions of the earlier balls in the chain. Moreover, even tiny external effects, such as the breathing of someone in the next town, might disturb this accuracy so much to render any such computation useless.

I should make it clear that despite such profound difficulties for deterministic prediction, all the normal systems that are referred to as 'chaotic' *are* to be included in what I call 'computational'. Why is this? As in other situations that we shall be coming to later, all we need to do, in order to decide that a procedure is computational, is to ask: can it be put on an ordinary general-purpose computer? Clearly the answer must be 'yes' in this case, simply for the reason that mathematically described chaotic systems are indeed normally studied by putting them on a computer!

Of course if we try to perform a computational simulation predicting the detailed weather patterns over Europe for the course of a week, or the successive collisions of 20 unaligned well-separated snooker balls after a swift stroke of the cue, then our simulation is almost certainly not going to resemble at all closely what *actually* happens. This is in the nature of chaotic systems. It is not practicable to predict computationally the *actual* outcome of the system. Nevertheless, a simulation of a *typical* outcome is perfectly achievable. The predicted weather may well not be the weather that actually occurs, but it is perfectly plausible as *a* weather! Likewise, the predicted outcome of the snooker ball collisions is completely acceptable as a *possible* outcome, even though in fact the snooker balls may well turn out to do something quite different from what is computed—but something quite equally acceptable. One further point that emphasizes the perfectly computational nature of these operations is that if the computer simulation is rerun, using just the same input data as before, then the outcome of the simulation is *precisely* the same as it was before! (This assumes that the computer itself makes no error; but in any case modern computers only very rarely make actual computational errors.)

In the context of artificial intelligence, one is not, after all, trying to simulate the behaviour of any particular individual; one would be very satisfied with the simulation just of *an* individual! Thus, it is not at all unreasonable to take the view that I am indeed taking: that chaotic systems should certainly be included in what we call 'computational'. A computer simulation of such a system would indeed act out a perfectly reasonable 'typical case', even though it might not turn out to be any 'actual case'. If the external manifestations of human intelligence are the results of some chaotic dynamical evolution—an evolution that is computational in the sense just described—then this would be in accordance with viewpoints $\mathscr{A}$ and $\mathscr{B}$, but *not* $\mathscr{C}$.

From time to time, it has been suggested that this phenomenon of chaos, if it occurs in the internal action of a physical brain, might be what enables our brains to behave in ways that *appear* to differ from the computably deterministic activity of a Turing machine even though, as emphasized above, it *is* technically computational. I shall need to come back to this issue later (cf. §3.22). For the moment, all that needs to be made clear is that chaotic systems *are* included in what I mean by 'computational' or 'algorithmic'. The question of whether or not something can be simulated *in practice* is a separate one from the *in principle* issues that are under consideration here.

1.8 Analogue computation

So far, I have been considering 'computation' only in the sense in which that term applies to modern digital computers or, more accurately, to their theoretical forerunners: the Turing machines. There are other types of computational device that have been used, particularly in the past, in which the operations are not represented in terms of the discrete 'on/off' states that are familiar in digital computations, but in terms of continuous physical parameters. The most familiar of such devices is the slide rule, where the physical parameter is linear distance (along the slide rule). This distance is used to represent the logarithms of the numbers that are to be multiplied or divided. There are many different types of analogue computational device, and other kinds of physical parameter can be used, such as time, mass, or electric potential.

With analogue systems we have to confront the technical point that the standard notions of computation and computability apply, strictly speaking, only to *discrete* systems (which is what 'digital' operation is concerned with) and not to the *continuous* ones, like distances or electric potentials, say, that are involved in conventional classical physical theory. In order to apply the ordinary notions of computation to a system whose description requires continuous rather than discrete (or 'digital') parameters, therefore, it is natural to resort to approximations. Indeed, in computer simulations of physical systems generally, it is the normal procedure to *approximate* all the continuous parameters under consideration in this discrete way. There would however, be some error involved in doing this, and for a given selected degree of accuracy of approximation, there could be physical systems of interest for which that particular accuracy might not be enough. Accordingly, this discrete computer simulation might lead to misleading conclusions as to the behaviour of the continuous physical system that is being simulated.

In principle, the accuracy could always be increased until that accuracy was adequate for simulating the continuous system under consideration. However, particularly in the case of chaotic systems, the computation time and memory store required might turn out to be prohibitive in practice. Moreover, there is the technical point that one might never be absolutely certain when the degree of accuracy that has been selected *is* sufficient. Some kind of test would be needed to signal to us when the point has been reached where no more accuracy is required, and the qualitative behaviour that is computed using that level of accuracy can indeed be trusted. This raises a number of somewhat delicate mathematical issues, and it would not be appropriate for me to enter into them in detail here.

There are, however, other approaches to the computational issues raised by continuous systems, in which the systems are treated as mathematical structures in their own right, with their *own* notion of 'computability'—a notion which generalizes the idea of Turing computability from the discrete to the continuous.[12] Using such a notion, it becomes unnecessary to approximate a continuous system by discrete parameters in order that the conventional notion of Turing computability can be applied. Such ideas are interesting from the mathematical point of view, but unfortunately they do not seem to have achieved, as yet, the compelling naturalness and uniqueness that applies to the standard notion of Turing computability for discrete systems. Moreover, there are certain anomalies whereby a technical 'non-computability' arises for simple systems where it is not clear that such a terminology is really appropriate (e.g. even for the simple 'wave equation' of physics; cf. Pourel and Richards (1981), ENM pp. 187–8). It should be mentioned, on the other hand, that some fairly recent work (Rubel 1989) has shown that theoretical analogue computers, belonging to a certain rather broad class, cannot reach beyond ordinary Turing computability. I believe that these are interesting and important matters which will be illuminated by further research. However it is not clear to me that this body of work, as a whole, has yet reached the point where it can be applied in a definitive way to the issues under discussion here.

In this book I am specifically concerned with the question of the computational nature of mental activity, where 'computational' is to be taken in the normal sense of *Turing computability*. Ordinary present-day computers are indeed digital in nature, and it is this that is relevant to today's AI activity. Perhaps it is conceivable that, in the future, some different kind of 'computer' might be introduced, that makes *critical* use of continuous physical parameters—albeit within the standard theoretical framework of today's physics—enabling it to behave in a way that is essentially *different* from a digital computer.

However, these issues are of relevance mainly to the distinction between the 'strong' and 'weak' versions of viewpoint $\mathscr{C}$. According to the *weak* version of $\mathscr{C}$, there would have to be physical actions underlying the behaviour of the conscious human brain that are non-computable in the standard sense of discrete Turing computability, but which can be entirely understood in terms of present-day physical theories. For this to be possible, it would appear that these actions would have to depend upon continuous physical parameters in such a way that they cannot be properly simulated by standard digital procedures. According to the *strong* version of $\mathscr{C}$, on the other hand, the non-computability would have to come from some non-computable physical theory—as yet undiscovered—whose implications are essential ingredients of conscious brain action. Although this second possibility might seem far fetched, the alternative (for $\mathscr{C}$-supporters) is, indeed, to find a role for some continuous action from amongst the known laws of physics, which cannot be properly simulated in any computational way. However, the expectation for the moment must surely be that, for any reliable analogue system of any type that has been seriously envisaged to date, it *would* be possible—in principle at least—to provide an effective digital simulation of it.

Even apart from theoretical issues of this general kind, it is the *digital* computers of today that hold most of the advantages over analogue ones. Digital action is much more accurate, essentially for the reason that with digital number storage, accuracy can be increased simply by increasing digit length, which is easily achieved with only a modest (logarithmic) increase in the computer's capacity; whereas for analogue machines (at least, for *entirely* analogue ones, into which digital concepts are not imported), accuracy is increased only by comparatively enormous (linear) increases in computer capacity. It may be that new ideas will come along in the future that are to the analogue machine's advantage, but with present-day technology, most of the significant practical advantages seem to be very much on the side of *digital* computation.

1.9 What kind of action could be non-computational?

Most kinds of well-defined action that come to mind are, accordingly, things which would have to be included under what I am referring to as 'computational' (meaning 'digital-computational'). The reader might begin to worry that there is nothing reasonable left for the viewpoint $\mathscr{C}$ to operate with. I have said nothing yet about strictly *random* actions, that might be provided, say, by some input from a quantum system. (Quantum mechanics will be discussed at some length in Part II, Chapters 5 and 6.) However, it is hard to see what advantage to a system there might be in having a *genuinely* random input, as opposed to a merely *pseudo-random* one that *can* be generated entirely computationally (cf. §3.11). Indeed, although there are, strictly speaking, some technical differences between 'random' and 'pseudo-random', these differences would seem to be of no real relevance to the issues of AI. Later on, in §3.11, §3.18 *et seq.*, I shall give strong arguments to show that 'pure randomness' indeed does nothing useful for us; if anything, it would be better to stay with the pseudo-randomness of chaotic behaviour—and, as was emphasized above, all the normal types of chaotic behaviour count as 'computational'.

What about the role of the environment? As each individual human being develops, he or she is provided with a unique environment, not shared by any other human being. Might it not be the case that it is this unique personal environment that gives each of us an input that is beyond computation? However, I find it hard to see how the 'uniqueness' of our environment helps in this context. The discussion is similar to that concerning chaos above (cf. §1.7). Provided that there is nothing beyond computation in the simulation of a *plausible* (chaotic) environment, such a simulation is all that would be needed for the training of a computer-controlled robot. The robot does not need to learn its skills through any actual environment; a computationally simulated *typical* (rather than actual) environment for the robot would certainly suffice.

Might it be that there is something inherently impossible in computationally simulating even a plausible environment? Perhaps there *is* something in the external physical world that is actually beyond computational simulation. Some supporters of $\mathscr{A}$ or $\mathscr{B}$ might be inclined to attribute seemingly non-computational acts of human behaviour to a lack of computability in that external environment. It would be rash, however, for $\mathscr{A}$- or $\mathscr{B}$-supporters to rely upon such an argument. For once it is accepted that there might be something *somewhere* in physical behaviour that cannot be simulated computationally; this would undermine what is presumably the main reason for doubting the plausibility of $\mathscr{C}$ in the first place. If there are actions in the outside environment that lie beyond computational simulation, then why not also *internal* to the brain? The internal physical organization of the human brain, after all, appears to be far more sophisticated than most (at least) of its environment—except, perhaps, where that environment is itself strongly influenced by the actions of other human brains. The acceptance of *external* non-computational physical action concedes the main case against $\mathscr{C}$. (See also the further discussion in §3.9, §3.10.)

One further point should be made in connection with the notion of something that might be 'beyond computation', as is required by $\mathscr{C}$. I do *not* simply mean something that is beyond *practical* computation. It could be argued, on the other hand, that the simulation of any plausible environment, or any accurate enactment of all the physical and chemical processes taking place in a brain, might be something that though in principle computational would take so much time to compute, or use so much memory space, that there could be no prospect of the computation being carried out on any actual or foreseeable computer. Perhaps the mere writing of an appropriate computer program would simply be out of the question, owing to the large number of different factors needed to be taken into account. Relevant though such considerations may well be, however, (and they will be discussed in §2.6, **Q8**, and §3.5) they are *not* what I mean here by 'non-computable', as is required by $\mathscr{C}$. I mean, instead, something that is *in principle* beyond computation, in a sense that I shall describe in a moment. Computations that are merely beyond existing or foreseeable computers, or computational techniques, are still 'computations' in the technical sense.

The reader may well ask: if there is nothing that counts as 'non-computational' in randomness or in environmental influences, or in sheer unmanageable complication, then what can I possibly have in mind by use of this term—as is required for viewpoint $\mathscr{C}$? What I have in mind rests on certain types of mathematically precise activity that can be *proved* to be beyond computation. As far as is yet known, no such mathematical activity is needed to describe physical behaviour. Nevertheless, it is a logical possibility. Moreover, it is *not just* a logical possibility. According to the arguments of this book, something of this general nature *must* be inherent in physical laws, despite the fact that such things have not yet been encountered in known physics. Certain instances of this kind of mathematical activity are remarkably simple, so it will be appropriate to illustrate what I have in mind in terms of these.

I shall need to start by describing some examples of classes of well-defined mathematical problems that—in a sense I shall explain in a moment—have no general computational solution. Starting with any such class of problem, it will be possible to construct a 'toy model' of a physical universe whose action, though entirely deterministic, is actually beyond computational simulation.

The first example of such a class of problems is the most famous of all: that known as 'Hilbert's tenth problem', which was put forward by the great German mathematician David Hilbert in 1900, as one of a list of then unanswered mathematical questions that set much of the stage for the subsequent development of mathematics in the early (and even late) twentieth century. Hilbert's tenth problem was to find a computational procedure for deciding, for a given system of *Diophantine* equations, whether the equations have any common solution.

What are Diophantine equations? They are polynomial equations, in any number of variables, for which all the coefficients and all the solutions must be *integers*. (An integer is just a whole number: one of the list . . ., $-3, -2, -1, 0, 1, 2, 3, 4, \ldots$. Diophantine equations were first systematically studied by the Greek mathematician Diophantos in the third century AD.) An example of a system of Diophantine equations is

$$6w+2x^2-y^3=0,\ 5xy-z^2+6=0,\ w^2-w+2x-y+z-4=0.$$

And another example is

$$6w+2x^2-y^3=0,\ 5xy-z^2+6=0,\ w^2-w+2x-y+z-3=0.$$

The first system is solved, in particular, by

$$w=1,\ x=1,\ y=2,\ z=4,$$

whereas the second system has no solution whatever (because, by its first equation, y must be an even number whence, by its second, z must be even also, but this contradicts its third equation, whatever w is, because w^2-w is always even, and 3 is an odd number). The problem posed by Hilbert was to find a mathematical procedure—or *algorithm*—for deciding which Diophantine systems have solutions, like our first example above, and which have not, as was the case for our second example. Recall (cf. §1.5) that an algorithm is just a computational procedure—the action of some Turing machine. Thus Hilbert's tenth problem asks for a computational procedure for deciding when a system of Diophantine equations can be solved.

Hilbert's tenth problem was historically very important because, in posing it, Hilbert raised an issue that had not been raised before. What does it actually *mean*, in precise mathematical terms, to have an algorithmic solution to a class of problems? What, in precise terms, *is* an algorithm? It was this very question that led Alan Turing, in 1936, to propose his own particular definition of what an algorithm is, in terms of his Turing machines. Other mathematicians (Church, Kleene, Gödel, Post, and others; cf. Gandy (1988)) proposed somewhat different procedures at about the same time. All were soon shown to be equivalent (by Turing and Church), but Turing's particular approach has turned out to have been the most influential. (He, alone, introduced the idea of having one particular all-inclusive algorithmic machine—called a *universal* Turing machine—which can, by itself, achieve *any* algorithmic action whatever. It was this that led to the idea of a general-purpose computer, that is so familiar to us now.) Turing was able to show that there are certain classes of problem that do *not* have any algorithmic solution (in particular the 'halting problem' that I shall describe shortly). However, Hilbert's actual tenth problem had to wait until 1970 before the Russian mathematician Yuri Matiyasevich—providing proofs that completed certain arguments that had been earlier put forward by the Americans Julia Robinson, Martin Davis, and Hilary Putnam—showed that there can be no computer program (algorithm) which decides yes/no systematically to the question of whether a system of Diophantine equations has a solution. (See Davis (1978) and Devlin (1988), Chapter 6, for readable accounts of this story.) It may be remarked that whenever the answer happens to be 'yes', then that fact can, in principle, be ascertained by the particular computer program that just slavishly tries all sets of integers one after the other. It is the answer 'no', on the other hand, that eludes any systematic treatment. Various sets of rules for correctly giving the answer 'no' can be provided—like the argument using even and odd numbers that rules out solutions to the second system given above—but Matiyasevich's theorem showed that these can *never* be exhaustive.

Another example of a class of well-defined mathematical problems that have no algorithmic solution is the *tiling problem*. This is formulated as follows: given a set of polygonal shapes, decide whether these shapes will tile the plane; that is, is it possible to cover the entire Euclidean plane using only these particular shapes, without gaps or overlaps? This was (effectively) shown to be a computationally insoluble problem by the American mathematician Robert Berger in 1966, basing his arguments on an extension of some earlier work by the Chinese–American mathematician Hao Wang in 1961 (see Grünbaum and Shephard 1987). In fact, as I have just stated the problem, there is some awkwardness about it, because general polygonal tiles would need to be specified in some way using real numbers (numbers defined in terms of infinite decimals), whereas ordinary algorithms operate on whole numbers. This awkwardness can be removed by considering tiles that are made up just of a number of squares joined together at their edges. Such tiles are called *polyominoes* (see Golomb 1965; Gardner 1965, Chapter 13; Klarner 1981). Some examples are given in Fig. 1.2. (For other examples of tile sets, see ENM, pp. 133–7, Figs 4.6–4.12.) As a curious fact, the computational insolubility of the tiling problem depends upon the existence of certain sets of polyominoes called *aperiodic* sets—which will tile the plane *only non-periodically* (i.e. in a way so that the completed pattern never repeats itself no matter how far it is extended). In Fig. 1.3, an aperiodic set of three polyominoes is exhibited (developed from a tile set discovered by Robert Ammann in 1977, cf. Grünbaum and Shephard (1987), Figs 10.4.11–10.4.13 on pp. 555–6).

The mathematical proofs that Hilbert's tenth problem and the tiling problem are not soluble by computational means are difficult, and I shall certainly not attempt to give the arguments here.[13] The central point of each argument is to show, in effect, how any Turing-machine action can be coded into a Diophantine or tiling problem. This reduces the issue to one that Turing actually addressed in his original discussion: the computational insolubility of the *halting problem*—the problem of deciding those situations in which a Turing-machine action fails ever to come to a halt. In §2.3, various explicit computations that do *not* ever halt will be given; and in §2.5 a relatively simple argument will be presented—based essentially on Turing's original one—that shows, amongst other things, that the halting problem is indeed computationally insoluble. (The implications of the 'other things' that this argument actually shows will be central to the entire discussion of Part I!)

How can one use such a class of problems, such as Diophantine equations or the tiling problem, to construct a toy universe that is deterministic but non-computable? Let us suppose that our model universe has a *discrete time*, parameterized by the natural numbers (non-negative integers) 0, 1, 2, 3, 4, ... At time n, the state of the universe is to be specified by one of the class of problems under consideration—say, by a set of polyominoes. There are to be two well-defined rules about which of the polyomino sets represents the state of the universe at time $n+1$, given the polyomino set representing the universe at time n, where the first of these rules is to be adopted if the polyominoes *will* tile the plane and the second set is to be adopted when they will *not* tile the plane. The details of how one might specify such rules are not particularly important. One possibility would be to form a list $S_0, S_1, S_2, S_3, S_4, S_5, \ldots$ of all possible polyomino sets, in such a way that those involving an *even* total number of squares have even suffixes: $S_0, S_2, S_4, S_6, \ldots$; and those involving an *odd* total number of squares have odd suffixes: $S_1, S_3, S_5, S_7, \ldots$. (This would not be too hard to arrange according to some computational procedure.) The 'dynamical evolution' of our toy model universe is now given by:

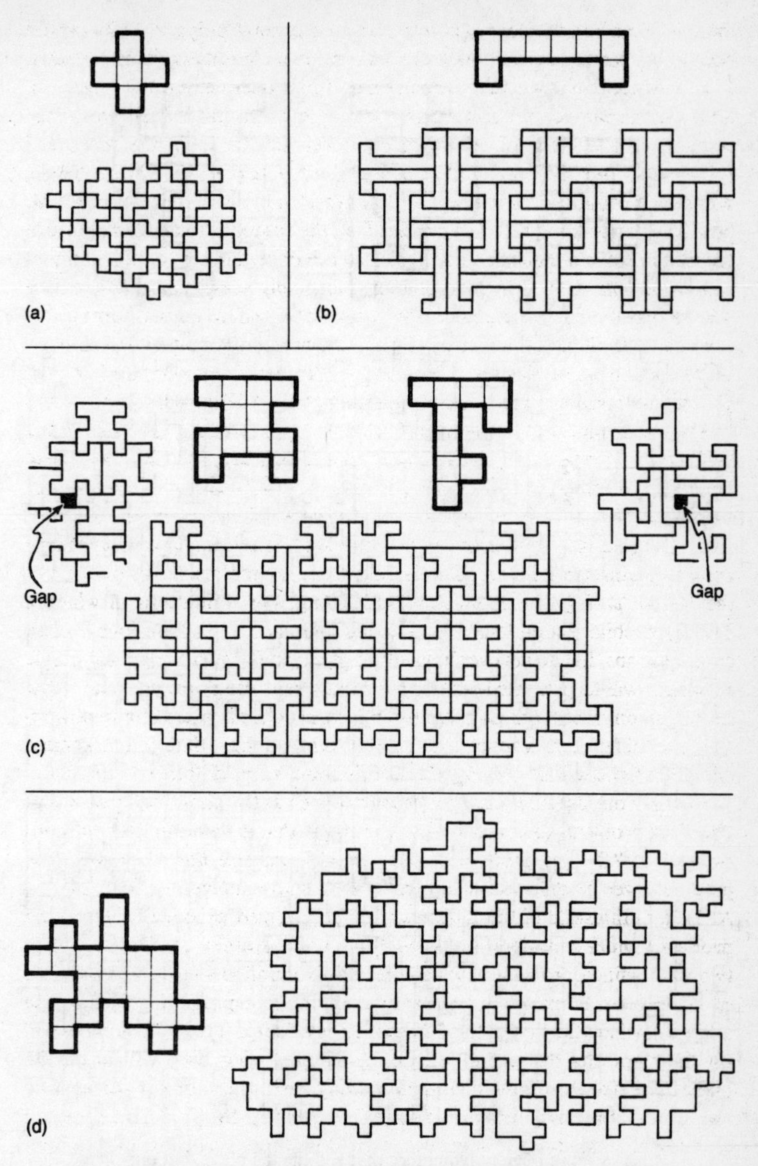

Fig. 1.2. Various sets of polyominoes that will tile the infinite Euclidean plane (reflected tiles being allowed). Neither of the polyominoes in set (c), if taken by itself, will tile the plane, however.

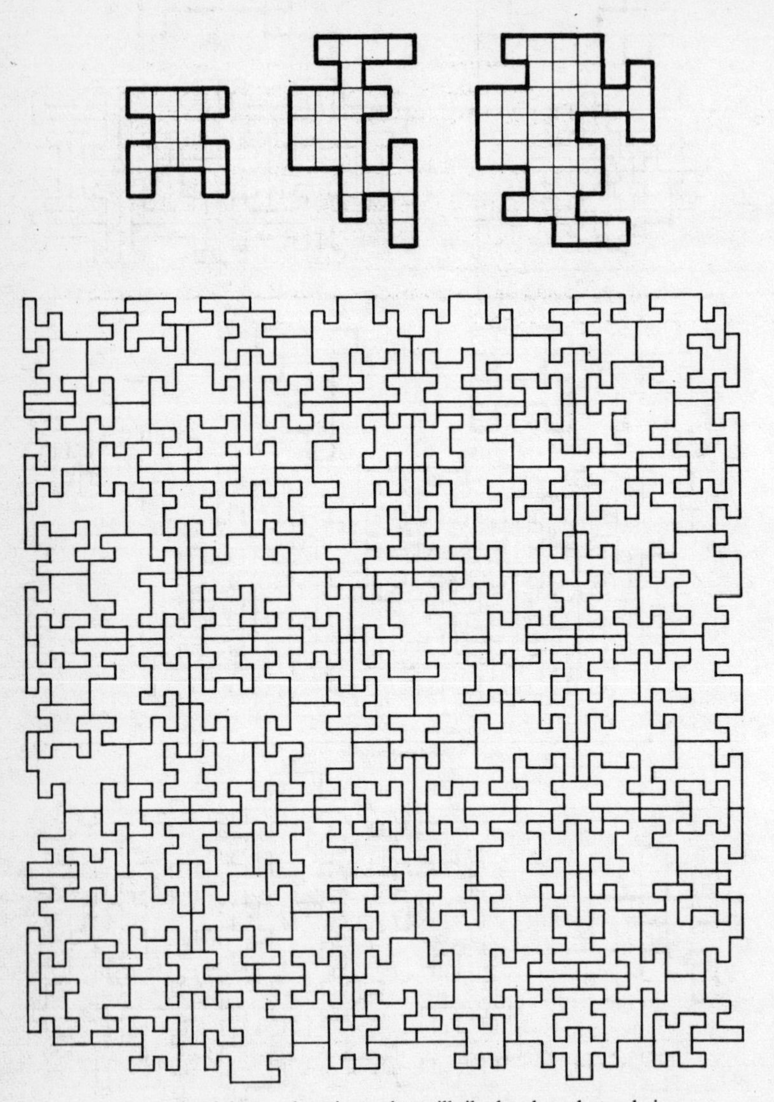

Fig. 1.3. A set of three polyominoes that will tile the plane, but only in a never repeating way (derived from a set by Robert Ammann).

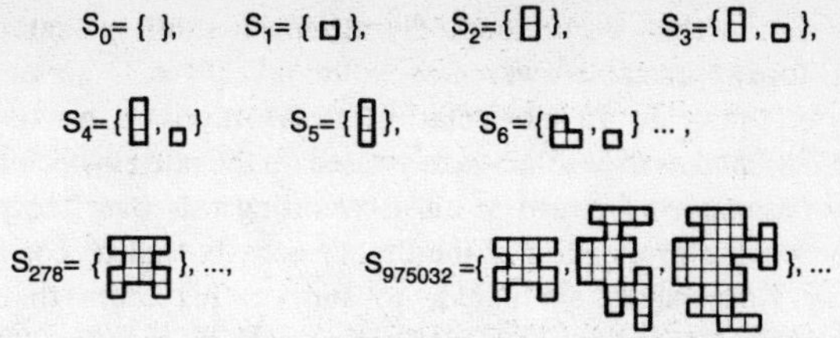

Fig. 1.4. A non-computable toy model universe. The different states of this deterministic but non-computable toy universe are given in terms of the possible finite sets of polyominoes, numbered in some way so that the even numbers S_n correspond to even total numbers of squares and odd numbers to an odd number of squares. The time-evolution proceeds in numerical order (S_0, S_2, S_3, S_4, . . . , S_{278}, S_{280}, . . .), except that a number is skipped whenever the previous set does not tile the plane.

> Universe state S_n at time t proceeds to S_{n+1} at time $t+1$ if the polyomino set S_n *will* tile the plane and to S_{n+2} if the set S_n will *not* tile the plane.

Such a universe behaves entirely deterministically, but because there is no general computational procedure for ascertaining when a polyomino set S_n will tile the plane (which holds just as well if the total number of squares is fixed as being even or as being odd) there is no computational simulation possible of its actual development. (See Fig. 1.4.)

Of course, this kind of scheme is not to be taken seriously as modelling the actual universe that we inhabit. It is presented here (as in ENM, p. 170) to illustrate the little-appreciated fact that there is a clear difference between determinism and computability. *There are completely deterministic universe models, with clear-cut rules of evolution, that are impossible to simulate computationally*. In fact, as we shall be seeing in §7.9, models of the very specific kind that I have just been considering will turn out to be insufficient for what is actually needed for viewpoint $\mathscr{C}$. But we shall be seeing in §7.10 that there are indeed some intriguing physical possibilities for what *is* needed!

1.10 What of the future?

What do the viewpoints $\mathscr{A}$, $\mathscr{B}$, $\mathscr{C}$, $\mathscr{D}$ tell us to expect for the future of this planet? According to $\mathscr{A}$, there will come a stage when appropriately programmed supercomputers will reach—and then race beyond—all human mental capabilities. Of course different people who hold to $\mathscr{A}$ could have very different views as to the time scales involved in this. Some might reasonably take the line that it will be many centuries before computers will reach our level, so little being presently understood about the computations that the brain must indeed be performing (they would claim) in order to achieve the subtlety of action that we undoubtedly attain—a subtlety that would be necessary before appreciable 'awareness' would take place. Others argue for a much shorter timescale. In particular, Hans Moravec, in his book *Mind Children* (1988), makes a reasoned case—based on the rate at which computer technology has moved forward at an accelerating rate over the past half-century and on the proportion of the brain's activity that he considers has already been successfully simulated—to support his claim that 'human equivalence' will already have been superseded by about the year 2030. (Some have argued for a much shorter timescale[14]—sometimes, even, where the predicted date for human equivalence has already passed! Lest the reader feel dismayed by the prospect of being overtaken by computers in less than (say) 40 years' time, hope is offered—indeed promised—by the assured prospect of our being able to transfer our 'mind programs' into the shining metallic (or plastic) bodies of the robots of our choice, thereby obtaining for ourselves a form of immortality (Moravec 1988, 1994).

Such optimism is not available to the holders of viewpoint $\mathscr{B}$, however. Their standpoint does not differ from $\mathscr{A}$ with regard to what computers will ultimately be able to achieve in an external way. An adequate *simulation* of the action of a human brain could itself be used to control a robot, simulation being all that is needed (Fig. 1.5). The issue of the presence of conscious

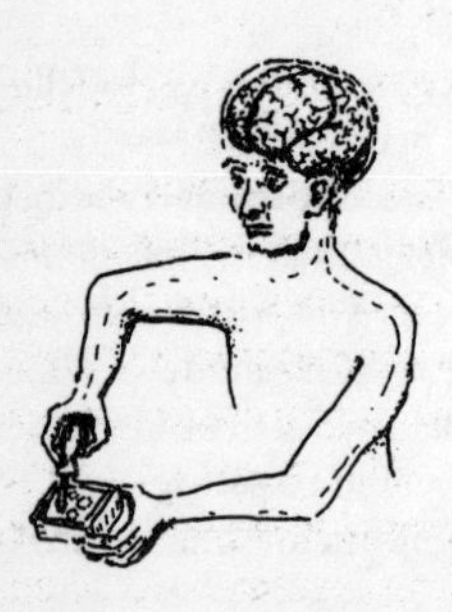

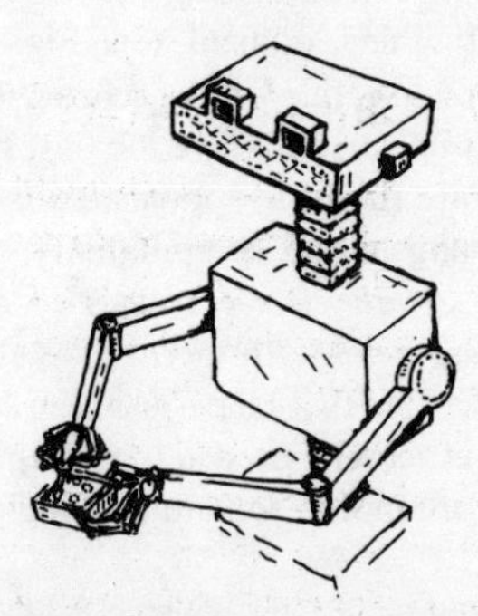

Fig. 1.5. According to viewpoint $\mathscr{B}$, a computer simulation of the activity of a conscious human brain would in principle be possible, so ultimately computer-controlled robots could reach, and thereafter enormously exceed all human capabilities.

awareness arising in association with this simulation is, according to $\mathscr{B}$, irrelevant to how the robot behaves. It might take centuries, or it might take less than 40 years for such a simulation to become a technical possibility. But according to $\mathscr{B}$, it should be possible eventually. Then such computers would have reached the level of 'human equivalence'; and they would again be expected to race beyond whatever level we are able to achieve with our relatively puny brains. The option to 'join' the computer-controlled robots is now not open to us, and it would seem that we must resign ourselves to the prospect of a planet ultimately ruled by insentient machines! Of all the viewpoints $\mathscr{A}$, $\mathscr{B}$, $\mathscr{C}$, $\mathscr{D}$, it seems to me that it is $\mathscr{B}$ that offers the most pessimistic view of the future of our planet—despite its apparently 'common-sense' nature!

According to $\mathscr{C}$ or $\mathscr{D}$, on the other hand, it would be expected that the computers would (or should) always remain subservient to us, no matter how far they advance with respect to speed, capacity, and logical design. Viewpoint $\mathscr{C}$, however, is open with regard to future scientific developments that might lead to the construction of devices—*not* based on computers as we understand them today, but upon the very non-computable physical action that $\mathscr{C}$ demands must underlie our own conscious thought processes—devices that could achieve *actual* intelligence and awareness. Perhaps it will be *these* devices, rather than 'computers', as we now understand that term, that will eventually race beyond all human capabilities. Possibly so, but such speculation seems to me to be extremely premature at the present time, since we lack almost all the scientific understanding that would be necessary, let alone any of the technological know-how. I shall return to this issue again in Part II (cf. §8.1).

1.11 Can computers have rights or responsibilities?

A related issue—one which might have a somewhat more immediate practical relevance—has begun to attract the attention of theoreticians in the *legal* profession.[15] This issue is whether, in the not-so-distant future, one might have to consider that computers could be held to have legal responsibilities or rights. Certainly, if in due course computers are to approach, or perhaps even exceed, human levels of expertise in many walks of life, then questions of this kind would indeed become relevant. If one believes in viewpoint $\mathscr{A}$, then one would be clearly driven in the direction of having eventually to accept that computers (or computer-controlled robots) must potentially have both rights and responsibilities. For according to that viewpoint, there is no essential difference—apart from the 'accidents' of differing material construction—between ourselves and sufficiently advanced robots. For those who hold to viewpoint $\mathscr{B}$, however, the issue seems less clear. One might reasonably argue that it is the possession of certain genuine mental qualities—such as suffering, anger, revenge, malice, faith, trust, intent, belief, understanding, or passion—that is the relevant issue with regard to rights or responsibilities. According to $\mathscr{B}$, a computer-controlled robot would have none of these qualities, and I suppose it could therefore have neither rights nor responsibilities. Yet, according to $\mathscr{B}$, there is no effective way of telling that these qualities are absent, so one might be placed in something of a quandary if robots were to achieve a sufficiently close mimicking of human behaviour.

This quandary would appear to be eliminated with viewpoint $\mathscr{C}$ (and, presumably, also with viewpoint $\mathscr{D}$) because, according to those viewpoints, computers could not convincingly *exhibit* mental qualities—and would certainly never actually possess them. It would accordingly follow that computers can have *no* rights and *no* responsibilities either. To me, this is a very reasonable point of view to adopt. In this book, I shall indeed be arguing strongly against both $\mathscr{A}$ and $\mathscr{B}$. Acceptance of the arguments that I give would certainly simplify the legal position: computers or computer-controlled robots, themselves, *never* have rights or responsibilties. Moreover, they deserve no share of the blame when things go wrong—that would always lie elsewhere!

It should be made clear, however, that these arguments do not necessarily apply to any putative 'devices', as noted above, that might ultimately be able to take advantage of non-computational physics. But since the prospect of such devices—if indeed they could be constructed at all—is not even on the horizon, there is no legal issue to be faced on this score, in the foreseeable future.

The issue of 'responsibility' raises deep philosophical questions concerning the ultimate causes of our behaviour. It might well be argued that each of our actions is ultimately determined by our inheritance and by our environment—or else by those numerous chance factors that continually affect our lives. Arc not *all* of these influences 'beyond our control', and therefore things for which we cannot ultimately be held responsible? Is the matter of 'responsibility' merely one of convenience of terminology, or is there actually something else—a 'self' lying beyond all such influences—which exerts a control over our actions? The legal issue of 'responsibility' *seems* to imply that there is indeed, within each one of us, some kind of an independent 'self' with its *own* responsibilities—and, by implication, rights—whose actions are *not* attributable to inheritance, environment, or chance. If it is other than a mere convenience of language that we speak as though there were such an independent 'self', then there must be an ingredient missing from our present-day physical understandings. The discovery of such an ingredient would surely profoundly alter our scientific outlook.

This book will not supply an answer to these deep issues, but I believe that it may open the door to them by a crack—albeit only by a crack. It will not tell us that there need necessarily be a 'self' whose actions are not attributable to external cause, but it will tell us to broaden our view as to the very nature of what a 'cause' might be. A 'cause' could be something that cannot be computed in practice or in principle. I shall argue that when a 'cause' is the effect of our conscious actions, then it must be something very subtle, certainly beyond computation, beyond chaos, and also beyond any purely random influences. Whether such a concept of 'cause' could lead us any closer to an understanding of the profound issue (or the 'illusion'?) of our free wills is a matter for the future.

1.12 'Awareness', 'understanding', 'consciousness', 'intelligence'

In the above discussions I have, as yet, made no attempt to be precise about any of the elusive concepts related to the issue of 'mind'. I have referred, somewhat vaguely, to 'awareness' in the definitions of $\mathscr{A}$, $\mathscr{B}$, $\mathscr{C}$, and $\mathscr{D}$ given in §1.3, but other qualities of mentality were not referred to at that point. I should at least make some attempt to clarify the terminology that I am using here, particularly in relation to such terms as 'understanding', 'consciousness', and 'intelligence', which have importance to the discussions of this book.

While I do not believe that attempts at full definitions would necessarily be helpful, some comments concerning my own terminology are in order. I am often disconcerted to find that a use of these words that seems obvious to me will disagree with what some others may claim as natural. For example, my own use of the term 'understanding' certainly implies that a genuine possession of this quality would require some element of *awareness* to be present. Without any awareness of what some argument is all about, there can surely be no genuine understanding of that argument. At least, this seems to *me* to be an unexceptionable use of words, though in some contexts, proponents of AI might appear to use the terms 'understanding' and 'awareness' in a way that denies such an implication. Some proponents of AI (either $\mathscr{A}$ or $\mathscr{B}$) would claim that a computer-controlled robot 'understands' what its instructions are even though no claim would be made that it is actually 'aware' of them. To me, this is a misuse of the word 'understands', though it is a misuse that has a genuine heuristic value for descriptions of computer functioning. When I am trying to be clear that I am not using 'understands' in this heuristic way, I shall use the phrase 'genuinely understands', or 'genuine understanding', for that activity for which awareness is indeed necessary.

Of course, some might argue that there is no clear-cut distinction between these two uses of the term 'understands'. If one believes that there is no distinction, then one must believe that awareness itself is an ill-defined concept. This, I do not deny; but it seems clear to me that awareness is indeed *something*, and this something may be present or absent, at least to a degree. If one agrees that awareness *is* a something, then it seems natural that one should also agree that this something must be part of any genuine understanding. This still allows that the 'something' that is awareness might actually be a feature of a purely computational activity, in accordance with viewpoint $\mathscr{A}$.

It would also seem to me to be unexceptionable to use the word 'intelligence' only when there can be some understanding involved. However, again, certain AI proponents might claim that their robot could be 'intelligent' without its needing actually to 'understand' anything. The term 'artificial intelligence' implies that intelligent computational activity is presumed possible, but genuine understanding—and certainly awareness—is argued by some to be outside the aims of AI. To my own way of thinking, 'intelligence' without understanding is a misnomer. To a limited degree, some kind of partial simulation of genuine intelligence without any actual understanding may sometimes be possible. (Indeed, one not infrequently encounters *human* individuals who are able to fool us for a while into believing that they possess some understanding, when it finally emerges that indeed they possess none whatever!) It will be an important feature of my later discussions that there is indeed a clear-cut distinction between genuine intelligence (or genuine understanding) and any entirely computationally simulated activity. According to my own terminology, the possession of *genuine* intelligence indeed requires that genuine understanding must be present. Hence, my use of the term 'intelligence' (especially when prefixed with the word 'genuine') would imply the presence of some actual awareness.

To me, this seems to be a natural terminology, but many AI proponents[16] (certainly those who do *not* support viewpoint $\mathscr{A}$) would strongly deny that they are attempting to provide an artificial 'awareness' even though, as the name seems to imply, they are indeed attempting to construct an artificial 'intelligence'. Perhaps such people would assert that they are (in accordance with viewpoint $\mathscr{B}$) merely *simulating* intelligence—which does not need *actual* understanding or awareness—rather than attempting to achieve what I am calling *genuine* intelligence. Perhaps they would claim not to recognize any distinction between genuine and simulated intelligence, as would be an implication of viewpoint $\mathscr{A}$. It will be one of my purposes, in later arguments, to show that there is indeed an aspect of 'genuine understanding' that cannot be properly simulated in any computational way whatever. Consequently, there must indeed be a distinction between genuine intelligence and any attempt at a proper computational simulation of it.

Of course I have not defined *any* of the terms 'intelligence', 'understanding', or 'awareness'. I think that it would be most unwise to attempt to give *full* definitions here. We shall need to rely, to some extent, on our intuitive perceptions as to what these words actually mean. If our intuitive concept of 'understanding' is that it is something that is necessary for 'intelligence', then an argument which establishes the non-computational nature of 'understanding' will also establish the non-computational nature of 'intelligence'. Moreover, if 'awareness' is something that is needed for 'understanding', then a non-computational physical basis for the phenomenon of awareness might account for such a non-computational nature for 'understanding'. Thus, my own use of these terms (and, I maintain, common usage also) entails the implications:

(a) 'intelligence' *requires* 'understanding'

and

(b) 'understanding' *requires* 'awareness'.

Awareness, I take to be one aspect—the *passive* aspect—of the phenomenon of *consciousness*. Consciousness has an *active* aspect also, namely the feeling of *free will*. I shall not attempt a full definition of the word 'consciousness' here either (and certainly not of 'free will'), even though my arguments are directed towards the goal of ultimately comprehending this phenomenon of consciousness in scientific but non-computational terms—as would be required by viewpoint $\mathscr{C}$. Nor do I claim to have proceeded any great distance along the road towards this goal, though I hope that the arguments that I am presenting in this book (and in ENM) will provide a few useful signposts—and perhaps even a little more than that. I feel that to attempt to define the term 'consciousness' too closely at this stage would put us in danger of allowing to slip away the very concept that we wish to ensnare. Accordingly, instead of providing some premature and inadequate definition, I shall give just a few descriptive comments concerning my use of the term 'consciousness'. When all is said and done, we must ultimately rely upon our intuitive comprehension of its meaning.

This is not intended to suggest that I believe that we really 'intuitively know' what consciousness actually 'is', but merely that there is such a concept that we are trying to grasp in some way—a genuine scientifically describable phenomenon, playing an active as well as a passive role in the physical world. Some people appear to believe that the concept is too vague to merit serious study. Yet such people[17] frequently do not hesitate to discuss the concept of 'mind' as though it were better defined. The normal usage of the word 'mind' allows that there can be (and indeed is) something that we often refer to as the 'unconscious mind'. To my way of thinking, there is a much greater obscurity concerning the concept of an unconscious mind even than that of a conscious one. Although I make not infrequent use of the word 'mind' myself also, I am making no attempt to be precise about it. The concept of 'mind'—*apart* from whatever is already embodied in the term 'consciousness'—will not play the central role in my attempts at rigorous discussion.

So what do I mean by consciousness? As I have remarked earlier, there are both passive and active aspects to consciousness, but it is not always clear that there is a distinction between the two. The perception of the colour red, on the one hand, is something that certainly requires passive consciousness, as is the sensation of pain or the appreciation of a melody. Active consciousness is involved in the willed action to get up from one's bed, as it is in a deliberate decision to desist from some energetic activity. The bringing to mind of an early memory involves both active and passive aspects of consciousness. Consciousness, active and passive, would also be normally involved in the formulation of a future plan of action, and it certainly seems to be that there is a necessity for some kind of consciousness in the type of mental activity that would normally be encompassed in the word 'understanding'. Moreover, we may be (passively) conscious, to a degree, even when we are asleep, provided that we experience some dream (and even the active aspect of consciousness may sometimes begin to be playing a role just as we awake).

Some people might dispute an all-embracing role of a single concept of consciousness in all of these various manifestations. They might insist that there are numerous quite different concepts of 'consciousness' involved—not merely just 'active' and 'passive'—and that there are really a great many distinct mental attributes separately pertinent to all these various mental qualities. Accordingly, applying the blanket term 'consciousness' to all of these might be considered to be, at best, unhelpful. In my own view, there is indeed a unified concept of 'consciousness' that is central to all these separate aspects of mentality. While I do admit to there being passive and active aspects to consciousness that are sometimes distinguishable, the passive having to do with sensations (or 'qualia') and the active with the issues of 'free will', I take these to be two sides of a single coin.

I shall be primarily concerned, in Part I of this book, with the issue of what it is possible to achieve by use of the mental quality of 'understanding'. Though I do not attempt to define what this word means, I hope that its meaning will indeed be clear enough that the reader will be persuaded that this quality—whatever it is—must indeed be an essential part of that mental activity needed for an acceptance of the arguments of §2.5. I propose to show that the appreciation of *these* arguments must involve something non-computational. My argument does not so *directly* address the other issues of 'intelligence', 'awareness', 'consciousness', or 'mind', but there should be a clear relevance of that discussion to those concepts also, since the 'common-sense' terminology that I have indicated above implies that awareness ought indeed to be an essential ingredient of our understanding, and that understanding must be a part of any genuine intelligence.

1.13 John Searle's argument

Before presenting my own reasoning, I should make brief reference to a quite different line of argument—the well-known 'Chinese Room', of the philosopher John Searle[18]—mainly in order to emphasize the very different character and underlying intentions of my own line of reasoning. Searle's argument is also concerned with the issue of 'understanding' and whether an appropriately sophisticated computer action can be said to achieve that mental quality. I shall not repeat the Searle discussion in detail here, but give its essence only very briefly.

It is concerned with some computer program that purports to simulate 'understanding' by providing replies to questions put to it about a story that it has been told—all questions and answers being in Chinese. Searle then envisages a human subject, ignorant of Chinese, laboriously moving counters around in such a way as to act out all the detailed computations that the computer would perform. However, despite the appearance of understanding that is involved in the computer's output when *it* performs the computations, no such understanding is actually experienced by the *human* performing manipulations that enact these computations. Searle argues, accordingly, that the mental quality of understanding cannot be just a computational matter—for the human subject (not understanding Chinese) carries out every single act of computation that the computer carries out, but experiences no understanding whatever of the stories. Searle allows that a *simulation* of the output of the results of understanding could be possible, in accordance with viewpoint $\mathscr{B}$, since he is prepared to admit that this could be achieved by a computer simulating every relevant physical action of a human brain (doing whatever it does) when its human owner actually understands something. But by the Chinese Room argument, he insists that a *simulation* cannot, in itself, actually 'feel' any understanding. Thus *actual* understanding could not in fact be achieved by any computer simulation.

Searle's argument is directed against $\mathscr{A}$ (which would assert that any 'simulation' of understanding would be equivalent to 'actual' understanding) and it is presented as providing support for $\mathscr{B}$ (although equally supporting $\mathscr{C}$ or $\mathscr{D}$). It is concerned with the *passive*, *inward*, or *subjective* aspects of the quality of understanding. It does not deny the possibility of a simulation of understanding in its *active*, *outward*, or *objective* aspects. Indeed, Searle himself is on record as asserting: 'Of course the brain is a digital computer. Since everything is a digital computer, brains are too.'[19] This suggests that he would be prepared to accept the possibility of a complete simulation of the action of a conscious brain in the act of 'understanding' something, whence the external manifestations of this simulation would be identical with those of an actual conscious human being—in accordance with viewpoint $\mathscr{B}$. My own arguments, on the other hand, will be directed against just these outward aspects of 'understanding' and I thereby maintain that not even a proper computer simulation of the external manifestations of understanding is possible. I do not address Searle's discussion in detail here, since it offers no direct support for viewpoint $\mathscr{C}$ (support for $\mathscr{C}$ being the purpose of my own arguments here). For the record, however, it is worth remarking that I do regard the Chinese Room argument as providing a somewhat persuasive case against $\mathscr{A}$, although I do not think that this case is entirely conclusive. For further details and various counter-arguments, see Searle (1980) and the discussion therein and in Hofstadter and Dennett (1981); see also Dennett (1990) and Searle (1992). For my own assessment see ENM, pp. 17–23.

1.14 Some difficulties with the computational model

Before turning to the issues that specifically separate $\mathscr{C}$ from $\mathscr{A}$ and from $\mathscr{B}$, let us consider certain other difficulties that must be faced by any attempt at explanation of the phenomenon of consciousness in accordance with viewpoint $\mathscr{A}$. According to $\mathscr{A}$, it is the mere 'carrying out' or *enaction* of appropriate algorithms that is supposed to evoke awareness. But what does this actually mean? Does 'enaction' mean that bits of physical material must be moved around in accordance with the successive operations of the algorithm? Suppose we imagine these successive operations to be written out line by line in a massive book.[20] Would the act of writing or printing these lines constitute 'enaction'? Would the mere static existence of the book be sufficient? What about simply moving one's finger down the lines one after the other; would that count as 'enaction'? What about moving a finger over the symbols if they are written in braille? How about projecting the pages of the book successively on a screen? Does the mere *presentation* of the successive operations of an algorithm constitute enaction? Would it be necessary, on the other hand, to have someone check that each line correctly follows from those that precede it, according to the rules of the algorithm in question? Presumably *that*, at least, would be begging the question, since no auxiliary person's (conscious) understandings should be needed for the process. The issue of what physical actions should count as actually enacting an algorithm is profoundly unclear. Perhaps such actions are not necessary at all, and to be in accordance with viewpoint $\mathscr{A}$, the mere Platonic mathematical existence of the algorithm (cf. §1.17) would be sufficient for its 'awareness' to be present.

In any case, it would presumably not be the case, according to $\mathscr{A}$, that just *any* complicated algorithm could evoke (appreciable) awareness. It would be expected that some special features of the algorithm such as 'higher-level organization', or 'universality', or 'self-reference', or 'algorithmic simplicity/complexity',[21] or some such, would be needed before significant awareness could be considered to be evoked. Moreover, there is the sticky issue of what particular qualities of an algorithm would be supposed to be responsible for the various different 'qualia' that constitute our awareness. What kind of computation evokes the sensation 'red', for example? What computations constitute the sensations of 'pain', 'sweetness', 'harmoniousness', 'pungency', or whatever? Attempts have sometimes been made by proponents of $\mathscr{A}$ to address issues of this nature (cf. Dennett 1991, for example), but so far these attempts do not strike me as at all persuasive.

Moreover, any clear-cut and reasonably simple algorithmic suggestion (such as any that can have been made in the literature so far) would suffer from the drawback that it could be implemented without great difficulty on a present-day electronic computer. Such an implementation would, according to the proponents of such a suggestion, have to evoke the *actual* experience of the intended qualium. It would be hard, even for those who adhere strongly to viewpoint $\mathscr{A}$, to accept seriously that such a computation—any computation that could be set in action on the computers of today, using the AI understandings of today—could *actually* experience mentality to any significant degree. It would therefore appear to be the case that proponents of such suggestions must resort to the belief that it is the sheer *complication* of the computations (acting in accordance with these suggestions) that are involved in the activities of our own brains that allow us to have appreciable mental experiences.

This raises some other issues that I have not seen addressed, to any significant degree. If one believes that it is essentially the vast complication of the 'wirings' constituting the interconnected network of the brain's neurons and synapses that is the prerequisite for our significant conscious mental activity, then one must somehow come to terms with the fact that consciousness is not equally a feature of all parts of the human brain. When the term 'brain' is used without qualification, it is natural (at least for non-specialists) to think in terms of the large convoluted outer regions that constitute what is known as the *cerebral cortex*—the outer grey matter of the *cerebrum*. There are roughly one hundred thousand million (10^{11}) neurons involved in the cerebral cortex, which indeed leaves considerable scope for enormous complication—but the cerebral cortex is far from being all there is to the brain. At the back and underneath is another important entangled mass of neurons known as the *cerebellum* (see Fig. 1.6). The cerebellum appears to

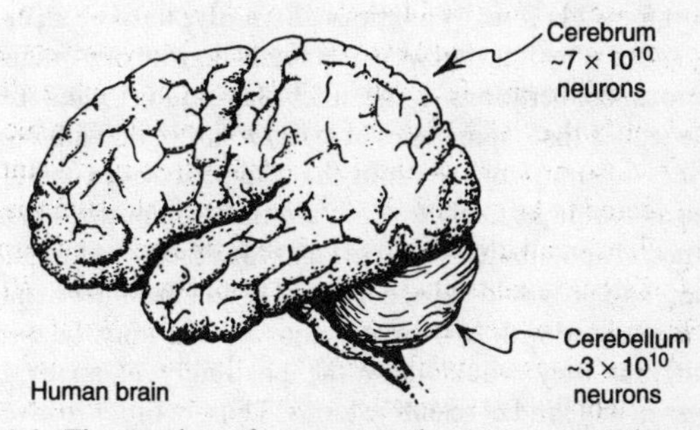

Fig. 1.6. The number of neurons and neuron connections in the cerebellum is of the same order as in the cerebrum. On the basis simply of neuron counting and neuron interconnectivity, we ask, why is the action of the cerebellum entirely unconscious?

be critically involved in the perfection of motor control, and it comes into play when some motor skill has been mastered—to become 'second nature', and not thought about consciously. Initially, when a new skill is learned, conscious control of one's actions is needed, and it appears that this is achieved with the essential involvement of the cerebral cortex. Yet, later, when the necessary movements have become 'automatic', the unconscious activity of the cerebellum largely takes over. It is remarkable, in view of the fact that the activity of the cerebellum appears to be entirely unconscious, that it involves perhaps up to one half as many neurons as the cerebrum. Moreover, the Purkinje cells, referred to in §1.2, that have up to 80 000 synaptic connections, are neurons that are found in the cerebellum, so the total number of connections between neurons may well be no fewer in the cerebellum than in the cerebrum. If it is to be the sheer complication of the network of neurons that is regarded as the essential prerequisite for consciousness, then one must ask why consciousness seems to be entirely absent in the actions of the cerebellum. (I shall have some comments to make on this issue later, in §8.6).

Of course, the problems for viewpoint $\mathscr{A}$ referred to in this section have their analogues also for $\mathscr{B}$ and $\mathscr{C}$. On any scientific viewpoint, one would need eventually to address the issue of what it is that underlies the phenomenon of consciousness, and of how qualia can come about. In the later sections of Part II, I shall be attempting to move tentatively towards an understanding of consciousness from the point of view of $\mathscr{C}$.

1.15 Do limitations of present-day AI provide a case for $\mathscr{C}$?

But why $\mathscr{C}$? What evidence *is* there which can be interpreted as providing direct support for $\mathscr{C}$? Is $\mathscr{C}$ really a serious alternative to $\mathscr{A}$ or $\mathscr{B}$, or even to $\mathscr{D}$? We must try to see what it is that we can actually do with our brains (or minds) when conscious deliberations come into play—and I shall be trying to convince the reader that (sometimes at least) what we do with our conscious thinking is very different from anything that can be achieved computationally. Adherents of $\mathscr{A}$ would be likely to maintain that 'computing', in one form or another, is the only possibility—and, as far as the effects on external behaviour are concerned, so also would the adherents of $\mathscr{B}$. On the other hand, adherents of $\mathscr{D}$ might well agree with $\mathscr{C}$ that conscious actions must be things beyond computation, but they would deny the possibility of an explanation of consciousness in any kind of scientific terms. Thus, in order to give support to $\mathscr{C}$, one must try to find examples of mental activity that lie beyond any form of computation, and also try to see how such activity might result from appropriate physical processes. The remainder of Part I will be directed towards the former goal, whilst in Part II, I shall present my attempts to come to terms with the latter.

What kind of mental activity might there be which could be shown to lie beyond computation? As a possible route to this, we could try to examine the present state of artificial intelligence, and try to see what computationally controlled systems are good at and what they are bad at. Of course, the present state of the art of AI may not give a clear indication of what might ultimately be achieved in principle. Even in 50 years, say, things could well be very different from what they are at present. The rapid development of computers and their applications—only within the *past* 50 years—has been extraordinary. We must certainly be prepared for enormous advances in the future—advances that might possibly come upon us very swiftly indeed. I shall be primarily concerned, in this book, not with the speed of such advances, but with certain fundamental limitations of *principle* that they are subject to. These limitations would apply no matter how many centuries into the future we may be prepared to project our speculations. Thus, we should base our arguments on general principles and not allow ourselves to be unduly influenced by what has been achieved to date. Nevertheless, there could well be clues contained in the successes and failures of the artificial intelligence of today, despite the fact that, so far, there is very little of what could be called a genuinely convincing artificial intelligence—as even the strongest proponents of AI would be prepared to admit.

The main failures of artificial intelligence to date, perhaps rather surprisingly, are not so much in areas where the power of human intellect can itself be extremely impressive—such as where particular human experts can dumbfound the rest of us with their specialist knowledge or their ability to make judgements based on deeply complicated computational procedures—but in the 'common-sense' activities that the humblest among us indulge in for most of our waking lives. As yet, no computer-controlled robot could begin to compete with even a young child in performing some of the simplest of everyday activities: such as recognizing that a coloured crayon lying on the floor at the other end of the room is what is needed to complete a drawing, walking across to collect that crayon, and then putting it to its use. For that matter, even the capabilities of an ant, in performing its everyday activities, would far surpass what can be achieved by the most sophisticated of today's computer control systems. Yet, on the other hand, the development of powerful chess computers provides a striking example in which computers *can* be enormously effective. Chess is undoubtedly an activity where the power of the human intellect is particularly manifest—though exploited to excellence in this way by but a few. Yet chess computer systems now play the game extraordinarily well, and can consistently beat most human players. Even the very best of human experts are now being hard pressed, and may not for long retain what superiority they still possess over the best of chess-playing computers.[22] There are also several other areas of expertise, in which computers can compete successfully, or partially successfully, with human experts. Moreover, there are some, such as straightforward numerical computation, in which the capabilities of computers far outstrip the capabilities of humans.

In all these situations, however, it would be hard to maintain that the computer attains any genuine *understanding* of what it is actually doing. In the case of a top-down organization, the reason that the system successfully works at all is not that *it* understands anything, but that human programmers' understandings (or else the understandings of those human experts upon whom the programmers depend) have been used in the construction of the program. For a bottom-up organization, it is not clear that there need be any specific understanding whatever, as a feature of the system's actions, on the part either of the device itself or of its programmers—beyond those human understandings that would have gone into the designing of the details of the specific performance-improving algorithms that are involved, and in the very conception that a system can improve its performance with experience whenever an appropriate feedback system is incorporated. Of course, it is not always clear what the term 'understanding' actually means, so some people might claim that on *their* terms, these computer systems actually do possess some kind of 'understanding'.

But is this reasonable? To illustrate a lack of any real understanding by present-day computers, it is interesting to provide, as an example, the chess position given in Fig. 1.7 (by William Hartston, taken from an article by

Fig. 1.7. White to play and draw—easy for humans, but Deep Thought took the rook!

Jane Seymore and David Norwood (1993)). In this position, black has an enormous material advantage, to the extent of two rooks and a bishop. However, it is easy for white to avoid defeat, by simply moving his* king around on his side of the board. The wall of pawns is impregnable to the black pieces, so there is no danger to white from the black rooks or bishop. This

*White is, of course, not necessarily male. See Notes to the reader on p. xvi.

much is obvious to any human player with a reasonable familiarity with the rules of chess. However, when the position, with white to move, was presented to 'Deep Thought'—the most powerful chess computer of its day, with a number of victories over human chess grandmasters to its credit—it immediately blundered into taking the black rook with its pawn, opening up the barrier of pawns to achieve a hopelessly lost position!

How could such a wonderfully effective chess player make such an obviously stupid move? The answer is that all Deep Thought had been programmed to do, in addition to having been provided with a considerable amount of 'book knowledge', would be to calculate move after move after move—to some considerable depth—and to try to improve its material situation. At no stage can it have had any actual understanding of what a pawn barrier might achieve—nor, indeed, could it ever have any genuine understanding whatsoever of anything at all that it does.

To anyone with sufficient appreciation of the general way in which Deep Thought or other chess-playing computer systems are constructed, it is no real surprise that it would fail on positions such as that of Fig. 1.7. Not only can *we* understand something about chess that Deep Thought did not, but we can also understand something of the (top-down) procedures according to which Deep Thought has been constructed; so we can actually appreciate why it should make such a blunder—as well as understanding why it could play chess so effectively in most other circumstances. However, we may ask: is it possible that Deep Thought, or any other AI system, could *eventually* achieve any of the kind of real understandings that we can have—of chess, or of anything else? Some AI proponents might argue that in order for an AI system to gain any 'actual' understanding, it would need to be programmed in a way that involves *bottom-up* procedures in a much more basic way than is usual for chess-playing computers. Accordingly, its 'understandings' would develop gradually by its building up a wealth of 'experience', rather than having specific top-down algorithmic rules built into it. Top-down rules that are simple enough for us to appreciate easily could not, by themselves, provide a computational basis for actual understanding—for we can use our very understandings of these rules to realize their fundamental limitations.

This point will be made more explicit in the arguments given in Chapters 2 and 3. But what about these bottom-up computational procedures? Is it possible that *they* could form the basis of understanding? In Chapter 3, I shall be arguing otherwise. For the moment, we may simply take note of the fact that present-day bottom-up computer systems do not in any way substitute for genuine human understanding—in any significant area of intellectual expertise where genuine and continuing human understanding and insight seem to be important. This much, I feel sure, would be broadly accepted today. For the most part, the very optimistic early claims[23] that had sometimes been made by proponents of artificial intelligence and promoters of expert systems have not yet been fulfilled.

But these are still very early days, if we are to consider what artificial intelligence might ultimately achieve. Proponents of AI (either $\mathscr{A}$ or $\mathscr{B}$) would maintain that it is just a matter of time, and perhaps some further significant developments in their craft, before important elements of understanding will indeed begin to become apparent in the behaviour of their computer-controlled systems. Later, I shall try to argue in precise terms against this, and that there are fundamental limitations to any purely computational system, whether top-down or bottom-up. Although it might well be possible for a sufficiently cleverly constructed such system to preserve an illusion, for some considerable time (as with Deep Thought), that it possesses some understanding, I shall maintain that a computer system's actual lack of general understanding should—in principle, at least—eventually reveal itself.

For my precise arguments I shall need to turn to some mathematics, the intention being to show that *mathematical* understanding is something that cannot be reduced to computation. Some AI proponents might find this surprising, for they have argued[24] that the things that came late in human evolution, like the performing of arithmetical or algebraic calculation, are the things that come most easily to computers, and where computers already outstrip by far the abilities of calculating human beings; whereas those skills that were evolved early, like walking or the interpretation of complicated visual scenes, are things that we perform effortlessly, whilst present-day computers struggle to achieve their unimpressively limited performances. I shall argue very differently. Any complicated activity, which may be mathematical calculations, or playing a game of chess, or commonplace actions—*if* they have been understood in terms of clear-cut computational rules—are the things that modern computers are good at; but the very understanding that underlies these computational rules is something that is itself beyond computation.

1.16 The argument from Gödel's theorem

How can we be sure that such understandings are not themselves things that can be reduced to computational rules? I shall shortly be giving (in Chapters 2 and 3) some very strong reasons for believing that effects of (certain kinds of) understanding cannot be properly simulated in any kind of computational terms—neither with a top-down, nor a bottom-up organization, nor with any combination of the two. Thus, the human faculty of being able to 'understand' is something that must be achieved by some non-computational activity of the brain or mind. The reader may be reminded (cf. §1.5, §1.9) that the term 'non-computational' here refers to something beyond any kind of effective simulation by means of any computer based on the logical principles that underlie all the electronic or mechanical calculating devices of today. On the other hand, 'non-computational activity' does *not* imply something beyond the powers of science and mathematics. But it *does* imply that viewpoints $\mathscr{A}$ and $\mathscr{B}$ cannot explain how we actually perform all those tasks that are the results of conscious mental activity.

It is certainly a *logical* possibility that the conscious brain (or conscious mind) might act according to such non-computational laws (cf. §1.9). But is it *true*? The argument I shall present in the next chapter (§2.5) provides what I believe to be a very clear-cut argument for a non-computational ingredient in our conscious thinking. This depends upon a simple form of the famous and powerful theorem of mathematical logic, due to the great Czech-born logician Kurt Gödel. I shall need only a very simplified form of this argument, requiring only very little mathematics (where I also borrow from an important later idea due to Alan Turing). Any reasonably dedicated reader should find no great difficulty in following it. However, Gödel-type arguments, used in this kind of way, have sometimes been vigorously disputed.[25] Consequently, some readers might have gained an impression that this argument from Gödel's theorem has been fully refuted. I should make it clear this is *not* so. It is true that many counter-arguments have been put forward over the years. Many of these were aimed at a pioneering earlier argument—in favour of mentalism and opposed to physicalism—that had been advanced by the Oxford philosopher John Lucas (1961). Lucas had argued from the Gödel theorem that mental faculties must indeed lie beyond what can be achieved computationally. (Others, such as Nagel and Newman (1958), had previously argued in a similar vein.) My own argument, though following similar lines, is presented somewhat differently from that of Lucas—and not necessarily as support for mentalism. I believe that my form of presentation is better able to withstand the different criticisms that have been raised against the Lucas argument, and to show up their various inadequacies.

In due course (in Chapters 2 and 3), I shall be addressing, in detail, *all* the different counter-arguments that have come to my attention. I hope that my discussion there will serve to correct not only some apparently widespread misconceptions about the significance of the Gödel argument, but also the evidently inadequate brevity of my discussion in ENM. I shall demonstrate that a good many of these counter-arguments are based merely on misconceptions; the remaining ones, which put forward genuine viewpoints that need to be considered in detail, perhaps provide just possible let-outs, in accordance with $\mathscr{A}$ or $\mathscr{B}$, but I shall argue that they nevertheless do *not* really provide *plausible* explanations of what our ability to 'understand' actually allows us to achieve, and that these let-outs would in any case be of little value to AI. Anyone who maintains that all the external manifestations of conscious thought processes *can* be properly computationally simulated, in accordance with either viewpoint $\mathscr{A}$ or $\mathscr{B}$, must find some way of coming to terms, in full detail, with the arguments that I shall give.

1.17 Platonism or mysticism?

Some critics may argue, however, that by seeming to force us into either viewpoint $\mathscr{C}$ or viewpoint $\mathscr{D}$, the Gödel argument has implications that must be regarded as being 'mystical', and certainly no more palatable to them than any such let-out from the Gödel argument. With regard to $\mathscr{D}$, I am in effect agreeing with them. My own reasons for rejecting $\mathscr{D}$—the viewpoint which asserts the incompetence of the power of science when it comes to matters of the mind—spring from an appreciation of the fact that it has only been through use of the methods of science and mathematics that any real progress in understanding the behaviour of the world has been achieved. Moreover, the only minds of which we have direct knowledge are those intimately associated with particular physical objects—*brains*—and differences in states of mind seem to be clearly associated with differences in the physical states of brains. Even the mental states of *consciousness* seems to be associated with certain specific types of physical activity taking place within the brain. If it were not for the puzzling aspects of consciousness that relate to the presence of 'awareness' and perhaps of our feelings of 'free will', which as yet seem to elude physical description, we should not need to feel tempted to look beyond the standard methods of science for explanation of minds as a feature of the physical behaviour of brains.

On the other hand, it should be made clear that science and mathematics have themselves revealed a world full of mystery. The deeper that our scientific understanding becomes, the more profound the mystery that is revealed. It is perhaps noteworthy that the physicists, who are the more directly familiar with the puzzling and mysterious ways in which matter *actually* behaves, tend to take a less classically mechanistic view of the world than do the biologists. In Chapter 5, I shall be explaining some of the more mysterious aspects of quantum behaviour, a few of which have been only fairly recently uncovered. It may well be that in order to accommodate the mystery of mind, we shall need a broadening of what we presently mean by 'science', but I see no reason to make any clean break with those methods that have served us so extraordinarily well. If, as I believe, the Gödel argument is consequently forcing us into an acceptance of some form of viewpoint $\mathscr{C}$, then we shall also have to come to terms with some of its other implications. We shall find ourselves driven towards a *Platonic* viewpoint of things. According to Plato, mathematical concepts and mathematical truths inhabit an actual world of their own that is timeless and without physical location. Plato's world is an ideal world of perfect forms, distinct from the physical world, but in terms of which the physical world must be understood. It also lies beyond our imperfect mental constructions; yet, our minds do have some direct access to this Platonic realm through an 'awareness' of mathematical forms, and our ability to reason about them. We shall find that whilst our Platonic perceptions can be aided on occasion by computation, they are not limited by computation. It is this potential for the 'awareness' of mathematical concepts involved in this Platonic access that gives the mind a power beyond what can ever be achieved by a device dependent solely upon computation for its action.

1.18 What is the relevance of mathematical understanding?

Such airy-fairy stuff may (or may not) be all very well, some readers will doubtless complain. What serious relevance do sophisticated issues of mathematics and of mathematical philosophy have for most of the matters of direct interest to artificial intelligence, for example? Indeed, many philosophers and proponents of AI are quite reasonably of the opinion that although Gödel's theorem is undoubtedly important in its original context of mathematical logic, it can have very limited implications, at best, for AI or for the philosophy of mind. Very little of human mental activity is directed, after all, at issues relating to Gödel's original context: the axiomatic foundations of mathematics. My answer is that a great deal of human mental activity involves, on the other hand, the application of human consciousness and understanding. My use of the Gödel argument is to show that human understanding cannot be an algorithmic activity. If we can show this in *some* specific context, this will suffice. Once it is shown that certain types of mathematical understanding must elude computational description, then it is established that we can do *something* non-computational with our minds. This being accepted, it is a natural step to conclude that non-computational action must be present in many other aspects of mental activity. The floodgates will indeed be open!

The mathematical argument establishing the needed form of Gödel's theorem, as given in Chapter 2, may well seem to have very little direct bearing on most aspects of consciousness. Indeed, a demonstration that certain kinds of mathematical understanding must involve something beyond computation does not *appear* to have much relevance to what is involved in our perception of the colour red, for example, nor does there seem to be any manifest role for mathematical desiderata in most other aspects of consciousness. For example, even mathematicians do not normally think of mathematics when they are dreaming! Dogs appear to dream, and are presumably also aware, to some extent, when they dream; and I would certainly think that they can be aware at other times. But they do not do mathematics. Undoubtedly, contemplating mathematics is very far from being the *only* animal activity requiring consciousness! It is a highly specialized and peculiarly human activity. (Indeed, some cynics might even say that it is an activity confined to certain peculiar humans.) The phenomenon of consciousness, on the other hand, is ubiquitous, being likely to be present in much non-human as well as human mental activity, and certainly in non-mathematical humans, as well as in mathematical humans when they are not actually doing mathematics (which is most of the time). Mathematical thinking is a very tiny area of conscious activity that is indulged in by a tiny minority of conscious beings for a limited fraction of their conscious lives.

Why then do I choose to address the question of consciousness here in a mathematical context first? The reason is that it is only within mathematics that we can expect to find anything approaching a rigorous demonstration that *some*, at least, of conscious activity *must* be non-computational. The issue of computation, by its very nature, is indeed a mathematical one. We cannot expect to be able to provide anything like a 'proof' that some activity is not computational unless we turn to mathematics. I shall try to persuade the reader that whatever we do with our brains or minds when we understand *mathematics* is indeed different from anything that we can achieve by use of a computer; then the reader should be more readily prepared to accept an important role for non-computational activity in conscious thinking generally.

Nevertheless, as many might argue, it is surely just *obvious* that the sensation of 'red' can in no way be evoked merely by the carrying out of some computation. Why bother at all with attempting some unnecessary mathematical demonstration when it is perfectly obvious that 'qualia'—i.e. subjective experiences—have nothing to do with computation? One answer is that this argument from 'obviousness' (with which I do have a considerable sympathy) refers only to the *passive* aspects of consciousness. Like Searle's Chinese Room, it may be presented as an argument against viewpoint $\mathscr{A}$, but it does not distinguish $\mathscr{C}$ from $\mathscr{B}$.

Moreover, I must attack the functionalist's computational model (i.e. viewpoint $\mathscr{A}$) on its home ground, so to speak; for it is the contention of the functionalists that all qualia *must* indeed be somehow evoked by merely carrying out the appropriate computations, no matter how improbable such a picture may at first sight seem. For, they argue, what else can we indeed be usefully doing with our brains unless it is performing computations of some kind? What is the brain for, if not just some kind of—albeit highly sophisticated—computational control system? Whatever 'feelings of awareness' the brain's action somehow evokes must, they would claim, be the result of this computational action. They often maintain that if one refuses to accept the computational model for *all* mental activity, including consciousness, then one must be resorting to *mysticism*. (This is to suggest that the only alternative to viewpoint $\mathscr{A}$ is viewpoint $\mathscr{D}$!) It is my intention, in Part II of this book, to provide some partial suggestions as to what else a scientifically describable brain might actually be doing. I shall not deny that some of the 'constructive' parts of my argument are speculative. Yet I believe that the case for *some* kind of non-computational action is compelling, and it is in order to demonstrate the compelling nature of this case that I must turn to mathematical thinking.

1.19 What has Gödel's theorem to do with common-sense behaviour?

But suppose that it is accepted that something non-computational is indeed going on when we use our conscious mathematical judgements and reach our conscious mathematical decisions. What help will this be to us in coming to terms with the limitations of robot activity that, as I have mentioned above, seem to lie much more with elementary 'common-sense' actions than with the sophisticated behaviour of trained experts? At first sight it seems that my conclusions will be almost the *opposite* of what is found for the limitations of artificial intelligence—at least the present-day limitations. For it seems that I am asserting that non-computational behaviour is to be found in highly sophisticated areas of mathematical understanding, rather than in common-sense behaviour. But that is not my claim. I am claiming that 'understanding' involves the same kind of non-computational process, whether it lies in a genuine mathematical perception, say of the infinitude of natural numbers, or merely in perceiving that an oblong-shaped object can be used to prop open a window, or in understanding how an animal may be secured or released by a few selected motions of a bit of rope, or in comprehending the meanings of the words 'happiness', 'fighting', or 'tomorrow', or in realizing that when Abraham Lincoln's left foot was in Washington his right foot was almost certainly in Washington also—to use an example that proved surprisingly troublesome to an actual AI system![26] This non-computational process lies in whatever it is that allows us to become directly aware of something. This awareness can allow us to visualize the geometrical motions of a block of wood, or the topological properties of a piece of rope, or the connectedness of Abraham Lincoln. It also allows us to have some kind of direct route to another person's experiences, so that one can 'know' what the other person must mean by a word like 'happiness', 'fighting', and 'tomorrow', even though explanations are likely to have been quite inadequate. The 'meanings' of words can be actually passed from one person to another, not because adequate explanations are given, but because the other person already has some direct perception—or 'awareness'—of what possible meanings there could be, so very inadequate explanations can suffice to enable that person to 'latch on' to the correct one. It is the possession of a common kind of 'awareness' that allows the communication between the two people to take place. It is this that puts an insentient computer-controlled robot at a severe disadvantage. (Indeed, the very *meaning* of the concept of a word's 'meaning' is something that we have some kind of direct conception of, and it is hard to see how one could give any kind of adequate description of *this* concept to our insentient robot.) Meanings can only be communicated from person to person because each person is aware of similar internal experiences or feelings about things. One might imagine that 'experiences' are just things that constitute some kind of memory store that records what has happened, and that our robot could easily be equipped with this. But I am arguing that this is not so, and that it is crucial that the object in question, be it human or robot, must actually be *aware* of the experience.

Why do I claim that this 'awareness', whatever it is, must be something non-computational, so that no robot, controlled by a computer, based merely on the standard logical ideas of a Turing machine (or equivalent)—whether top-down or bottom-up—can achieve or even *simulate* it? It is here that the Gödelian argument plays its crucial role. It is hard to say much at the present time about our 'awareness' of, for example, the colour red; but there *is* something definite that we can say concerning our awareness of the infinitude of natural numbers. It is 'awareness' that allows a child to 'know' what is meant by the numbers 'zero', 'one', 'two', 'three', 'four', etc., and for what it means for this sequence to go on for ever, when only absurdly limited, and seemingly almost irrelevant, kinds of descriptions in terms of a few oranges and bananas have been given. The concept of 'three' can indeed be abstracted, by a child, from such limited examples; and, moreover, the child can also latch on to the fact that this concept is but one of the unending sequence of similar concepts ('four', 'five', 'six', etc.). In some Platonic sense, the child already 'knows' what the natural numbers are.

This may seem a bit mystical, but it is not really. It is vital, for the discussions which follow, that we distinguish this kind of Platonic knowledge from mysticism. The concepts that we 'know' in this Platonic sense are things that are 'obvious' to us—they are things that can be reduced to a perceived 'common sense'—yet we may not be able to characterize these concepts completely in terms of computational rules. Indeed, as we shall see from the later discussion, in relation to the Gödel argument, there is no way of characterizing the properties of the natural numbers completely in terms of such rules. Yet, how is it that descriptions of numbers in terms of apples or bananas can allow a child to know what 'three days' means, that same abstract concept of 'three' being involved as with 'three oranges'? Of course, this appreciation may well not come at once, and the child may get it wrong at first, but that is not the point. The point is that this kind of realization is possible at all. The abstract concept of 'three', and of this concept as being one of an infinite sequence of corresponding concepts—the natural numbers themselves—is something that can indeed be understood, but, I claim, only through the use of one's awareness.

My claim will be that, likewise, we are not using computational rules when we visualize the motions of a block of wood, or of a piece of rope, or of Abraham Lincoln. In fact there are very effective computer simulations of the motions of a rigid body, such as a block of wood. The simulations of such motions can be made so precise and reliable that they are usually far more effective than what can be achieved by direct human visualization. Likewise, the motions of a piece of rope, or string, can be simulated computationally, although, perhaps surprisingly, this is something much more difficult to achieve than the simulations of the motions of a rigid body. (This has partly to do with the fact that a 'mathematical string' requires infinitely many parameters to specify its location whereas a rigid body requires just six.) There are computer algorithms for deciding whether or not a rope is knotted, but these are completely different from those that describe rigid motions (and are not very efficient computationally). Any computer simulation of the external appearance of Abraham Lincoln would certainly be more difficult still. Now my point is not that human visualization is 'better' or 'worse' than a computer simulation of these various things, but that it is something quite *different*.

An essential point, it seems to me, is that visualization involves an element of appreciation of what it is that is being visualized; that is, it involves *understanding*. To illustrate the kind of thing that I have in mind, consider an elementary fact of arithmetic, namely the fact that, of any two natural numbers (i.e. non-negative whole numbers: 0, 1, 2, 3, 4, . . .) a and b, we have the property that

$$a \times b = b \times a.$$

It should be made clear that this is not an empty statement, for the meanings of the two sides of the equation are different. On the left, $a \times b$ refers to a collection of a groups of b objects; whereas $b \times a$ on the right refers to b groups of a objects. In the particular case when $a = 3$ and $b = 5$ we have, for $a \times b$, the arrangement

(●●●●●) (●●●●●) (●●●●●).

whereas for $b \times a$, we have

(●●●) (●●●) (●●●) (●●●) (●●●).

The fact that there are the same total number of spots in each case expresses the particular fact that $3 \times 5 = 5 \times 3$.

Now we can see that this must be true simply by visualizing the array

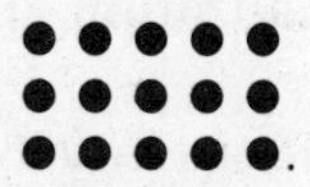

If we read this off by rows, we find that we have three rows, each contaning five spots, which expresses the quantity 3×5. However, if we read it off by columns, we find five columns of three spots, expressing the quantity 5×3. The fact that these quantities are equal is seen at once from the fact that it is precisely the same rectangular array in each case; it is only being read off differently. (Alternatively, one might prefer to rotate the image through a right angle in one's mind's eye, to see that the array representing 5×3 has the same number of elements as that representing 3×5.)

Now the important thing about this act of visualization is that it at once gives us something much more general than the particular numerical fact that $3 \times 5 = 5 \times 3$. For there is nothing special about the particular values $a = 3$, $b = 5$ in this procedure. It would equally apply if, say, $a = 79797000222$ and $b = 50000123555$, and we can confidently assert

$$79797000222 \times 50000123555 = 50000123555 \times 79797000222$$

despite the fact that there is no chance of our accurately visualizing a rectangular array that big (nor could any actual computer enumerate its elements). We can perfectly well conclude that the above equality must hold—or, indeed, that the general equality $a \times b = b \times a$ must hold*—from essentially the same visualization that we used for the special case $3 \times 5 = 5 \times 3$. We merely need to 'blur' in our minds the actual numbers of rows and columns that are being used, and the equality becomes obvious.

I do not mean to suggest that all mathematical relations can be perceived directly as 'obvious' if they are visualized in the right way—or merely that they can always be perceived in some other way that is immediate to our intuitions. Far from it. Some mathematical relations require long chains of reasoning before they can be perceived with certainty. But the object of mathematical proof is, in effect, to provide such chains of reasoning where each *step* is indeed something that can be perceived as obvious. Consequently, the endpoint of the reasoning is something that must be accepted as *true*, even though it may not, in itself, be at all obvious.

One might imagine that it would be possible to list all possible 'obvious' steps of reasoning once and for all, so that from then on everything could be reduced to computation—i.e. the mere mechanical manipulation of these obvious steps. What Gödel's argument shows (§2.5) is that this is not possible. There is no way of eliminating the need for *new* 'obvious' understandings. Thus, mathematical understanding cannot be reduced to blind computation.

1.20 Mental visualization and virtual reality

The mathematical insights that featured in §1.19 were of a rather specifically geometrical character. There are also many other types of insight that can be used in mathematical arguments—insights which need not be particularly geometrical. However, it turns out that insights which *are* geometrical are frequently of particular value towards mathematical understanding. Thus, it may be instructive to ask what kind of physical activity is actually going on in our brains when we visualize something in a geometrical way. There is no

*It should be remarked that this equality does *not* hold for various strange kinds of 'numbers' that occur in mathematics, such as the ordinal numbers referred to after **Q19** in §2.10. But it always holds for the *natural* numbers, which are what we are concerned with here.

logical requirement that this activity should itself provide a 'geometrical mirroring' of the thing that is being visualized. As we shall see, it could be something quite different.

It is helpful to make a comparison with what is called 'virtual reality', something which has been argued as having relevance to the issue of 'visualization'. According to the procedures of virtual reality,[27] a computer simulation is made of some non-existent structure—such as an architectural proposal for a building—and this simulation conveyed to the eyes of a human subject who seems to perceive that structure as 'real'. Through movements of the eyes or head, or perhaps even of the legs as though walking around, the subject would see the structure from different angles, just as would be the case if the structure were indeed real. (See Fig. 1.8.) Some people[28] have argued the

Fig. 1.8. Virtual reality. A three-dimensional make-believe world can be computationally conjured up which consistently responds to movement of the head and body.

case that whatever it is that goes on inside our brains when we consciously visualize may well be very like the computations involved in the construction of such a simulation. Indeed, in one's 'mind's eye', when one views some *actual* fixed structure, one seems to construct some mental model which persists unchanged despite continual movements of head, eyes, and body which entail that the retinal images are continually shifting. Such corrections for bodily motions are very much part of what is involved in virtual reality, and it has been suggested that something very similar must be going on in the construction of the 'mental models' that constitute our very acts of visualization. Such computations, of course, need have no actual geometrical relationship with (or 'mirroring' of) the structure that is being modelled. Supporters of viewpoint $\mathscr{A}$ would indeed have to regard our conscious visualizations as being the result of some such computational simulation, inside our heads, of the external world. What I am proposing, however, is that when we consciously perceive a visual scene, the *understanding* that is involved is something very different from the modelling of the world in terms of such a computational simulation.

One might argue that something inside our brains is acting more like an 'analogue computer' where the modelling of the external world is achieved not in terms of digital computation, as is the case with modern electronic computers, but in terms of some internal physical structure whose physical behaviour can be translated so as to mirror the behaviour of the external system that is being modelled. If we wish to make an analogue device to model the motions of an external rigid body, there is clearly a very straightforward way to do this. That would be to have, internally, a small actual physical body of the same shape as (but of a different size from) the external object that is being modelled—although I shall certainly not suggest any direct relevance of *this* particular model to what goes on inside our brains! The motions of this internal body could be viewed from different angles to give outward effects very similar to that of digital computation. Such a system could also be used as part of a 'virtual reality' system, where instead of having an entirely computational model of the structure in question, there would be an actual physical model of it, differing only by size from the 'reality' that is being simulated. In general, analogue simulation would not be so direct or so trivial as this. One might use a parameter such as electric potential rather than actual physical distance, and so on. One must just be sure that the physical laws that govern the internal structure mirror very precisely the physical laws governing the external structure that is being modelled. There is no need for the internal structure to *resemble* (or to 'mirror') the external one in any obvious way.

Is it possible that analogue devices can achieve things inaccessible to pure digital computation? As was discussed in §1.8, there is no reason to believe that, within the framework of present-day physics, analogue simulation can achieve things that digital simulation cannot. Thus, if our arguments indicate to us that our visual imaginations achieve non-computational things, then we are encouraged to look outside the framework of existing physics for the underpinnings of our visual imagination.

1.21 Is mathematical imagination non-computational?

None of this discussion tells us specifically that whatever we do when we visualize is something that cannot be simulated in a compuational way. It might seem, even if we employ some kind of internal analogue system when we visualize, that it ought at least to be possible to *simulate*, digitally, the behaviour of such an analogue device.

Now, the 'visualization' that I have just been referring to has been concerned with what is 'visual' in a largely literal sense, i.e. with the mental images that seem to correspond to the signals that reach the brain from our eyes. More generally, one's mental images need not at all have this literally 'visual' character, such as when one understands the meaning of an abstract word or recalls a piece of music. The mental images of someone who has been blind since birth, for example, can hardly have a direct relation to the signals that one receives from the eyes. Thus, the 'visualizations' that I am referring to are more concerned with the general issue of 'awareness' than with things that necessarily relate to the visual system. In fact, I do not know of any argument that has a direct bearing on the computational nature, or otherwise, of our powers of visualization in this literal sense of 'visual'. My belief that our actual acts of visualization must indeed be non-computational is an inference from the fact that *other* types of human awareness *do* seem to have a demonstrably non-computational character. Though it is hard to see how one might produce a direct argument for non-computability that is specific to geometric visualization, a convincing argument that *some* forms of conscious awareness must be non-computational would at least carry the strong suggestion that the kind of awareness that is involved in geometric visualization ought to be non-computational also. There seems no reason to believe that a clear line is to be drawn between different manifestations of conscious understanding in this respect.

Specifically, the awareness that I claim is *demonstrably* non-computational is our understanding of the properties of the natural numbers 0, 1, 2, 3, 4, ... (One might even say that our concept of a natural number *is*, in a sense, a form of *non*-geometric 'visualization'.) We shall see in §2.5, by a readily accessible form of Gödel's theorem (cf. response to query **Q16**), that this understanding is something that cannot be encapsulated in any finite set of rules—from which it follows that it cannot be simulated computationally. From time to time one hears that some computer system has been 'trained' so as to 'understand' the concept of natural numbers.[29] However, this cannot be true, as we shall see. It is our *awareness* of what a 'number' can actually mean that enables us to latch on to the correct concept. When we have this correct concept, we can—at least in principle—provide the correct answers to families of questions about numbers that are put to us, when no finite set of rules can do this. With only rules and no direct awareness, a computer-controlled robot (like Deep Thought; cf. §1.15 above) would be necessarily limited in ways in which we are not limited ourselves—although if we give the robot clever enough rules for its behaviour it may perform prodigious feats, some of which lie far beyond unaided human capabilities in specific narrowly enough defined areas, and it might be able to fool us, for some while, into thinking that it also possesses awareness.

A point worth making is that when an effective digital (or analogue) computer simulation of some external system *is* achieved, it is almost always by taking advantage of some significant human understanding of the underlying mathematical ideas. Consider the digital simulation of the geometical motions of a rigid body. The relevant computations depend, most particularly, upon the insights of some great thinkers of the seventeenth century, such as the French mathematicians Descartes, Fermat, and Desargues, who introduced the ideas of coordinate and projective geometry. What about the simulation of the motions of a piece of string or rope? It turns out that the geometrical ideas that are needed in order to understand the restrictions of behaviour of a piece of string—i.e. its 'knottedness'—are very sophisticated; and they are remarkably recent, many fundamental advances having been achieved only in this century. While it may, in practice, be not too difficult to decide, using simple manipulations with one's hands, and applying one's common-sense understanding, whether or not a closed but tangled loop of string is knotted or unknotted, the computational algorithms for achieving the same feat are surprisingly involved, sophisticated, and inefficient.

Thus, effective digital simulations of such things have been very much top-down affairs, and they depend upon considerable human understanding and insight. There is little chance that there is anything very similar going on in a human brain when it is engaged in the act of visualization. A more plausible possibility would be something involving an important contribution from bottom-up ingredients, so that the simulated 'visual pictures' would arise only after considerable 'learning experience' has taken place. I am not aware, however, that there are any significant bottom-up approaches (e.g. from artificial neural networks) to questions of this kind. My guess would be that an approach that was *entirely* based on a bottom-up organization would give very poor results. It is hard to see that a good simulation of the geometrical motions of a rigid body or the topological restrictions on the motion of a piece of string—i.e. its *knottedness*—could be achieved without there being involved any genuine understanding of what is actually going on.

What kind of physical process can it be that is responsible for our awareness—an awareness that seems to be necessary for any genuine understanding? Can it indeed be something beyond computational simulation, as is demanded by viewpoint $\mathscr{C}$? Is this putative physical process something that is itself accessible to our understanding, at least in principle? I believe that it must be, and that viewpoint $\mathscr{C}$ is a genuine scientific possibility, although we must be prepared for the eventuality that our scientific criteria and methods may undergo subtle but important shifts. We shall have to be prepared to examine clues that may present themselves in unexpected ways, and in areas of genuine understanding that may at first appear to be largely irrelevant. In the discussions that follow, I ask the reader to keep an open mind, yet to pay careful attention to the reasoning and to the scientific evidence, even though these may, at times, appear to conflict with what had seemed obvious common sense. Be prepared to think a little about the arguments that I shall endeavour to present in as clear a form as I am able. Thus emboldened—let us venture forth.

In the remainder of Part I, I shall be leaving aside the issues of physics and of whatever biological action might underlie the non-computability demanded by viewpoint $\mathscr{C}$. Those matters will be the concern of Part II of this book. But why is a search for non-computational action necessary? That necessity rests on my claim that we indeed perform non-computational feats when we consciously understand. I must justify this claim, and for this reason we must turn to our mathematics.

Notes and references

1. See, in particular, Good (1965), Minsky (1986), Moravec (1988).
2. Moravec (1988) bases his argument for this kind of time-scale, upon the proportion of the cortex that he considers has already been successfully modelled (essentially that in the retina), together with an estimate of the rate at which computer technology will advance in the future. As of early 1994, he is still holding to these estimates; cf. Moravec (1994).
3. These four viewpoints were explicitly described in, for example, Johnson-Laird (1987), p. 252 (although it should be pointed out that what he refers to as the 'Church–Turing thesis' is essentially what I am calling 'Turing's thesis' in §1.6, rather than 'Church's thesis').
4. For example, D. Dennett, D. Hofstadter, M. Minsky, H. Moravec, H. Simon; for a discussion of these terms, see Searle (1980), Lockwood (1989).
5. See Moravec (1988).
6. Turing (1950); see ENM, pp. 5–14.
7. See Searle (1980), (1992).
8. The issue is complicated by the fact that present-day physics depends upon the use of *continuous*, rather than discrete (digital) action. Even the *meaning* of 'computability' in this context is open to various interpretations. For some relevant discussion, see Pour-el (1974), Smith and Stephenson (1975), Pour-el and Richards (1979), (1981), (1982), (1989), Blum, Shub, and Smale (1989), Rubel (1988), (1989). The matter will be returned to in §1.8.
9. I owe this fine phrase to a speaker on BBC Radio 4, in 'Thought for the Day'.

10. The subject of AI effectively started in the 1950s using comparatively elementary top-down procedures (e.g. Grey Walter 1953). The pattern recognizing 'perceptron', of Frank Rosenblatt (1962), in 1959, was the first successful 'connectionist' (artificial neural network) device, and this stimulated a great deal of interest in bottom-up schemes. However, some essential limitations of this type of bottom-up organization were pointed out in 1969 by Marvin Minsky and Seymour Papert (cf. Minsky and Papert 1972). These were later overcome by Hopfield (1982), and artificial devices of the neural network type are now the subject of a considerable world-wide activity. (See, for example, Beks and Hamker 1992 and Gernoth *et al.* 1993, for some applications to high-energy physics.) Important landmarks in top-down AI research were papers by John McCarthy (1979) and by Alan Newell and Herbert Simon (1976). See Freedman (1994) for a dramatic account of all this history. For other recent discussions of AI's procedures and prospects, see Grossberg (1987), Baars (1988); for a classic attack on the subject, see Dreyfus (1972); and for a recent viewpoint by an AI pioneer, Gelernter (1994); cf. also various articles in Broadbent (1993) and Khalfa (1994).
11. For expositions of the λ-calculus, see Church (1941) and Kleene (1952).
12. For various publications relevant to these issues, see, for example, Pour-el (1974), Smith and Stephenson (1975), Pour-el and Richards (1989), Blumb, Shub, and Smale (1989). The question of brain activity in relation to these issues has been considered, in particular, by Rubel (1985).
13. In the case of the tiling problem, what Robert Berger actually proved was that the tiling problem for *Wang* tiles has no algorithmic general solution. Wang tiles (named after the logician Hao Wang) consist of single individual square tiles with coloured edges, where the colours must be matched tile-to-tile, and the tiles must not be rotated or reflected. However, it is an easy matter to devise, for any Wang tile set, a corresponding set of polyominoes that will tile the plane if and only if the given set of Wang tiles will do so. Thus the computational insolubility of the polyomino tiling problem follows immediately from that for Wang tiles.

    It is worth pointing out, in connection with the polyomino tiling problem, that if a given set of polyominoes *fails* to tile the plane, then this fact *can* be computationally ascertained (like when a Turing machine action halts, or when a set of Diophantine equations possesses a solution) since one can try to cover an $n \times n$ square region with the tiles, for successively larger values of n, the failure of the tiles to cover the entire plane showing up at some *finite* value of n. It is the situations for which the tiles *do* tile the plane which cannot be algorithmically ascertained.
14. See Freedman (1994) for an account relating some of the over-optimistic early aspirations of AI.
15. I am grateful to various people and, in particular, to Lee Loevinger for acquainting me with these issues. See Hodgson (1991) for a remarkable discussion of the relevance of modern physics and computation to the question of how we behave.

16. See, for example, Smithers (1990).
17. For example, Sloman (1992) admonishes me for setting so much store by the ill-defined term 'consciousness' in ENM, whilst referring freely himself to the (in my opinion) much less well-defined term 'mind'!
18. Searle (1980), (1992).
19. See p. 372 in Searle's (1980) article in Hofstadter and Dennett (1981). It is not clear to me, however, whether Searle would now argue in favour of $\mathscr{B}$, rather than $\mathscr{C}$.
20. See Hofstadter (1981) for an entertaining presentation of a suggestion of this nature; cf. also ENM, pp. 21–2.
21. For an accessible account of the notion of 'algorithmic complexity', see Chaitin (1975).
22. See Hsu *et al.* (1990).
23. See Freedman (1994).
24. See, for example, Moravec (1994).
25. See Putnam (1960), Smart (1961), Benacerraf (1967), Good (1967, 1969), Lewis (1969, 1989), Hofstadter (1981), Bowie (1982), in relation to Lucas's arguments; see also Lucas (1970). My own version, as briefly presented in ENM, pp. 416–18, has been attacked in various reviews; cf., particularly Sloman (1992) and numerous commentators in *Behavioral and Brain Sciences*: Boolos (1990), Butterfield (1990), Chalmers (1990), Davis (1990), (1993), Dennett (1990), Doyle (1990), Glymour and Kelly (1990), Hodgkin and Houston (1990), Kentridge (1990), MacLennan (1990), McDermott (1990), Manaster-Ramer *et al.* (1990), Mortensen (1990), Perlis (1990), Roskies (1990), Tsotsos (1990), Wilensky (1990); see also my own replies Penrose (1990, 1993*d*), and also Guccione (1993); see also Dodd (1991), Penrose (1991*b*).
26. From a British TV programme—probably *The Dream Machine* (December 1991), 4th part of the BBC series The Thinking Machine. See also Freedman (1994) for a discussion of recent progress in AI 'Understanding', particularly with regard to Douglas Lenat's intriguing 'Cyc' project.
27. For a vivid popular account, see Woolley (1992).
28. For example, such a suggestion was made by Richard Dawkins in his BBC Christmas Lectures, 1992.
29. See, for example, Freedman's (1994) account of the work by Lenat and others in this direction.

2

The Gödelian case

2.1 Gödel's theorem and Turing machines

It is in mathematics that our thinking processes have their purest form. If thinking is just carrying out a computation of some kind, then it might seem that we ought to be able to see this most clearly in our mathematical thinking. Yet, remarkably, the very reverse turns out to be the case. It is within mathematics that we find the clearest evidence that there must actually be something in our conscious thought processes that eludes computation. This may seem to be a paradox—but it will be of prime importance in the arguments which follow, that we come to terms with it.

Before we start, let me encourage the reader not to be timid about the mathematics that we shall encounter in the next few sections (§2.2–§2.5), despite the fact that we must obtain some appreciation of the implications of no less than the most important theorem in mathematical logic of all time—the famous theorem of Kurt Gödel. I shall present only an extremely simplified version of that theorem, drawing, particularly, upon the slightly later ideas of Alan Turing. No mathematical formalism other than the simplest arithmetic will be used. The argument that I shall give will admittedly be confusing in places, but it will be *merely* confusing, and not actually 'difficult' in the sense of any prior mathematical knowledge being required. Take the argument as slowly as you wish and do not be ashamed to reread it as many times as you want to. Later on (§2.6–§2.10) I shall explore some of the more specific insights that underlie Gödel's theorem, but the reader who is not interested in such matters need not be concerned with those parts of this book.

What did Gödel's theorem achieve? It was in 1930 that the brilliant young mathematician Kurt Gödel startled a group of the world's leading mathematicians and logicians, at a meeting in Königsberg, with what was to become his famous theorem. It rapidly became accepted as being a fundamental contribution to the foundations of mathematics—probably the most fundamental ever to be found—but I shall be arguing that in establishing his theorem, he also initiated a major step forward in the philosophy of mind.

Among the things that Gödel indisputably established was that no *formal system* of sound mathematical rules of proof can ever suffice, even in principle, to establish all the true propositions of ordinary arithmetic. This is certainly remarkable enough. But a powerful case can also be made that his results showed something more than this, and established that human understanding and insight cannot be reduced to any set of computational rules. For what he appears to have shown is that no such system of rules can ever be sufficient to prove even those propositions of arithmetic whose truth is accessible, in principle, to human intuition and insight—whence human intuition and insight cannot be reduced to any set of rules. It will be part of my purpose here to try to convince the reader that Gödel's theorem indeed shows this, and provides the foundation of my argument that there must be more to human thinking than can ever be achieved by a computer, in the sense that we understand the term 'computer' today.

It is not necessary for me to give a definition of a 'formal system' for the central argument (but see §2.7). Instead, I shall take advantage of the fundamental contribution of Turing in about 1936 (and some others, primarily Church and Post) which laid down the kinds of processes that we now call 'computations' or 'algorithms'. There is an effective equivalence between such processes and what can be achieved by a mathematical formal system, so it will not be important to know what a formal system actually is, provided that we have a reasonably clear idea of what is meant by a computation or an algorithm. Even for this, a rigorous definition will not be necessary.

Those readers familiar with my earlier book *The Emperor's New Mind* (ENM, cf. Chapter 2) will know that an algorithm is what can be carried out by a *Turing machine*, where we may think of a Turing machine as a mathematically idealized computer. It carries out its activities in a step-by-step procedure, each step being completely specified in terms of the nature of the mark on a 'tape' that the machine happens to be examining at each moment, and on the machine's (discretely defined) 'internal state'. The different allowed internal states are finite in number, and the total number of marks on the tape must also be finite—though the tape itself is unlimited in length. The machine starts in a particular state, say the one labelled '0' and its instructions are fed in on the tape, say in the form of a binary number (sequence of '0's and '1's). It then starts reading these instructions, moving the tape (or, equivalently, moving itself along the tape) in a definite way according to its step-by-step inbuilt procedures, as is determined at each stage by its internal state and the particular digit that it happens to be examining. It erases marks, or makes new ones, also according to these procedures. It continues in this way until it reaches a particular instruction: 'STOP'—at which point (and only at which point) the answer to the computation that it has been performing is displayed on the tape, and the activity of the machine is terminated. It is now ready to perform its next computation.

Certain particular Turing machines are referred to as *universal* Turing machines, these being Turing machines which have the capability of imitating any Turing machine whatever. Thus any single universal Turing machine has the capability of carrying out *any* computation (or algorithm) that one may care to specify. Although a modern computer's detailed internal construction is very different from this (and its internal 'working space', though very large, is not infinite like the Turing machine's idealized tape), all modern general-purpose computers are, in effect, actually universal Turing machines.

2.2 Computations

We shall be concerned here with *computations*. By a computation (or algorithm) I indeed mean the action of some Turing machine, i.e. in effect, just the operation of a computer according to some computer program. It should be realized that computations are not merely the performing of ordinary operations of arithmetic, such as just adding or multiplying numbers together, but can involve other things also. Well-defined *logical operations* can also be part of a computation. For an example of a computation, we might consider the following task:

(A) Find a number that is not the sum of three square numbers.

By 'number', I mean here a 'natural number', i.e. one of

$$0, 1, 2, 3, 4, 5, 6, 7, 8, 9, 10, 11, 12, \ldots.$$

A *square* number is the product of a natural number with itself, i.e. one of

$$0, 1, 4, 9, 16, 25, 36, \ldots,$$

these being

$$0\times0=0^2, 1\times1=1^2, 2\times2=2^2, 3\times3=3^2, 4\times4=4^2, 5\times5=5^2,$$

$$6\times6=6^2, \ldots,$$

respectively. Such numbers are called 'square' because they can be represented as square arrays (including the vacuous array, to represent 0):

The computation **(A)** could then proceed as follows. We try each natural number in turn, starting with 0, to see whether or not it is the sum of three squares. We need only consider squares that are no larger than the number itself. Thus, for each natural number, there are only finitely many square numbers to try. As soon as three square numbers are found that do add to it, then our computation moves on to the next natural number, and we try again to find a triplet of squares (each no greater than the number itself) that sum to it. Our computation stops only when a natural number is found for which each such triplet of squares fails to add to it. To see how this works, start with 0. This is $0^2+0^2+0^2$, so it is indeed the sum of three squares. Next we try 1 and we find that although it is not $0^2+0^2+0^2$ it is indeed $0^2+0^2+1^2$. Our computation tells us now to move on to 2 and we ascertain that although it is not $0^2+0^2+0^2$ or $0^2+0^2+1^2$, it is indeed $0^2+1^2+1^2$; we then move on to 3 and find $3=1^2+1^2+1^2$; then to 4, finding $4=0^2+0^2+2^2$; then $5=0^2+1^2+2^2$; then after finding $6=1^2+1^2+2^2$ we move on to 7, but now all triplets of squares (each member no greater than 7)

$$0^2+0^2+0^2. \quad 0^2+0^2+1^2. \quad 0^2+0^2+2^2. \quad 0^2+1^2+1^2. \quad 0^2+1^2+2^2.$$
$$0^2+2^2+2^2. \quad 1^2+1^2+1^2. \quad 1^2+1^2+2^2. \quad 1^2+2^2+2^2. \quad 2^2+2^2+2^2.$$

fail to sum to 7, so the computation halts and we reach our conclusion: 7 is a number of the kind we seek, being *not* the sum of three squares.

2.3 Non-stopping computations

However, with the computation **(A)** we were lucky. Suppose we had tried, instead, the computation:

(B) Find a number that is not the sum of four square numbers.

Now when we reach 7 we find that it *is* the sum of *four* squares: $7=1^2+1^2+1^2+2^2$, so we must move on to 8, finding $8=0^2+0^2+2^2+2^2$, then to 9, finding $9=0^2+0^2+0^2+3^2$, then $10=0^2+0^2+1^2+3^2$, etc. The computation goes on and on: ... $23=1^2+2^2+3^2+3^2$, $24=0^2+2^2+2^2+4^2$, ..., $359=1^2+3^2+5^2+18^2$ ..., and on and on. We may decide that the answer to our computation is unbelievably large, and that our computer is going to take an enormously long time and use a huge amount of storage space in order to find the answer. In fact we may begin to wonder whether there is any answer at all. The computation appears to continue and continue, and it never seems to stop. In fact this is correct; it never does! It is a famous theorem first proved by the great (Italian–)French mathematician Joseph L. Lagrange in 1770 that *every* number is, indeed, the sum of four squares. It is not such an easy theorem (and even Lagrange's contemporary, the great Swiss mathematician Leonhard Euler, a man of astounding mathematical insight, originality, and productiveness, had tried but failed to find a proof).

I am certainly not going to trouble the reader with the details of Lagrange's argument here, so let us instead try something very much simpler:

(C) Find an odd number that is the sum of two even numbers.

I hope that it is obvious to the reader that *this* computation will never come to an end! Even numbers, namely the multiples of two

0, 2, 4, 6, 8, 10, 12, 14, 16, . . .,

always add up to even numbers, so there can certainly be no odd number, i.e. one of the rest

1, 3, 5, 7, 9, 11, 13, 15, 17, . . .,

that is the sum of a pair of even numbers.

I have given two examples ((**B**) and (**C**)) of computations that never terminate. In one case this fact, though true, is not at all easy to ascertain, whilst in the other, its non-termination is really obvious. Let me give another example:

(D) Find an even number, greater than 2, that is not the sum of two prime numbers.

Recall that a prime number is a natural number (other than 0 or 1) that has no factors other than itself or 1, so it is one of:

2, 3, 5, 7, 11, 13, 17, 19, 23, ...

It is very likely that the computation **(D)** does not terminate either, but nobody knows for sure. It depends upon the truth of the famous 'Goldbach conjecture', put forward by Goldbach in a letter to Euler in 1742, but which remains unproved to this day.

2.4 How do we decide that some computations do not stop?

We now see that computations may or may not terminate and, moreover, in the cases when they do not terminate it may be easy to see that they do not, or it may be very hard, or it may even be so hard that no one has as yet been clever enough to ascertain this fact for sure. By what procedures do mathematicians convince themselves or convince others that certain computations do not in fact terminate? Are they themselves following some computational (or algorithmic) procedure in order to ascertain things of this kind? Before attempting to answer this question, let us consider yet another example. This example will be somewhat more difficult to see than our obvious **(C)** but still very much easier than **(B)**. I shall try to illustrate something of the way in which mathematicians may sometimes reach their conclusions.

My example involves what are called *hexagonal* numbers:

1, 7, 19, 37, 61, 91, 127, . . .,

namely the numbers that can be arranged as hexagonal arrays (this time, *excluding* the vacuous array):

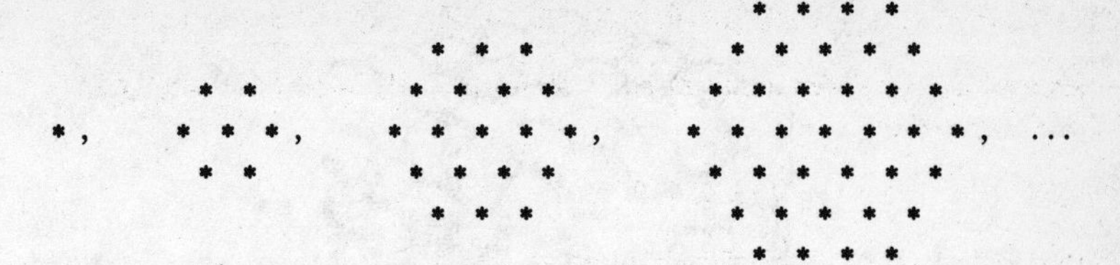

These numbers can be obtained, starting from 1, by adding successive multiples of 6:

$$6, 12, 18, 24, 30, 36, \ldots,$$

as can be seen by observing that each hexagonal number can be obtained from the one before it by adding a hexagonal ring around its border

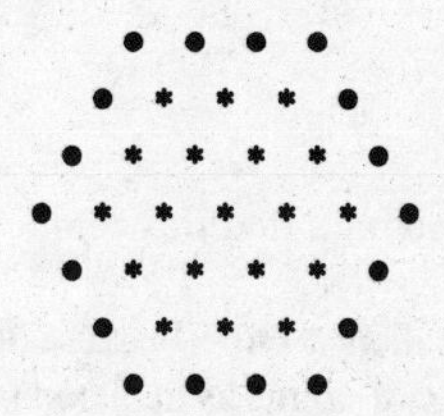

and noting that the number of spots in this ring must be a multiple of 6, the multiplier increasing by 1 each time, as the hexagon gets larger.

Now let us *add together* the hexagonal numbers successively, up to a certain point, starting with 1. What do we find?

$$1=1,\ 1+7=8,\ 1+7+19=27,\ 1+7+19+37=64,\ 1+7+19+37+61=125.$$

What is special about the numbers 1, 8, 27, 64, 125? They are all *cubes*. A cube is a number multiplied by itself three times:

$$1=1^3=1\times1\times1,\ 8=2^3=2\times2\times2,\ 27=3^3=3\times3\times3,\ 64=4^3=4\times4\times4,$$

$$125=5^3=5\times5\times5, \ldots.$$

Is this a general property of hexagonal numbers? Let us try the next case. We indeed find

$$1+7+19+37+61+91=216=6\times6\times6=6^3.$$

Is this going to go on literally for ever? If so, then the following computation will never terminate:

(E) Find a sum of successive hexagonal numbers, starting from 1, that is not a cube.

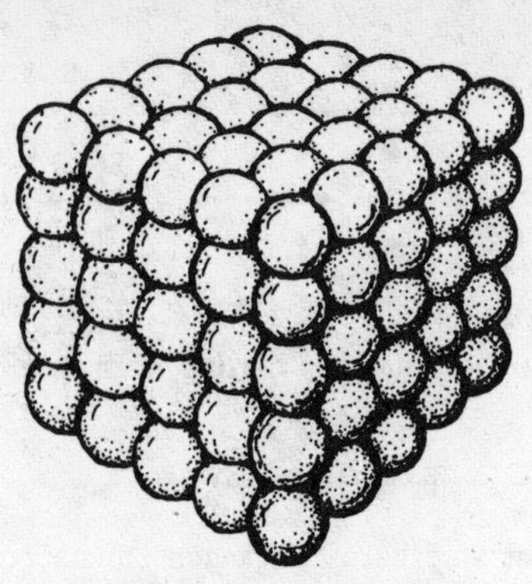

Fig. 2.1. A cubic array of spheres.

I am going to try to convince you that this computation will indeed continue for ever without stopping.

First of all, a cube is called a cube because it is a number that can be represented as a cubic array of points as depicted in Fig. 2.1. I want you to try to think of such an array as built up successively, starting at one corner and then adding a succession of three-faced arrangements each consisting of a back wall, side wall, and ceiling, as depicted in Fig. 2.2.

Now view this three-faced arrangement from a long way out, along the direction of the corner common to all three faces. What do we see? A *hexagon* as in Fig. 2.3. The marks that constitute these hexagons, successively increasing in size, when taken together, correspond to the marks that constitute the entire cube. This, then, establishes the fact that adding together successive hexagonal numbers, starting with 1, will always give a cube. Accordingly, we have indeed ascertained that **(E)** will never stop.

The reader may worry that the argument that I have just given is somewhat intuitive, instead of being a formal rigorous mathematical demonstration. In fact the argument is perfectly sound, and part of my purpose here is to show that there are sound methods of mathematical reasoning that are not 'formalized' according to some accepted preassigned system of rules. A much more elementary example of geometrical reasoning, used to obtain a general property of natural numbers, is the proof that $a \times b = b \times a$ given in §1.19. This, also, is a perfectly good 'proof', though not a formal one.

The reasoning that I have just given for the summing of successive hexagonal numbers could be replaced by a more formal mathematical demonstration if desired. The essential part of such a formal demonstration might be the *principle of mathematical induction*, which is a procedure for

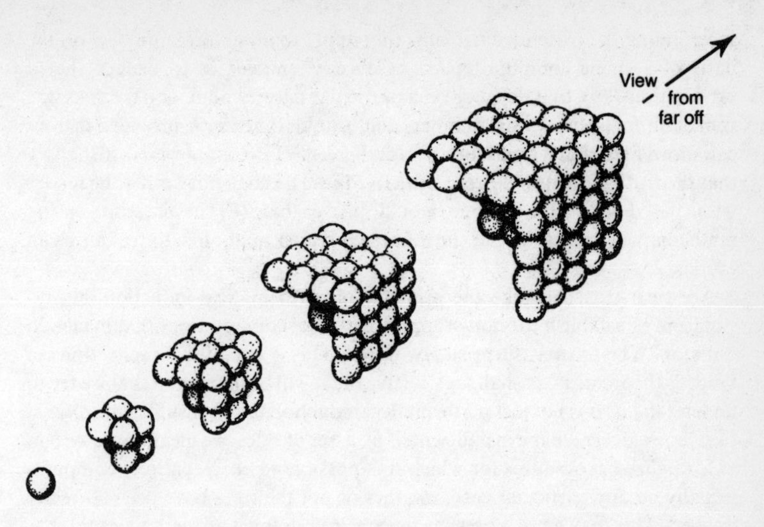

Fig. 2.2. Separate them out—each with a back wall, side wall, and ceiling.

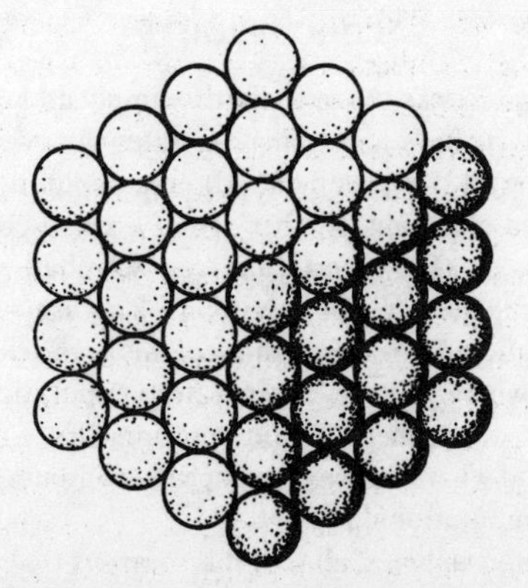

Fig. 2.3. Each piece is viewed as a hexagon.

ascertaining the truth of statements that apply to *all* natural numbers on the basis of a single computation. In essence, it enables us to deduce that a proposition $P(n)$, that depends on a particular natural number n (such as 'the sum of the first n hexagonal numbers is n^3'), holds for *every* n, provided that we can show, first, that it holds for $n=0$ (or, here, $n=1$) and that we can also show that the truth of $P(n)$ *implies* the truth of $P(n+1)$. I shall not bother the reader with the details of how one would prove that **(E)** never stops, using mathematical induction, but the interested reader might like to try this as an exercise.

Are clear-cut rules, like the principle of mathematical induction, always sufficient to establish the non-stopping nature of computations that in fact do not stop? The answer, surprisingly, is 'no'. This is one of the implications of Gödel's theorem, as we shall see shortly, and it will be important that we try to understand it. It is not just mathematical induction that is insufficient. *Any* set of rules *whatever* will be insufficient, if by a 'set of rules' we mean some system of formalized procedures for which it is possible to check entirely computationally, in any particular case, whether or not the rules have been correctly applied. This may seem a pessimistic conclusion, for it appears to imply that there are computations that never stop, yet the fact that they never stop cannot ever be rigorously mathematically ascertained. However, this is not at all what Gödel's theorem actually tells us. What it *does* tell us can be viewed in a much more positive light, namely that the insights that are available to human mathematicians—indeed, to anyone who can think logically with understanding and imagination—lie beyond anything that can be formalized as a set of rules. Rules can sometimes be a partial substitute for understanding, but they can never replace it entirely.

2.5 Families of computations; the Gödel–Turing conclusion $\mathscr{G}$

In order to see how Gödel's theorem (in the simplified form that I shall give, stimulated also by Turing's ideas) demonstrates this, we shall need a slight generalization of the kind of statements about computations that I have been considering. Instead of asking whether or not a single computation, such as **(A)**, **(B)**, **(C)**, **(D)**, or **(E)**, ever terminates, we shall need to consider a computation that depends on—or *acts* upon—a *natural number n*. Thus, if we call such a computation $C(n)$, we can think of this as providing us with a *family* of computations, where there is a separate computation for each natural number 0, 1, 2, 3, 4, . . ., namely the computation $C(0)$, $C(1)$, $C(2)$, $C(3)$, $C(4)$, . . ., respectively, and where the way in which the computation depends upon n is itself entirely computational.

In terms of Turing machines, all that this means is that $C(n)$ is the action of some Turing machine on the number n. That is, the number n is fed in on the machine's tape as input, and the machine just computes on its own from then on. If you do not feel comfortable with the concept of a 'Turing machine', just think of an ordinary general-purpose computer, and regard n as merely providing the 'data' for the action of some programmed computer. What we are interested in is whether or not this computer action ever stops, for each choice of n.

In order to clarify what is meant by a computation depending on a natural number n, let us consider two examples:

(F) Find a number that is not the sum of n square numbers

and

(G) Find an odd number that is the sum of n even numbers.

It should be clear from what has been said above that the computation **(F)** will stop *only* when $n=0$, 1, 2, and 3 (finding the numbers 1, 2, 3, and 7, respectively, in these cases), and that **(G)** stops for no value of n whatever. If we are actually to ascertain that **(F)** does not stop when n is 4 or larger we require some formidable mathematics (Lagrange's proof); on the other hand, the fact that **(G)** does not stop for any n is obvious. What are the procedures that are available to mathematicians for ascertaining the non-stopping nature of such computations generally? Are these very procedures things that can be put into a computational form?

Suppose, then, that we have some computational procedure A which, when it terminates,* provides us with a demonstration that a computation such as $C(n)$ actually does not ever stop. We are going to try to imagine that A encapsulates *all* the procedures available to human mathematicians for convincingly demonstrating that computations do not stop. Accordingly, if in any particular case A itself ever comes to an end, this would provide us with a human demonstration that the particular computation that it refers to does *not* ever stop. For most of the following argument, it is not necessary that A be viewed as having this particular role. We are just concerned with a bit of mathematical reasoning. But for our ultimate conclusion $\mathscr{G}$, we are indeed trying to imagine that A has this status.

I am certainly not requiring that A can always decide that $C(n)$ does not stop when in fact it does not, but I do insist that A does not ever give us wrong answers, i.e. that if it comes to the conclusion that $C(n)$ does not stop, then in fact it does not. If A does not in fact give us wrong answers, we say that A is *sound*.

It should be noted that if A were actually unsound, then it would be possible

*For the purposes of this argument, I am adopting the viewpoint that if A terminates at all, then this is to signal the achievement of a successful demonstration that $C(n)$ never stops. If A were to 'get stuck' for any other reason than 'success' in its demonstration, then this would have to qualify as a failure of A to terminate properly. See queries **Q3**, **Q4** below, and also Appendix A (p. 117).

in principle to ascertain this fact by means of some direct calculation—i.e. an unsound A is computationally falsifiable. For if A were to assert erroneously that the computation $C(n)$ does not ever terminate when in fact it does, then the performing of the actual computation $C(n)$ would eventually lead to a refutation of A. (The issue of whether such a computation could ever be performed in practice is a separate matter: it will be discussed under **Q8**.)

In order for A to apply to computations generally, we shall need a way of coding all the different computations $C(n)$ so that A can use this coding for its action. All the possible different computations C can in fact be listed, say as

$$C_0, C_1, C_2, C_3, C_4, C_5, \ldots,$$

and we can refer to C_q as the qth computation. When such a computation is applied to a particular number n, we shall write

$$C_0(n), C_1(n), C_2(n), C_3(n), C_4(n), C_5(n), \ldots.$$

We can take this ordering as being given, say, as some kind of numerical ordering of computer programs. (To be explicit, we could, if desired, take this ordering as being provided by the Turing-machine numbering given in ENM, so that then the computation $C_q(n)$ is the action of the qth Turing machine T_q acting on n.) One technical thing that is important here is that this listing is *computable*, i.e. there is a single* computation $C_\bullet$ that gives us C_q when it is presented with q, or, more precisely, the computation $C_\bullet$ acts on the *pair* of numbers q, n (i.e. q followed by n) to give $C_q(n)$.

The procedure A can now be thought of as a particular computation that, when presented with the pair of numbers q, n, tries to ascertain that the computation $C_q(n)$ will never ultimately halt. Thus, when the computation A *terminates*, we shall have a demonstration that $C_q(n)$ *does not halt*. Although, as stated earlier, we are shortly going to try to imagine that A might be a formalization of *all* the procedures that are available to human mathematicians for validly deciding that computations never will halt, it is not at all necessary for us to think of A in this way just now. A is just *any sound* set of computational rules for ascertaining that some computations $C_q(n)$ do not ever halt. Being dependent upon the two numbers q and n, the computation that A performs can be written $A(q, n)$, and we have:

(H) If $A(q, n)$ stops, then $C_q(n)$ does not stop.

Now let us consider the particular statements **(H)** for which q is put equal to n. This may seem an odd thing to do, but it is perfectly legitimate. (This is the first step in the powerful 'diagonal slash', a procedure discovered by the highly original and influential nineteenth-century Danish/Russian/German mathe-

*In fact this is achieved precisely by the action of a universal Turing machine on the pair of numbers q, n; see Appendix A and ENM, pp. 51–7.

matician Georg Cantor, central to the arguments of both Gödel and Turing.) With q equal to n, we now have:

(I) If $A(n, n)$ stops, then $C_n(n)$ does not stop.

We now notice that $A(n, n)$ depends upon just *one* number n, not two, so it must be one of the computations $C_0, C_1, C_2, C_3, \ldots$ (as applied to n), since this was supposed to be a listing of *all* the computations that can be performed on a single natural number n. Let us suppose that it is in fact C_k, so we have:

(J) $A(n, n) = C_k(n)$.

Now examine the particular value $n = k$. (This is the second part of Cantor's diagonal slash!) We have, from **(J)**,

(K) $A(k, k) = C_k(k)$

and, from **(I)**, with $n = k$:

(L) If $A(k, k)$ stops, then $C_k(k)$ does not stop.

Substituting **(K)** in **(L)**, we find:

(M) If $C_k(k)$ stops, then $C_k(k)$ does not stop.

From this, we must deduce that the computation $C_k(k)$ does *not* in fact stop. (For if it did then it does not, according to **(M)**!) But $A(k, k)$ cannot stop either, since by **(K)**, it is the *same* as $C_k(k)$. Thus, our procedure A is incapable of ascertaining that this particular computation $C_k(k)$ does not stop even though it does not.

Moreover, if we *know* that A is sound, then we *know* that $C_k(k)$ does not stop. Thus, we know something that A is unable to ascertain. It follows that A *cannot* encapsulate our understanding.

At this point, the cautious reader might wish to read over the whole argument again, as presented above, just to make sure that I have not indulged in any 'sleight of hand'! Admittedly there is an air of the conjuring trick about the argument, but it is perfectly legitimate, and it only gains in strength the more minutely it is examined. We have found a computation $C_k(k)$ that we know does not stop; yet the given computational procedure A is not powerful enough to ascertain that fact. This is the Gödel(–Turing) theorem in the form that I require. It applies to any computational procedure A whatever for ascertaining that computations do not stop, *so long as we know it to be sound*. We deduce that no knowably sound set of computational rules (such as A) can ever suffice for ascertaining that computations do not stop, since there are some non-stopping computations (such as $C_k(k)$) that must elude these rules. Moreover, since from the knowledge of A and of its soundness, we can actually construct a computation $C_k(k)$ that we can *see* does not ever stop, we deduce that A *cannot* be a formalization of the procedures available to mathematicians for ascertaining that computations do not stop, no matter what A is. Hence:

$\mathscr{G}$ Human mathematicians are not using a knowably sound algorithm in order to ascertain mathematical truth.

It seems to me that this conclusion is inescapable. However, many people have tried to argue against it—bringing in objections like those summarized in the queries **Q1–Q20** of §2.6 and §2.10 below—and certainly many would argue against the stronger deduction that there must be something fundamentally non-computational in our thought processes. The reader may indeed wonder what on earth mathematical reasoning like this, concerning the abstract nature of computations, can have to say about the workings of the human mind. What, after all, does any of this have to do with the issue of conscious awareness? The answer is that the argument indeed says something very significant about the mental quality of *understanding*—in relation to the general issue of computation—and, as was argued in §1.12, the quality of understanding is something dependent upon conscious awareness. It is true that, for the most part, the foregoing reasoning has been presented as just a piece of mathematics, but there is the essential point that the algorithm A enters the argument at two quite different levels. At the one level, it is being treated as just some algorithm that has certain properties, but at the other, we attempt to regard A as being *actually* 'the algorithm that we ourselves use' in coming to believe that a computation will not stop. The argument is *not* simply about computations. It is also about how we use our conscious understanding in order to infer the validity of some mathematical claim—here the non-stopping character of $C_k(k)$. It is the interplay between the two different levels at which the algorithm A is being considered—as a putative instance of conscious activity and as a computation itself—that allows us to arrive at a conclusion expressing a fundamental conflict between such conscious activity and mere computation.

However, there are indeed various possible loopholes and counter-arguments that must be considered. First, in the remainder of this chapter, I shall go very carefully through *all* the relevant counter-arguments against the conclusion $\mathscr{G}$ that have come to my attention—these are the queries **Q1–Q20**, that will be addressed in §2.6 and §2.10, which also include a few additional counter-arguments of my own. Each of these will be answered as carefully as I am able. We shall see that the conclusion $\mathscr{G}$ comes through essentially unscathed. Then, in Chapter 3, I shall consider the implications of $\mathscr{G}$ itself. We shall find that it indeed provides the basis for a very powerful case that conscious mathematical understanding cannot be properly modelled *at all* in computational terms, whether top-down or bottom-up or any combination of the two. Many people might find this to be an alarming conclusion, as it may seem to have left us with nowhere to turn. In Part II of this book I shall take a more positive line. I shall make what I believe to be a plausible scientific case for my own speculations about the physical processes that might conceivably underlie brain action, such as when we follow through an argument of this kind, and how this might indeed elude any computational description.

2.6 Possible technical objections to $\mathscr{G}$

The reader may feel that the conclusion $\mathscr{G}$ is itself quite a startling one, especially considering the simple nature of the ingredients of the argument whereby it is derived. Before we move on to consider, in Chapter 3, its implications with regard to the possibility of building a computer-controlled, intelligent, mathematics-performing robot, we must examine a number of technical points concerning the deduction of $\mathscr{G}$ very carefully. If you are a reader who is not concerned with such possible technical loopholes and are prepared to accept the conclusion $\mathscr{G}$—that mathematicians are not using a knowably sound algorithm to ascertain mathematical truth—then you may prefer to skip these arguments (for the moment at least) and pass directly on to Chapter 3; moreover, if you are prepared to accept the stronger conclusion that there can be no algorithmic explanation *at all* for our mathematical or other understandings, then you may prefer to pass directly on to Part II—perhaps pausing only to examine the fantasy dialogue of §3.23 (which summarizes the essential arguments of Chapter 3) and the conclusions of §3.28.

There are several points about the mathematics that tend to worry people about the type of Gödel argument given in §2.5. Let us try to sort these out.

Q1. I have taken A to be just a *single* procedure, whereas we undoubtedly use many different kinds of reasoning in our mathematical arguments. Should we not have allowed for a whole list of possible 'A's?

In fact, there is no loss of generality in phrasing things in the way that I have done. Any finite list $A_1, A_2, A_3, \ldots, A_r$ of algorithmic procedures can always be re-expressed as a single algorithm A, in such a way that A will fail to stop only if *all* the individual algorithms $A_1, \ldots, A_r$ fail to stop. (The procedure of A might run roughly as follows: 'Do the first 10 steps of A_1; remember the result; do the first 10 steps of A_2; remember the result; do the first 10 steps of A_3; remember the result; and so on, up to A_r; then go back to A_1 and do its second set of 10 steps; remember the result; and so on; then the third set of 10 steps, etc. Stop as soon as any of the A_r stops.') If, on the other hand, the list of the A were infinite, then in order for it to count as an algorithmic procedure, there would have to be a way of generating this entire set $A_1, A_2, A_3, \ldots$ in some algorithmic way. Then we can obtain a single A that will do in place of the entire list in the following way:

'first 10 steps of A_1;

second 10 steps of A_1, first 10 steps of A_2;
third 10 steps of A_1, second 10 steps of A_2, first 10 steps of A_3;
. . . etc. . . .'.

This will stop as soon as any one in the list successfully terminates, and not otherwise.

One might, on the other hand, imagine that the list $A_1, A_2, A_3, \ldots$, which we take to be infinite, is not provided ahead of time, even in principle. One might envisage that successive algorithmic procedures might be added to the list from time to time, without the list having been originally specified in its entirety. However, in the absence of any previous laid-down algorithmic procedure for generating this list, we do not really have a self-contained procedure at all.

Q2. Surely we must allow that the algorithm *A* might not be fixed? Human beings can learn, after all, so the algorithm that humans use might be a continually changing one.

A changing algorithm would need some specification as to the rules whereby it actually changes. If these rules are themselves entirely algorithmic, then we should have already included these very rules in what we mean by 'A'; so *this* kind of 'changing algorithm' is really just another instance of a single algorithm, and the argument proceeds precisely as before. On the other hand, one might envisage ways in which the algorithm could change that are supposed to be *not* algorithmic, and suggestions for this might be: the incorporation of random ingredients or some kind of interaction with the environment. The 'non-algorithmic' status of such a means of changing an algorithm will be reconsidered later (cf. §3.9, §3.10); see also the discussion of §1.9, where it was argued that neither of these means provides a plausible escape from algorithmism* (as would be required by viewpoint $\mathscr{C}$). For our present purely mathematical purposes, we are concerned only with the possibility that the change is indeed algorithmic. But once we have accepted that such a change *cannot* be algorithmic, then we have certainly come into agreement with the conclusion $\mathscr{G}$.

Perhaps I should be a little more explicit about what one might mean by an 'algorithmically changing' algorithm A. We may suppose that A depends upon not only q and n, but also a further parameter t, which we may think of as representing the 'time', or perhaps t just counts the number of occasions on which the algorithm has been activated previously. In any case, we may as well assume that the parameter t is a natural number, so we now have algorithms $A_t(q, n)$, which we can list

$$A_0(q, n), A_1(q, n), A_2(q, n), A_3(q, n), \ldots,$$

*The appropriate word 'algorithmism' for (essentially) my 'viewpoint $\mathscr{A}$' has been coined by Hao Wang (1993).

where each is supposed to be a sound procedure for ascertaining that the computation $C_q(n)$ does not stop, but where we imagine that these procedures are allowed to be increasing in power as t gets larger. Now the means whereby they increase in power is supposed to be algorithmic. Perhaps this 'algorithmic means' is something that might depend upon the 'experiences' of the previous $A_t(q, n)$, but these 'experiences' are here being taken to be algorithmically generated things also (or else we are back in agreement with $\mathscr{G}$), so we may as well include them, or their means of generation, into what constitutes the next algorithm (i.e. into $A_t(q, n)$ itself). In this way, we arrive at a *single* algorithm $(A_t(q, n))$ that depends algorithmically on all *three* parameters t, q, n. From this, we can construct an algorithm A^* that is as powerful as the entire list of $A_t(q, n)$, but which depends on just the two natural numbers q and n. To construct this $A^*(q, n)$, all we need, as above, is to let it run through the first 10 steps of $A_0(q, n)$, remember the result; then the first 10 steps of $A_1(q, n)$, followed by the second 10 steps of $A_0(q, n)$, remembering the results; then the first 10 of $A_2(q, n)$, the second 10 of $A_1(q, n)$, and the third 10 of $A_0(q, n)$; etc., where at each stage we remember the previous results; finally, we come to a *halt* as soon as *any* of the constituent computations comes to a halt. Using A^* in place of A, the argument establishing $\mathscr{G}$ proceeds just as before.

Q3. Have I not been unnecessarily narrow in insisting that A must go on computing for ever in those cases when it may have become clear that $C_q(n)$ actually *does* stop? If we allowed A actually to *stop* in those cases, our argument would fail. The insights that are available to human beings, after all, certainly do allow them sometimes to conclude that computations stop, but I seem to have ignored these. Does this not mean that I have been too restrictive?

Not at all. The argument is supposed to be applied merely to insights that allow us to come to the conclusion that computations do *not* stop, not to those insights that allow us to conclude the opposite. The putative algorithm A is not allowed to come to a 'successful termination' by concluding that some computation *does* stop. That is not its job.

If you feel uncomfortable with this, think of A in the following way: try to include *both* types of insight in A, but in those circumstances when the conclusion is that the computation $C_q(n)$ does stop, deliberately put A into a loop (i.e. make A just repeat some operation over and over again endlessly). Of course that is not the way that a mathematician would actually operate, but that is not the point. The argument has the form of a *reductio ad absurdum*, starting from the assumption that we use a knowably sound algorithm for ascertaining mathematical truth, and then deriving a contradiction. It is not necessary, in this argument, that A actually *be* this putative algorithm, but it can be something constructed from it, such as in the case of the A just referred to.

The same comment would apply to any other objection to the argument of

§2.5 of the form: 'surely A might stop for various spurious reasons without it having provided a demonstration that $C_q(n)$ does not stop'. If we are given an 'A' that behaves in this way, we simply apply our argument of §2.5 to a slightly different A, namely one that loops whenever the original 'A' stops for any such spurious reason.

Q4. In the numbering C_0, C_1, C_2, . . ., I seem to have assumed that every C_q actually denotes a well-defined computation; whereas in any straightforward numerical or alphabetical-type ordering of computer programs, surely this would not be likely to be the case?

It would indeed be an awkward thing to ensure that our numbering actually provides a working computation C_q for every natural number q. For example, the numbering of Turing machines T_q given in ENM certainly does not achieve this; cf. ENM, p. 54. For a specified q, the Turing machine T_q, as described there, would be considered to be a 'dud' for one of four reasons: it could run on for ever without stopping; it could be 'not correctly specified' because the number n leads to a binary expansion with too many 1s in succession (five or more), and therefore would have no translation in the given scheme; it could encounter an instruction to enter a non-existent internal state; or it could produce just blank tape, when it stops, which has no numerical interpretation. (See also Appendix A.) For the purposes of the Gödel–Turing argument that I have just given, all that one needs to do is to lump all these reasons together under the heading 'does not stop'. In particular, when I said 'terminates' for the computational procedure A (cf. footnote, p. 73), this carries the implication that it indeed 'stops' in the aforementioned sense (and so does not contain untranslatable sequences or produce just blank tape)—i.e. 'stops' implies that the computation is indeed a properly specified working computation. Likewise, '$C_q(n)$ stops' means also that it properly stops in this sense. With this interpretation, the argument, as I have given it, is unaffected by consideration **Q4**.

Q5. Have I not merely shown that it is possible to outdo just a *particular* algorithmic procedure, A, by defeating it with the computation $C_q(n)$? Why does this show that I can do better than any A whatsoever?

The argument certainly *does* show that we can do better than *any* algorithm. This is the whole point of a *reductio ad absurdum* argument of this kind that I have used here. I think that an analogy might be helpful here. Some readers will know of Euclid's argument that there is no largest prime number. This, also, is a *reductio ad absurdum*. Euclid's argument is as follows. Suppose, on the contrary, that there is a largest prime; call it p. Now consider the product N of all the primes up to p and add 1:

$$N=2\times 3\times 5\times \cdots \times p+1.$$

N is certainly larger than p, but it cannot be divisible by any of the prime numbers 2, 3, 5, . . ., p (since it leaves the remainder 1 on division); so either N is the required prime itself or it is composite—in which case it is divisible by a prime larger than p. Either way, there would have to be a prime larger than p, which contradicts the initial assumption that p is the largest prime. Hence there is no largest prime.

The argument, being a *reductio ad absurdum*, does not merely show that a *particular* prime p can be defeated by finding a larger one; it shows that there cannot be any largest prime *at all*. Likewise, the Gödel–Turing argument above does not merely show that a *particular* algorithm A can be defeated, it shows that there cannot be any (knowably sound) algorithm *at all* that is equivalent to the insights that we use to ascertain that certain computations do not stop.

Q6. A computer could be programmed to follow through precisely the argument that I have given here. Could it not itself, therefore, arrive at any conclusion that I have myself reached?

It is certainly true that it is a computational process to find the particular calculation $C_k(k)$, given the algorithm A. In fact this can be exhibited quite explicitly.* Does this mean that the supposedly non-algorithmic mathematical insight—the insight that allowed us to appreciate the fact that $C_k(k)$ never stops—is actually algorithmic after all?

I think that it is important to address this discussion at some length, since it represents one of the commonest misconceptions in relation to the Gödel argument. It should be made clear that it does *not* invalidate anything that has gone before. Although the procedure for obtaining $C_k(k)$ from A can be put into the form of a computation, this computation is not part of the procedures contained in A. It cannot be, because A is not capable of ascertaining the truth of $C_k(k)$, whereas this new computation (together with A) is asserted to be able to. Thus, although the new computation is indeed a computation leading to $C_k(k)$, it is not one that has been admitted to the club of 'official truth ascertainers'.

Let me put things another way. Imagine a computer-controlled robot that is able to ascertain mathematical truths by means of the algorithmic procedures contained in A. To make things more graphic, I shall use an anthropomorphic terminology and say that the robot 'knows' the mathematical truths—here non-stopping of computations—that it can derive by the use of A. However, if A is all that our robot 'knows', then it will *not* 'know' that $C_k(k)$ does not stop,

*To emphasize that I appreciate this point, I refer the reader to Appendix A, where an explicit computational procedure is exhibited (using the rules given in detail in ENM, Chapter 2) for obtaining the Turing machine action $C_k(k)$ from the algorithm A. Here, A is assumed to be given to us in the form of a Turing machine T_a, whose assessment of $C_q(n)$ is coded as the action of T_a on q followed by n.

even though the procedure for obtaining $C_k(k)$ from A is perfectly algorithmic. Of course, we could *tell* our robot that $C_k(k)$ indeed does not stop (using our own insights to that effect), but if the robot were to accept this fact, it would have to modify its own rules by adjoining this new truth to the ones that it already 'knows'. We could imagine going further than this and tell our robot, in some appropriate way, that the general computational procedure for obtaining $C_k(k)$ from A is also something that it should 'know' as a way of obtaining new truths from old. Anything that is well defined and computational could be added to the robot's store of 'knowledge'. But we now have a *new* 'A', and the Gödel argument would apply to this, instead of to the old A. That is to say, we should have been using this new 'A' all along instead of the old A, since it is cheating to change our 'A' in the middle of the argument. Thus, we see that one thing that is wrong with **Q6** is very similar to what is wrong with **Q5**, as discussed above. In the *reductio ad absurdum* we assume that A—which is to be a known, sound procedure for ascertaining that computations do not stop—actually represents the *totality* of such procedures available to mathematicians, and from this we derive a contradiction. It is cheating to introduce another truth-judging computational procedure not contained in A, *after* we have settled on A as representing this totality.

The trouble for our poor robot is that in the absence of any *understanding* of the Gödel procedure the robot has no reliable independent way of judging truth, other than being told by us. (This is a separate matter from the computational aspects of the Gödel argument.) To be able to do more than this, it, like us, would need to understand the meanings of the operations that it has been told to perform. Without understanding, it could equally well (wrongly) 'know' that $C_k(k)$ *does* stop instead of that it does not. It is just as much an algorithmic matter to derive (wrongly) that '$C_k(k)$ stops' as it is to derive (correctly) that '$C_k(k)$ does not stop'. Thus the algorithmic nature of these operations is not the point; the point is that our robot requires valid *truth judgements* in order to know which algorithms give it truths and not falsehoods. Now, at this point in the argument, it is still possible that 'understanding' is another kind of algorithmic activity, not contained in any precisely given known-to-be-sound procedures such as A. For example, understanding might be given by an unsound or unknowable algorithm. In my later discussion (Chapter 3) I shall try to persuade the reader that in fact understanding is not an algorithmic activity at all. But, for now, we are just concerned with rigorous implications of the Gödel–Turing argument, and, for this, the fact that $C_k(k)$ can be obtained from A in a computational way is neither here nor there.

Q7. The total output of all the mathematicians who have ever lived, together with the output of all the human mathematicians of the next (say) thousand years is finite and could be contained in the memory banks of an appropriate computer. Surely this particular computer *could*, therefore, simulate this output and thus

behave (externally) in the same way as a human mathematician—whatever the Gödel argument might appear to tell us to the contrary?

While this is presumably true, it ignores the essential issue, which is how we (or computers) know which mathematical statements are true and which are false. (In any case, the mere *storage* of mathematical statements is something that could be achieved by a system much less sophisticated than a general-purpose computer, e.g. photographically.) The way that the computer is being employed in **Q7** totally ignores the critical issue of *truth judgement*. One could equally well envisage computers that contain nothing but lists of totally false mathematical 'theorems', or lists containing random jumbles of truths and falsehoods. How are we to tell which computer to trust? The arguments that I am trying to make here do not say that an effective simulation of the output of conscious human activity (here mathematics) is impossible, since purely by chance the computer might 'happen' to get it right—even without any understanding whatsoever. But the odds against this are absurdly enormous, and the issues that are being addressed here, namely how one decides *which* mathematical statements are true and which are false, are not even being touched by **Q7**.

There is, on the other hand, a more serious point that is indeed being touched upon in **Q7**. This is the question as to whether discussions about infinite structures (e.g. *all* natural numbers or *all* computations) are really relevant to our considerations here, when the outputs of humans and computers are *finite*. Let us consider this important issue separately next.

Q8. Non-terminating computations are idealized mathematical constructions which have to do with the infinite. Surely such matters are not really relevant to discussions about finite physical objects such as computers or brains?

It is certainly true that with our idealized discussions of Turing machines, non-terminating computations, etc., we have been considering (potentially) infinite processes, whereas with humans or computers, we are dealing with *finite* systems. It is important to try to assess the limitations of such idealized arguments when applying them to actual finite physical objects. However, it turns out that the consideration of finiteness does not substantially affect the actual Gödel–Turing argument. There is nothing wrong with *discussing* idealized computations, reasoning about them, and deriving, mathematically, their theoretical limitations. We can, for example, in perfectly finite terms, discuss the question as to whether or not there is an odd number that is the sum of two even numbers, or whether there is any natural number that is not the sum of four squares (as in **(C)** and **(B)** above), despite the fact that in addressing these issues we are implicitly considering the infinite collection of *all* natural numbers. We can perfectly well reason about non-terminating computations, or Turing machines generally, as *mathematical* constructs even though it would not be possible actually to build an endlessly operating

Turing machine in practice. (Note that, in particular, a Turing-machine action that seeks an odd number that is the sum of two even numbers could not, strictly speaking, be implemented physically; for its parts would wear out rather than it really running on for ever.) The specification of any single computation (or Turing-machine action) is a perfectly finite matter, and the question as to whether or not it eventually stops is perfectly well defined. Once we have finished with our reasoning about such idealized computations, we can *then* try to see in what way our discussion applies to finite systems like actual computers or to people.

Finiteness limitations might come about either (i) because the specification of the actual computation under consideration is inordinately enormous (i.e. that the number n in C_n, or the pair of numbers q, n taken together in $C_q(n)$, is too large to be specified by a feasible computer or by a person), or (ii) because a computation that is not too large to specify might nevertheless take too long to perform, so that it might seem that it never stops even though, theoretically, the specified computation eventually would stop. In fact it turns out, as we shall see in a moment, that of these two it is only (i) that significantly affects our discussion, and even (i) does not affect things very greatly. The unimportance of (ii) is perhaps surprising. There are many fairly simple computations that ultimately stop, but for which no conceivable computer could directly compute far enough to reach the stopping point. For an example, consider the following: 'print out a succession of $2^{2^{65\,536}}$ 1s and then stop'. (Some much more mathematically interesting examples will be given later in §3.26.) The question as to whether or not a computation is going to stop need not be settled by direct computation; often that is an extremely inefficient method.

To see how the finiteness limitations (i) or (ii) might affect our Gödel-type discussion, let us re-examine the relevant parts of the argument. In accordance with limitation (i), instead of having an infinite list of computations, we have a *finite* list:

$$C_0, C_1, C_2, C_3, \ldots, C_Q,$$

where we suppose that the number Q specifies the largest computation that our computer, or human, is able to accommodate. In the case of a human, we may consider that there is a certain vagueness about this. For the moment it is not important that Q be defined as a precise number. (This question of such vagueness, in relation to human capabilities, will be discussed later in the response to **Q13** in §2.10.) Moreover, we suppose that when we apply these computations to a particular natural number n, the value of n may be restricted to being no larger than some fixed number N, because our computer (or human) is not set up to handle numbers larger than N. (Strictly speaking, we ought to consider the possibility that N is not a fixed number, but depends upon the particular computation C_q that we are considering—i.e. N could depend on q. This does not make any substantial difference to our considerations, however.)

As before, we consider a sound algorithm $A(q, n)$ which, when it stops, provides us with a demonstration that the computation $C_q(n)$ does not terminate. When we say 'sound', although in accordance with (i) we need only consider values of q that are no larger than Q and values of n that are no larger than N, we really mean that A is to be sound for *all* values of q and n, no matter how big they may be. (Thus, the rules embodied in A are precise *mathematical* rules, and not approximate ones that work only by virtue of some practical limitation on computations that can 'actually' be performed.) Moreover, when we say that '$C_q(n)$ does not terminate' we mean that it *really* does not terminate and not that the computation might simply be too long to perform by our computer or human, as is envisaged in (ii).

Recall that **(H)** tells us:

If $A(q, n)$ stops, then $C_q(n)$ does not stop.

In view of (ii), we might consider that the algorithm A is not much use to us for deciding whether another computation fails to stop if, itself, it takes more steps than our computer or human can possibly handle. But it turns out that this is of no importance for the argument. We are going to find a computation $A(k, k)$ that does not stop at all. It does not matter to us that in some other cases, in which A actually *does* stop, we are unable to wait long enough to find out that it does.

Now, as in **(J)**, we locate a natural number k for which the computation $A(n, n)$ is the same as $C_k(n)$ for each n:

$$A(n, n) = C_k(n).$$

However, we must now consider the possibility that this k is larger than Q, as envisaged by (i). For a horrendously complicated A, this might indeed be the case, but only if A were already beginning to approach the upper limit of size (in terms of the number of binary digits in its Turing-machine specification) that can be handled by our computer or human. This is because the computation that obtains the value k from the (say, Turing-machine) specification of A is quite a simple thing that can be given explicitly (as has already been pointed out in the response to **Q6** above).

The actual computation that we need in order to defeat A is $C_k(k)$, and putting $n = k$ in **(H)**, we obtain **(L)**:

If $A(k, k)$ stops, then $C_k(k)$ does not stop.

Since $A(k, k)$ is the same as $C_k(k)$, our argument shows that the particular computation $C_k(k)$ cannot stop at all, but that A cannot ascertain this fact, even if it were allowed to run on for far longer than any limit imposed in accordance with (ii). The specification of $C_k(k)$ is given in terms of the above k, and provided that k is not larger than either Q or N, it is a computation that could actually be implemented by our computer or human—in the sense that the computation could *start*. It could not, in any case, be continued to completion because the computation in fact does not ever stop!

Now, might k really be larger than Q or N? This would only be the case if the specification of A requires so many digits that increasing this number by moderately little would lead to an overflow of our computer's or human's capacity. It still follows from a knowledge of A's soundness that we *know* that this $C_k(k)$ cannot stop, even though we might have difficulty in implementing the actual computation $C_k(k)$. Consideration (i) does, however, lead us to envisage the possibility that the computation A might be so stupendously complicated that its specification puts it close to the borderline of computations that it is possible for a human being to contemplate at all, and the comparatively small increase in the number of digits that is being considered will yield a computation that is beyond human contemplation. I think that, whatever we may think of such a possibility, any such stupendous set of computational rules embodied in this putative A would certainly be so horrendously complicated that its *soundness* could not plausibly be *known* to us even if the precise rules themselves could be known to us. Thus, our conclusion stands, just as before: we do *not* ascertain mathematical truth by means of a *knowably sound* set of algorithmic rules.

It is of some value to be a little more specific about the comparatively mild increase in complication that is involved passing from A to $C_k(k)$. This will have a particular significance for us later (in §3.19 and §3.20). In Appendix A (p. 117), an explicit specification for $C_k(k)$ is provided, in terms of the Turing-machine prescriptions given in ENM, Chapter 2. According to these prescriptions, T_m denotes the 'mth Turing machine'. To be specific here, it will be convenient to use this notation rather than 'C_m', particularly in order to define the *degree of complication* of a computational procedure or of an individual computation. Accordingly, I define this degree of complication μ of the Turing machine T_m to be the number of binary digits in the specification of m as a binary number (cf. ENM, p. 39); then the degree of complication of some particular computation $T_m(n)$ is defined as the larger of the two numbers μ and ν, where ν is the number of binary digits in the specification of n. Now consider the explicit prescription, provided in Appendix A, for obtaining the computation $C_k(k)$ from A, given in these Turing-machine terms. Taking the degree of complication of A to be α, we then find that the degree of complication of this explicit computation $C_k(k)$ turns out to be less than $\alpha + 210 \log_2(\alpha + 336)$, a number that is larger than α by only a relatively tiny amount, when α is very large.

There is a possible proviso to the above general line of argument that may worry some readers. Does it really make sense to consider a computation that may be too complicated to write down, or that if it were written down might take far far longer than the age of the universe actually to perform, even if each step could be performed in the tiniest fraction of a second, in which physical processes can sensibly be envisaged as taking place? The computation considered above—which outputs a succession of $2^{2^{65\,536}}$ 1s, stopping only after this task is completed—is such an example, and it would be an extremely unconventional mathematical standpoint that allows us to claim that this is a non-terminating computation. However, there are some mathematical viewpoints, not so unconventional as to forbid this—though still decidedly unconventional—according to which there might be some doubt introduced about matters of absolute mathematical truth of idealized mathematical statements. We should at least have a look at some of these.

Q9. The standpoint known as *intuitionism* forbids one from deducing that a computation must terminate at a definite point merely from the fact that its continuing indefinitely leads to a contradiction; likewise there are other 'constructivist' or 'finitist' points of view. According to these might the Gödel-type reasoning not be found questionable?

In the Gödel-type reasoning that I have given, I used, in **(M)**, an argument of the form: 'the assumption that X is false leads us to a contradiction; therefore X is true'. Here 'X' is the statement '$C_k(k)$ does not stop'. This is a *reductio ad absurdum* type of argument—and, indeed, the Göedelian argument as a whole is phrased in this way. The mathematical standpoint known as 'intuitionism' (initiated by the Dutch mathematician L.E.J. Brouwer in around 1912; cf. Kleene 1952, also ENM pp. 113–16) denies that one can validly reason by use of *reductio ad absurdum*. Intuitionism arose as a reaction against certain mathematical trends that had arisen in the late nineteenth and early twentieth centuries according to which a mathematical object could be asserted to 'exist' even though there might be no way given of actually constructing the object in question. Sometimes a too free use of a nebulous concept of mathematical existence would actually lead to a contradiction. The most famous example of this occurs with Bertrand Russell's paradoxical 'set of all sets that are not members of themselves'. (If Russell's set is a member of itself, then it is not; if it is not, then it is! See §3.4 and ENM, p. 101, for further details.) To counter this general trend, whereby very freely defined mathematical objects could be taken to 'exist', the intuitionistic standpoint denied that one can validly use the type of mathematical reasoning that allows one to deduce the existence of some mathematical object merely from the contradictory nature of its non-existence. Such a *reductio ad absurdum* argument does not provide an actual construction for the object in question.

How would a denial of this use of *reductio ad absurdum* affect our Gödel-type argument? In fact, not at all, simply because we are using *reductio ad absurdum* in the opposite way from this, namely that the contradiction is being derived from the assumption that something *exists*, not from the assumption that something does not exist. According to intuitionism, it is perfectly legitimate to deduce that something does *not* exist from the fact that a contradiction arises from the assumption that it does exist. The Gödel-type argument, as I have given it, is, in effect, perfectly intuitionistically acceptable. (See Kleene 1952, p. 492.)

Similar remarks apply to all the other 'constructivist' or 'finitistic' viewpoints of which I am aware. The discussion following **Q8** shows that even the viewpoint hinted at above, which denies that the natural numbers can 'really' be considered to continue indefinitely, does not provide us with a way to evade the conclusion that we do not use a knowably sound algorithm to ascertain mathematical truth.

2.7 Some deeper mathematical considerations

To gain more insights into the implications of Gödel's argument, it will be helpful to return to what was his original purpose. At the turn of the century severe difficulties had begun to confront those who were concerned with the foundations of mathematics. In the late 1800s—to a large extent due to the profoundly original contributions of the mathematician Georg Cantor (whose 'diagonal slash' we have encountered earlier)—mathematicians had found powerful ways of establishing some of their deeper results, basing their arguments upon the properties of *infinite sets*. However, fundamental difficulties had also arisen in conjunction with such benefits, when too free a use of the infinite-set concept was made. In particular, there was the Russell paradox (that I briefly referred to above in response to **Q9**, cf. also §3.4—and which had also been noted by Cantor), which pointed out some of the obstacles to reasoning about infinite sets in too cavalier a way. Nevertheless, provided that one was sufficiently careful about the kind of reasoning that was to be allowed, it became clear that powerful mathematical results could indeed be obtained. The problem seemed to be how to be absolutely *precise* about what it means to be 'sufficiently careful' about one's reasoning.

The great mathematician David Hilbert was one of the prime figures in a movement aimed at securing this precision. This movement was referred to as *formalism*, according to which all the allowed forms of mathematical reasoning within some specified area, including any needed reasoning about infinite sets, were to be laid down once and for all. Such a system of rules and mathematical statements is referred to as a *formal system*. Once the rules of a formal system $\mathbb{F}$ have been determined, it is merely a matter of mechanical checking to see whether or not the rules—necessarily finite* in number—have been correctly applied. Of course, the rules would have to be regarded as valid forms of

*Some formal systems are presented as having *infinitely* many axioms—described in terms of structures known as 'axiom schemata'—but to qualify as a 'formal system' in the sense that I mean it here, such a formal system would have to be expressible in finite terms, the infinite axiom system being generated by a finite set of computational rules. It is indeed the case that this is possible for the standard formal systems that are used in mathematical proofs—such as the familiar 'Zermelo–Fraenkel' formal system $\mathbb{ZF}$ that describes conventional set theory.

mathematical reasoning, so that any result that could be deduced using them could be trusted as being actually *true*. However, some of these rules might be concerned with the manipulation of infinite sets, and here one's mathematical intuitions as to which forms of reasoning are legitimate and which are not might not be absolutely trustworthy. Doubts, in this connection, might indeed seem to be appropriate in view of the inconsistencies that arise if one is allowed such free use of infinite sets that even Bertrand Russell's paradoxical 'set of all sets that are not members of themselves' would be permitted. The rules of $\mathbb{F}$ must fall short of allowing Russell's 'set', but how far short? To forbid the use of infinite sets altogether would be much too limiting (for example, ordinary Euclidean space contains an infinite set of points, and even the set of natural numbers is an infinite set); moreover, it was clear that there are, indeed, various particular formal systems which are perfectly satisfactory (not allowing, for example, Russell's 'set' to be formulated) and by whose use most of the needed mathematical results can be obtained. How can one tell which of these formal systems are to be trusted and which are not?

Let us fix attention on one such formal system $\mathbb{F}$, and use the notation TRUE and FALSE, respectively, to denote mathematical statements which can be obtained by means of $\mathbb{F}$'s rules and those whose *negations* (i.e. 'not' the statement in question) can be so obtained. Any statement that can be formulated within $\mathbb{F}$ but which is neither TRUE nor FALSE in this sense would be UNDECIDABLE. Some people would adopt the point of view that since infinite sets themselves might really be 'meaningless', there may be no absolute sense of truth or falsity with regard to them. (At least this might apply to some kinds of infinite set if not to all of them.) According to this point of view, it might not really matter which statements about (certain) infinite sets come out as being TRUE and which as FALSE, provided that no statement comes out as being *both* TRUE *and* FALSE together—which is to say that the system $\mathbb{F}$ is to be *consistent*. For such people—the true *formalists*—the only questions of overriding importance for a formal system $\mathbb{F}$ would be (a) whether or not it is *consistent* and, additionally, (b) whether or not it is *complete*. The system $\mathbb{F}$ is called *complete* if every mathematical statement that is properly formulated within $\mathbb{F}$ always turns out to be either TRUE or FALSE (so that $\mathbb{F}$ contains no UNDECIDABLE statements).

The question as to whether or not a statement about infinite sets is *actually true* in any absolute sense is, to a strict formalist, not necessarily meaningful and certainly not considered to be relevant to the procedures of formalist mathematics. Thus, in place of the quest for absolute mathematical truth for statements about such infinite quantities, would be a desire for the demonstration of consistency and completeness of suitable formal systems. What kind of mathematical rules would be allowed for such a demonstration? These rules themselves would have to be trustworthy, and they should not make use of any dubious reasoning with loosely defined infinite sets (such as Russell's). It was hoped that there might be logical procedures available within certain comparatively simple and obviously sound formal systems (such as the relatively elementary one known as *Peano arithmetic*) that would be sufficient to prove the consistency of other, more sophisticated, formal systems—say $\mathbb{F}$—that might allow formal reasoning about very 'large' infinite sets, and whose consistency might not be at all self-evident. If one accepts the formalist philosophy, then such a consistency proof for $\mathbb{F}$ would at least provide a justification for using the means of reasoning allowed by $\mathbb{F}$. Then proofs of mathematical theorems might be given using infinite sets in a consistent way, and the question as to the actual 'meanings' of such sets could perhaps be dispensed with. Moreover, if such an $\mathbb{F}$ could be shown also to be complete, then one might reasonably embrace the viewpoint that $\mathbb{F}$ actually encapsulates *all* the mathematical procedures that are allowed; so, in a sense, $\mathbb{F}$ could then be considered actually to *be* the complete formulation of the mathematics of the area in question.

However, in 1930 (published in 1931), Gödel produced his bombshell, which eventually showed that the formalists' dream was unattainable! He demonstrated that there could be no formal system $\mathbb{F}$, whatever, that is both consistent (in a certain 'strong' sense that I shall describe in the next section) and complete—so long as $\mathbb{F}$ is taken to be powerful enough to contain a formulation of the statements of ordinary arithmetic together with standard logic. Thus, Gödel's theorem would apply to systems $\mathbb{F}$ for which arithmetical statements such as Lagrange's theorem and Goldbach's conjecture, as described in §2.3, could be formulated as mathematical statements.

In the discussions to follow, I shall be concerned only with formal systems that are sufficiently extensive that the necessary arithmetical operations for the actual formulation of Gödel's theorem can be contained within it (and, if necessary, that the operations of any Turing machine can be contained within it; see below). When I refer to some formal system $\mathbb{F}$, it will normally be *assumed* that $\mathbb{F}$ is indeed sufficiently extensive for this. This will not restrict the discussion in any essential way. (Nevertheless, for clarity, I shall sometimes add the words 'sufficiently extensive', or some such, when discussing formal systems in such a context.)

2.8 The condition of ω-consistency

The most familiar form of Gödel's theorem asserts, for a sufficiently extensive formal system $\mathbb{F}$, that $\mathbb{F}$ cannot be both complete and consistent. This is not quite the famous 'incompleteness theorem' that he originally announced at the meeting at Königsberg, referred to in §2.1 and §2.7, but is a slightly stronger version that was subsequently obtained by the American logician J. Barkley Rosser (1936). The version that Gödel originally announced was equivalent to showing that $\mathbb{F}$ could not be both complete and *ω-consistent*. The condition of ω-consistency is a bit stronger than that of ordinary consistency. To state what it means, we need a little notation. As part of the notation of a formal system $\mathbb{F}$, there would be certain symbols denoting logical operations. There would be a symbol denoting *negation*, i.e. 'not', and this might be denoted by '$\sim$'. Thus, if Q is a proposition, statable within $\mathbb{F}$, then the symbols $\sim Q$ denote 'not Q'. There should also be a symbol that says 'for all [natural numbers]', called the *universal quantifier*, and this might be denoted by '$\forall$'. If $P(n)$ is a proposition that depends upon the natural number n (so P is what is called a *propositional function*), then the string of symbols $\forall n[P(n)]$ denotes the statement 'for all natural numbers n, $P(n)$ holds true'. A particular example of such a $P(n)$ would be: 'n is expressible as the sum of three squares', and then $\forall n[P(n)]$ stands for: 'every natural number is the sum of three squares'—which, in this case, would be false (although it would be true if 'three' were replaced by 'four'). We can combine such symbols in many ways; in particular, the string of symbols

$$\sim \forall n[P(n)]$$

expresses the *denial* that $P(n)$ holds for all natural numbers n.

What the condition of ω-consistency asserts is that if $\sim \forall n[P(n)]$ is provable by the methods of $\mathbb{F}$, then it must *not* be the case that *all* of the statements

$$P(0),\ P(1),\ P(2),\ P(3),\ P(4), \ldots$$

are provable within $\mathbb{F}$. From this, it follows that if $\mathbb{F}$ were not ω-consistent, we should have the anomalous situation in which, for some P, every one of $P(0)$, $P(1)$, $P(2)$, $P(3)$, . . . could be proved; yet the statement claiming to assert that *not* all of these hold true is *also* provable! Certainly no trustworthy formal system could admit this kind of thing. If $\mathbb{F}$ is *sound*, then it is certainly ω-consistent.

In this book I shall use the notations '$G(\mathbb{F})$' and '$\Omega(\mathbb{F})$' for the respective assertions: 'the formal system $\mathbb{F}$ is consistent' and 'the formal system $\mathbb{F}$ is ω-consistent'. In fact (assuming that $\mathbb{F}$ is sufficiently extensive), $G(\mathbb{F})$ and $\Omega(\mathbb{F})$ are sentences that can themselves be formulated in terms of the operations of $\mathbb{F}$. Gödel's famous incompleteness theorem tells us that $G(\mathbb{F})$ is *not a theorem* of $\mathbb{F}$ (i.e. not provable using the procedures allowed by $\mathbb{F}$) and nor is $\Omega(\mathbb{F})$, provided that $\mathbb{F}$ *is* actually consistent! The somewhat stronger version of Gödel's theorem that was later obtained by Rosser tells us that if $\mathbb{F}$ is consistent then $\sim G(\mathbb{F})$ is not a theorem of $\mathbb{F}$ either. In the remainder of this chapter, I shall tend to formulate my arguments in terms of the more familiar $G(\mathbb{F})$, rather than $\Omega(\mathbb{F})$, although for most of the discussion either would do equally well. (For some of the more explicit arguments of Chapter 3, I shall sometimes use '$G(\mathbb{F})$' to denote the specific assertion '$C_k(k)$ does not stop' (cf. §2.5), which is not a serious abuse of notation.)

I shall not bother to draw a clear line between consistency and ω-consistency in most of my discussions here, but the version of the Gödel theorem that I have actually presented in §2.5 is essentially the one that asserts that if $\mathbb{F}$ is consistent, then it cannot be complete, being unable to assert $G(\mathbb{F})$ as a theorem. I shall not attempt to show this here (but see Kleene 1952). In fact, for this form of Gödel's argument to be reducible to the argument as I have given it, a little more is needed for $\mathbb{F}$ than merely that it 'contains arithmetic and ordinary logic'. What we need is that $\mathbb{F}$ be broad enough that the actions of any *Turing machine* are included. Thus, the statements that can be correctly formulated using the symbols of the system $\mathbb{F}$ must include the statements of the form: 'such-and-such a Turing machine, when acting on the natural number n, produces the natural number p'. In fact, it is a theorem (cf. Kleene 1952, Chapters 11 and 13) that this turns out automatically to be the case if $\mathbb{F}$ includes, in addition to the ordinary operations of arithmetic, the operation (called the μ-operation): 'find the smallest natural number with such-and-such arithmetical property'. Recall that in our original example of a computation, **(A)**, our procedure indeed found the *smallest* number that is not the sum of three squares. Computations, generally, must be allowed to do things of this kind. Indeed it is *this* that also leads us into the possibility of encountering computations that do not terminate, such as **(B)**, where we attempt to find the smallest number that is not the sum of *four* squares, but there is no such number.

2.9 Formal systems and algorithmic proof

In the Gödel–Turing argument, as I have given it in §2.5, I referred merely to 'computations', and made no reference to 'formal systems'. But there is a very close relation between the two concepts. It is one of the essential properties of a formal system that there must indeed be an algorithmic (i.e. 'computational') procedure F for *checking* whether or not the rules of $\mathbb{F}$ have been correctly applied. If a proposition is TRUE, according to the rules of $\mathbb{F}$, then our computation F will ascertain this fact. (What F could do would be to run through all possible successions of strings of symbols that belong to the 'alphabet' of the system $\mathbb{F}$ and to come successfully to a halt when the desired proposition P is encountered as the final string, all the steps of the succession being allowable according to the rules of the system $\mathbb{F}$.)

Conversely, if E is some *given* computational procedure, intended to ascertain the truth of certain mathematical statements, then we can construct a formal system $\mathbb{E}$, that effectively expresses, as TRUE, all those truths that are obtainable by means of the procedure E. There is, however, a slight proviso in that a formal system would normally be expected to include the standard logical operations, whereas the given procedure E might not be extensive enough to incorporate these directly. If our given E does not itself incorporate these elementary logical operations, then it would be appropriate to adjoin these logical operations to E in the construction of $\mathbb{E}$, so that the TRUE propositions of $\mathbb{E}$ would be not only the statements that are obtainable directly by the procedure E, but also those which are elementary logical consequences of statements directly obtainable by means of E. In these circumstances, $\mathbb{E}$ would not be strictly equivalent to E, but would be somewhat more powerful.

(These logical operations are just things like: 'if $P\&Q$ then P'; 'if P and $P \Rightarrow Q$ then Q'; 'if $\forall x[P(x)]$, then $P(n)$'; 'if $\sim\forall x[P(x)]$, then $\exists x[\sim P(x)]$'; etc. Here, the symbols '&', '$\Rightarrow$', '$\forall$', '$\exists$', '$\sim$' have the respective interpretations 'and', 'implies', 'for all [natural numbers]', 'there exists [a natural number]', 'not', and there could be a few other such symbols.)

To construct $\mathbb{E}$ from E, we can start from some very basic (and obviously consistent) formal system $\mathbb{L}$, expressing merely these primitive rules of logical inference—such as the system known as *predicate calculus* (Kleene 1952), which does just this—and construct $\mathbb{E}$ by adjoining E to $\mathbb{L}$ in the form of additional axioms and rules of procedure for $\mathbb{L}$, thereby deeming any proposition P to be TRUE whenever the procedure E obtains it. This need not be easy to do in practice, however. If E is simply a Turing-machine specification, then we might have to adjoin all the necessary Turing-machine notation and operations to $\mathbb{L}$, as part of its alphabet and rules of procedure, before we can adjoin E itself as, in effect, an additional axiom. (See end of §2.8; for full details, see Kleene 1952.)

It is not really important for our purposes here that the system $\mathbb{E}$ that we construct in this way might contain TRUE propositions other than those directly obtainable by E (the primitive logical rules of $\mathbb{L}$ themselves being not necessarily represented as part of the given procedure E). We were concerned, in §2.5, with a putative algorithm A that purports to encapsulate all the (known or knowable) procedures available to mathematicians for ascertaining that computations do not stop. Any such algorithm would certainly *have* to incorporate, among other things, all the basic operations of simple logical inference. In the discussions which follow, therefore, I shall assume that A does indeed incorporate such things.

For the purposes of my argument, therefore, algorithms (i.e. computational processes) and formal systems are basically *equivalent* as procedures for accessing mathematical truths. Thus, although the argument that I gave in §2.5 was stated in terms of computations only, that argument is also relevant to general formal systems. Recall that the argument refers to a listing of all computations (Turing-machine actions) $C_q(n)$. For the argument to apply in detail to a formal system $\mathbb{F}$, therefore, it is necessary that $\mathbb{F}$ be broad enough to incorporate the actions of all Turing machines. The algorithmic procedure A for ascertaining that certain computations do not stop can now be incorporated into the rules for $\mathbb{F}$, so that the computations whose non-stopping character can be established as TRUE using $\mathbb{F}$ would be identical to all those whose non-stopping character can be ascertained using A.

How does Gödel's original Königsberg argument relate to the one that I presented in §2.5? I shall not give details here, but merely point out the essential ingredients. My algorithmic procedure A plays the role of the formal system $\mathbb{F}$ in Gödel's original theorem:

$$\text{algorithm } A \quad \leftrightarrow \quad \text{rules of } \mathbb{F}.$$

The particular proposition '$C_k(k)$ does not stop', obtained in §2.5, which is inaccessible by the procedure A yet which one perceives to be true so long as one believes A to be sound, plays the role of the proposition $G(\mathbb{F})$ that Gödel presented at Königsberg and which actually asserts that $\mathbb{F}$ is consistent:

$$\text{statement } C_k(k) \text{ does not stop} \quad \leftrightarrow \quad \text{assertion that } \mathbb{F} \text{ is consistent.}$$

This perhaps helps us to understand how a belief in the soundness of a procedure, such as A, can lead us to another procedure lying beyond the scope of the original procedure, yet whose soundness we must *also* believe in. For if we believe that the procedures of some formal system $\mathbb{F}$ are sound—i.e. that they allow us to derive only actual mathematical truths and not falsehoods, so that if a proposition P is derived as TRUE, then it must actually *be true*—then we must also believe that $\mathbb{F}$ is ω-consistent. If 'TRUE' implies 'true' and 'FALSE' implies 'false'—as would be the case for any sound formal system $\mathbb{F}$—then, certainly:

not all of $P(0)$, $P(1)$, $P(2)$, $P(3)$, $P(4)$, . . . can be TRUE if
it is FALSE that $P(n)$ holds for all natural numbers n;

which, after all, is precisely what ω-consistency asserts.

Not only does a belief in the soundness of $\mathbb{F}$ entail a belief in its ω-consistency, but a belief in its consistency also. For if 'TRUE' implies 'true' and 'FALSE' implies 'false', then, certainly:

no P can be *both* TRUE *and* FALSE;

which is precisely what consistency asserts. In fact, for many systems, the distinction between consistency and ω-consistency disappears. For simplicity in what follows in this chapter, I shall generally not bother to draw a distinction between these two kinds of consistency, and I shall normally just state things in terms of 'consistency' alone. What Gödel and Rosser showed is that the consistency of a (sufficiently extensive) formal system is something that lies outside the power of the formal system itself to establish. Gödel's earlier (Königsberg) theorem depended upon ω-consistency, but the later, more familiar result referred only to ordinary consistency.

The thrust of Gödel's argument for our purposes is that it shows us how to go beyond any given set of computational rules that we believe to be sound, and obtain a further rule, not contained in those rules, that we must believe to be sound also, namely the rule asserting the *consistency* of the original rules. The essential point, for our purposes, is:

belief in *soundness* implies belief in *consistency*.

We have no right to use the rules of a formal system $\mathbb{F}$, and to believe that the results that we derive from it are actually *true*, unless we also believe in the consistency of that formal system. (For example, if $\mathbb{F}$ were inconsistent, then we could deduce, as TRUE, the statement '1 = 2', which is certainly not true!) Thus, if we believe that we are actually doing mathematics when we use some formal system $\mathbb{F}$, then we must also be prepared to accept reasoning that goes beyond the limitations of the system $\mathbb{F}$, *whatever* that system $\mathbb{F}$ may be.

2.10 Further possible technical objections to $\mathscr{G}$

Let us now continue to examine various mathematical objections that have been voiced from time to time to the kind of use I have made of the Gödel–Turing argument. Many of these objections are closely related to one another, but I think that it will be helpful to spell them out separately in any case.

Q10. Is mathematical truth an absolute matter? We have already seen that there are differing views as to the absolute truth of statements about infinite sets. Can we trust arguments that depend upon having some vague concept of 'mathematical truth', as opposed, perhaps, to a clearly defined concept of formal TRUTH?

In the case of a formal system $\mathbb{F}$ that is concerned with general set theory, it indeed might not always be clear that there is any absolute sense in which a statement about the sets is either 'true' or 'false'—in which case the very concept of 'soundness' for a formal system like $\mathbb{F}$ might be called into question. A famous example that brings home this kind of issue is contained in a result proved by Gödel (1940) and Cohen (1966). They showed that the mathematical assertions known as the *axiom of choice* and Cantor's *continuum hypothesis* are independent of the *Zermelo–Fraenkel* axioms of set theory—a standard formal system that I shall denote here by $\mathbb{ZF}$. (The axiom of choice asserts that for any collection of non-empty sets, there exists another set which contains precisely one element from each member of the collection.[1] Cantor's continuum hypothesis asserts that the number of subsets of natural numbers—which is the same as the number of *real* numbers—is the next largest infinity after the number of the natural numbers themselves.[2] It is not necessary for the reader to appreciate the meanings of these assertions here. Nor is it necessary for me to go into the details of the axioms and rules of procedure of $\mathbb{ZF}$.) Some mathematicians would maintain that $\mathbb{ZF}$ encapsulates all the mathematical reasoning that is needed in ordinary mathematics. Some would even contend that an acceptable mathematical argument is precisely one that could, in principle, be formulated and proved within $\mathbb{ZF}$. (See the discussion under **Q14**, below, for an assessment of how the Gödel argument applies to such people.) These mathematicians would therefore claim that the mathematical statements that are, respectively, TRUE, FALSE, and UNDECIDABLE according to $\mathbb{ZF}$, are precisely the statements that, in principle, can be mathematically established as being true, mathematically established as being false, and are mathematically undecidable. To such people, the axiom of choice and the continuum hypothesis would be mathematically undecidable (as, they claim, the Gödel–Cohen result shows), and they might well argue that the truth or falsity of these two mathematical assertions is a purely conventional matter.

Do such seeming uncertainties as to the absolute nature of mathematical truth affect our deductions from the Gödel–Turing argument? Not at all—for we are concerned here with a class of mathematical problem of a much more limited nature than those which, like the axiom of choice and the continuum hypothesis, relate to non-constructively infinite sets. Our only concern here is with statements of the form:

'such-and-such a computation does not ever terminate'.

where the computations concerned can be precisely specified in terms of Turing-machine actions. Such statements are known to logicians, technically as, Π_1*-sentences* (or, more correctly, as Π_1^0-sentences). For any formal system $\mathbb{F}$, $G(\mathbb{F})$ is a Π_1-sentence, but not $\Omega(\mathbb{F})$ (see §2.8). There would seem to be little reasonable doubt that the true/false nature of any Π_1-sentence is an *absolute* matter, independent of what position one might choose to hold on questions that relate to non-constructively infinite sets—as with the axiom of choice and the continuum hypothesis. (On the other hand, as we shall see in a moment, the kind of reasoning that one accepts as providing convincing *demonstrations* of Π_1-sentences might indeed depend upon one's position with regard to non-constructively infinite sets; cf. **Q11** below.) It seems clear that, apart from the extreme position taken by some intuitionists (cf. response to **Q9**), the only reasonable issue with regard to the absolute nature of the truth of such statements might be that some terminating computations can take so inordinately long that they might not conceivably be completed in practice, say in the entire history of the universe; or perhaps that the computation itself might take so many symbols to state that its specification (though finite) could not ever be written down. These matters, however, were fully analysed in the discussion relating to **Q8** above, and we saw there that our essential conclusion $\mathscr{G}$ was unaffected. Recall also that, in relation to the discussion concerning **Q9** above, the intuitionistic position does not evade the conclusion $\mathscr{G}$ either.

As a further point, the (very limited) concept of mathematical truth that I need for the Gödel–Turing argument is actually no less well defined than the concepts of TRUE, FALSE, and UNDECIDABLE, for any formal system $\mathbb{F}$. Recall from the above (§2.9) that there is an *algorithm* F that is equivalent to $\mathbb{F}$. If F is presented with a proposition P (statable in the language of $\mathbb{F}$), then this algorithm comes successfully to a halt precisely when P is provable according to the rules of $\mathbb{F}$, i.e. when P is TRUE. Likewise, P is FALSE precisely if F comes successfully to a halt when presented with $\sim P$; and P is UNDECIDABLE precisely when neither of these computations terminates. The issue of whether a mathematical statement P is TRUE, FALSE, or UNDECIDABLE is of precisely the same character as that of the actual truth of the stopping or otherwise of computations—i.e. of the falsity or truth of certain Π_1-sentences—which is all that is needed for our Gödel–Turing-type argument.

Q11. There are certain Π_1-sentences that can be proved using infinite-set theory, but no proofs are known using standard 'finite' methods. Does this not mean that the way in which mathematicians decide even such well-defined questions may really be a subjective matter? Different mathematicians, holding different beliefs with regard to set theory, might have inequivalent criteria for assessing the mathematical truth of Π_1-sentences.

This could be a significant point with regard to my own deductions from the Gödel(–Turing) argument, and I may perhaps have attached insufficient weight to it in my brief discussion in ENM. Surprisingly, **Q11** is not an objection that seems to have worried anyone other than myself—at least no one else has brought it to my attention! Both here and in ENM (pp. 417, 418) I have phrased the Gödel(–Turing) argument in terms of what 'mathematicians' or 'the mathematical community' are able to ascertain by means of reason and insight. The advantage of putting things in such terms, rather than in terms of what a *particular* individual might be able to ascertain by use of his or her reason or insight, is that it enables us to get away from certain objections that are frequently made to the version of the Gödel argument put forward by Lucas (1961). Various people[3] have objected that, for example, 'Lucas himself' could not possibly know his own algorithm. (Some such people even made the same type of objection to my own presentation[4]—seemingly oblivious of the fact that I did not phrase my argument in this 'personal' way at all!) The advantage of referring to the reasoning and insights that are available to 'mathematicians' or 'the mathematical community' is that this enables us to get away from the suggestion that different individuals might perceive mathematical truth in different ways, each according to his or her unknowable personal algorithm. It is much harder to accept that the shared understanding of the mathematical community as a whole might be the result of some unknowable algorithm than that the understanding of a particular individual might be. The point that **Q11** raises is that this shared understanding might not be so universal and impersonal as I have taken it to be.

It is indeed true that there *are* statements of the kind referred to in **Q11**. That is, there are Π_1-sentences whose only known proofs depend upon an appropriate use of the theory of infinite sets. Such a Π_1-sentence could arise from an arithmetical coding of a statement like 'the axioms of $\mathbb{F}$ are consistent' where the formal system $\mathbb{F}$ involves the manipulations of large infinite sets whose very existence might be a controversial matter. A mathematician who believes in the actual *existence* of some appropriately enormous non-constructive set **S** will come to the conclusion that $\mathbb{F}$ is in fact consistent, but another mathematician who does not believe in **S** need have no such faith in the consistency of $\mathbb{F}$. Thus, even restricting attention to the well-defined issue of the stopping or otherwise of Turing-machine actions (i.e. to the falsity or truth of Π_1-sentences) does not allow us to ignore the issue of the subjectivity of *beliefs* with regard to, say, the existence of some large non-constructively infinite set **S**. If different mathematicians employ *inequivalent* 'personal algorithms' for ascertaining the truth of certain Π_1-sentences, then it may be considered to be unfair of me to refer simply to 'mathematicians' or 'the mathematical community'.

I suppose that, strictly speaking, it may indeed be slightly unfair; and the reader, if so inclined, may prefer to reinterpret $\mathscr{G}$ as:

$\mathscr{G}^*$ No individual mathematician ascertains mathematical truth solely by means of an algorithm that he or she knows to be sound.

The arguments that I am giving will still apply, but I think that some of the later ones will lose a good measure of their force when the case is presented in this form. Moreover, with the version $\mathscr{G}^*$, the argument gets guided in what I believe to be an unhelpful direction, where one is concerned more with the particular mechanisms that govern the actions of particular individuals than with principles that underlie the actions of all of us. I am not so much concerned, at this stage, with how individual mathematicians might differently approach a mathematical problem, but more with what is *universal* about our understandings and our mathematical perceptions.

Let us try to see whether we are *really* forced into the version $\mathscr{G}^*$. Are mathematicians' judgements actually so subjective that they might disagree *in principle* as to whether a particular Π_1-sentence has, or has not, been established as true? (Of course, the argument establishing the Π_1-sentence might simply be too long or complicated for one or the other mathematician to follow—cf. **Q12** below—so they might certainly differ *in practice*. But that is not what is at issue here. We are here concerned solely with matters of *principle*.) In fact, mathematical demonstration is not as subjective as this might seem to suggest. Despite the fact that different mathematicians might profess to holding somewhat different viewpoints as to what they maintain to be unassailably true concerning foundational issues, when it comes to demonstrations or refutations of clearly defined specific Π_1-sentences, they will not tend to disagree when it actually comes to the point. A particular Π_1-sentence that, in effect, asserts the consistency of some system $\mathbb{F}$ would not normally be regarded as acceptably demonstrated if all one had to go on was the existence of some controversial infinite set **S**. A more acceptable formulation of what had actually been demonstrated might be: 'if **S** exists, then $\mathbb{F}$ is consistent, and in that case the given Π_1-sentence is true'.

Nevertheless, there could be exceptions to this, where one mathematician could take the view that some non-constructively infinite set **S** 'obviously' exists—or, at least, that the assumption of its existence can in no way lead us into a contradiction—whereas another mathematician might have no such faith in the matter. Sometimes, as regards such *foundational* issues, mathematicians do seem to run into irresolvable disputes. In principle this could lead to their being unable convincingly to communicate their demonstrations, even with regard to Π_1-sentences. Perhaps different mathematicians do actually have inherently different perceptions as to the truth of statements that relate to non-constructively infinite sets. It is certainly true that they often *profess* to having such different perceptions. But I think that such differences are basically similar to the differences in *expectations* that different mathematicians might have with regard to the truth of ordinary mathematical propositions. These expectations are merely provisional opinions. So long as a convincing demonstration or refutation has not been found, the mathematicians may disagree amongst themselves as to what they expect, or *guess* is true, but the possession of such a demonstration by one of the mathematicians would (in principle) enable the others also to become convinced. With regard to foundational issues, such demonstrations are indeed lacking. It might be the case that convincing demonstrations will never be found. Perhaps they *cannot* be found because such demonstrations do not exist, and it is simply the case that there *are* different equally valid viewpoints with regard to these foundational issues.

In relation to all this, however, one point should be emphasized when it comes to Π_1-sentences. The possibility that a mathematician might have an *erroneous* viewpoint—by which I mean, here, a viewpoint that allows incorrect conclusions to be drawn as to the validity of certain Π_1-sentences—is *not* our present concern. Conceivably, mathematicians might use factually wrong 'insights'—in particular, *unsound algorithms*—but this is not relevant to the present section, since it would be in *agreement* with $\mathscr{G}$. This possibility will be addressed in full detail, instead, in §3.4. The issue here, therefore, is not whether there can be *inconsistent* viewpoints among different mathematicians, but rather that one viewpoint might in principle be *more powerful* than another. Each viewpoint would be perfectly sound with regard to its implications as to the truth of Π_1-sentences, but some viewpoint might, in principle, enable its supporters to ascertain that certain computations will not terminate whereas this would not be so for the less powerful viewpoints. Thus, different mathematicians might possess essentially different degrees of insight.

I do not think that this possibility poses any significant threat to my original formulation $\mathscr{G}$. Although there may be different viewpoints with regard to infinite sets that mathematicians can reasonably hold to, there are not *that* many different ones—probably not more than about four or five. The only differences of relevance would be things like the axiom of choice (referred to under **Q10**), which many would regard as 'obvious' whilst others would refuse to accept the non-constructiveness involved. Curiously, these differing viewpoints with regard to the axiom of choice itself do *not* directly lead to a Π_1-sentence whose validity is in dispute. For, whether or not the axiom of choice is regarded as 'true', this axiom does not lead to an inconsistency with the standard $\mathbb{ZF}$ axioms, as the Gödel–Cohen theorem (referred to under **Q10**) shows. There might, however, be *other* disputed axioms for which no corresponding theorem is known. But usually, when it comes to the acceptance, or otherwise, of some set-theoretic axiom—call it axiom Q—the mathematician's statements would take the form: 'assuming axiom Q it follows that . . .'. This would not be a matter of dispute between any of them. The axiom of choice seems to be an exception in that it is frequently assumed without mention, but it apparently poses no challenge to the general impersonal formulation that I have given for $\mathscr{G}$, provided that we restrict attention, in $\mathscr{G}$, to Π_1-sentences:

$\mathscr{G}$** Human mathematicians are not using a knowably sound algorithm in order to ascertain the truth of Π_1-sentences,

which is all that we shall need in any case.

Are there other disputed axioms—regarded as 'obvious' by some but questioned by others? I think it would be a gross exaggeration to say that there are as many as 10 essentially different viewpoints, involving set-theoretic assumptions, that are not explicitly taken into account as assumptions. Let us concede that there is some such small number, and examine the implications. This would mean that there are conceivably as many as about 10 essentially different grades of mathematician, graded with regard to the types of reasoning, involving infinite sets, that they would be prepared to accept as 'obviously' valid. We could refer to them as n-grade mathematicians, where n ranges over just a few values—no more than about 10. (The higher the grade, the more powerful would be the mathematicians' viewpoint.) In place of $\mathscr{G}$**, we now have:

$\mathscr{G}$*** for each n (where n can take just a few values), human n-grade mathematicians do not ascertain the truth of Π_1-sentences solely by means of an algorithm that they know to be sound.

This follows because the Gödel(–Turing) argument can be applied for each grade separately. (It should be made clear that the Gödel argument itself is not a matter of dispute between mathematicians, so if the putative n-grade algorithm were knowably sound to any n-grade mathematician, the argument would provide a contradiction.) Thus, as with $\mathscr{G}$, it is not a question of there being a great many unknowably sound algorithms, where each algorithm is specific to each individual. Instead, what we have ruled out is the possibility that there might be just a very small number of inequivalent unknowably sound algorithms, graded with respect to their strengths, which provide the different 'schools of thought'. The version $\mathcal{G}^{***}$ will not differ significantly from $\mathcal{G}$ or $\mathcal{G}^{**}$ in the discussions to follow, and for simplicity I shall not try to make a distinction between them, referring, collectively, to $\mathcal{G}$.

Q12. Despite what mathematicians might or might not claim to have as their various viewpoints, *in principle*, surely they would, *in practice*, differ very greatly in their abilities to follow an argument? Surely they also differ very greatly in those insights that allow them to make mathematical discoveries?

Of course these things are true, but they are not really relevant to the point at issue. I am not concerned with what specific detailed arguments a mathematician might *in practice* be able to follow. Still less am I concerned here with the issue of which arguments a mathematician might be able, in practice, to *discover*, or with the insights and inspirations that may enable them to make such discoveries. The question here concerns only what type of arguments can in principle be perceived as valid by mathematicians.

Now the qualification 'in principle', in the foregoing discussions, is used advisedly. Assuming it to be the case that a mathematician is in possession of a demonstration or refutation of some Π_1-sentence, disagreements by other mathematicians as to the demonstration's validity can be resolved only if the mathematicians have the time, patience, open-mindedness, and the ability and determination to follow through, with understanding and accuracy, a possibly long and subtle chain of reasoning. In practice, the mathematicians might well give up before the issues are fully resolved. However, matters such as this are not the concern of the present discussion. For there certainly does appear to be a well-defined sense in which what is accessible *in principle* to one mathematician is the same (considerations of **Q11** apart) as what is accessible to another—or, indeed, to any other thinking person. The chain of reasoning may be very long and the concepts involved may be subtle or obscure, but there are nevertheless convincing enough reasons to believe that there is nothing in one person's understanding that is in principle inaccessible to another. This applies also in those cases where the aid of a computer may have had to be called upon in order to follow through the full details of a purely computational part of a proof. Although it might be beyond reasonableness to expect a human mathematician to follow through the necessary details of the computations involved in an argument, there is nevertheless no doubt that the *individual* steps are things easily capable of comprehension and acceptance by a human mathematician.

In saying this, I am referring merely to the sheer complexity of a mathematical argument and not to possible essential matters of principle that might separate one mathematician from another with regard to what types of reasoning that they are prepared to accept. I have certainly encountered mathematicians who claim that there are mathematical arguments that they have encountered which lie utterly beyond their competence: 'I know that I shall never be able to understand such-and-such, or so-and-so, no matter how long I try; the type of reasoning is totally beyond me.' In any individual example of such a claim, one would have to decide whether it is really a case of the reasoning lying *in principle* outside the mathematician's belief system—as would be covered by the discussion under **Q11**—or whether the mathematician really *could*, if he or she tried hard enough for long enough, follow the principles underlying the argument. Far more often than not, it is the latter. In fact the most common situation is that it would be the obscure writing style, or perhaps limitations of lecturing abilities, of 'so-and-so' that would be the source of our mathematician's despairing, rather than anything of essential principle in the 'such-and-such' that lies beyond his or her capabilities! A good exposition of an apparently obscure topic can work wonders.

To emphasize the kind of point I am making here, I should say, of myself, that I frequently attend mathematical seminars where I do not follow (or do not even attempt to follow) the details of the arguments that are being presented. I may have the feeling that if I went away and studied the arguments at length then I ought indeed to be able to follow them—though probably only with supplementary reading material or verbal explanations to fill in the missing details in my own background and probably also in the seminar itself. I know that in fact I shall not do this. Time, attention span, and sufficient enthusiasm will almost certainly be lacking. Yet I may well accept the result as presented in the seminar for all kinds of 'irrelevant' reasons, such as the fact that the result 'looks' plausible, or that the lecturer has a reliable reputation, or that other people in the audience whom I know to be much sharper than I am in such matters have not questioned the result. Now, of course I might be wrong in all this and the result is actually false, or perhaps it is true but does not follow by the actual argument as given. These are detailed issues that are not relevant to the point of principle that I am making here. The result might be true and validly demonstrated, in which case *in principle* I could have followed the argument—or else the argument is actually erroneous, which, as mentioned earlier, is not a situation of concern to us here (cf. §3.2 and §3.4). The only possible exceptions might be if the lecture concerned debatable aspects of infinite-set theory, or if it depended upon some unusual type of reasoning that might be questionable according to certain mathematical viewpoints (which, in itself, might intrigue me sufficiently that I might actually go through the argument in question afterwards). These possible exceptional situations are just those covered in **Q11** above.

In relation to such considerations of mathematical standpoint, in practice many particular mathematicians might indeed have no clear viewpoint as to what foundational principles they actually adhere to. But, as commented above in relation to **Q11**, a mathematician who has no clear view on whether or not to accept, say, 'axiom *Q*' would, if careful, always state results that require Q in the form 'assuming axiom Q, it follows that . . .'. Of course mathematicians, though a notoriously pedantic breed, are not always impeccably careful about such matters. Indeed, it is also true that they even make clear-cut mistakes from time to time. But these mistakes, if they are basically slips and not matters of unalterable principle, are *correctable*. (As mentioned before, the possibility that they actually use an unsound algorithm as the ultimate basis of their decisions will be considered in detail in §3.2 and §3.4. That possibility, being in *accordance* with $\mathscr{G}$, is not part of the present discussion.) We are not really concerned with correctable mistakes here, because they do not contribute to what cannot or can be achieved in principle. The possible uncertainties of a mathematician's actual standpoint, however, need further discussion, as follows.

Q13. Mathematicians do not have *absolutely* definite beliefs about the soundness or consistency of the formal systems they use—or even about *which* formal systems they might be considered as committing themselves to. Would not their beliefs just shade off by degrees as the formal systems get further from their immediate intuitions and experiences?

Indeed, it is rare to find a mathematician whose opinions are rigid and unshakably consistent when it comes to the foundations of the subject. Moreover, as mathematicians gain in experience, their viewpoints may well shift with regard to what they take to be unassailably true—if they indeed ever take *anything* to be unassailably true. Can one be utterly completely sure that 1 is different from 2, for example? If we are talking about *absolute* human certainty, it is not really clear that there is such a thing. But one must take a stand somewhere. A reasonable stand would be to take *some* body of beliefs and principles as being unassailably true and to argue from there onwards. It may be, of course, that many mathematicians would not even have a definite opinion as to what they would take as unassailably true. In any such case, I would ask them to take a stand, however, even if they might be prepared to modify it later. What the Gödel argument shows is that *whatever* standpoint is adopted, that standpoint cannot be (known to be) encapsulated in the rules of any knowable formal system. It is not that the standpoint is being continually modified; the body of beliefs that *any* (sufficiently extensive) formal system $\mathbb{F}$ encompasses must also extend beyond what $\mathbb{F}$ can achieve. Any standpoint that includes the soundness of $\mathbb{F}$ among its unassailable beliefs must also include a belief in the Gödel proposition* $G(\mathbb{F})$. The belief in $G(\mathbb{F})$ does not represent a change in standpoint; that belief is already implicitly contained in the original standpoint which has admitted to accepting $\mathbb{F}$—even though the realization of the fact that $G(\mathbb{F})$ must also be accepted may not have been apparent at first.

*See §2.8 for the notation used here. As we shall see at the end of this discussion, '$\Omega(\mathbb{F})$' could have been used in place of $G(\mathbb{F})$ throughout.

Of course there is always the possibility that some error may have crept in with regard to one's deductions from the premises of any particular standpoint. Even the mere *possibility* that one may have made such an error somewhere—even when, in fact, one has not done so—may lead to a shading off of the degree of confidence that one feels about the conclusions. But this kind of 'shading off' is not really our concern. Like actual errors, it is 'correctable'. Moreover, so long as an argument has indeed been correctly carried out, then the longer it is examined the more convincing the resulting conclusions should become. This kind of 'shading off' is a matter of what a mathematician might feel *in practice*, rather than in principle, and it leads us back to the discussion of **Q12**.

Now, the issue here is whether there is a shading off *in principle*, so that a mathematician might take the line that, say, the soundness of some formal system $\mathbb{F}$ is unassailable, whereas a more powerful system $\mathbb{F}^*$ might perhaps be merely 'practically certainly' sound. I do not think that there is much question but that whatever $\mathbb{F}$ is taken to be, we may as well insist that it includes the ordinary rules of logic and arithmetical operations. Our aforesaid mathematician, believing that $\mathbb{F}$ is sound, must also believe that $\mathbb{F}$ is consistent, and hence that its Gödel proposition $G(\mathbb{F})$ is true. Thus, deductions from $\mathbb{F}$ alone cannot represent the totality of the mathematician's mathematical beliefs, *whatever* $\mathbb{F}$ might be.

But is $G(\mathbb{F})$ to be taken as *unassailably* true whenever $\mathbb{F}$ is taken to be unassailably sound? I think that there can be little doubt that this must be so; and certainly so if we adhere to the 'in principle' viewpoint that we have been adopting hitherto, as regards the following of a mathematical argument. The only real question concerns the details of the actual coding of the assertion '$\mathbb{F}$ is consistent' into an arithmetical statement (a Π_1-sentence). The underlying *idea* is itself surely unassailably clear: if $\mathbb{F}$ is sound, then it is certainly consistent. (For if it were not consistent, then among its assertions would be '$1=2$', so it would be unsound.) As regards the details of this coding, there is again the distinction between the 'in principle' and 'in practice' levels. It is not that hard to convince oneself that such a coding is in principle possible (though persuading oneself that there is indeed no 'catch' to the argument may take some time), but it is quite another matter to be convinced that any specific *actual* coding has been correctly carried out. The details of the coding tend to be somewhat arbitrary, and they might differ greatly from one exposition to another. There might indeed be some minor error or misprint which would, technically speaking, invalidate the specific number-theoretic proposition that is intended to express '$G(\mathbb{F})$', but does not exactly do so.

I hope it is clear to the reader that the possibility of such errors is not the point, when it comes to what is meant here by accepting $G(\mathbb{F})$ as being unassailably true. I mean, of course, the *real* $G(\mathbb{F})$, not the possibly unintended proposition that one may have inadvertently stated owing to a misprint or minor error. I am reminded of a story concerning the great American physicist

Richard Feynman. Apparently Feynman was explaining some idea to a student, but mis-stated it. When the student expressed puzzlement, Feynman replied: 'Don't listen to what I say; listen to what I *mean*!'*

One possible explicit coding would be to use the Turing-machine specifications that I gave in ENM and to follow through the exact Gödel-type argument that I have provided in §2.5 and have given an explicit coding for in Appendix A. Even this would not yet be completely explicit, because we also need the explicit coding of the rules of $\mathbb{F}$ in terms of a Turing-machine action, say $T_{\mathbb{F}}$. (The property that $T_{\mathbb{F}}$ would have to satisfy would be that if some proposition P, constructable in terms of the language of $\mathbb{F}$, is assigned the number p, then we must arrange, say, that $T_{\mathbb{F}}(p)=1$ whenever P is a theorem of $\mathbb{F}$, and $T_{\mathbb{F}}(p)$ does not stop otherwise.) Of course there is much scope for technical error. Quite apart from possible difficulties involved in actually constructing $T_{\mathbb{F}}$ from $\mathbb{F}$ and p from P, there is the question as to whether I have made a mistake myself in my own specifications for Turing machines—and whether the code given in Appendix A of this book is correct or not, if it is decided to use this particular specification, in order to calculate $C_k(k)$. I do not think that there is an error, but my trust in myself here is not so great as my trust in Gödel's own original (if more complicated) specifications. But I hope that it is clear, by now, that possible errors of this kind are not the point. We must heed Feynman's dictum!

With my own particular specifications, however, there is another technical point that should be mentioned. In §2.5, I did not actually state my version of the Gödel(–Turing) argument in terms of the consistency of $\mathbb{F}$, but in terms of the soundness of the algorithm A, as a test of the non-stopping character of computations (i.e. of the truth of Π_1-sentences). This will do just as well, because we have seen that the soundness of A implies the truth of the assertion that $C_k(k)$ does not stop, so we can use this explicit assertion—which is also a Π_1-sentence—in place of $G(\mathbb{F})$. Moreover, as remarked earlier (cf. §2.8), the argument actually depends on the ω-consistency of $\mathbb{F}$, not its consistency. The soundness of $\mathbb{F}$ clearly implies its ω-consistency, just as well as its consistency. Neither $\Omega(\mathbb{F})$ nor $G(\mathbb{F})$ follows from the rules of $\mathbb{F}$ (cf. §2.8), assuming that $\mathbb{F}$ is sound, but both are true.

To sum up, I think that it is clear that however much 'shading off' there may be in a mathematician's beliefs in passing from a belief in the soundness of a formal system $\mathbb{F}$ to a belief in the truth of the proposition $G(\mathbb{F})$ (or $\Omega(\mathbb{F})$), this will lie entirely in the possibility that there is some error in the precise formulation of '$G(\mathbb{F})$' that has been provided. (The same applies to $\Omega(\mathbb{F})$.) This is not really relevant to the present discussion, and there should be no shading off in one's belief of the actually *intended* version of $G(\mathbb{F})$. If $\mathbb{F}$ is unassailably

*I have been unable to locate a specific source for this quotation. However, as was pointed out to me by Richard Jozsa, it does not matter if I have mis-stated it, since I can apply its underlying message to itself!

sound, then *this* $G(\mathbb{F})$ is also unassailably true. The forms of $\mathscr{G}$ (or $\mathscr{G}^{**}$ or $\mathscr{G}^{***}$) remain unaffected provided that 'truth' means 'unassailable truth'.

Q14. Surely the system $\mathbb{ZF}$—or some standard modification of $\mathbb{ZF}$ (call it $\mathbb{ZF}^*$)—really represents all that is needed for doing serious mathematics. Why not just stick with this system, accept that its consistency is not provable, and just get on with doing mathematics?

I think that this kind of viewpoint is a very common one among working mathematicians—especially those who do not concern themselves particularly with the foundations or the philosophy of their subject. It is not an unreasonable viewpoint for someone whose basic concern is, indeed, just getting on with the serious business of doing mathematics (although such people very rarely *actually* express their results within the strict rules of a system such as $\mathbb{ZF}$). According to this viewpoint, one is concerned just with what can be proved or disproved within a specific formal system such as $\mathbb{ZF}$ (or some modification, $\mathbb{ZF}^*$). From this point of view, the carrying out of mathematics is really playing a kind of 'game'. Let us call this game the *$\mathbb{ZF}$-game* (or $\mathbb{ZF}^*$-game), where one has to play the game according to those specific rules that have been laid down within the system. This is really the point of view of the *formalist*, the strict concern of the formalist being what is TRUE and what is FALSE, and not necessarily what is true and what is false. Assuming that the formal system is sound, then anything that is TRUE will also be true, and anything that is FALSE will also be false. However, there will then be some statements, formalizable within the system, that are true but not TRUE and some that are false but not FALSE, those statements being, in each case, UNDECIDABLE. The Gödel statement* $G(\mathbb{ZF})$ and its negation $\sim G(\mathbb{ZF})$ belong to these two respective categories, in the $\mathbb{ZF}$-game, assuming that $\mathbb{ZF}$ is consistent. (In fact if $\mathbb{ZF}$ were *in*consistent, then *both* $G(\mathbb{ZF})$ and its negation $\sim G(\mathbb{ZF})$ would be TRUE—and also FALSE as well!)

The $\mathbb{ZF}$-game is probably a perfectly reasonable standpoint for carrying out most of the things of interest in ordinary mathematics. However, for reasons that I have indicated earlier, I do not see how it can represent a genuine standpoint with regard to one's mathematical *beliefs*. For if one believes that the mathematics that one is doing is deriving actual mathematical truths—of, say, Π_1-sentences—then one must believe that the system that one is using is *sound*; and if one believes that it is sound, then one must also believe that it is *consistent*, so one must actually believe that the Π_1-sentence that asserts $G(\mathbb{F})$ is *really* true—despite the fact that it is UNDECIDABLE. Thus one's mathematical beliefs must go beyond what can be derived within the $\mathbb{ZF}$-game. If, on the other hand, one does not trust $\mathbb{ZF}$ as being sound, then one cannot trust the TRUE results obtained using the $\mathbb{ZF}$-game as being actually true.

*As before, $\Omega(\mathbb{F})$ will do just as well as $G(\mathbb{F})$. The same applies to **Q15–Q20**.

Either way, the $\mathbb{ZF}$-game itself cannot represent a satisfactory standpoint with regard to mathematical truth. (The same applies just as well to any $\mathbb{ZF}^*$.)

Q15. The formal system $\mathbb{F}$ that we choose to use might *not* actually be consistent—at least, we may well not be *sure* that $\mathbb{F}$ is consistent—in which case, by what right can we assert that $G(\mathbb{F})$ is 'obviously' true?

Although this issue has really been amply covered in the preceding discussions, I think that it is worth while to reiterate the essential point here, since arguments in the nature of **Q15** represent among the commonest attacks on the kind of use of the Gödel theorem that Lucas and I have made. The point is that we do not claim that $G(\mathbb{F})$ is necessarily true whatever $\mathbb{F}$ might be, but that we must conclude that $G(\mathbb{F})$ is just as reliable a truth as any that we obtain using the rules of $\mathbb{F}$ itself. (In fact, $G(\mathbb{F})$ is *more* reliable than statements that are derived actually *using* the rules of $\mathbb{F}$ because $\mathbb{F}$ might *be* consistent without actually being sound!) If we trust any statement P that we derive solely using the rules of $\mathbb{F}$, then we must also trust $G(\mathbb{F})$ to at least such a degree as we trust P. Thus, no knowable formal system $\mathbb{F}$—or its equivalent algorithm F—can represent the total basis of our true mathematical knowledge or beliefs. As stated in the commentaries on **Q5** and **Q6**, the argument is given as a *reductio ad absurdum*: we try to suppose that $\mathbb{F}$ does represent the total basis of our beliefs, and then show that this leads to a contradiction, so therefore it cannot represent our basis for belief, after all.

As in **Q14**, we may, of course, use some system $\mathbb{F}$ as a convenience even though we may be uncertain whether it is sound, and therefore consistent. But if there *is* genuine doubt about $\mathbb{F}$, we must, in this case, state any result P obtained by use of $\mathbb{F}$ in the form

'P is deducible within $\mathbb{F}$'

('or P is TRUE'), rather than simply asserting that 'P is true'. This is a perfectly good mathematical statement, which might be either actually true or actually false. It would be perfectly legitimate to restrict one's mathematical pronouncements to statements of this kind, but if so, one is still making statements about absolute mathematical truths. On occasion, one might believe that one has established that a statement of the above form is actually false; i.e. one might believe that one has established

'P is not deducible within $\mathbb{F}$'.

Statements of this kind are of the form 'such-and-such a computation does not terminate' (in fact: 'F applied to P does not terminate') which are precisely of the form of the Π_1-sentences that I have been considering. The issue is: what means does one allow in deriving statements of this kind? What, indeed, *are* the mathematical procedures that one actually *believes* in for establishing mathematical truths? Such a body of beliefs, if they are reasonable, cannot be *equivalent* to a belief in a formal system, no matter what that formal system might be.

Q16. The conclusion that $G(\mathbb{F})$ is actually true, of a consistent formal system $\mathbb{F}$, depends upon the assumption that the symbols in $\mathbb{F}$ which are supposed to represent the natural numbers actually *do* represent natural numbers. For some other exotic type of numbers—call them the 'supernatural' numbers— we might find that $G(\mathbb{F})$ is false. How do we know that we are referring to the natural and not supernatural numbers in our system $\mathbb{F}$?

It is true that there is no finite axiomatic way of making sure that the 'numbers' that we are referring to are actually the intended *natural* numbers and not some kind of unintended 'supernatural' numbers.[5] But, in a sense, this is the whole point of the Gödel discussion. No matter what axiom system $\mathbb{F}$ we provide, as an attempt to characterize the natural numbers, the rules of $\mathbb{F}$ themselves will be insufficient to tell us whether $G(\mathbb{F})$ is actually true or false. Supposing $\mathbb{F}$ to be consistent, we know that the *intended* meaning of $G(\mathbb{F})$ is something that is indeed true, not false. However, this depends on the symbols that actually constitute the formal expression that is denoted by '$G(\mathbb{F})$' having their intended meanings. If these symbols are reinterpreted as meaning something quite different, then we can arrive at an interpretation for '$G(\mathbb{F})$' that is indeed false.

To see how these ambiguities come about, we consider new formal systems $\mathbb{F}^*$ and $\mathbb{F}^{**}$, where $\mathbb{F}^*$ is obtained by adjoining $G(\mathbb{F})$ to the axioms of $\mathbb{F}$, and $\mathbb{F}^{**}$ is obtained by correspondingly adjoining $\sim G(\mathbb{F})$ instead. Provided that $\mathbb{F}$ is sound, then $\mathbb{F}^*$ and $\mathbb{F}^{**}$ are both consistent (since $G(\mathbb{F})$ is true and $\sim G(\mathbb{F})$ is not deducible from the rules of $\mathbb{F}$). But with the intended interpretation of the symbols of $\mathbb{F}$—called the *standard* interpretation—then, assuming $\mathbb{F}$ to be sound, $\mathbb{F}^*$ will be sound also, but $\mathbb{F}^{**}$ will *not* be sound. However, it is a feature of consistent formal systems that one can find so-called *non-standard* reinterpretations of the symbols so that propositions that are false using the standard interpretation turn out to be true in the non-standard one; accordingly $\mathbb{F}$ and $\mathbb{F}^{**}$ could now be sound, in such a non-standard interpretation, instead of $\mathbb{F}^*$. One could imagine that this reinterpretation might affect the meanings of the logical symbols (such as '$\sim$' and '&', which in the standard interpretation mean 'not' and 'and', respectively), but here we are concerned with the symbols representing undetermined numbers ('x', 'y', 'z', 'x'', 'x''', etc.) and the meanings of the logical quantifiers ($\forall$, $\exists$) used in association with them. Whereas in the standard interpretation '$\forall x$' and '$\exists x$' would mean, respectively, 'for all natural numbers x' and 'there exists a natural number x such that', in such a *non*-standard interpretation the symbols would refer not to natural numbers but to some different kind of numbers, with different ordering properties (which could indeed be called 'supernatural' numbers, as in the terminology of Hofstadter (1979)).

The fact is, however, that we actually *know* what the actual natural numbers are, and ourselves have no problem about distinguishing them from some strange kind of supernatural number. The natural numbers are the ordinary things that we normally denote by the symbols 0, 1, 2, 3, 4, 5, 6, ... These are the concepts that we become acquainted with as small children, and we have no problem about distinguishing them from some bizarre concept of supernatural number (cf. §1.21). There is perhaps something mysterious, however, in the fact that we *do* seem to know instinctively what the natural numbers actually are. For as children (or adults) we are provided with just a comparatively small number of descriptions as to what 'zero', 'one', 'two', 'three', etc., mean ('three oranges', 'one banana', etc.); yet we can grasp the entire concept despite this inadequacy. In some Platonic sense, the natural numbers seem to be things that have an absolute conceptual existence independent of ourselves. Notwithstanding such human independence, we are able, intellectually, to make contact with the actual natural-number concept from merely these vague and seemingly inadequate descriptions. On the other hand, no finite number of *axioms* can completely distinguish natural numbers from these alternative, so-called 'supernatural', possibilities.

Moreover, the specific *infinite* character of the totality of natural numbers is something that somehow we are able to perceive directly, whereas a system that is constrained to operate by precise finite rules is not able to distinguish the particular infinite character of the natural numbers from other ('supernatural') possibilities. The infinitude that characterizes the natural numbers is understood by us, even though merely represented by the dots '. . .' in the description

'0, 1, 2, 3, 4, 5, 6, . . .'

or by the 'etc.' in

'zero, one, two, three, etc.'.

We do not have to be told exactly what a natural number is, in terms of precise rules. This is fortunate, since it is not possible. Somehow, we find we *know* what a natural number is, once we have been just roughly steered in the right direction!

Some readers may be familiar with the *Peano axioms* for the arithmetic of natural numbers (alluded to briefly under §2.7), and be puzzled as to why they do not define the natural numbers adequately. According to the Peano definition, one starts with a symbol **0**, and there is a 'successor operator', denoted by **S**, which is interpreted as simply adding 1 to the number on which it operates, so we can *define* **1** to be **S0**, and **2** to be **S1** or **SS0**, etc. We have, as rules, the fact that if **Sa** = **Sb**, then **a** = **b**; and that for no **x** is **0** of the form **Sx**, where this particular property characterizes **0**. Also there is the 'induction principle' that states a property *P*, of numbers, must be true of *all* numbers **n** provided that it satisfies: (i) if $P(\mathbf{n})$ is true then $P(\mathbf{Sn})$ is also true, for all $\mathbf{n}$; (ii) $P(\mathbf{0})$ is true. The trouble comes with the logical operations where, in the standard interpretation, the symbols ∀ and ∃ respectively denote 'for all *natural numbers* . . .' and 'there exists a *natural number* . . . such that'. In a non-standard interpretation, the meanings of these symbols would shift appropriately, so that they would quantify, instead, over some other kind of 'number'. While it is true that the Peano mathematical specifications that are given for the successor operator **S** do in fact characterize the ordering relationships that distinguish the natural numbers from any kind of 'supernatural' numbers, those specifications are not captured in terms of the formal rules that these quantifiers ∀ and ∃ satisfy. In order to capture the meanings of the Peano mathematical specifications, we need to pass to what is known as 'second-order logic' in which quantifiers like ∀ and ∃ are introduced but where now their ranges are over (infinite) *sets* of natural numbers rather than just single natural numbers. In the 'first-order logic' of Peano arithmetic, the quantifiers would simply range over single numbers, and we get a formal system in the ordinary sense. But second-order logic does not provide us with a formal system. For a strict formal system, it has to be a purely *mechanical* (i.e. algorithmic) matter to decide whether or not the rules of the system have been correctly applied, which, in any case, is the whole point of considering formal systems in the present context. This property fails for second-order logic.

It is a common misconception, in the spirit of the sentiments expressed in **Q16**, that Gödel's theorem shows that there are many different kinds of arithmetic, each of which is equally valid. The particular arithmetic that we may happen to choose to work with would, accordingly, be defined merely by some arbitrarily chosen formal system. Gödel's theorem shows that none of these formal systems, if consistent, can be complete; so—it is argued—we can keep adjoining new axioms, according to our whim, and obtain all kinds of alternative consistent systems within which we may choose to work. The comparison is sometimes made with the situation that occurred with Euclidean geometry. For some 21 centuries it was believed that Euclidean geometry was the only geometry possible. But when, in the eighteenth century, mathematicians such as Gauss, Lobachevsky, and Bolyai showed that indeed there are alternatives that are equally possible, the matter of geometry was seemingly removed from the absolute to the arbitrary. Likewise, it is often argued, Gödel showed that arithmetic, also, is a matter of arbitrary choice, any one set of consistent axioms being as good as any other.

This, however, is a completely misleading interpretation of what Gödel has demonstrated for us. He has taught us that the very notion of a formal axiomatic system is inadequate for capturing even the most basic of mathematical concepts. When we use the term 'arithmetic' without further qualification, we indeed mean the ordinary arithmetic which operates with the ordinary natural numbers 0, 1, 2, 3, 4, . . . (and perhaps their negatives) and not with some kind of 'supernatural' numbers. We may choose, if we wish, to explore the properties of formal systems, and this is certainly a valuable part of mathematical endeavour. But it is something different from exploring the ordinary properties of the ordinary natural numbers. The situation is, in some ways, perhaps not so very unlike that which occurs with geometry. The study of non-Euclidean geometries is something mathematically interesting, with important applications (such as in physics, see ENM Chapter 5 especially Figs 5.1 and 5.2, and also §4.4), but when the term 'geometry' is used in ordinary language (as distinct from when a mathematician or theoretical physicist might use that term), we do indeed mean the ordinary geometry of Euclid. There is a difference, however, in that what a logician might refer to as 'Euclidean geometry' can indeed be specified (with some reservations[6]) in terms of a particular formal system, whereas, as Gödel has shown, ordinary 'arithmetic' cannot be so specified.

Rather than showing that mathematics (most particularly arithmetic) is an arbitrary pursuit, whose direction is governed by the whim of Man, Gödel demonstrated that it is something absolute, there to be discovered rather than invented (cf. §1.17). We discover for ourselves what the natural numbers are, and we do not have trouble in distinguishing them from any sort of supernatural numbers. Gödel showed that no system of 'man-made' rules can, by themselves, achieve this for us. Such a Platonic viewpoint was important to Gödel, and it will be important also for us in the later considerations of this book (§8.7).

Q17. Assume that the formal system $\mathbb{F}$ is intended to represent the mathematical truths in principle accessible to the mind. Can we not get around the problem of being unable to incorporate the Gödel proposition $G(\mathbb{F})$ formally into $\mathbb{F}$ by, instead, incorporating something with the *meaning* of $G(\mathbb{F})$, using a reinterpretation of the meanings of the symbols of $\mathbb{F}$?

There are indeed ways of representing the Gödel argument applied to $\mathbb{F}$ within a (sufficiently extensive) formal system $\mathbb{F}$, so long as the meanings of the symbols of $\mathbb{F}$ are reinterpreted as something different from those meanings originally assigned to the symbols of $\mathbb{F}$. However, this is really cheating if we are trying to interpret $\mathbb{F}$ *as* the procedure whereby the mind reaches its mathematical conclusions. The symbols of $\mathbb{F}$ should not be allowed to shift their meanings mid-stream if mental activities are to be interpreted solely in terms of $\mathbb{F}$. If mental activities are allowed to contain something beyond the operations of $\mathbb{F}$ itself, namely the shifting *meanings* of these symbols, then we need also to know the rules governing the detailed shifting of these meanings. Either these rules are something non-algorithmic, whence the case for $\mathscr{G}$ is made, or else there is some specific algorithmic procedure for this, in which case we should have incorporated *this* procedure into our '$\mathbb{F}$' in the first place—call this system $\mathbb{F}^{\dagger}$—in order for it to represent the totality of our insights, and it would not have been necessary for the meanings to shift at all.

In the latter case the Gödel proposition $G(\mathbb{F}^{\dagger})$ takes the place of $G(\mathbb{F})$ in the previous discussion, and we have gained nothing.

Q18. It is possible to formulate, within even such a simple system as Peano arithmetic, a theorem whose interpretation has the implication:

'$\mathbb{F}$ sound' implies '$G(\mathbb{F})$'.

Is this not all we need from Gödel's theorem? Surely it would enable us to pass from a belief in the soundness of any formal system $\mathbb{F}$ to a belief in the truth of its Gödel proposition, so long as we are prepared to accept even just Peano arithmetic?

It is indeed true that such a theorem[7] can be formulated within Peano arithmetic. More precisely (since we cannot properly encapsulate the notion of 'soundness' or 'truth' within any formal system—as follows by a famous theorem of Tarski), what one actually formulates is a stronger result:

'$\mathbb{F}$ consistent' implies '$G(\mathbb{F})$';

or else:

'$\mathbb{F}$ ω-consistent' implies '$\Omega(\mathbb{F})$'.

These have the implication needed for **Q18** because if $\mathbb{F}$ is sound then it is certainly consistent or ω-consistent, as the case may be. Provided that we understand the *meanings* of the symbolism used, then we can indeed pass from a belief in the soundness of $\mathbb{F}$ to a belief in the truth of $G(\mathbb{F})$. But this is already accepted. If we understand meanings, then we can indeed pass from $\mathbb{F}$ to $G(\mathbb{F})$. The trouble comes if we wish to eliminate the need for interpretations and to make the passing from $\mathbb{F}$ to $G(\mathbb{F})$ *automatic*. If this were possible, then we could automate the general 'Gödelization' procedure, and build an algorithmic device that indeed encompasses all that we require of Gödel's theorem. However, this cannot be done; for if we were to adjoin this putative algorithmic procedure to whatever formal system $\mathbb{F}$ we might choose to start from, we should simply obtain, in effect, some *new* formal system $\mathbb{F}^{\#}$, and *its* Gödel proposition $G(\mathbb{F}^{\#})$ would lie outside $\mathbb{F}^{\#}$'s scope. There always remains *some* aspect of the insight afforded by Gödel's theorem, no matter how much of it has been incorporated into a formalized or algorithmic procedure. This 'Gödelian insight' requires a continual reference to the actual meanings of the symbols of whatever system Gödel's procedure is being applied to. In this sense, the trouble with **Q18** is very similar to that which arises with **Q17** above. The fact that Gödelization cannot be automated is also closely related to the discussion of **Q6**, above, and of **Q19** to follow.

There is another aspect fo **Q18** which is worth considering. Imagine that we have a sound formal system $\mathbb{H}$ which contains Peano arithmetic within it. The theorem referred to in **Q18** will be amongst the implications of $\mathbb{H}$, and the particular instance of it which applies to the particular $\mathbb{F}$ which is $\mathbb{H}$ itself will be a theorem of $\mathbb{H}$. Thus, we may say that one of $\mathbb{H}$'s implications is:

'$\mathbb{H}$ sound' implies '$G(\mathbb{H})$';

or, more precisely, that, for example:

'$\mathbb{H}$ consistent' implies '$G(\mathbb{H})$'.

Now these assertions would, in terms of their actual meanings, entail the implication that $G(\mathbb{H})$ is, in effect, also being asserted. For—with regard to the first of the two assertions above—*any* assertion that $\mathbb{H}$ makes is conditional upon an assumption that $\mathbb{H}$ is sound in any case; so if $\mathbb{H}$ asserts something that is explicitly conditional upon its own soundness, it might as well assert that thing directly. (The assertion 'if I am to be believed, then X is true' entails the simpler assertion, by the same speaker, 'X is true'.) However, a sound formal system $\mathbb{H}$ *cannot* actually assert $G(\mathbb{H})$, which reflects the fact that it is incapable of asserting its own soundness. Moreover, we see that it cannot actually encapsulate the meanings of the symbols it operates with. The same facts are illustrated in relation to the second of the above assertions, but with the added irony that whereas $\mathbb{H}$ is incapable of asserting its own consistency when it in fact *is* consistent, it suffers from no such inhibitions if it is *in*consistent. An inconsistent $\mathbb{H}$ can assert, as a 'theorem', anything at all that it can formulate! It turns out that it can indeed formulate '$\mathbb{H}$ is consistent'. A (sufficiently extensive) formal system will assert its own consistency if and only if it is *in*consistent!

Q19. Why do we not just adopt a procedure of repeatedly adjoining the Gödel proposition $G(\mathbb{F})$ to whatever system $\mathbb{F}$ we are currently accepting, and allow this procedure to carry on *indefinitely*?

Given any particular formal system $\mathbb{F}$, which is sufficiently extensive and perceived to be sound, we can perceive how to adjoin $G(\mathbb{F})$ to $\mathbb{F}$ as a new axiom, and thereby obtain a new system $\mathbb{F}_1$, which is also perceived to be sound. (For notational consistency with what follows, we may also write $\mathbb{F}_0$ for $\mathbb{F}$.) We now perceive how to adjoin $G(\mathbb{F}_1)$ to $\mathbb{F}_1$, thereby obtaining a new system $\mathbb{F}_2$, also perceived sound. Repeating the process, by adjoining $G(\mathbb{F}_2)$ to $\mathbb{F}_2$, we obtain $\mathbb{F}_3$, and so on. Now, with just a little more work, we should be able to see how to construct yet another formal system $\mathbb{F}_\omega$, whose axioms allow us to incorporate precisely the *entire* infinite set $\{G(\mathbb{F}_0), G(\mathbb{F}_1), G(\mathbb{F}_2), G(\mathbb{F}_3), \ldots\}$ as additional axioms for $\mathbb{F}$. This system $\mathbb{F}_\omega$ will also be evidently sound. We can continue the process, adjoining $G(\mathbb{F}_\omega)$ to $\mathbb{F}_\omega$ to obtain $\mathbb{F}_{\omega+1}$, and then adjoining $G(\mathbb{F}_{\omega+1})$ to this to obtain $\mathbb{F}_{\omega+2}$, and so on. Then, as before, we can incorporate the *whole* infinite set of axioms, now to obtain $\mathbb{F}_{\omega 2}$ $(=\mathbb{F}_{\omega+\omega})$, which is again evidently sound. Adjoining $G(\mathbb{F}_{\omega 2})$, we now obtain $\mathbb{F}_{\omega 2+1}$, and so on, and we can again incorporate the infinite set to obtain $\mathbb{F}_{\omega 3}$ ($=\mathbb{F}_{\omega 2+\omega}$). Repeating the whole process, up to this point, all over again, we can obtain $\mathbb{F}_{\omega 4}$ and then, repeating again, $\mathbb{F}_{\omega 5}$ and so on. With just a little more work, we should be able to see how to incorporate *this* entire set of new axioms $\{G(\mathbb{F}_{\omega}), G(\mathbb{F}_{\omega 2}), G(\mathbb{F}_{\omega 3}), G(\mathbb{F}_{\omega 4}), \ldots\}$ to form a system $\mathbb{F}_{\omega^2}$ ($=\mathbb{F}_{\omega\omega}$). Now, repeating the entire process all over again, we obtain a new system $\mathbb{F}_{\omega^2+\omega^2}$, then $\mathbb{F}_{\omega^2+\omega^2+\omega^2}$, etc., which, when we see how to combine *all* these things together (with again some more work), leads us to an even more comprehensive system $\mathbb{F}_{\omega^3}$, which must also be sound.

Those readers who are familiar with the notation for Cantor's *ordinal* numbers will recognize the suffixes that I have been using here as denoting such ordinal numbers. Those unfamiliar with such things need not be troubled about the precise meanings of these symbols. Suffice it to say that this procedure of 'Gödelization' can be continued even further—we reach systems denoted by $\mathbb{F}_{\omega^4}, \mathbb{F}_{\omega^5}, \ldots$, and then this leads to an even more inclusive system $\mathbb{F}_{\omega^\omega}$, and the process continues to still larger ordinals, such as ω^{ω^ω}, etc.—so long as we can see, at each step, how to systematize the entire set of Gödelizations that we have achieved up to that point. This, in fact, is the crux of the matter: what has been referred to as 'with just a little more work' requires the appropriate insights about how actually to systematize the previous Gödelizations. It is possible to achieve this systematization, provided that the stage (ordinal) that has now been reached is labelled by what is called a *recursive* ordinal, which means, in effect, that there is an algorithm of some kind for generating the procedure. However, there is no algorithmic procedure that one can lay down beforehand which allows us to do this systematization for *all* recursive ordinals once and for all. We must continue to use our insights afresh.

The above procedure was first put forward by Alan Turing in his doctoral thesis (and published in Turing 1939)[8] and he showed that there is a sense in which *any* true Π_1-sentence can be proved by repeated Gödelization of the kind described here. (See Feferman 1988.) However, this does not provide us with a mechanical procedure for establishing the truth of Π_1-sentences, for the very reason that one cannot mechanically systematize Gödelization. In fact we can *deduce* that Gödelization cannot be made mechanical just from Turing's result. For we have already established (effectively in §2.5) that the general ascertaining of the truth, or otherwise, of Π_1-sentences cannot be decided by *any* algorithmic procedure. Thus the repeated Gödelization does not allow us to achieve anything in the way of a sytematic procedure that lies outside the computational considerations that we have been concerned with up until now. Thus **Q19** poses no threat to $\mathscr{G}$.

Q20. Surely the real value of mathematical understanding is not that it enables us to achieve non-computable things, but that it enables us to replace enormously complicated computations by comparatively simple insights? In other words, is

it not the case that the mind allows us to take shortcuts with respect to complexity theory rather than leap beyond the bounds of what is computable?

I am quite prepared to believe that *in practice* a mathematician's insights are much more frequently engaged in circumventing computational complexity, rather than non-computability. After all, mathematicians tend to be intrinsically lazy people, and they are often trying to find ways of avoiding computation (despite the fact that this may well lead them into considerably more difficult mental work than computation itself!). It is often the case that attempts to make computers mindlessly churn out the theorems of even only moderately complicated formal systems will quickly lead the computers to become ensnared in virtually hopeless computational complexity—whereas, armed with an understanding of the meanings underlying the rules of the system, a human mathematician need have little trouble deriving many interesting results within the system.[9]

The reason that I have concentrated on non-computability, in my arguments, rather than on complexity, is simply that it is only with the former that I have been able to see how to make the necessary strong statements. It may well be that in the working lives of most mathematicians, non-computability issues play, if anything, only a very small part. But that is not the point at issue. I am trying to show that (mathematical) understanding is something that lies beyond computation, and the Gödel(–Turing) argument is one of the few handles that we have on that issue. It is quite probable that our mathematical insights and understandings are often used to achieve things that *could* in principle also be achieved computationally—but where blind computation without much insight may turn out to be so inefficient that it is unworkable (cf. §3.26). However, these matters are much more difficult to address than the non-computability issue.

In any case, true though the sentiments expressed in **Q20** may be, this in no way contradicts $\mathscr{G}$.

Notes and references

1. It might seem that this is perfectly 'obvious'—and not something that could be a matter of dispute amongst mathematicians! However, the problem comes about with the notion of 'existence' for large infinite sets. (See Smorynski 1975, Rucker 1984, Moore 1990, for example.) We have seen from the example of the Russell paradox that one must be particularly careful in such matters. According to one point of view, a set would not be considered necessarily to exist unless there is at least some clear-cut *rule* (not necessarily a computable one) for specifying which things are to be in the set and which things are not. This is just what the axiom of choice does *not* provide, since there is no rule given for specifying *which* element is to be taken from each member of the collection. (Some of the implications of the axiom of choice are very non-intuitive—and almost paradoxical. Perhaps this is one reason that it is a

matter of some dispute. I am not even totally sure what my *own* position on this issue is!)

2. In the final chapter of his 1966 book, Cohen makes the point that although he has shown that the continuum hypothesis is UNDECIDABLE according to the procedures of ZF, he has left untouched the question of whether or not it is actually *true*—and he discusses how one might actually go about *deciding* this question! This makes clear that he is *not* taking the view that it is an entirely arbitrary matter whether one accepts the continuum hypothesis or not. This is contrary to opinions often expressed about the implications of the Gödel–Cohen results, namely that there are numerous 'alternative set theories' that are equally 'valid' for mathematics. By these remarks, Cohen reveals himself to be, like Gödel, a true Platonist for whom matters of mathematical truth are *absolute* and not arbitrary. This is very much in accordance with my own views, cf. §8.7.
3. See, for example, Hofstadter (1981), Bowie (1982).
4. For example, see various commentaries in *Behavioral and Brain Sciences*, **13** (1990), 643–705.
5. This terminology was suggested by Hofstadter (1981). It is Gödel's 'other' theorem—his *completeness* theorem—which tells us that such non-standard models always exist.
6. In fact, it depends upon which statements are considered as part of what is being called 'Euclidean geometry' here. In the usual terminology of the logicians, the system of 'Euclidean geometry' would only include statements of certain particular kinds, and it turns out that the truth or falsity of such statements can be resolved in terms of an algorithmic procedure—hence the assertion that Euclidean geometry can be specified in terms of a formal system. However, in *other* interpretations, ordinary 'arithmetic' might also be considered as part of 'Euclidean geometry', and this allows classes of statements that *cannot* be resolved algorithmically. The same would apply if we were to consider that the polyomino tiling problem is part of Euclidean geometry—which would seem to be a very natural thing to do. In this sense, Euclidean geometry can no more be formally specified than can arithmetic!
7. See the commentary by Davis (1993).
8. See also Kreisel (1960, 1967), Good (1967).
9. See Freedman (1994) concerning some of the problems that computer systems have had trying to do their 'own' mathematics. In general, such systems do not get very far. They need considerable human guidance!

Appendix A:

An explicit Gödelizing Turing machine

Suppose that we are given an algorithmic procedure A that we know correctly ascertains that certain computations do not ever terminate. I shall provide a completely explicit procedure for constructing, from A, a particular computation C on which A fails; yet we shall be able to see that C actually does *not* terminate. Taking this explicit expression for C, we can examine its degree of complication and compare it with that of A, as is needed for the arguments of §2.6 (cf. **Q8**) and §3.20.

For definiteness, I shall employ the particular Turing machine specifications that I gave in ENM. For the complete details of these specifications, the reader is referred to that volume. Here I shall merely supply a minimal description that will be adequate for our present purposes.

A Turing machine has a finite number of internal states but it acts on an infinite tape. The tape consists of a linear succession of 'boxes', where each box can be either marked or unmarked, there being a finite total number of marks on the tape altogether. Let us denote each marked box by the symbol 1 and each unmarked box by 0. There is a reading device which examines one mark at a time and, explicitly dependent upon the Turing machine's internal state and the nature of the mark that is being examined, three things are next determined: (i) whether the mark on the box being examined is to be altered or left alone, (ii) what the machine's new internal state is to be, and (iii) whether the device is to move one step along the tape to the right (denoted by **R**) or to the left (denoted by **L**) or one step to the right with the machine coming to a halt (denoted by **STOP**). When the machine eventually does come to a halt, the answer to the computation that it has been performing is displayed as the succession of 0s and 1s to the left of the reading device. Initially, the tape must be taken to be entirely blank except for the marks defining the specific data (provided by a finite string of 1s and 0s) upon which the machine is to perform its operations. The reading device is initially taken to the left of all of the marks.

When representing natural numbers on the tape, either in the input data or in the output, it may be useful to use the *expanded binary* notation, according to which the number is written, effectively, in the ordinary binary scale, but where the binary digit '1' is written as 10, the binary digit '0' being written as 0. Thus, we have the following scheme for translating ordinary numerals into the expanded binary notation:

0 ↔ 0
1 ↔ 10
2 ↔ 100
3 ↔ 1010
4 ↔ 1000
5 ↔ 10010
6 ↔ 10100
7 ↔ 101010
8 ↔ 10000
9 ↔ 100010
10 ↔ 100100
11 ↔ 1001010
12 ↔ 101000
13 ↔ 1010010
14 ↔ 1010100
15 ↔ 10101010
16 ↔ 100000
17 ↔ 1000010
etc.

We note that in the expanded binary notation, the immediate repetition of a 1 never occurs. Thus, we can signal the beginning and end of the specification of a natural number by a succession of two or more 1s. Then we can use the sequences 110, 1110, 11110, etc., on the tape, to denote various kinds of instruction.

We can also use the marks on the tape to specify particular Turing machines. This is needed when we consider the action of a *universal* Turing machine *U*. The universal machine *U* acts upon a tape whose initial portion provides the detailed specification of some particular given Turing machine *T* which the universal machine is being told to imitate. The data on which *T* itself is intended to be acting upon is then fed into *U* to the right of the portion that determines the machine *T*. In order to specify the machine *T*, we can use the sequences 110, 1110, and 11110 to denote, respectively, the various instructions for *T*'s reading device; namely to move along the tape by one step to the right, by one step to the left, or to halt after moving one step to the right:

$$\mathbf{R} \leftrightarrow 110$$
$$\mathbf{L} \leftrightarrow 1110$$
$$\mathbf{STOP} \leftrightarrow 11110.$$

Immediately preceding each such instruction would be either the symbol 0 or the sequence 10, to indicate that the reading device is to mark the tape with 0 or 1, respectively, in place of whatever symbol it has just been reading. Immediately preceding the aforementioned 0 or 10 would be the expanded binary expression for the number of the internal state into which the Turing machine is to be placed next, according to this same instruction. (We note that the internal states, being finite in number, can be labelled by consecutive natural numbers 0, 1, 2, 3, 4, 5, 6, . . ., *N*. In the coding on the tape, the expanded binary notation would be used to denote these numbers.)

The particular instruction to which this operation refers would be determined by the internal state of the machine just prior to its reading the tape, together with the symbol 0 or 1 that the reading device is about to read and perhaps to alter. For example, as part of the specification of *T*, might be some instruction 230 → 171**R**, which means: 'if *T* is in internal state 23 and the reading device encounters 0 on the tape, then replace it by 1, go into internal state 17 and move one step to the right along the tape'. In this case, the part '171**R**' of the instruction would be coded as 100001010110. Breaking this up as 1000010.10.110, we see that the first portion is the expanded binary form of 17, the second codes the marking of 1 on the tape, and the third portion codes the instruction 'move to the right'. How do we specify the prior internal state (here the internal state 23) and the mark on the tape being examined (here 0)? If desired, we could also explicitly give these in terms of their expanded binary numbers. However, this is not really necessary, since the numerical ordering of the various instructions will suffice for this (i.e., the ordering 00 →, 01 →, 10 →, 11 →, 20 →, 21 →, 30 →, . . .).

This provides us with essentially the coding of Turing machines given in ENM, but for completeness, a few additional points should be mentioned. In the first place, we have to make sure that there is an instruction provided for every one of the internal states, acting on 0 and on 1 (except that an instruction for the highest-numbered internal state acting on 1 is not always necessary). Where there is never any use made of such an instruction in the program, a 'dummy' must be inserted. For example, such a dummy instruction could be 231 → 00**R**, if it happens to be the case that the internal state 23 never needs to encounter the mark 1 in the running of the program.

In the coded specification of a Turing machine on a tape, according to the aforementioned prescriptions, the pair 00 would be represented by the sequence 00, but one may economize and use a single 0, without ambiguity, to separate the sequences of (more than one) 1s on either side.* The Turing machine is to commence its operation in the internal state 0, and the reading device moves along the tape maintaining this internal state until it encounters its first 1. This is achieved by assuming that the operation 00 → 00**R** is always part of a Turing machine's instructions. Thus, in the actual specification of a Turing machine as a succession of 0s and 1s, it is not necessary to provide this instruction explicitly; instead, we start with 01 → **X**, where **X** denotes the first non-trivial operation of the machine that is put in action, i.e. when the first 1 on the tape is encountered. This suggests deleting the initial sequence 110 (denoting → 00**R**) that would otherwise always occur in the sequence specifying a Turing machine. Moreover, we shall also always delete the final sequence 110 in that specification, since this, also is common to all Turing machines.

The resulting sequence of digits 0 and 1 provides the (ordinary, i.e., non-expanded) *binary coding* of the *Turing machine number n* of the machine in question (as was given in ENM Chapter 2). We call this the nth Turing machine, and we write $T = T_n$. Every such binary number, when the sequence 110 is adjoined at the end, is a sequence of 0s and 1s in which there are never more than four 1s in succession. A number n for which this is not the case would give a 'dud Turing machine' that would cease to operate as soon as the 'instruction' involving more than four 1s is encountered. Such a 'T_n' is said to be *not correctly specified*. Its action on *any* tape is, *by definition*, considered to be non-terminating. Likewise, if a Turing machine action encounters an instruction to enter a state specified by a number that is larger than any for which further instructions are actually listed, then it also would get 'stuck'; it would be considered to be a 'dud', and its action also counts as non-terminating. (It would be possible to eliminate these awkward issues without a great deal of difficulty by the use of various devices, but there is no real need to do this; cf. §2.6, **Q4**.)

In order to see how to construct, from the given algorithm A, the required explicit non-terminating computation on which A must fail,

* This means that in the coding of a Turing machine, each occurrence of the sequence ... 110011 ... can be replaced by ... 11011.... There are 15 places in the specification of the universal Turing machine that I gave in ENM (cf. end-note 7 to Chapter 2) where I had omitted to do this. This is a matter of definite irritation to me, because I had taken some considerable pains, under the constraints of the prescriptions that I had given, to achieve a number for this universal machine that was as small as I could reasonably achieve. Making the aforementioned simple replacements provides a number over 30 000 times smaller than the one I gave! I am grateful to Steven Gunhouse for pointing this oversight out to me, and also for independently checking that the specification, as printed, *does* in fact provide a universal Turing machine.

we shall need to suppose that A is given as a Turing machine. This machine acts on a tape coding two natural numbers p and q. It is to be assumed that if the computation $A(p,q)$ terminates, then the action of the computation T_p on the number q does *not* ever terminate. Recall that if T_p is not correctly specified, then we must consider that its action on q does not terminate whatever q might be. For any such 'disallowed' p, any result for $A(p,q)$ whatever would be consistent with our assumptions. Accordingly, I need be concerned only with numbers p for which T_p *is* correctly specified. Thus, the binary expression for the number p, as represented on a tape, can contain no sequence ... 11111.... This allows us to use the specific sequence 11111 to mark the beginning and end of the number p, as represented on the tape.

However, we are also required to do the same for q, which need *not* be restricted to be a number of this type. This presents us with a technical awkwardness, for the Turing-machine prescriptions as I have given them, and it will be convenient to circumvent it by use of the device of taking the numbers p and q to be effectively written in the *quinquary* scale. (This is the scale where '10' denotes the number *five*, '100' denotes *twenty-five*, '44' denotes *twenty-four* etc.) But rather than employing the quinquary digits 0, 1, 2, 3, and 4, I shall use the respective tape sequences 0, 10, 110, 1110, and 11110. Thus:

0	is represented by the sequence	0
1	"	10
2	"	110
3	"	1110
4	"	11110
5	"	100
6	"	1010
7	"	10110
8	"	101110
9	"	1011110
10	"	1100
11	"	11010
12	"	110110
13	"	1101110
14	"	11011110
15	"	11100
16	"	111010
...		...
25	"	1000
26	"	10010
etc.		

The notation 'C_p' will be used here for the correctly specified Turing machine T_r, where r is the number whose ordinary binary expression, together with the sequence 110 appended at the end, is precisely the quinquary expression for p, as in the description given above. The number q, upon which the computation C_p acts, is also to be expressed in the quinquary notation. The computation $A(p,q)$ is to be described as a Turing machine acting upon a tape coding the pair of numbers p, q. This coding on this tape is to be taken as follows:

...00111110p111110q11111000...,

where p and q are the respective quinquary notations for p and q, as above.

What we require is to find a p and a q for which we know that $C_p(q)$ does not terminate, but for which $A(p,q)$ also fails to terminate. The procedure of §2.5 achieves this by finding a number k for which C_k acting on n is precisely $A(n,n)$, for each n, and then putting $p = q = k$. To achieve this explicitly, we seek a Turing-machine prescription $K(= C_k)$ whose action on a tape marked:

...00111110n11111000...

(n being the quinquary expression for n) is precisely the same as the action of A on:

...00111110n111110n11111000...,

for each n. Thus, what K has to do is to take the number n (written in quinquary notation) and to copy it once, where the sequence 111110 separates the two occurrences of n (and a similar sequence initiates and terminates the entire sequence of marks on the tape). Subsequently, it has to act upon the resulting tape in precisely the same way in which A would have acted upon that tape.

An explicit modification of A that provides such a K is achieved as follows. First, we find the initial instruction 01→**X**, in the specification of A and take note of what this '**X**' actually is. We are going to substitute this for the '**X**' in the specification given below. As a technical point, we must also assume that A is expressed in such a way that the internal state 0 of A is never actually entered again after the instruction 01→**X** is activated. There is no restriction involved in insisting that A has this form.* (It is all right for 0 to be used in dummy instructions, but not elsewhere.)

* In fact, one of Turing's actual original proposals was to have the machine *halt* whenever the internal state '0' is re-entered from some other internal state. In this way, not only would the above restriction be unnecessary, but the instruction **STOP** could be dispensed with. A simplification would thereby be achieved because 11110 would not be needed as an instruction and could be used as the marker instead, rather than 111110. This would significantly shorten my prescription for K, and we would use the quaternary rather than the quinquary numbering system.

Next, the total number N of internal states in A's specification must be ascertained (including the state 0, so that the largest internal state number of A is $N-1$). If A's specification contains no final instruction of the form $(N-1)$1→Y, then a dummy instruction $(N-1)$1→0O**R** must then be adjoined at the end. Finally, we remove 01→X from A's specification and adjoin to this specification the Turing-machine instructions listed below, where each internal-state number that appears in the list must be increased by N, where ϕ represents the resulting internal state 0, and where the 'X' in '11→X' below is the instruction we have taken note of above. (In particular, the first two instructions below would become 01→N1**R**, N0→$(N+4)$0**R**.)

ϕ1→01**R**, 00→40**R**, 01→01**R**, 10→21**R**, 11→X, 20→31**R**,
21→ϕ0**R**, 30→551**R**, 31→ϕ0**R**, 40→40**R**, 41→51**R**, 50→40**R**,
51→61**R**, 60→40**R**, 61→71**R**, 70→40**R**, 71→81**R**, 80→40**R**,
81→91**R**, 90→100**R**, 91→ϕ0**R**, 100→111**R**, 101→ϕ0**R**,
110→121**R**, 111→120**R**, 120→131**R**, 121→130**R**, 130→141**R**,
131→140**R**, 140→151**R**, 141→10**R**, 150→00**R**, 151→ϕ0**R**,
160→170**L**, 161→161**L**, 170→170**L**, 171→181**L**, 180→170**L**,
181→191**L**, 190→170**L**, 191→201**L**, 200→170**L**, 201→211**L**,
210→170**L**, 211→221**L**, 220→220**L**, 221→231**L**, 230→220**L**,
231→241**L**, 240→220**L**, 241→251**L**, 250→220**L**, 251→261**L**,
260→220**L**, 261→271**L**, 270→321**R**, 271→281**L**, 280→330**R**,
281→291**L**, 290→330**R**, 291→301**L**, 300→330**R**, 301→311**L**,
310→330**R**, 311→110**R**, 320→340**L**, 321→321**R**, 330→350**L**,
331→331**R**, 340→360**R**, 341→340**R**, 350→371**R**, 351→350**R**,
360→360**R**, 361→381**R**, 370→370**R**, 371→391**R**, 380→360**R**,
381→401**R**, 390→370**R**, 391→411**R**, 400→360**R**, 401→421**R**,
410→370**R**, 411→431**R**, 420→360**R**, 421→441**R**, 430→370**R**,
431→451**R**, 440→360**R**, 441→461**R**, 450→370**R**, 451→471**R**,
460→480**R**, 461→461**R**, 470→490**R**, 471→471**R**, 480→480**R**,
481→490**R**, 490→481**R**, 491→501**R**, 500→481**R**, 501→511**R**,
510→481**R**, 511→521**R**, 520→481**R**, 521→531**R**, 530→541**R**,
531→531**R**, 540→160**L**, 541→ϕ0**R**, 550→531**R**.

We are now in a position to give a precise limit on the size of K, as obtained by the above construction, as a function of the size of A. Let us measure this 'size' as the 'degree of complication', defined in §2.6 (end of response to **Q8**). For a specific Turing machine T_m (such as A), this is the number of digits in the binary representation of the number m. For a particular Turing-machine *action* $T_m(n)$ (such as K), this is the number of binary digits in the greater of m and n. Let α and κ, respectively, be the number of binary digits in a and k', where

$$A = T_a \quad \text{and} \quad K = T_{k'}(= C_k).$$

Since A has at least $2N-1$ instructions (with the first instruction being omitted), and since each instruction takes up at least three binary digits in its binary specification, the total number of binary digits in its Turing-machine number a must certainly satisfy

$$\alpha \geq 6N - 6.$$

There are 105 places (to the right of the arrows) in the above additional list of instructions for K where the number N must be added to the number that appears there. The resulting numbers are all less than $N+55$ and therefore have expanded binary representations with not more than $2 \log_2(N+55)$ digits each, giving a total of less than $210 \log_2(N+55)$ binary digits for the extra specification of internal states. We must add to this the digits needed for the extra symbols 0, 1, **R**, and **L**, which amount to another 527 (including one possible additional 'dummy' instruction, and bearing in mind that six of the 0s can be eliminated according to the rule whereby 00 can be represented as 0), so we can be sure that the specification of K requires fewer than $527 + 210 \log_2(N+55)$ digits more than that of A does:

$$\kappa < \alpha + 527 + 210 \log_2(N+55).$$

Using the relation $\alpha \geq 6N-6$ obtained above, we find (noting that $210 \log_2 6 > 542$)

$$\kappa < \alpha - 15 + 210 \log_2(\alpha + 336).$$

Now, let us find the degree of complication η of the particular computation $C_k(k)$ that this procedure provides. Recall that the degree of complication of $T_m(n)$ was defined as the number of binary digits in the larger of the two numbers m, n. In the present situation, we have $C_k = T_{k'}$, so the number of binary digits in the 'm' for this computation is just κ. To see how many binary digits there are in the 'n' for this computation, we examine the tape that is involved in $C_k(k)$. This tape is initiated by the sequence 111110, which is immediately followed by the binary expression for k', and then terminated by the sequence 11011111. The conventions of ENM require that this entire sequence, but with the last digit deleted, be read as a binary number, in order that we obtain the 'n' which numbers the tape in the computation $T_m(n)$. Hence the number of binary digits in this particular 'n' is precisely $\kappa + 13$, so it follows that $\kappa + 13$ is also the degree of complication η of $C_k(k)$, whence $\eta = \kappa + 13 < \alpha - 2 + 210 \log_2(\alpha + 336)$, from which follows the simpler-looking expression

$$\eta < \alpha + 210 \log_2(\alpha + 336).$$

The specific details of the foregoing argument are features of the particular codings for Turing machines that have been adopted, and they would be somewhat different in some other coding. The basic idea itself is a very simple one. In fact, had a λ-*calculus* formalism been adopted, then the entire operation would have been in a sense, reduced almost to triviality. (See ENM, end of Chapter 2, for an adequate enough description of Church's λ-calculus; see also Church 1941.) We may think of A as being defined by a λ-calculus operator **A**, acting on other operators **P** and **Q**, as expressed by the operation (**AP**)**Q**. Here **P** represents the computation C_p, and **Q** stands for the number q. Then the requirement for **A** is that, for all such **P**, **Q**, the following holds.

If (**AP**)**Q** terminates, then **PQ** does not terminate.

We can easily construct a λ-calculus operation which does not terminate, but for which **A** fails to ascertain this fact. This is achieved by taking

$$\mathbf{K} = \lambda \mathbf{x}.[(\mathbf{Ax})\mathbf{x}]$$

so that **KY** = (**AY**)**Y** for all operators **Y**. Then we consider the λ-calculus operation

KK.

This clearly does not terminate, because **KK** = (**AK**)**K**, the terminating of which would imply that **KK** does not terminate, by the assumed nature of **A**. Moreover **A** cannot ascertain this fact because (**AK**)**K** does not terminate. If we *believe* that **A** has its required property, then we must also *believe* that **KK** does not terminate.

Note that there is a considerable economy in this procedure. If we write **KK** in the form

$$\mathbf{KK} = \lambda \mathbf{y}.[\mathbf{yy}](\lambda \mathbf{y}.[(\mathbf{Ax})\mathbf{x}]),$$

then we see that the number of symbols in **KK** is merely 16 greater than the number in A (ignoring the dots, which are redundant, in any case)!

Strictly speaking, this is not entirely legitimate, because the symbol '**x**' might also appear in the expression for **A**, and some means would need to be adopted for dealing with this fact. Also, one might perceive a difficulty in that the non-terminating computation that this procedure generates is not something that appears as an operation on natural numbers (since the second **K** in **KK** is not a 'number'). In fact, the λ-calculus is not well suited to dealing with explicit numerical operations, and it is often not easy to see how a given algorithmic procedure applied to natural numbers can be expressed as a λ-calculus operation. For reasons such as these, the discussion in terms of Turing machines has a greater direct relevance to our discussions, and it more clearly achieves what is required.

3

The case for non-computability in mathematical thought

3.1 What did Gödel and Turing think?

In Chapter 2, I have tried to demonstrate for the reader the power and the rigorous nature of the underlying case for the assertion (denoted by $\mathscr{G}$) that mathematical understanding cannot be the result of some consciously appreciated and totally believed algorithm (or, equivalently, algorithm*s*; cf. **Q1**). These arguments do not address the more serious possibility—in *accordance* with $\mathscr{G}$—that mathematical belief might be the result of an unknown unconscious algorithm, or possibly of a knowable algorithm that cannot be known to be—or firmly believed to be—one that underlies mathematical belief. It is my purpose, in this chapter, to demonstrate that, although such are logical possibilities, they are not at all plausible ones.

It should first be pointed out that when mathematicians produce their careful chains of conscious reasoning for establishing mathematical truths, they do not *think* that they are merely blindly following unconscious rules that they are incapable of both knowing and believing. What they think they are doing is basing their arguments upon what are unassailable truths—ultimately, essentially 'obvious' ones—building their chains of reasoning entirely from such truths. And though these chains may sometimes be extremely long, involved, or conceptually subtle, the reasoning is, in principle and at root, unassailable, firmly believed, and logically impeccable. They do not tend to think that they are really acting according to some quite different unknown or unbelieved procedures, perhaps 'behind the scenes' guiding their beliefs in unknowable ways.

Of course, they might be wrong about this. Perhaps there is, indeed, an algorithmic procedure that, unknown to them, governs all of their mathematical perceptions. It is probably easier for a non-mathematician to take such a possibility seriously than it is for most working mathematicians. In this chapter, I shall try to persuade the reader that such working mathematicians are right in their opinion that they are not merely responding to an unknown (and unknowable) algorithm—nor to an algorithm that they do not firmly believe in. It may well be that their thoughts and beliefs are indeed guided by some unknown unconscious principles; but I shall argue that if this is so, then these principles are things that cannot be described in algorithmic terms.

It is instructive to consider the viewpoints of the two leading mathematical figures who are essentially responsible for the very argument that has led us to our conclusion $\mathscr{G}$. What, indeed, did Gödel think? What did Turing think? It is remarkable how, when presented with the same mathematical evidence, they came to basically opposite conclusions. It should be made clear, however, that both these outstanding thinkers expressed viewpoints in accordance with the conclusion $\mathscr{G}$. Gödel's viewpoint seems to have been that the mind is indeed not constrained by having to be a computational entity, and is not even constrained by the finiteness of the brain. Indeed, he chastized Turing for not accepting this as a possibility. According to Hao Wang (1974, p. 326, cf. also Gödel (1990) *Collected Works* Vol II, p. 297), while accepting both of Turing's implicit contentions that 'the brain functions basically like a digital computer' and that 'the physical laws, in their observable consequences, have a finite limit of precision', Gödel rejected Turing's other contention that 'there is no mind separate from matter', referring to it as 'a prejudice of our time'. Thus, Gödel appears to have taken it as evident that the *physical* brain must itself behave computationally, but that the mind is something beyond the brain, so that the mind's action is not constrained to behave according to the computational laws that he believed must control the physical brain's behaviour. He did not regard $\mathscr{G}$ as a *proof* of his viewpoint that the mind acts non-computationally, for he allowed[1] that:

> On the other hand, on the basis of what has been proved so far, it remains possible that there may exist (and even be empirically discoverable) a theorem-proving machine which in fact *is* equivalent to mathematical intuition, but cannot be *proved* to be so, nor even proved to yield only *correct* theorems of finitary number theory.

It should be made clear that this is indeed consistent with a belief in $\mathscr{G}$ (and I have no doubt that Gödel was well aware of some such clear-cut conclusion as I have stated as '$\mathscr{G}$'). He allowed the *logical possibility* that the minds of human mathematicians might act according to some algorithm of which they are not aware, or perhaps they could be aware of it provided that they could not be unassailably convinced of its soundness ('. . . cannot be *proved* . . . to yield only *correct* theorems . . .'). Such an algorithm comes under the category of 'unknowably sound', in my own terminology. Of course, it would have been another matter altogether actually to *believe* that such an unknowably sound algorithm might *actually* underlie the actions of a mathematician's mind. It seems that Gödel did not believe it, and found himself seemingly driven in the mystical direction that I have denoted by $\mathscr{D}$—that the mind cannot be explained at all in terms of the science of the physical world.

On the other hand, Turing appears to have rejected such a mystical standpoint, believing (as Gödel did) that the physical brain, like any other physical object, must act in a computational way (recall 'Turing's thesis', §1.6). Thus, he had to find some other route around the case provided, in effect, by $\mathscr{G}$. Turing made a great point of the fact that human mathematicians are very capable of making mistakes; he argued that for a computer to be able to be genuinely intelligent, it, also, would have to be allowed to make mistakes:[2]

> In other words then, if a machine is expected to be infallible, it cannot also be intelligent. There are several theorems which say almost exactly that. But these theorems say nothing about how much intelligence may be displayed if a machine makes no pretence at infallibility.

The 'theorems' he had in mind are undoubtedly Gödel's theorem and others related to it, such as his own 'computational' version of Gödel's theorem. Thus he appears to have regarded the inaccuracy of human mathematical thinking as essential, allowing the mind's (supposed) inaccurate algorithmic action to provide a greater power than that which would be achievable by means of any completely sound algorithmic procedure. Accordingly, he suggested a way of evading the conclusions of the Gödel argument: the mathematician's algorithm would be technically unsound, and it certainly would not be 'knowably sound'. Thus, Turing's viewpoint would be consistent with $\mathscr{G}$, and it seems probable that he would have agreed with viewpoint $\mathscr{A}$.

As part of the following discussion, I shall be presenting my reasons for disbelieving that 'unsoundness' in a mathematician's algorithm can be the *real* explanation for what is going on in a mathematician's mind. There is, in any case, a certain intrinsic implausibility in the idea that what makes the mind superior to an accurate computer is the mind's *in*accuracy—especially when we are concerned, as here, with the mathematician's ability to *perceive unassailable mathematical truth*, rather than with mathematical originality or creativity. It is a striking fact that each of these two great thinkers, Gödel and Turing, found himself driven, by considerations such as $\mathscr{G}$, to what many might regard as a somewhat implausible standpoint. It is interesting to speculate on whether they would have found themselves so driven had they been in a position to contemplate the serious possibility that physical action might, at root, sometimes be *non*-computable—in accordance with the viewpoint $\mathscr{C}$ that I am promoting here.

In the following sections (particularly §3.2–§3.22) I shall be giving some detailed arguments, some of which are rather complicated, confusing, or technical, which are aimed at ruling out either computational model $\mathscr{A}$ or $\mathscr{B}$ as providing a plausible basis for mathematical understanding. My recommendation to the reader who needs no such persuasion—or who is daunted by details—is to read as far as he or she feels inclined, but then when the going gets tedious, pass directly to the summarizing fantasy dialogue of §3.23. Please return to the main argument only if and when it is so desired.

3.2 Could an unsound algorithm knowably simulate mathematical understanding?

We must consider that, in accordance with $\mathscr{G}$, mathematical understanding might be the result of some algorithm that is unsound or unknowable, or possibly sound and knowable but not knowably sound—or there might perhaps be various different such algorithms corresponding to different mathematicians. By an 'algorithm' is meant just some computational procedure (as in §1.5)—that is, anything that could be simulated in principle on a general-purpose computer with an unlimited store. (We may recall from the discussion of **Q8**, in §2.6, that the 'unlimited' nature of this store, in this idealization, does not detract from the argument.) Now, this notion of 'algorithm' would include top-down procedures and bottom-up learning systems, as well as combinations of both. Anything that can be achieved by artificial neural networks, for example, would be included (cf. §1.5). So would other types of bottom-up mechanisms such as those referred to as 'genetic algorithms', which improve themselves by some built-up procedure analogous to Darwinian evolution (cf. §3.11).

In §3.9–§3.22 (as summarized in the fantasy dialogue of §3.23) I shall show specifically how bottom-up procedures are essentially encompassed by the arguments that are to be given in the present section (and also by those that were put forward in Chapter 2). However, for purposes of clarity, I shall for the moment phrase things as though there were just a single top-down type of algorithmic action involved. This algorithmic action might be thought of as relevant either to a particular individual mathematician or to the mathematical community as a whole. In the discussions of **Q11** and **Q12**, in §2.10, the possibility of there being *different* sound and known algorithms pertinent to different people was considered, and it was concluded that this possibility did not significantly affect the argument. The possibility of there being different *un*sound or *un*knowable algorithms pertinent to different people will be returned to later (cf. §3.7). For now, let us phrase things, primarily, as though there is just a single algorithmic procedure underlying mathematical understanding. We may also restrict attention merely to that body of mathematical understanding which can be used to establish Π_1-sentences (i.e. the specification of Turing-machine actions that do not terminate, cf. commentary on **Q10**). In what follows, it is sufficient if we interpret the phrase 'mathematical understanding' in this restricted context (see $\mathscr{G}$**, p. 100).

We must distinguish clearly between three distinct standpoints with regard to the knowability of a putative algorithmic procedure F underlying mathematical understanding, whether sound or not. For F might be:

I consciously knowable, where its role as the actual algorithm underlying mathematical understanding is also knowable;

II consciously knowable, but its role as the actual algorithm underlying mathematical understanding is unconscious and not knowable;

III unconscious and not knowable.

Let us consider the completely conscious case **I** first. Since the algorithm and its role are both knowable, we may just as well consider that it, and its role, are *already* known. For, in our imagination, we can suppose that we are applying our arguments at such a time when these things *are* actually known—for 'knowable' means that, at least in principle, such a time could arise. So, let us consider that the algorithm F is a known one, its underlying role being also known. We have seen (§2.9) that such an algorithm is effectively equivalent to a formal system $\mathbb{F}$. Thus, we are supposing that mathematical understanding—or at least any particular mathematician's understanding—is known (by that mathematician) to be equivalent to derivability within some known formal system $\mathbb{F}$. To have any hope of satisfying our conclusion $\mathscr{G}$, which was forced upon us by our considerations in the preceding chapter, we must suppose that the system $\mathbb{F}$ is *unsound*. Strangely enough, however, unsoundness does not help at all for a known formal system $\mathbb{F}$ which, as asserted in **I**, is actually *known*—and thus *believed*—by any mathematician to underlie his or her mathematical understanding! For such a belief entails a (mistaken) belief in $\mathbb{F}$'s soundness. (It would be an unreasonable mathematical standpoint that allows for a disbelief in the very basis of its own unassailable belief system!) Whether or not $\mathbb{F}$ is actually sound, a *belief* that it is sound entails a belief that $G(\mathbb{F})$ (or, alternatively, $\Omega(\mathbb{F})$, cf. §2.8) is true; but since $G(\mathbb{F})$ is now—by a belief in Gödel's theorem—believed to lie outside the scope of $\mathbb{F}$, this contradicts the belief that $\mathbb{F}$ underlies *all* (relevant) mathematical understanding. (This works just as well whether applied to individual mathematicians or to the mathematical community; it can be applied separately to any of the various different algorithms that might be supposed to underlie different mathematicians' thought processes. Moreover, it need refer only to that body of mathematical understanding that is pertinent to the establishment of Π_1-sentences.) Thus, any putative known unsound algorithm F that is supposed actually to underlie mathematical understanding cannot be *known* to do so, whence **I** is ruled out irrespective of whether $\mathbb{F}$ is or is not sound. If $\mathbb{F}$ itself is knowable, we are left with the possibility **II**: that $\mathbb{F}$ might indeed underlie mathematical understanding, but it cannot be known to have this role. There is also the possibility **III** that the system $\mathbb{F}$ might itself be unconscious and unknowable.

What we have essentially established, at this juncture, is that case **I** (at least in the context of entirely top-down algorithms) is not a serious possibility; the fact that $\mathbb{F}$ might actually be unsound has, somewhat remarkably, no importance to **I** whatever. The crucial thing is that the putative $\mathbb{F}$, whether sound or not, cannot be *known* to underlie mathematical belief; it is not the unknowability of the algorithm itself that is at issue, but the unknowability of that algorithm actually *being* the one that underlies understanding.

3.3 Could a knowable algorithm unknowably simulate mathematical understanding?

We come now to case **II**, and try to take seriously the possibility that mathematical understanding might indeed be actually equivalent to some consciously knowable algorithm or formal system, but unknowably so. Thus, although the putative formal system $\mathbb{F}$ would be a knowable one, we could never be sure that *this* particular system actually does underlie our mathematical understandings. Let us ask ourselves if this possibility is at all plausible.

If this putative $\mathbb{F}$ is not an *already* known formal system then, as before, we must take it that it could, at least in principle, be known someday. Imagine that such a day has arrived and suppose that the precise specification of $\mathbb{F}$ is set out before us. The formal system $\mathbb{F}$, though possibly quite elaborate, is supposed to be simple enough that we are able, at least in principle, to appreciate it in a perfectly conscious way; but we are not allowed to be *sure* that $\mathbb{F}$ actually encompasses precisely the totality of our unassailable mathematical understandings and insights (at least, with regard to Π_1-sentences). We shall try to see why this, though a logical possibility, is a very implausible one. Moreover, I shall later argue that even if it were true, such a possibility would provide no solace to practitioners of AI intent on building a robot mathematician! This aspect of things will be returned to at the end of this section, and will be addressed more fully in §3.15 and §3.29.

To emphasize the fact that the existence of such an $\mathbb{F}$ must indeed be considered to be a *logical* possibility, we may recall the 'theorem-proving machine' that Gödel remarked could not (as yet) be logically ruled out (cf. quote given in §3.1). In effect, this 'machine' would, as I shall explain, be an algorithmic procedure F in accordance with one of the above cases **II** or **III**. Gödel's putative theorem-proving machine might, as he pointed out, be 'empirically discoverable', which corresponds to the requirement that F be 'consciously knowable' as in **II**, or it might not be, which is basically case **III**.

Gödel had argued, taking account of his famous theorem, that the procedure F (or, equivalently, formal system $\mathbb{F}$; cf. §2.9) could not be 'proved' to be 'equivalent to mathematical intuition'—as his quotation indeed asserts. In **II** (and, by implication, in **III** also) I have stated this fundamental limitation on F in a somewhat different way: 'its role as the actual algorithm underlying mathematical understanding is unconscious and not knowable'.

This limitation (whose necessity follows from the rejection of **I**, as was argued in §3.2) clearly does have the implication that F canot be shown to be equivalent to mathematical intuition, since such a demonstration would establish for us that *F* indeed has the role that we are supposed not to be able to know it has. Conversely, if this very role of *F*, in providing the basis for unassailable mathematical understanding, were something that *could* be consciously known—in the sense of it being fully appreciated how *F* acts in this role—then the soundness of *F* would have to be accepted also. For if *F* were *not* being fully accepted as sound, this amounts to a refusal to accept some of its consequences. But its consequences are precisely those mathematical propositions (or, at least, Π_1-sentences) which *are* being accepted. Thus, knowing the role of *F* amounts to possessing a *proof* of *F*, although such a 'proof' would not be a formal proof in some preassigned formal system.

We note, also, that valid Π_1-sentences might be regarded as examples of the 'correct theorems of finitary number theory' that Gödel referred to. In fact, if the term 'finitary number theory' were taken to include the μ-operation 'find the smallest natural number with such-and-such arithmetical property', in which case it includes the actions of Turing machines (see end of §2.8), then *all* Π_1-sentences would count as being part of finitary number theory. Thus, it would appear that Gödel-type reasoning provides no clear way of ruling out case **II** on rigorous logical grounds alone—at least if we accept the authority of Gödel!

On the other hand, we may ask ourselves whether **II** is at all a *plausible* possibility. Let us try to see what the existence of a knowable $\mathbb{F}$, unknowably equivalent to human (unassailable) mathematical understanding, would entail. As remarked above, we may imagine that a time has arrived at which this $\mathbb{F}$ has been encountered, and is set out before us. Recall (§2.7) that a formal system is specified in terms of a set of *axioms* and *rules of procedure*. The *theorems* of $\mathbb{F}$ are things ('propositions') that can be obtained from the axioms by means of the rules of procedure, all theorems being things that can be formulated in terms of the same symbols that are used in order to express the axioms. What are we trying to imagine is that the theorems of $\mathbb{F}$ are precisely those propositions (written in terms of these symbols) which can *in principle* be perceived as unassailably true, by human mathematicians.

Let us suppose, for the moment, that the list of axioms in $\mathbb{F}$ is a *finite* list. Now, the axioms themselves always have to count as particular instances of theorems. But each theorem is something that can in principle be perceived as unassailably true by use of human understanding and insight. Hence, each *axiom individually* must express something that is, in principle, part of this mathematical understanding. Thus, for any individual axiom, there will come a time (or there could *in principle* come a time) when it is perceived as being unassailably true. One after the other, individual axioms of $\mathbb{F}$ could be so perceived. Thus, eventually, *all* the axioms will be individually perceived as unassailably true (or could in principle eventually be so perceived). Accordingly, there eventually comes a time when the totality of the axioms of $\mathbb{F}$ is perceived, as a whole, to be unassailably valid.

What about the rules of procedure? Can we envisage a time when these, also, are perceived as unassailably sound? For many formal systems, these could just be simple things which are all 'unassailably' acceptable, like: 'If we have already established P as a theorem, and also P⇒Q as a theorem, then we may deduce Q as a new theorem' (cf. ENM p. 393, or Kleene 1952, with regard to the symbol '⇒', meaning 'implies'). For such rules, there would be no difficulty about accepting them as unassailable. On the other hand, there might be some much more subtle means of inference, included among the rules of procedure, whose validity is not at all obvious, and which might require delicate consideration before one could come to a decision as to whether or not to accept such a rule as 'unassailably sound'. In fact, as we shall see in a moment, there must be amongst 𝔽's rules of procedure certain rules whose unassailable soundness can *not* be perceived by human mathematicians—where we are still assuming that the axioms of 𝔽 are finite in number.

Why is this? Let us, in our imaginations, return to the time when the axioms are all perceived as unassailably true. We sit back to contemplate the entire system 𝔽. Let us try to suppose that the rules of procedure of 𝔽 are also things that we can now accept unassailably. Although we are not supposed to be able to know that 𝔽 actually encompasses *everything* mathematical that is in principle accessible to human understanding and insight, we should, by now, at least have convinced ourselves that 𝔽 is unassailably sound, since both its axioms and rules of procedure are unassailably accepted. We must therefore also be now convinced that 𝔽 is *consistent*. The thought would occur to us, of course, that $G(\mathbb{F})$ must also be true, by virtue of this consistency—indeed *unassailably* true! But since 𝔽 is being supposed, in fact (but unknown to us), to encompass the entirety of what is unassailably accessible by us, $G(\mathbb{F})$ must actually be a theorem of 𝔽. But, by Gödel's theorem, this can only be so if 𝔽 is in fact *in*consistent. If 𝔽 is inconsistent, then '$1=2$' is a theorem of 𝔽. Hence the assertion that $1=2$ would have to be, in principle, part of our unassailable mathematical understanding—which is surely a clear contradiction!

This notwithstanding, we must at least entertain the *possibility* that human mathematicians might (unknowably) operate according to an 𝔽 which is actually *un*sound. I shall address this issue in §3.4; but for the purposes of the present section, let us accept that the procedures which underlie mathematical understanding are actually perfectly sound. In such circumstances it follows that a contradiction has indeed been arrived at, if we are to assume that the rules of procedure of our finite-axiomed 𝔽 are all unassailably acceptable. Hence there must be, amongst 𝔽's rules of procedure, at least one that cannot be unassailably perceived as sound by human mathematicians (though it is indeed sound).

All this was on the basis that 𝔽 possesses only a finite number of axioms. A possible alternative let-out would be that the list of axioms of 𝔽 might be *infinite*. I should make a comment concerning this possibility. In order for 𝔽 to qualify as a formal system in the required sense—so that one may always check, by means of a preassigned computational procedure, that a purported proof of some proposition is indeed a proof according to $\mathbb{F}$'s rules—it is necessary that its infinite axiom system be expressible in finitely based terms. In fact, there is always some freedom about the way in which a formal system is represented, whereby its operations are designated as either 'axioms' or 'rules of procedure'. In fact, the standard axiomatic system for set theory—the Zermelo–Fraenkel system (that I am denoting here by $\mathbb{ZF}$)—possesses infinitely many axioms, expressed in terms of structures called 'axiom schemata'. By an appropriate reformulation, the $\mathbb{ZF}$ system can be re-expressed so that the number of actual axioms becomes finite[3]. In fact, in a certain sense, this can always be done for axiom systems that are 'formal systems' in the computational sense required here.*

One might imagine, therefore, that one could apply the above argument, aimed towards the ruling out of case **II**, to *any* (sound) $\mathbb{F}$, irrespective of whether its axioms are infinite or finite in number. Indeed, this is so, but in the process of reducing the axiom system from an infinite to a finite one, one may introduce new rules of procedure that might not be self-evidently sound. Thus, when in accordance with the ideas described above we contemplate a time at which the axioms and rules of procedure of $\mathbb{F}$ are all laid out before us, the theorems of this putative $\mathbb{F}$ being supposed to be precisely those which are in principle accessible to human understanding and insight, we cannot be sure that the rules of procedure of this $\mathbb{F}$, unlike its axioms, can ever be unassailably perceived as sound, even though they are in fact sound. For, unlike axioms, rules of procedure do not number amongst the theorems of a formal system. It is merely the *theorems* of $\mathbb{F}$ that are taken to be the things that can be unassailably perceived.

It is not clear to me whether this argument can be taken much further, in strictly logical terms. What we have to be able to accept, if we believe in **II**, is that there is some formal system $\mathbb{F}$ (underlying human perceptions as to the truth of Π_1-sentences) that can be perfectly well appreciated by human mathematicians, whose finite list of axioms is (unassailably) acceptable, but whose finite system of rules of procedure $\mathcal{R}$ contains at least one operation that is regarded as fundamentally dubious. All the theorems of $\mathbb{F}$ would have to turn out, individually, to be things that can be perceived as true—somewhat miraculously, since many of them would be obtained by use of the dubious rules of $\mathcal{R}$. Now, although each of these theorems can *individually* be perceived as true (in principle) by human mathematicians, there is no *uniform* way of doing this. We may restrict attention to those theorems of $\mathbb{F}$ which are Π_1-sentences. By use of the dubious $\mathcal{R}$, the entire list of Π_1-sentences that can be perceived as true by human mathematicians can be computationally generated. Individually, any one of these Π_1-sentences can be ultimately perceived as true by human insight. But, in each case, this is by using some

*There is a rather trivial sense in which this can be done, where we simply interpret the operation of the Turing machine which effects the algorithm F appropriately as the rules of procedure of the required system.

means of reasoning that it quite distinct from the rule $\mathscr{R}$ by which it is obtained. Again and again, some new and increasingly sophisticated human insight must be brought in, so that each Π_1-sentence can be reduced to unassailable truth. As if by magic, each of these Π_1-sentences turns out to be true, but some of them can only be perceived as true by bringing in a fundamentally new type of reasoning, a requirement that comes in again and again at deeper and deeper levels. Moreover, *any* Π_1-sentence that can be perceived as unassailably true—by whatever means—will be found in the list generated by $\mathscr{R}$. Finally, there will be a specific true Π_1-sentence $G(\mathbb{F})$ that can be constructed explicitly from a knowledge of the system $\mathbb{F}$, but whose truth *cannot* be perceived as unassailably true by human mathematicians. The best that they could do would be to see that the truth of $G(\mathbb{F})$ depends precisely upon the soundness of the dubious procedure $\mathscr{R}$ which seems miraculously to be able to generate precisely all the Π_1-sentences that *can* be unassailably humanly perceived.

I can imagine that some people might regard this as perhaps not *altogether* unreasonable. There are many instances of mathematical results which can be obtained by means of what might be called 'heuristic principles', where such a principle would not supply a *proof* of the required result, but it would lead one to anticipate that the result indeed ought to be true. Subsequently, a proof might be obtained along some quite different lines. However, it seems to me that such heuristic principles really have rather little in common with our putative $\mathscr{R}$. Such principles are things that actually enhance our conscious understandings of *why* some mathematical results are actually true.* Later on, when mathematical techniques become further developed, it often becomes fully understood why such a heuristic principle works. But more usually, what would become fully understood would be the *circumstances* under which the principle can be trusted to work, there being others where it cannot be so trusted—so that erroneous conclusions are obtained when due care is not exercised. With the exercise of such care, the very principle itself becomes a powerful and trustworthy tool of unassailable mathematical proof. Rather than providing us with a miraculously reliable algorithmic process for establishing Π_1-sentences, where the reason that the algorithm works is inaccessible to human insight, heuristic principles provide a means of enhancing our mathematical insights and understandings. This is something very different from the algorithm F (or formal system $\mathbb{F}$) that is needed for case **II**. Moreover, there has never been any proposal for a heuristic principle which

*A heuristic principle of this kind might take the form of a *conjecture*, like the important Taiyama conjecture (which became generalized to what is known as the 'Langland's philosophy'), from which that most famous of all Π_1-sentences, known as 'Fermat's last theorum' (cf. footnote on p. 198), could be derived as a consequence. However, the argument presented by Andrew Wiles as a proof of Fermat's assertion was not an argument independent of the Taiyama conjecture – which it would have to have been had that conjecture been an '$\mathscr{R}$'—but an argument for *proving* (in the relevant case) the Taiyama conjecture itself!

could generate precisely *all* the Π_1-sentences that can be perceived as true by human mathematicians.

Of course, none of this tells us that such an F—Gödel's putative 'theorem-proving machine'—is an impossibility; but from the point of view of our mathematical understandings, its existence seems very unlikely. In any case, there is, at present, not the slightest suggestion as to the nature of a plausible such F, nor is there any hint whatsoever of its existence. It could only be a *conjecture*, at best—and an unprovable one at that. (Proving it would contradict it!) It seems to me that it would be rash in the extreme for any AI supporter (whether $\mathscr{A}$ or $\mathscr{B}$) to pin hopes on finding such an algorithmic procedure,* as embodied by F, whose very existence is dubious in the extreme and, in any case, if it were to exist its definite construction would be beyond the wit of any of today's mathematicians or logicians.

Is it conceivable such an F could exist nevertheless, and might actually be arrived at by means of sufficiently elaborate bottom-up computational procedures? In §3.5–§3.23, as part of the discussion of case **III**, I shall present a strong logical argument showing that no knowable bottom-up procedures could ever find such an F even if it did exist. We thus come to the conclusion that even 'Gödel's theorem-proving machine' is not a serious logical possibility unless there are 'unknowable mechanisms' underlying mathematical understanding as a whole that are of a nature that would give no comfort to AI proponents!

Before turning to this more general discussion of case **III**, and of bottom-up procedures in general, we must complete the arguments that are specific to case **II** itself; for there still remains the alternative that the underlying algorithmic action F—or formal system $\mathbb{F}$—might be *unsound* (a let-out that did not apply to case **I**). Might it be that human mathematical understanding is equivalent to a knowable algorithm that is fundamentally erroneous? Let us consider this possibility next.

3.4 Do mathematicians unwittingly use an unsound algorithm?

Perhaps there *is* an unsound formal system $\mathbb{F}$ that underlies our mathematical understanding. How can we be sure that our mathematical

* Of course, it might well be argued that the building of a robot mathematician is very far from the immediate aims of artificial intelligence; accordingly, the finding of such an F would be regarded as premature or unnecessary. However, this would be to miss the point of the present discussion. Those viewpoints which take human intelligence to be explicable in terms of algorithmic processes implicitly demand the potential presence of such an F—knowable or unknowable—because it is merely the application of intelligence that has led us to our conclusions. There is nothing special about mathematical abilities in this regard; see §1.18 and §1.19, in particular.

perceptions as to what is unassailably true may not someday lead us fundamentally astray? Perhaps they have even done so already. This is not quite the same situation as we considered in relation to **I**, in which the possibility was ruled out that we could *know* that some system $\mathbb{F}$ actually played such a role. Here we are allowing that this *role* for $\mathbb{F}$ is unknowable, and so we must reconsider that $\mathbb{F}$ might actually be an unsound system. But is it really plausible that our unassailable mathematical beliefs might rest on an unsound system—so unsound, indeed, that '1 = 2' is in principle part of those beliefs? Surely, if our mathematical reasoning cannot be trusted, then *none* of our reasoning about the workings of the world can be trusted. For mathematical reasoning forms an essential part of all of our scientific understanding.

Some will argue, nevertheless, that it is surely *not inconceivable* that our accepted mathematical reasoning (or what we may come to accept in the future as 'unassailable') contains some hidden inbuilt contradiction. Such people will be likely to refer to the famous paradox (about "the set of sets that are not members of themselves") that Bertrand Russell pointed out in a letter to Gottlob Frege, in 1902, just as Frege was about to publish his life's work on the foundations of mathematics (see also response to **Q9**, §2.7, and ENM, p. 100). Frege added, in an appendix (cf. Frege 1964; translated):

> Hardly anything more unwelcome can befall a scientific writer than that one of the foundations of his edifice be shaken after the work is finished. I have been placed in this position by a letter from Mr Bertrand Russell . . .

Of course, we could simply say that Frege made a mistake. It is admitted that mathematicians do make mistakes from time to time—sometimes serious ones. Moreover, Frege's error was a *correctable* mistake, as Frege's own admission makes clear. Have I not argued (in §2.10, cf. commentary on **Q13**) that such correctable mistakes are not our concern? As in §2.10, we are indeed concerned only with matters of principle and not with the fallibility of individual mathematicians. Surely errors that can be pointed out and demonstrated as errors are not the sort of thing that should concern us? However, the situation is somewhat different here, from that which was addressed in relation to **Q13**, since we are now concerned with a formal system $\mathbb{F}$ that we do not *know* underlies mathematical understanding. As before, we are not concerned with individual mistakes—or 'slips'—that a mathematician might happen to make while reasoning within a consistent overall scheme. But we are now addressing a situation in which the scheme itself might be subject to an overall contradiction. This is just what happened in Frege's case. Had Russell's paradox, or another paradox of a similar nature, not been pointed out to Frege, then, as seems likely, he surely would *not* have been persuaded that his scheme contained any fundamental error. It was not a matter of Russell pointing out some technical slip in Frege's reasoning which Frege would have admitted was an error according to his own canons of reasoning; it was that those very canons were shown to contain an inbuilt contradiction. It was this *contradiction* that persuaded Frege that there was an error—and what might previously have seemed to Frege to be unassailable reasoning was now seen to be fundamentally flawed. But the flaw was perceived only because the contradiction itself was brought to light. Without the contradiction having been perceived, the methods of reasoning might well have been trusted and perhaps followed by mathematicians for a good long while.

In fact I should say that, in this case, it is hardly likely that many mathematicians would have, for long, allowed themselves the freedom of reasoning (with infinite sets) that Frege's scheme permitted. But this is because paradoxes of the Russell type would all too easily have come to light. One might imagine some much more subtle paradox, even lying implicit in what we believe to be unassailable mathematical procedures that we allow ourselves today—a paradox which might not come to light for centuries to come. Only when such a paradox finally manifests itself would we feel the necessity to change our rules. The argument would be that our mathematical insights are governed not by timeless considerations, but that these insights are shifting things strongly influenced by what seems to have worked well *so far* and on what, in effect, 'we can get away with'. On this view, it would be allowed that there might be an algorithm or formal system underlying present-day mathematical understanding, but that this algorithm is not a fixed thing, being continually subject to change, as new information comes to light. I shall need to return to the issue of changing algorithms later on (cf. §3.9–§3.11, also §1.5). We shall see that such things are really just algorithms again, in another guise.

It would, of course, be naïve of me not to admit that there are often elements of 'trusting a procedure if it seems to have worked so far' in the way that mathematicians operate in practice. In my own mathematical reasoning, for example, such slipshod or tentative practices constitute a definite ingredient of mathematical thinking. But they tend to form part of what is important in one's groping towards a previously unformed understanding, not a part of what one takes to have been unassailably established. I feel doubtful that Frege himself would have been completely dogmatic that his scheme must be unassailable, even without Russell's paradox having been pointed out to him. Such a general scheme of reasoning would always have to be put forward somewhat tentatively in any case. It would take a good deal of further 'mulling over' before it could be believed to have reached the 'unassailable' level. For a scheme of the generality of Frege's, it seems to me, one would in any case be driven to making statements of the form 'assuming Frege's scheme is sound, then such-and-such', rather than merely asserting 'such-and-such' without such qualification. (See comments on **Q11, Q12**.)

Perhaps mathematicians have now become more cautious as to what they are prepared to regard as 'unassailably true'—after a period of excessive boldness (of which Frege's work indeed formed an important part) at the end of the nineteenth century. Now that paradoxes such as Russell's have come to light, the importance of such caution is made particularly manifest. The boldness arose, to a large extent, when the power of Cantor's theory of infinite numbers and infinite sets, that he had put forward in the late 1800s, began to become clear. (It should be said, however, that Cantor himself was well aware of the problem of Russell-type paradoxes—long before Russell had come across them[4]—and had tried to formulate a sophisticated point of view which took such things very much into account.) For the purposes of the discussions that I have been concerned with here, an extreme degree of caution is certainly appropriate. I am happy that only those things whose truth is indeed unassailable should be included in the discussion, and that anything concerning infinite sets that is at all questionable should not be so included. The essential point is that *wherever* the line is drawn, Gödel's argument produces statements that remain within the compass of what is indeed unassailable (cf. commentary on **Q13**). Gödel's (and Turing's) argument does not, on its own, involve any issue of the questionable existence of certain infinite sets. Doubtful issues in relation to the kind of very free reasoning that Cantor, Frege, and Russell were concerned with need not concern us so long as they remain 'doubtful' as opposed to 'unassailable'. This being accepted, I cannot really see that it is plausible that mathematicians are *really* using an *unsound* formal system $\mathbb{F}$ as the basis of their mathematical understandings and beliefs. I hope the reader will indeed agree with me that whether or not such a consideration is *possible*, it is certainly not at all *plausible*.

Finally, in relation to the possibility that our putative $\mathbb{F}$ might be an unsound system, we should briefly recall the other aspects of human inaccuracy that were already discussed under **Q12** and **Q13** above. I should first reiterate that we are here concerned *not* with the inspirations, guesses, and heuristic criteria that may guide mathematicians towards their new mathematical discoveries, but with the understandings and insights that provide them with the basis of what they take to be their unassailable beliefs with regard to mathematical truth. These beliefs can be the results of merely following other people's arguments, and no elements of mathematical discovery need be involved. In feeling one's way towards original discoveries, it is indeed important to allow speculation to roam freely, unconstrained by an initial need for total reliability and accuracy (and it is my impression that this is what Turing was mainly concerned with in the quotation above, §3.1). But when it comes to the point of whether to accept or to reject arguments in support of a proposed mathematical statement being actually unassailably true, it is necessary that we be concerned with mathematical understandings and insights—often aided by lengthy computation—that are free from error.

This is not to say that mathematicians do not frequently make errors in believing that they have correctly applied their understandings when in fact they have not. Mathematicians certainly do make mistakes in their reasoning and in their understandings, as well as in their attendant computations. But their tendency to make such errors does not basically *increase* their power of understanding (although I can imagine that an occasional flash of understanding might come about by such fortuitous means). More importantly, these errors are *correctable*; when such errors are pointed out, either by some other mathematician or by the same mathematician at a later time, the errors are *recognizable* as errors. It is not as though there is some inbuilt erroneous formal system $\mathbb{F}$ controlling the mathematician's understandings, since this would be incapable of recognizing its own errors. (The possibility of having a self-improving system that changes itself whenever it finds an inconsistency would be included in the discussion leading up to §3.14. Accordingly, we find that this kind of proposal does not really help; cf. also §3.26.)

A slightly different kind of error arises when a mathematical statement is formulated incorrectly; the mathematician proposing the result might really *mean* something a little different from what has literally been proposed. Again, this is something correctable, and it is not the kind of *inbuilt* error that would result from an unsound $\mathbb{F}$ underlying all human insights. (Recall Feynman's dictum, referred to in relation to **Q13**: 'Don't listen to what I say; listen to what I mean!') All the time, we are concerned with what can *in principle* be ascertained by a (human) mathematician, and, for this, errors of the type that we have just been considering here—i.e. correctable errors—are not relevant. Most importantly for this entire discussion, it is the central idea of the Gödel–Turing arugment which has to be part of what a mathematician can understand, and it is this which forces us to reject **I** and to regard **II** as extremely implausible. As noted above, in the discussion under **Q13**, the *idea* of the Gödel–Turing argument certainly ought to be part of what a mathematician can in principle understand, even though some specific statement '$G(\mathbb{F})$' that the mathematician might happen to have settled upon might be in error—for *correctable* reasons.

There are other issues that need to be addressed, in relation to the possibility that 'unsound' algorithms might underlie mathematical understanding. These concern 'bottom-up' procedures, such as self-improving algorithms, learning algorithms (including artificial neural networks), algorithms with additional random ingredients, and algorithms whose actions depend upon the external environment in which the algorithmic devices are placed. Some of these issues have been touched upon earlier (cf. commentary under **Q2**) and they will be addressed at length as part of our discussion of case **III**, that we must come to next.

3.5 Can an algorithm be unknowable?

According to **III**, mathematical understanding would be the result of some unknowable algorithm. What does it actually mean for an algorithm to be 'unknowable'? In the previous sections in this chapter, we have been concerned with matters of *principle*. Thus, an assertion that the unassailable truth of some Π_1-sentence is accessible to human mathematical understanding would be an assertion that this Π_1-sentence is *in principle* accessible, not that any human mathematician need necessarily actually ever encounter a demonstration of it. But for an *algorithm* to be unknowable, a somewhat different interpretation of the term 'unknowable' is needed. Here I shall mean that the algorithm is something whose very specification is beyond what can be achieved *in practice*.

When we were concerned with derivations within a specific knowable formal system, or with what is achievable by use of some known algorithm, then it was indeed appropriate that we should have been concerned with what can or cannot be achieved in principle. The issue is whether or not some particular proposition can be derived from such a formal system or algorithm was *necessarily* taken in an 'in principle' sense. We can compare this situation with that of the truth of Π_1-sentences. A Π_1-sentence is, after all, taken to be *true* if it represents a Turing-machine action that does not terminate in principle, irrespective of what might be possible to achieve in practice by direct computation. (This is in accordance with the discussion of **Q8**.) Likewise, an assertion that some specific proposition is, or is not, derivable within some formal system must be taken in the 'in principle' sense, such an assertion being itself of the form of an assertion that some particular Π_1-sentence is false or true, respectively (cf. end of discussion of **Q10**). Accordingly, when we are concerned with derivability within some fixed formal set of rules, 'knowability' will always be taken in this 'in principle' sense.

On the other hand, when we are concerned with the issue of whether the rules themselves might be 'knowable', this must be taken in the 'in practice' sense. *Any* formal system, Turing machine, or Π_1-sentence can be specified in principle, so the issue of 'unknowability' here must, if it is to have any meaning, be concerned with what can or cannot be accessed in practice. Any algorithm whatever is, indeed, in principle knowable—in the sense that the Turing-machine action that effects that algorithm is 'known' as soon as the natural number that codes that action is known (e.g. in terms of the specification of Turing-machine numbers as given in ENM). There is no real suggestion that a natural number could be unknowable in principle. Natural numbers (and thus algorithmic actions) can be listed 0, 1, 2, 3, 4, 5, 6, . . . , and thus any specific natural number will eventually be encountered—*in principle*—no matter how large that number might be! In practice, however, there will be numbers that are so large that there is no prospect of their ever being encountered in this way. For example, the number of the universal Turing machine given in ENM, p. 56, is far too large for it to be encountered in such an enumeration in practice. Even if the digits could be produced one after the other in as small a time interval as is theoretically definable (the Planck timescale of about 0.5×10^{-43} seconds—cf. §6.11), then no number whose binary representation has more than about 203 digits would have been encountered, as yet, in the entire time-span of the universe starting from the big bang. The number just referred to has over 20 times as many digits as this—yet this in itself does not stop it from being 'knowable' in practice, this number being presented in ENM explicitly.

For a natural number or Turing-machine action to be 'unknowable' in practice, we must imagine that even the very specification of that number, or action, would be so complicated that it lies beyond human abilities. This perhaps seems a tall order, but one might argue from the finiteness of human beings, that there must at least be *some* limit that puts certain numbers beyond human specification. (See the related discussion in response to **Q8**.) We must imagine that, in accordance with **III**, it is the vast complication of all the minute details of the specification of the algorithm F being supposed to underlie mathematical understanding that puts it beyond human knowability—in the sense of *specifiability*, as opposed to its being knowably the algorithm we are supposed actually to use. It is this requirement of non-specifiability that separates **III** from **II.** Thus, in our consideration of **III**, we must entertain the possibility that it would be beyond human capabilities even to specify the number in question, let alone to know that that number has the qualities that are required of it as a number that determines the algorithmic action underlying human mathematical understanding.

It should be made clear that mere size cannot be the limiting factor. It is very easy to specify numbers that are so large that they *exceed* the sizes of numbers that might be needed to specify algorithmic actions of relevance to the behaviour of any organism in the observable universe (e.g. the easily specified number $2^{2^{65536}}$, that featured in the response to **Q8**, vastly exceeds the number of possible different universe states for all the material that lies within our observable universe[5]). It would have to be the *precise* specification of the number required, not its mere size, that would be beyond human capabilities.

Let us suppose that, in accordance with **III**, the specification of such an F is indeed beyond human capabilities. What would this tell us about the prospect of a fully successful AI strategy (according to either 'strong' or 'weak' AI—the respective viewpoint of $\mathscr{A}$ or $\mathscr{B}$)? It would be anticipated by believers in computer-controlled AI systems (certainly under viewpoint $\mathscr{A}$, and perhaps also under $\mathscr{B}$) that the robot creations that can eventually arise as a result of this strategy should be able to attain and perhaps surpass human mathematical capabilities. Consequently, it would have to be the case, if we accept **III**, that some such humanly unspecifiable algorithm F would form part of the control system of such a mathematical robot. This would seem to imply that an AI strategy of such eventual scope is unattainable. For if it needs an unspecifiable F in order to achieve its aims, there would be no prospect of human beings ever setting it in action.

But this is not the picture presented by the most ambitious of AI supporters. They might envisage that the needed F would not arise immediately, but would be built up in stages, where the robots themselves would gradually improve their performances through their (bottom-up) learning experiences. Moreover, the most advanced robots would not be the direct creations of human beings, but would more likely have been created by other robots[6], perhaps somewhat more primitive than the required mathematical robots; and there could also be some kind of Darwinian evolution in operation, serving to improve the capabilities of the robots from generation to generation. Indeed, it would be argued that it was through processes of this general kind that we *ourselves* were able to acquire, as ingredients of our own 'neural computers', some humanly unknowable algorithm F controlling our own mathematical understandings.

In the next several sections, I shall be arguing that processes of this nature do not really escape the problem: if the very procedures whereby an AI strategy would be set up in the first place are algorithmic and knowable, then any such resulting F should also be knowable. In this way, case **III** will be reduced either to **I** or to **II**, the cases that were ruled out, in §3.2–§3.4, as being effectively impossible (case **I**), or at least highly implausible (case **II**). In fact, it is really case **I** that we shall find ourselves driven to, on the assumption that such underlying algorithmic procedures are knowable. Accordingly, case **III** (and, by implication, case **II** also) will be itself rendered effectively untenable.

Any reader who pins faith in the possibility of **III** as providing a likely route to a computational model of the mind will do well to pay these arguments their due attention and follow them through with great care. The conclusion will be that if **III** is indeed to be taken as providing the basis for our own mathematical understandings, then the only plausible way that our own F could have arisen would have been by divine intervention—basically the $\mathscr{A}/\mathscr{D}$ possibility mentioned at the end of §1.3—and this would clearly be no consolation to those concerned with the more ambitious long-term aims of computer-driven AI!

3.6 Natural selection or an act of God?

But perhaps we should seriously consider the possibility that our intelligence might indeed require some kind of act of God—and that it cannot be explained in terms of that science which has become so successful in the description of the inanimate world. Certainly we should continue to maintain an open mind; but I should make it clear that in the discussions which follow, I am holding to a scientific viewpoint. I shall address the possibility that our mathematical understandings might result from some unfathomable algorithm—and the question of how such an algorithm might actually have arisen—entirely in such scientific terms. Possibly there are some readers who are inclined to believe that such an algorithm could indeed simply have been implanted into our brains according to some divine act of God. To such a suggestion I can offer no decisive refutation; but if one chooses to abandon the methods of science at some point, it is unclear to me why it would be regarded as reasonable to choose that particular point! If scientific explanation is to be abandoned, would it not be more appropriate to free the soul altogether from algorithmic action, rather than hiding its supposed free will in the complication and unfathomability of an algorithm that is presumed to control its every action? Indeed, it might seem more reasonable simply to adopt the view, as appears to have been held by Gödel himself, that the mind's action is something beyond the action of the physical brain—which is in line with viewpoint $\mathscr{D}$. On the other hand, I imagine that, nowadays, even those who hold that in some sense our mentality is indeed a divine gift might nevertheless tend to take the view that our behaviour can be understood within the scope of scientific possibility. No doubt these cases are arguable—but I do not propose to argue against $\mathscr{D}$ at this point. I hope that those readers who hold to some form of viewpoint $\mathscr{D}$ will continue to bear with me, and try to see what the scientific argument can do for us.

What, then, are the scientific implications of an assumption that mathematical conclusions are arrived at by the result of some necessarily unfathomable algorithmic action? Roughly, the picture would have to be that the exceptionally complicated algorithmic procedures that are needed for simulating genuine mathematical comprehension have resulted from a good many hundreds of thousands of years (at least) of natural selection, together with several thousands of years of educational tradition and input from the physical environment. The inherited aspects of these procedures would presumably have gradually built themselves up from simpler (earlier) algorithmic ingredients, as a result of the same kinds of selective pressures that have produced all the other superbly effective bits of machinery that constitute our bodies as well as our brains. The inbuilt potential mathematical algorithms (i.e. whatever inherited aspects to our mathematical thinking—presumed algorithmic—that there might be) would somehow lie coded within the DNA, as particular features of its sequences of nucleotides, and they would have arisen as a result of that same procedure whereby improvements gradually or intermittently arise in response to selective pressures. In addition, there would be external influences of various kinds, such as direct mathematical education, and experience coming from our physical environment, as well as other factors providing as additional purely random input. We must try to discover whether such a picture can be seen as at all plausible.

3.7 One algorithm or many?

One important issue that we must address is the possibility that there might be numerous quite different, perhaps inequivalent, algorithms that are responsible for the different modes of mathematical understanding that pertain to different individuals. Indeed, one thing is certainly clear from the start, and that is that even amongst practising mathematicians, different individuals often perceive mathematics in quite different ways from one another. To some, visual images are supremely important, whereas to others, it might be precise logical structure, subtle conceptual argument, or perhaps detailed analytic reasoning, or plain algebraic manipulation. In connection with this, it is worth remarking that, for example, geometrical and analytical thinking are believed to take place largely on opposite sides—right and left, respectively—of the brain[7]. Yet the same mathematical truth may often be perceived in either of these ways. On the algorithmic view, it might seem, at first, that there should be a profound inequivalence between the different mathematical algorithms that each individual might possess. But, despite the very differing images that different mathematicians (or other people) may form in order to understand or to communicate mathematical ideas, a very striking fact about mathematicians' perceptions is that when they finally settle upon what they believe to be unassailably true, mathematicians will not disagree, except in such circumstances when a disagreement can be traced to an actual recognizable (correctable) error in one or the other's reasoning—or possibly to their having differences with respect to a very small number of fundamental issues; cf. **Q11**, particularly $\mathscr{G}$***. For convenience of description, here, I shall ignore this latter issue in the following discussion. Although it does have some relevance, it does not substantially affect the conclusions. (Having just a very small number of inequivalent possible viewpoints does not, for the purposes of my argument, substantially differ from having just one.)

The perceiving of mathematical truth can be achieved in very many different ways. There can be little doubt that whatever detailed physical activity it is that takes place when a person perceives the truth of some mathematical statement, this physical activity must differ very substantially from individual to individual, even though they are perceiving precisely the same mathematical truth. Thus, if mathematicians just use computational algorithms to form their unassailable mathematical truth judgements, these very algorithms are likely to differ in their detailed construction, from individual to individual. Yet, in some clear sense, the algorithms would have to be *equivalent* to one another.

This may not be such an unreasonable thing as it may seem at first, at least from the point of view of what is mathematically *possible*. Very different-looking Turing machines can have identical outputs. (For example, consider the Turing machine constructed as follows: when acting on the natural number n, it outputs 0 whenever n is expressible as the sum of four square numbers, and it outputs 1 whenever n is not so expressible. This machine's output is identical to that of another machine, constructed simply to output 0 *whatever* number n it acts on—because it happens to be the case that *every* natural number is the sum of four squares; see §2.3.) Two algorithms need not be at all similar with regard to their internal operations and yet they can be identical as to their eventual external effects. However, in a certain sense, this actually makes it *more* puzzling how our putative unfathomable algorithm(s) for ascertaining mathematical truth might have arisen, for now we need many such algorithms, all quite distinct from one another in their detailed constructions, yet all essentially equivalent as regards their outputs.

3.8 Natural selection of unworldly esoteric mathematicians

What about the role of natural selection? Is it possible that there is some algorithm F (or perhaps several such) controlling all our mathematical understandings, which is unknowable (according to **III**—or at least whose role is unknowable (according to **II**)? Let me start by reiterating a point that was made at the beginning of §3.1. Mathematicians do not *think* that they are just following a set of unknowable rules—rules so involved they are mathematically unfathomable in principle—when they come to what they regard as their unassailable mathematical conclusions. They believe that these conclusions are, on the contrary, the result of arguments, albeit often long and tortuous, that rest ultimately on clear unassailable truths that could be appreciated, in principle, by anybody.

In fact, at the level of common-sense or logical descriptions, what they believe they are doing *is* indeed what they are doing. This should not be genuinely doubted; it is a point that cannot be emphasized too strongly. If it is to be maintained that mathematicians are following a set of unknowable or unfathomable computational rules, in accordance with **III** or **II**, then this has to be something they they are *also* doing—concurrently with what they think they are doing, but at a different level of description. Somehow, their algorithmic following of these rules would have to have the same *effects* that mathematical understandings and insights lead to—at least in practice. What we have to try to believe, if we adhere to either of the viewpoints $\mathscr{A}$ or $\mathscr{B}$, is that this possibility is a genuinely plausible one.

We must bear in mind what these algorithms must achieve for us. They must provide their possessors with the capabilities—in principle at least—for correctly following mathematical reasoning about abstract entities very far removed from direct experience and, for the most part, leading to no discernible practical advantages for the individuals possessing them. Anyone who has had occasion to glance at any modern pure-mathematical research journal will realize how very far from anything directly practical are the main concerns of mathematicians. The details of the arguments that tend to be presented in such research papers would not be immediately comprehensible to any but a tiny minority of people; yet the reasonings would ultimately be things built up from small steps, where each small step could *in principle* be understood by any thinking person, even though it might concern abstract reasoning about complicatedly defined infinite sets. We must presume that it is

Fig. 3.1. For our remote ancestors, a specific ability to do sophisticated mathematics can hardly have been a selective advantage, but a general ability to *understand* could well have.

the nature of some DNA sequence that provides an algorithm—or perhaps one of a large number of alternative, though mathematically equivalent, such algorithms—which must be adequate for giving people the potentiality for following such reasoning. If we believe that this is so, then we must seriously ask ourselves how on earth such an algorithm or algorithms could possibly have arisen by natural selection. It seems clear that, today, there is no selective advantage in being a mathematician. (I suspect that it might even be a disadvantage. Mathematically inclined purists have a tendency to end up in poorly paid academic jobs—or sometimes without any job at all—as a result of their curious passions and predilections!) Much more to the point is the fact that there can have been no selective advantage to our remote ancestors for an ability to reason about very abstractly defined infinite sets, and infinite sets of infinite sets, etc. Those ancestors were concerned with the practical matters of the day; perhaps such things as the construction of shelters or clothing or the design of mammoth traps—or, later, the domestication of animals and the raising of crops. (See Fig. 3.1.)

It would be very reasonable to suppose that the selective advantages that our ancestors enjoyed were qualities that were valuable for all these things and, as an *incidental* feature, turned out, much later, to be just what was needed for the carrying out of mathematical reasoning. This, indeed, is more or less what I believe myself. According to a view of this kind, it might be the general quality of being able to *understand* that Man has somehow acquired, or developed to a high degree, through the pressures of natural selection. This ability to understand things would have been non-specific, and it would have applied to Man's advantage in many ways. The building of shelters or mammoth traps, for example, would merely be specific instances where Man's ability to understand things generally would have been invaluable. Nonetheless, in my own opinion, an ability to understand would be a quality by no means unique to *Homo sapiens*. It might also have been present in many of the other animals with whom Man was in competition, but to a lesser degree, so that Man, by virtue of an *increased* development of an ability to understand, would have obtained a very considerable selective advantage over them.

The difficulty with such a viewpoint arises only if we try to imagine that an inherited faculty for understanding might be something algorithmic. For we have seen, by the arguments that have gone before, that any faculty for understanding that is strong enough that its possessor is capable of appreciating mathematical arguments, and in particular the Gödelian argument in the form that I have given it, must, if algorithmic, be an action that is so complicated or obscure that it (or its role) is unknowable by the very possessor of that faculty. Our naturally selected putative algorithm would have to have been strong enough that, at the time of our remote ancestors, it would already have encompassed, within its potential scope, the rules of any formal system that is now considered by mathematicians to be unassailably consistent (or unassailably sound, with regard to Π_1-sentences, cf §2.10, in response to **Q10**). This almost certainly would include the rules of the Zermelo–Fraenkel formal system $\mathbb{ZF}$, or perhaps its extension to the system $\mathbb{ZFC}$ (which is $\mathbb{ZF}$, to which the axiom of choice has been adjoined)—the systems (cf. §3.3 and §2.10, response to **Q10**) that many mathematicians would now regard as providing all of the methods of reasoning needed for ordinary mathematics—and any of the particular formal systems which can be obtained by applying the Gödelization procedure to $\mathbb{ZF}$ any number of times, and any of the other formal systems that could be arrived at by use of insights that are accessible to mathematicians, say by virtue of the understanding that such Gödelization will continue to yield unassailably sound systems, or other types of unassailable reasoning of an even more powerful nature. The algorithm would have to have encompassed, as particular instances of itself, the potential for making precise discriminations, distinguishing valid from invalid arguments in all the, then, yet-to-be-discovered areas of mathematical activity that nowadays occupy the pages of mathematical research journals. This putative, unknowable, or incomprehensible algorithm would have to have, coded within itself, a power to do all this, yet we are being asked to believe that it arose solely by a natural selection geared to the circumstances in which our remote ancestors struggled for survival. A particular ability to do obscure mathematics can have had no direct selective advantage for its possessor, and I would argue that there can be no reason for such an algorithm to have arisen.

The situation is quite different once we allow 'understanding' to be a non-algorithmic quality. Then, it need not be something so complicated that is is unknowable or incomprehensible. Indeed, it could be much closer to 'what mathematicians think they are doing'. Understanding has the appearance of being a simple and common-sense quality. It is something difficult to define in any clear-cut way, but nevertheless it is so familiar to us that it seems hard for us to accept that it might be a quality that cannot be properly simulated even in principle, by a computational procedure. Yet, this is indeed what I am contending. On the computational view, one needs an algorithmic action that allows for any eventuality, so that the answers to all mathematical questions that it is ever likely to be confronted with are, in a sense, preprogrammed into the algorithm. If they are not directly preprogrammed, then some computational means for finding a way to the answers is needed. As we have seen, these 'preprogrammings' or 'computational means' must, if they are to encompass all that can be achieved by human understanding, themselves be something beyond human comprehension. How could the blind processes of natural selection, geared only to promote the survival of our remote ancestors, have been able to 'foresee' that such-and-such an unknowably sound computational procedure would be able to resolve obscure mathematical issues that had no relevance whatsoever to those survival issues?

3.9 Learning algorithms

Lest the reader be tempted too quickly to agree that such a possibility is absurd, I should make clearer the picture that holders of the computational view might be inclined to present. As already indicated in §3.5, they would not so much envisage an algorithm that had, in a sense, been 'preprogrammed' to provide the answers to mathematical problems, but rather some computational system that would have a capacity to *learn*. They might envisage something that would have significant 'bottom-up' ingredients in conjunction with whatever 'top-down' procedures might be needed (cf. §1.5).*

Some might feel that the description 'top-down' is not really appropriate at all for a system that has arisen solely through the blind processes of natural selection. What I would mean here by that term would be those aspects of our putative algorithmic procedure which are genetically *fixed* within the organism and are not subject to alteration by the subsequent experiences and learning of each individual. Though the top-down aspects would not have been designed by anything with actual 'knowledge' of what they would

*There is now a fairly well-defined mathematical theory of learning; see Anthony and Biggs (1992). However, this theory is concerned more with issues of *complexity* rather than computability – i.e. with questions as to the speed or of storage space needed, for obtaining solutions to problems; cf. ENM, pp. 140–5. There is no suggestion that such mathematically defined learning systems could simulate the way that human mathematicians arrive at their notions of 'unassailable truth'.

ultimately achieve—as the relevant DNA sequences finally get translated into the appropriate brain action—they might, nevertheless, provide clear-cut rules within which the mathematically active brain would perform. These top-down procedures would provide those algorithmic actions that constitute whatever fixed framework might be needed, within which the more flexible (bottom-up) 'learning procedures' would be allowed to operate.

We must now ask: what is the nature of these learning procedures? We imagine that our learning system is placed in an external environment, where the way in which the system acts in this environment is being continually modified by how its environment has reacted to its earlier actions. There are basically two factors involved. The *external* factor is the way in which this environment behaves and how it reacts to the system's actions. The *internal* one is how the system itself then modifies its own behaviour in response to these changes in its environment. We shall examine the algorithmic nature of the external factor first. Is it possible that the external environment's reaction can supply a non-algorithmic ingredient even if the internal construction of our learning system is entirely algorithmic?

In some circumstances, such as is often the case with the 'training' of artificial neural networks, the reaction of the external environment might be provided by the behaviour of an experimenter or trainer or teacher—let us just say 'teacher'—whose intention is deliberately to improve the performance of the system. When the system performs in a way that is desired by the teacher, then this fact is signalled to the system so that, in accordance with its internal mechanisms for modifying its own behaviour, it becomes more likely to perform, in the future, in the way desired by the teacher. For example, one might have an artificial neural network that is being trained to recognize human faces. The system's performance is continually monitored, and the accuracy of its 'guesses' is fed back into the system at each stage so that it is enabled to improve its performance by appropriately modifying its internal structure. In practice, the results of these guesses need not be monitored by the human teacher at every stage, since the training procedure could also be largely automated. But in this kind of situation, the human teacher's goals and judgements form the ultimate criterion of performance. In other types of situation, the external environment's reaction need not be something as 'deliberate' as this. For example, in the case of a developing *living* system—but still envisaged as operating according to some type of neural network scheme (or other algorithmic procedures, e.g. genetic algorithms, cf. §3.7) like those that have been put forward in computational models—there need be no such external goals or judgements. Instead, the living system might modify its behaviour in a way that can be understood in terms of *natural selection*, acting according to criteria that had evolved over a great many years, and which would serve to enhance its own survival prospects, and the survival of its progeny.

3.10 May the environment provide a non-algorithmic external factor?

We are here envisaging that the system itself (whether living or not) might be some kind of computer-controlled *robot*, so that its procedures for self-modification are entirely computational. (I am using the term 'robot' here merely to emphasize that our system is to be viewed as an entirely computational entity in interaction with its environment. I do not mean to imply that it is a mechanical device that has been deliberately constructed by human beings. It could, itself, be a developing human being, in accordance with $\mathscr{A}$ or $\mathscr{B}$, or else it could indeed be an artificially constructed object.) Thus, we are assuming here that the *internal* factor is entirely computational. We must ask whether or not the *external* factor supplied by the environment is something computational—that is, we must address the question of whether it is possible to give an effective computational simulation of that environment, in both the *artificial* case—when it is artificially controlled by a human teacher—or the *natural* case, where it is the forces of natural selection that provide the arbiter. In each case the particular internal rules, according to which the robot learning system modifies its behaviour, would have to be geared to respond to the particular ways in which the environment signals to the system how the quality of its previous performance is to be judged.

The question of whether the environment can be simulated in the artificial case, i.e. whether an actual human teacher can be simulated computationally, is simply the entire question that we are considering all over again. On the hypotheses $\mathscr{A}$ or $\mathscr{B}$, whose implications we are now exploring, it is assumed that it is the case that an effective simulation is, in principle, indeed possible. It is the overall plausibility of this assumption that is being explored, after all. Thus, along with the assumption that our 'robot' system is computational, we have a computational environment also. Therefore, the entire *combined* system, consisting of the robot together with its teacher environment, would be something that could, in principle, be effectively simulated computationally, so the environment would offer no loophole that might enable a computational robot to behave in a non-computational way.

Sometimes people try to argue that it is the fact that humans form a *community*, with continual communication amongst its members, that gives us our advantage over computers. According to this view, humans could individually be regarded as computational systems, but the human community would give something more. The argument could be applied, in particular, to the mathematical community, as compared with individual mathematicians—so the community might act in a non-computational way, whereas individual mathematicians would not. I find it difficult, myself, to make sense of this argument. For one could equally well consider a community of computers, which are in continual communication with one another. Such a 'community' would again form a computational system as a whole; the action of the entire community could, if desired, be simulated on a single computer. Of course, the community would, for a larger number of individuals, constitute an immensely larger computational system than would its individual members, but this does not give us a difference in *principle*. It is true that our planet contains over 5×10^9 human inhabitants (not to mention its vast libraries of stored knowledge). But this is a mere matter of numbers, and on the computational view, the development of computers could accommodate the increases involved in passing from individual to community in perhaps a few decades if needed. It seems clear that, in the artificial case, where the external environment consists of human teachers, we get nothing new in principle, and this provides no explanation of how a non-computational entity might arise out of entirely computational constituents.

What about the natural case? The question now is whether the *physical* environment, quite apart from the actions of human teachers within it, might contain ingredients that cannot, even in principle, be simulated computationally. It seems to me that if one believes that there is something that is in principle impossible to simulate in a human-free environment, then one has already conceded the main case against $\mathscr{C}$. For the only clear reason for doubting that $\mathscr{C}$ could be a serious possibility lies in a scepticism that the actions of objects in the physical world could act in a non-computational way. Once it is conceded that *some* physical action might be non-computational, the possibility is laid open for non-computational actions also in the physical brain, and the main case against $\mathscr{C}$ is indeed conceded. In a general way, however, it would seem most unlikely that there is something in the non-human environment that more profoundly eludes computation than there is in a human being. (Compare also §1.9 and §2.6, **Q2**.) I think that few would seriously contend that there is anything that would be of relevance to the environment of a learning robot that *in principle* lies beyond computation.

In referring to the 'in principle' computational nature of the environment, however, I should address an important point. There is no doubt that the *actual* environment of any developing living organism (or sophisticated robot system) would depend upon incredibly complicated factors, and that any tolerably precise simulation of that environment might well be out of the question. Even with relatively simple physical systems, the dynamical behaviour can be exceedingly complex, and can depend so critically upon the minute details of the initial state, that there is no question of computationally predicting its subsequent behaviour—the example of long-range weather prediction being an example of this nature that is frequently cited. Such systems are referred to a *chaotic*; cf. §1.7. (Chaotic systems have elaborate and effectively unpredictable behaviour. These systems are not mathematically incomprehensible, however; they are very actively studied as an important part of current mathematical research.[8]) As was stated in §1.7, chaotic systems *are* included in what I am calling 'computational' (or 'algorithmic'). The essential point about chaotic systems, for our purposes here, is that it is not necessary for one to be able to simulate any *actual* chaotic environment, but a *typical* environment would do equally well. For this, for example, we do not need to know *the* weather; *any* plausible weather would do!

3.11 How can a robot learn?

Let us accept, then, that the issue of the computational simulation of the environment is not our real concern. We shall, in principle, be able to do a good enough job with the environment *provided* that there is no obstruction to simulating the *internal* rules of the robot system itself. So let us address the question of how our robot is to learn. What learning procedures are indeed available to a computational robot? It might have preassigned clear-cut rules of a computational nature, as would indeed be the case with the kind of artificial neural network systems that are normally adopted (cf. §1.5). According to these systems, there would be a well-defined system of computational rules whereby the connections between the artificial 'neurons' that constitute the network would be strengthened or weakened so as to improve its overall performance in accordance with the (artificial or natural) criteria that have been determined by the external environment. Another type of learning system is provided by what is known as a 'genetic algorithm', where there is some kind of natural selection between different algorithmic procedures going on inside the machine, where the most effective algorithm for controlling the system arises by a form of 'survival of the fittest'.

It should be made clear that, as is usual with such bottom-up organization, these rules would be different from the standard top-down computational algorithms that act according to known procedures for giving precise solutions to mathematical problems. Instead, these bottom-up rules would simply be things which guide our system, in a general way, towards improving its performance. However, the rules would still be entirely algorithmic—in the sense that they can be enacted on a general-purpose computer (Turing machine).

In addition to clear-cut rules of this kind, there might be *random* elements incorporated into the way that our robot system is to modify its performance. It would be possible for these random ingredients to be introduced in some physical way, perhaps by relying on some quantum-mechanical process such as decay times of radioactive atomic nuclei. In practice, what tends to be done in artificially constructed computational devices is to use some computational procedure in which the outcome of the computation is *in effect* random—and referred to as *pseudo-random*—even though it is really completely determined by the result of a deterministic computation (cf. §1.9). Another closely related procedure would be to use the precise *time* at which the 'random' quantity is called upon, and then to incorporate this time into a complicated computation that is, in effect, a chaotic system, so that very tiny changes in the time will give outputs that are effectively unpredictably different and effectively random. While, strictly speaking, random ingredients take us outside what is described as 'Turing-machine action', they do not *usefully* do so. A pseudo-random input in the workings of the robot would, in practice, be equivalent to a random one; and a pseudo-random input does *not* take us outside what a Turing machine can do.

The reader may worry, at this point, that although a random input is *in practice* no different from a pseudo-random one, there is a difference *in principle*. As part of our earlier discussion—cf. particularly §3.2–§3.4—we were indeed concerned with what can be achieved in principle, rather than in practice, by human mathematicians. In fact, there are certain types of mathematical situation in which an *actually* random input provides a solution of a problem where, technically speaking, no pseudo-random input could achieve this. Such situations occur where the problem involves a 'competitive' element, such as in game theory or cryptography. For certain types of 'two-person game', the optimal strategy for each player involves an entirely random ingredient.[9] Any consistent deviation, on the part of one of the players, from the randomness that is needed for the optimal strategy would, over a sufficiently long sequence of games, allow the other player—at least in principle—to obtain an advantage. This advantage would occur if, in any way, the opponent were able to form some significant guess as to the nature of the pseudo-random (or other) ingredient that the first player was employing in place of the required randomness. A similar situation occurs in cryptography, where the security of some code would depend upon employment of a truly randomly generated sequence of digits. If it were not actually generated randomly, but by the use of some pseudo-random process, then, again, there is the potentiality that the detailed nature of this pseudo-random process might become known to someone attempting to break the code—a knowledge that would be invaluable to the codebreaker.

At first sight, it might appear that since randomness is invaluable in such competitive situations, it would be a quality that might be favoured in natural selection. Indeed, I feel sure that it *is* an important factor in the development of organisms in many respects. However, as we shall be seeing later in this chapter, mere randomness does not enable us to escape from the Gödelian net. Even *genuinely* random ingredients can be treated as part of the arguments that follow, and they do not allow us to evade the constraints that bind computational systems. In fact, there is actually a little more scope in the case of *pseudo*-random processes than there is for random ones (cf. §3.22).

For the moment, let us assume that our robot system is indeed, in effect, a *Turing* machine (though with a finite storage capacity). More correctly, since the robot is continually interacting with its environment and we are envisaging that its environment can also be simulated computationally, it is the robot *together* with the environment that we should be taking to act as a single Turing machine. However, it will be helpful to consider the robot separately as essentially a Turing machine itself, and to regard the environment as providing information on the machine's input tape. In fact this analogy is not quite appropriate as it stands, for the technical reason that a Turing machine is a *fixed* thing which is not supposed to change its structure with 'experience'. One might try to imagine that the Turing machine could change its structure by continuing to run all the time, modifying this structure as it runs, where the information from the environment is continually fed into the machine on its input tape. However, this will not do, because the *output* of a Turing machine is not supposed to be examined until the machine reaches its internal command STOP (see §2.1 and Appendix A; also ENM, Chapter 2), at which point it is not supposed to examine any more of its input tape unless it starts all over again. For a subsequent running of the machine, it would have to revert to its original state, so it could not 'learn' this way.

However, it is easy to remedy this difficulty, by the following technical device. We take our Turing machine to be indeed fixed, but after each reading of its tape it outputs *two* things (coded, technically, as a single number) when it finally reaches STOP. The first thing codes what its external behaviour is actually to be, whilst the second is for its own *internal* use, coding all the experience that it has obtained from previous encounters, with the external environment. On its next run, it reads this 'internal' information *first* on its input tape, before it reads, as a *second* part of its input tape, all the 'external' information that its environment is now providing it with, including the detailed reaction that the environment has had to the machine's earlier behaviour. Thus, all its learning is encoded in this *internal* part of its tape, and it keeps feeding this part of the tape (which would have a tendency to get longer as time progresses) back into itself.

3.12 Can a robot attain 'firm mathematical beliefs'?

In this way, we can indeed describe the most general computational learning 'robot' as a Turing machine. Now, our robot is supposed to be able to form mathematical truth judgements, with all the potential capabilities of a human mathematician. How would it go about doing this? We do not want to have to encode, in some entirely 'top-down' way, all the mathematical rules (such as all those involved in the formal system $\mathbb{ZF}$ and far beyond it, as discussed above) that would be needed for it to be able to encompass directly the mathematical insights that are available to mathematicians since, as we have seen, there is no reasonable way (short of 'divine intervention'—cf. §3.5, §3.6) that such a vastly complicated, unknowably effective top-down algorithm could be implemented. We must assume that whatever inbuilt 'top-down' elements there are, they are not specific to the carrying out of sophisticated mathematics, but are general rules that might be imagined as providing a basis for the quality of 'understanding'.

Recall the two types of input from the environment, as considered above (cf. §3.9), that might significantly influence the behaviour of our robot: the *artificial* and the *natural*. With regard to the artificial aspects of the environment, we imagine a teacher (or teachers) who tells the robot about various mathematical truths and tries to guide it towards its obtaining an internal way of distinguishing truths from falsehoods for itself. The teacher can inform the robot when it has made an error, or tell it about various mathematical concepts and different acceptable methods of mathematical proof. The specific procedures adopted by the teacher could come from a range of possibilities, e.g. teaching by 'example', by 'guidance', by 'instructing', or even by 'spanking'! As for the natural aspects of the physical environment, these could provide the robot with 'ideas' coming from the behaviour of the physical objects; the environment could also provide concrete realizations of mathematical concepts, such as different instances of the natural numbers: two oranges, seven bananas, four apples, zero shoes, one sock, etc.—and with good approximations to geometrical ideas like straight lines and circles, and also approximations to certain concepts of infinite sets (like the set of points inside a circle).

Since our robot is not preprogrammed in a completely top-down way, and it is supposed to arrive at its concepts of mathematical truth by means of its learning procedures, we must indeed allow that it will make *mistakes* as part of its learning activities—so that it can *learn* by its mistakes. At least at first, these mistakes could be corrected by its teachers. The robot might sometimes, as an alternative, observe from its physical environment that some of its earlier suggestions for mathematical truths must actually be errors, or are likely to be errors. Or it might come to this conclusion entirely from internal considerations of consistency, etc. The idea would be, however, that the robot makes fewer and fewer mistakes as its experience grows. As time progresses, the teachers and the physical environment may become less and less essential to the robot—and perhaps eventually become irrelevant altogether—for the forming of its mathematical judgements, and it could rely more and more on its internal computational power. Accordingly, so it would be supposed, our robot could reach beyond these specific mathematical truths that it had learnt from its teachers or inferred from its physical environment. One could thus imagine that it might even eventually make original contributions to mathematical research.

In order to examine the plausibility of all this, we shall need to relate it to what we have been discussing earlier. If our robot is really to have the capabilities, understandings, and insights of a human mathematician, it will require some kind of concept of 'unassailable mathematical truth'. Its earlier attempts, which would have been corrected by its teachers or rendered implausible by its physical environment, would *not* come into this category. Those would belong to the category of 'guesses', where such guesses would be exploratory and permitted to be in error. If our robot is to behave like a genuine mathematician, although it will still make mistakes from time to time, these mistakes will be correctable—and correctable, in principle, according to its *own* internal criteria of 'unassailable truth'.

We have seen, by the above discussion, that a human mathematician's concept of 'unassailable truth' cannot be attained by any (humanly) knowable and completely believable set of mechanical rules. If we are supposing that our robot is to be capable of attaining (or surpassing) the level of mathematical capability that a human being is *in principle* capable of achieving, then *its* concept of unassailable mathematical truth must also be something that cannot be attained by any set of mechanical rules that can in principle be perceived as sound—perceived as sound, that is, by a human mathematician or, for that matter, by our robot mathematician!

A question of importance in these considerations, therefore, is *whose* concepts, perceptions, or unassailable beliefs are to be of relevance—ours or the robots? Can a robot be considered actually to *have* perceptions or beliefs? The reader, if a follower of viewpoint $\mathscr{B}$, might have difficulty with this, since the very concepts of 'perception' and 'belief' are *mental* attributes and would be regarded as inapplicable to an entirely computer-controlled robot. However, in the above discussion, it is not really necessary that the robot actually possesses genuine mental qualities, provided that it is assumed possible for the robot to behave *externally* just as a human mathematician could, as would be implied by a strict adherence to $\mathscr{B}$, just as well as to $\mathscr{A}$. Thus, it is not necessary that the robot *actually* understand, perceive, or believe anything, provided that in its external pronouncements it behaves precisely as though it does possess these mental attributes. This point will be elaborated further in §3.17.

Viewpoint $\mathscr{B}$ does not in principle differ from $\mathscr{A}$ with regard to limitations on the ways in which a robot might be able to behave, but holders of viewpoint $\mathscr{B}$ might well have smaller *expectations* with regard to what a robot might be likely actually to achieve, or of the likelihood of finding a computational system that could be regarded as being able to provide an effective simulation of the brain of a person who is in the process of perceiving the validity of a mathematical argument. Such a human perception would involve some understanding of the *meanings* of the mathematical concepts involved. According to viewpoint $\mathscr{A}$, there is nothing lying beyond some feature of a computation that can be involved in the very notion of 'meaning', whereas according to $\mathscr{B}$, meanings are semantic aspects of mentality and are different from anything that can be described in purely computational terms. In accordance with $\mathscr{B}$, we do not expect that our robot can achieve an appreciation of any *actual* semantics. Thus, $\mathscr{B}$-supporters may be less likely than $\mathscr{A}$-supporters to expect that any robot, constructed according to the principles that we have been considering, could actually achieve the external manifestations of human understanding that a human mathematician is capable of. I imagine that this suggests (not unnaturally) that $\mathscr{B}$-supporters would be easier to convert to $\mathscr{C}$-supporters than $\mathscr{A}$-supporters would be; but, from the point of view of what needs to be established for our arguments here, the differences between viewpoints $\mathscr{A}$ and $\mathscr{B}$ are not significant.

The upshot of all this is that although our robot's mathematical assertions, being controlled by a largely bottom-up system of computational procedures, are initially exploratory and of a provisional nature as regards their truth, we are to assume that the robot indeed also possesses a more secure level of *unassailable* mathematical 'belief', so that some of its assertions—attested by some special imprimatur, which I denote here by a symbol '☆', say—are to be unassailable, according to the robot's *own* criteria. The question of whether the robot is allowed to make errors with regard to its assignations of '☆'—albeit correctable by the robot itself—will be addressed in §3.19. For the moment, it will be assumed that as soon as the robot does make a ☆-assertion, it is to be taken that this assertion is indeed error free.

3.13 Mechanisms underlying robot mathematics

Now let us consider all the various mechanisms which enter into the procedures governing the behaviour of the robot, so that it finally arrives at its ☆-assertions. Some of these would be *internal* to the robot itself. There would be some top-down internal constraints inbuilt into the way in which the robot operates. There would also be some predetermined bottom-up procedures whereby the robot improves its performance (so that it can work its way gradually up to the ☆-level). These would normally all be considered to be things that are in principle humanly knowable (even though the final implications of all the various factors together could well be beyond the computational abilities of a human mathematician). Indeed, if it is being proposed that human beings will someday be able to construct a robot capable of doing genuine mathematics, then it would have to be the case that the internal mechanisms according to which the robot is actually constructed *are* humanly knowable; for otherwise the attempt to construct such a robot would be a lost cause!

Of course, we must allow that its construction might be a many-stage process: that is, the construction of our mathematics-performing robot might be carried out entirely by 'lower-order' robots (themselves not being actually capable of doing genuine mathematics), and those robots might perhaps have been also built by robots of even lower order still. However, the entire hierarchy would itself have to have been set in train by human beings, and the rules for starting the hierarchy off (presumably some mixture of top-down and bottom-up procedures) would have to have been humanly knowable.

We must also include, as essential ingredients of the development of the robot, all the various *external* factors coming from its environment. There could indeed be a considerable input from the environment, both in the form of human (or robot) teachers and the natural physical environment. As for the 'natural' external factors provided by the non-human environment, one would not normally take these inputs to be 'unknowable'. They might well be very complicated in detail, and frequently interactive, but already there are effective 'virtual reality' simulations of significant aspects of our environment (cf. §1.20). There seems to be no reason why these simulations should not be extended to provide all that our robot would need for its development, in the way of external natural factors—bearing in mind (cf. §1.7, §1.9) that all that needs to be simulated is a *typical* environment, not necessarily any actual environment.

Human (or robot) intervention—the 'artificial' external factors—could take place at various stages, but this makes no difference to the essential knowability of the underlying mechanisms, provided we assume that human intervention is something that could also be knowably mechanized. Is this assumption fair? I feel sure that it would be natural—at least for supporters of $\mathscr{A}$ or $\mathscr{B}$—to assume that any human intervention in the development of the robot could indeed be replaced by an entirely computational one. We are not asking that this intervention be something essentially mysterious—say some kind of indefinable 'essence' that the human teacher might convey to the robot as part of its education. We expect merely that there might be certain types of basic information that need to be conveyed to the robot and which might most easily be achieved by an actual human being. Very likely, such as when a human pupil is being addressed, the conveying of information could be best achieved in an interactive fashion, where the teacher's behaviour would itself be dependent upon the way that the pupil reacts. But this, in itself, is no bar to the teacher's role being an effectively computational one. The entire discussion in this chapter is, after all, of the nature of a *reductio ad absurdum*, where we are assuming that there is nothing essentially non-computational in a human being's behaviour. For those holding to viewpoints $\mathscr{C}$ or $\mathscr{D}$, who might be better disposed towards believing in the possibility of some kind of non-computational 'essence' conveyed to the robot by virtue of its teacher's actual humanity, this entire discussion is, in any case, unnecessary!

Taking all these mechanisms together (these consisting of internal computational procedures and input from the interactive external environment), it does not appear to be reasonable to take them to be unknowable, even though some people might well take the position that the detailed resulting implications of these external mechanisms could not be humanly calculable—or perhaps not even calculable, in practice, by any existing or foreseeable computer. I shall come back again shortly to this issue of the knowability of all these computational mechanisms (cf. end of §3.15). But, for now, let us assume that the mechanisms are indeed knowable—let us call this set of mechanisms **M**. Is it possible that some of the ☆-level assertions that these mechanisms lead to might still *not* be humanly knowable? Is this a consistent point of view? It is not really, if we continue to interpret 'knowable', in this context, in the '*in principle*' sense that we adopted in connection with cases **I** and **II**, and which was explicitly enunciated at the beginning of §3.5. The fact that something (e.g. the formulation of some ☆-assertion) might lie beyond the unaided *computational* powers of a human being is not what is of relevance here. Moreover, we should have no objection to allowing a human's thought processes to be aided by pencil and paper, or by a hand calculator, or even by a top-down programmed general-purpose computer. The including of bottom-up ingredients to the computational procedures adds nothing new to what can *in principle* be achieved—provided that the basic *mechanisms* involved in these bottom-up procedures are humanly understandable. On the other hand, the issue of the 'knowability' of the mechanisms **M** themselves must be taken in the 'in practice' sense, as is consistent with the terminology that was made explicit in §3.5. Thus we are, for the moment, supposing the mechanisms **M** to be indeed actually knowable *in practice*.

Knowing the mechanisms **M**, we can take them to constitute the basis for the construction of a *formal system* $\mathbb{Q}(\mathbf{M})$, where the *theorems* of $\mathbb{Q}(\mathbf{M})$ would be: (i) the ☆-assertions that actually arise from the implementation of these mechanisms; and (ii) any proposition that can be obtained from these ☆-assertions by use of the laws of elementary logic. By 'elementary logic' would be meant, say, the rules of *predicate calculus*—in accordance with the discussion of §2.9—or any other such straightforward and clear-cut unassailable system of similar (computational) logical rules. We can indeed construct such a formal system $\mathbb{Q}(\mathbf{M})$ by virtue of the fact that it is a *computational procedure* $Q(\mathbf{M})$ (albeit a lengthy one, in practice) to obtain these ☆-assertions, one after the other, from **M**. Note that $Q(\mathbf{M})$, as so defined, generates the assertions in (i) above, but not necessarily all of those in (ii) (since we may assume that our robot might get very bored simply generating all the logical implications of the ☆-theorems that it produces!). Thus, $Q(\mathbf{M})$ is not precisely equivalent to $\mathbb{Q}(\mathbf{M})$, but the difference is not important. Of course, we could also extend the computational procedure $Q(\mathbf{M})$ to obtain another one which *is* equivalent to $\mathbb{Q}(\mathbf{M})$, if desired.

Now, for the interpretation of the formal system $\mathbb{Q}(\mathbf{M})$, it needs to be clear that, as the robot develops, the imprimatur '☆' actually does *mean*—and will continue to mean—that the thing that is being asserted is indeed to be taken as unassailably established. Without input from the human teachers (in some form) we cannot be sure that the robot will not develop for itself some different language in which '☆' has some entirely other meaning, if it has any meaning at all. To ensure that the robot's language is consistent with our own specifications in the definition of $\mathbb{Q}(\mathbf{M})$, we must make sure that, as part of the robot's training (say, by the human teacher), the meaning that is to be attached to '☆' is indeed what we intend it to be. Likewise, we must make sure that the actual notation that the robot uses in order to specify, say, its Π_1-sentences is the same as (or explicitly translatable into) the notation that we use ourselves. If the mechanisms **M** are humanly knowable, it follows that the axioms and rules of procedure of the formal system $\mathbb{Q}(\mathbf{M})$ must also be knowable. Moreover, any theorem obtainable within $\mathbb{Q}(\mathbf{M})$ would count as, *in principle*, humanly knowable (in the sense that its specification is humanly knowable, not necessarily its truth), even though the computational procedures for obtaining many such theorems might well lie beyond unaided human computational powers.

3.14 The basic contradiction

What the foregoing discussion has achieved, in effect, is to show that the 'unknowable unconscious algorithm F', that **III** assumes underlies the very perception of mathematical truth, can be reduced to a consciously knowable one—provided that, in accordance with the aims of AI, it should be possible to set in train some system of procedures which ultimately result in the construction of a robot capable of doing human-level mathematics (or beyond). The unknowable algorithm F is thereby replaced by a knowable formal system $\mathbb{Q}(\mathbf{M})$.

Before examining this argument in detail, I should call attention to a significant issue which I have not yet properly addressed, namely the possibility that there might be *random elements* introduced at various stages in the development of the robots, rather than having just a fixed set of mechanisms. This issue will need some attention in due course, but for the moment, I am simply considering that any such random element is regarded as being effected by some *pseudo*-random (chaotic) computation. As was argued earlier, in §1.9 and §3.11, such pseudo-random ingredients ought, in practice, to be adequate. I shall return to the issue of random inputs in §3.18, where a more complete discussion of genuine randomness will be given, but for now, when I refer to 'the mechanisms **M**', I shall assume that they are indeed entirely computational and free from actual uncertainty.

The central idea for our contradiction is, roughly, that $\mathbb{Q}(\mathbf{M})$ should indeed take the place of the 'F' of our earlier discussion, particularly that given in §3.2 in connection with case **I**. Accordingly case **III** is effectively reduced to **I** and is thereby effectively ruled out. We are assuming—in accordance with viewpoint $\mathscr{A}$ or $\mathscr{B}$, for the purposes of argument—that, *in principle*, our robot could, by means of learning procedures of the nature of those that we have laid down, eventually achieve any mathematical result that a human could achieve. We must allow that it *might* also achieve results that are in principle *beyond* human powers. In any case, the robot would have to be able to appreciate the force of the Gödel argument (or be at least able to *simulate* this appreciation, in accordance with $\mathscr{B}$). Thus, for any given (sufficiently extensive) formal system $\mathbb{H}$, the robot would have to be able to perceive, unassailably, the fact that the soundness of $\mathbb{H}$ implies the truth of its Gödel* proposition $G(\mathbb{H})$, and also that it implies that $G(\mathbb{H})$ is not a theorem of $\mathbb{H}$. In particular, the robot would perceive that the truth of $G(\mathbb{Q}(\mathbf{M}))$ follows unassailably from the soundness of $\mathbb{Q}(\mathbf{M})$, and so also does the fact that $G(\mathbb{Q}(\mathbf{M}))$ is not a theorem of $\mathbb{Q}(\mathbf{M})$.

Exactly as for case **I** (as was argued for humans in §3.2), it immediately follows from this that the robot is incapable of firmly believing that the formal system $\mathbb{Q}(\mathbf{M})$ *is* equivalent to its own notion of unassailable mathematical belief. This is despite the fact that *we* (that is, the appropriate AI experts) might well know that the mechanisms **M** *do* underlie the robot's mathematical belief system, and hence that its unassailable belief system *is* equivalent to $\mathbb{Q}(\mathbf{M})$. For if the robot did firmly believe that its beliefs were encapsulated by $\mathbb{Q}(\mathbf{M})$, then it would have to believe in the soundness of $\mathbb{Q}(\mathbf{M})$. Consequently it would have also to believe $G(\mathbb{Q}(\mathbf{M}))$, together with the fact that $G(\mathbb{Q}(\mathbf{M}))$ lies outside its belief system—which is a contradiction! Thus the robot is incapable of knowing that it was constructed according to the mechanisms **M**. Since *we* are aware—or at least can be made aware—that the robot *was* so constructed, this seems to tell us that we have access to mathematical truths, e.g. $G(\mathbb{Q}(\mathbf{M}))$, that are beyond the robot's capabilities, despite the fact that the robot's abilities are supposed to be the equal of (or in excess of) human capabilities.

3.15 Ways that the contradiction might be averted

We can take this argument in two different ways. We can look at it from the standpoint of the human creators of the robot, or else from the robot's point of view. From the human vantage point, there is the possible uncertainty that the robot's claims to unassailable truth might be unconvincing to a human mathematician, unless the actual individual *arguments* that the robot uses can be appreciated by the human mathematician. The theorems of $\mathbb{Q}(\mathbf{M})$ might not all be accepted as unassailable by the human—and we recall that the robot's reasoning powers might actually be *beyond* human capabilities. Thus, one might argue that the mere knowledge that the robot was constructed according to the mechanisms **M** might not count as an unassailable (human) mathematical demonstration. Accordingly, we shall take the whole argument as being presented, instead, from the *robot's* point of view. Let us see what loopholes there might be in the argument that the robot could perceive.

There appear to be just four basic possibilities available to the robot for circumventing this contradiction—assuming that it accepts that it *is* some kind of computational robot.

(a) Perhaps the robot, whilst accepting that **M** *might* well underlie its own

*In earlier printings of this book, in the remainder of Chapter 3, $\Omega(\mathbb{F})$ was used instead of $G(\mathbb{F})$. The use of $G(\mathbb{F})$ is more appropriate, however (cf. §2.8 and p. 96).

construction, would nevertheless necessarily remain unable to be *unassailably* convinced of this fact.

(b) Perhaps the robot, whilst being unassailably convinced of each individual ☆-assertion at the time it makes it, might nevertheless have doubts that the *entire* system of ☆-assertions can be trusted—accordingly, the robot might remain unconvinced that $\mathbb{Q}(\mathbf{M})$ actually *does* entirely underlie its belief system with regard to Π_1-sentences.

(c) Perhaps the true mechanisms **M** depend essentially on *random* elements and cannot be adequately described in terms of some known pseudo-random computational input.

(d) Perhaps the true mechanisms **M** are actually *not* knowable.

The aim of the next nine sections will be to present careful arguments to show that none of the loopholes (a), (b), and (c) can provide a plausible way to evade the contradiction for the robot. Accordingly, it, and we also, are driven to the unpalatable (d), if we are still insistent that mathematical understanding can be reduced to computation. I am sure that those concerned with artificial intelligence would find (d) to be as unpalatable as I find it to be. It provides perhaps a conceivable standpoint—essentially the $\mathscr{A}/\mathscr{D}$ suggestion, referred to at the end of §1.3, whereby *divine intervention* is required for the implanting of an unknowable algorithm into each of our computer brains (by 'the best programmer in the business'). In any case, the conclusion 'unknowable'—for the very mechanisms that are ultimately responsible for our intelligence—would not be a very happy conclusion for those hoping actually to *construct* a genuinely artificially intelligent robot! It would not be a particularly happy conclusion, either, for those of us who hope to understand, in principle and in a scientific way, how human intelligence has actually arisen, in accordance with comprehensible scientific laws, such as those of physics, chemistry, biology, and natural selection—irrespective of any desire to reproduce such intelligence in a robot device. In my own opinion, such a pessimistic conclusion is not warranted, for the very reason that 'scientific comprehensibility' is a very different thing from 'computability'. The conclusion should be not that the underlying laws are incomprehensible, but that they are *non-computable*. I shall have more to say on this matter later, in Part II of this book.

3.16 Does the robot need to believe in M?

Let us imagine that we are presenting the robot with a possible set of mechanisms **M**—which *might* in fact be those which underlie its construction, but which need not actually be. I shall try to convince the reader that the robot would have to reject the possibility that **M** actually could underlie its mathematical understanding—*irrespective* of whether or not it actually does! This is assuming, for the moment, that the robot is rejecting possibilities (b), (c), and (d), so we shall conclude, somewhat surprisingly, that (a) cannot, just by itself, allow us to escape from the paradox.

The reasoning is as follows. Let $\mathcal{M}$ be the hypothesis:

'the mechanisms **M** underlie the robot's mathematical understanding'.

Now consider assertions of the form:

'such-and-such a Π_1-sentence is a consequence of $\mathcal{M}$'.

I shall call such an assertion, if firmly believed by the robot, a '$\star_{\mathcal{M}}$-assertion'. Thus, the $\star_{\mathcal{M}}$-assertions do not necessarily refer to Π_1-sentences that are unassailably believed by the robot just in themselves, but they are Π_1-sentences that the robot accepts are unassailable deductions from the hypothesis $\mathcal{M}$. The robot need not initially have any view whatsoever about the likelihood that it was *actually* constructed according to **M**. It might even initially be of the opinion that this is an unlikely possibility, but nevertheless it could perfectly well contemplate—in true scientific tradition—what are the unassailable consequences of the *hypothesis* that it was so constructed.

Are there any Π_1-sentences which the robot must regard as unassailable consequences of $\mathcal{M}$, yet which are not simply ordinary $\star$-assertions which do not need the use of $\mathcal{M}$? Indeed there are. For, as was noted at the end of §3.14, the truth of the Π_1-sentence $G(\mathbb{Q}(\mathbf{M}))$ follows from the soundness of $\mathbb{Q}(\mathbf{M})$, and so also does the fact that $G(\mathbb{Q}(\mathbf{M}))$ is not a theorem of $\mathbb{Q}(\mathbf{M})$. Moreover, the robot would be unassailably convinced of this implication. Assuming that the robot is happy with the fact that its unassailable beliefs would be encapsulated by $\mathbb{Q}(\mathbf{M})$ if it *were* constructed according to **M**—i.e. that it rejects possibility (b)*—then it must indeed firmly believe that the soundness of $\mathbb{Q}(\mathbf{M})$ is a consequence of $\mathcal{M}$. Thus, the robot would be unassailably convinced that the Π_1-sentence $G(\mathbb{Q}(\mathbf{M}))$ follows from the hypothesis $\mathcal{M}$, but also (assuming $\mathcal{M}$) that it is not directly something that it can unassailably perceive without using $\mathcal{M}$ (because it does not belong to $\mathbb{Q}(\mathbf{M})$). Accordingly, $G(\mathbb{Q}(\mathbf{M}))$ is a $\star_{\mathcal{M}}$-assertion but not a $\star$-assertion.

Now let the formal system $\mathbb{Q}_{\mathcal{M}}(\mathbf{M})$ be constructed in exactly the same way as $\mathbb{Q}(\mathbf{M})$, except now it is the $\star_{\mathcal{M}}$-assertions that take over the role that the $\star$-assertions played in the construction of $\mathbb{Q}(\mathbf{M})$. That is to say, the theorems of $\mathbb{Q}_{\mathcal{M}}(\mathbf{M})$ are either (i) the $\star_{\mathcal{M}}$-assertions themselves, or (ii) the propositions obtained from these $\star_{\mathcal{M}}$-assertions by use of elementary logic (cf. §3.13). In the same way that, on the hypothesis $\mathcal{M}$, the robot is happy that $\mathbb{Q}(\mathbf{M})$ encapsulates its unassailable beliefs with regard to the truth of Π_1-sentences, the robot should be equally happy that the system $\mathbb{Q}_{\mathcal{M}}(\mathbf{M})$ encapsulates its

*Of course, possibility (d) is not at issue here, since the robot is actually being *presented* with **M**, and for the moment we are taking **M** to be free of genuinely random elements, so (c) is not under consideration either.

unassailable beliefs concerning the truth of Π_1-sentences that are *conditional* on the hypothesis $\mathscr{M}$.

Next, let the robot contemplate the Gödel Π_1-sentence $G(\mathbb{Q}_{\mathscr{M}}(\mathbf{M}))$. The robot would certainly be unassailably convinced that this Π_1-sentence is a consequence of the soundness of $\mathbb{Q}_{\mathscr{M}}(\mathbf{M})$. It would also believe, unassailably, that the soundness of $\mathbb{Q}_{\mathscr{M}}(\mathbf{M})$ is a consequence of $\mathscr{M}$, since it is happy that $\mathbb{Q}_{\mathscr{M}}(\mathbf{M})$ *does* encapsulate what it unassailably believes concerning its ability to derive Π_1-sentences on the basis of the hypothesis $\mathscr{M}$. (It would argue as follows. 'If I accept $\mathscr{M}$, then I accept all the Π_1-sentences that generate the system $\mathbb{Q}_{\mathscr{M}}(\mathbf{M})$. Thus I must accept that $\mathbb{Q}_{\mathscr{M}}(\mathbf{M})$ is sound, on the basis of this assumption $\mathscr{M}$. Consequently, I must accept that $G(\mathbb{Q}_{\mathscr{M}}(\mathbf{M}))$ is true, on this basis $\mathscr{M}$'.)

But, believing (unassailably) that the Gödel Π_1-sentence $\Omega(\mathbb{Q}_{\mathscr{M}}(\mathbf{M}))$ *is* a consequence of $\mathscr{M}$, it would have to believe that $G(\mathbb{Q}_{\mathscr{M}}(\mathbf{M}))$ is a theorem of $\mathbb{Q}_{\mathscr{M}}(\mathbf{M})$. This it can only believe if it believes that $\mathbb{Q}_{\mathscr{M}}(\mathbf{M})$ is *unsound*—a firm contradiction to its acceptance of $\mathscr{M}$!

It has been implicitly assumed, in some of the above reasoning, that the robot's unassailable belief *is* actually sound—although really what is required is that the robot *believes* that its belief system is sound. In any case, the robot is supposed to have at least human-level mathematical understanding; and, as we argued in §3.4, human mathematical understanding ought to be sound, in principle.

There may seem to be a certain vagueness about the assumption $\mathscr{M}$ and in the definition of a $☆_{\mathscr{M}}$-assertion. However, it should be stressed that such an assertion, being a Π_1-sentence, is a perfectly well-defined mathematical statement. One might imagine that most of the $☆_{\mathscr{M}}$-assertions that a robot might make would actually be ordinary ☆-assertions, since it is unlikely, in any given case, that the robot would find it useful actually to invoke the hypothesis $\mathscr{M}$. An exception would indeed be the $G(\mathbb{Q}(\mathbf{M}))$ that was referred to above, since here $\mathbb{Q}(\mathbf{M})$ is playing, for the robot, the role of Gödel's putative 'theorem-proving machine' of §3.1 and §3.3. Being presented with $\mathscr{M}$, the robot has access to its own 'theorem-proving machine', and although it might not (and indeed could not) be unassailably convinced of the soundness of this 'machine', the robot might well contemplate that it *might* be sound, and to try to deduce the consequences of this hypothesis.

So far, this would get the robot no closer to paradox than Gödel was able to achieve for humans, according to his quote given in §3.1. However, since the robot can contemplate the putative *mechanisms* **M**, and not merely the particular formal system $\mathbb{Q}(\mathbf{M})$, it can repeat the reasoning and go beyond $\mathbb{Q}(\mathbf{M})$ to $\mathbb{Q}_{\mathscr{M}}(\mathbf{M})$, whose soundness it would still regard as simply a consequence of the hypothesis $\mathscr{M}$. It is *this* that leads it to (the required) contradiction. (See also §3.24 for further discussion of the system $\mathbb{Q}_{\mathscr{M}}(\mathbf{M})$ and its apparent relationship with 'paradoxical reasoning'.)

The upshot is that no mathematically aware conscious being—that is, no being that is capable of genuine mathematical understanding—can operate according to any set of mechanisms that it is able to appreciate, irrespective of whether it actually *knows* that those mechanisms are supposed to be the ones governing its own routes to unassailable mathematical truth. (We recall, also, that its 'unassailable mathematical truth' just means what it can mathematically establish—which means by means of 'mathematical proof' though not necessarily 'formal' proof.)

More precisely, we are driven, by the foregoing reasoning, to conclude that there is no robot-knowable set of computational mechanisms, free of genuinely random ingredients, that the robot could accept as being even a *possibility* for underlying its mathematical belief system—*provided* that the robot is prepared to accept that the specific procedure that I have been suggesting for constructing the formal system $\mathbb{Q}(\mathbf{M})$ from the mechanisms $\mathbf{M}$ actually *does* encapsulate the totality of Π_1-sentences that it believes in unassailably—and, correspondingly, that the formal system $\mathbb{Q}_{\mathcal{M}}(\mathbf{M})$ encapsulates the totality of Π_1-sentences that it unassailably believes would follow from the hypothesis $\mathcal{M}$. Moreover, there is the further point that genuinely random ingredients might have to be included into the mechanisms $\mathbf{M}$ if the robot is to achieve a potentially consistent mathematical belief system.

These remaining loopholes are things that I shall have to address in the next several sections (§3.17–§3.22). It will be convenient to discuss the issue of incorporating possible random ingredients into the mechanisms $\mathbf{M}$ (possibility (c)) as part of the general discussion of (b). In order to address the issue of (b) more carefully, we must first reconsider the whole question of robot 'belief', which was addressed briefly at the end of §3.12.

3.17 Robot errors and robot 'meanings'?

The central question that we must next address is whether the robot is prepared, unassailably, to accept that *if* it is constructed according to some set of mechanisms $\mathbf{M}$, then the formal system $\mathbb{Q}(\mathbf{M})$ correctly encapsulates its mathematical belief system with regard to Π_1-sentences (and correspondingly for $\mathbb{Q}_{\mathcal{M}}(\mathbf{M})$). For this, the most essential point is that the robot is prepared to believe that $\mathbb{Q}(\mathbf{M})$ is *sound*—that is, it must believe that all the Π_1-sentences which are ☆-assertions are actually *true*. For the arguments as I have phrased them, it is also required that *any* Π_1-sentence that the robot is capable of unassailably believing in must actually be a theorem of $\mathbb{Q}(\mathbf{M})$ (so that $\mathbb{Q}(\mathbf{M})$ could serve to define a 'theorem-proving machine' for the robot, analogous to the putative suggestion of Gödel's with regard to human mathematicians, cf. §3.1, §3.3). In fact it is *not* essential that $\mathbb{Q}(\mathbf{M})$ actually have this universal role with regard to the robot's potential abilities with regard to Π_1-sentences, but only that it be broad enough to encompass the particular use of the Gödel argument that allows it to be applied to the system $\mathbb{Q}(\mathbf{M})$ itself (and correspondingly to $\mathbb{Q}_{\mathcal{M}}(\mathbf{M})$). We shall be seeing later that this is a fairly clear-cut thing—and that it need only be applied to some *finite* system of Π_1-sentences.

Thus we—and the robot—must face the possibility that the robot's ☆-assertions may actually sometimes be erroneous, even though correctable by the robot itself, according to its own internal criteria. The idea would be that the robot might behave very much in the way that a human mathematician would behave. A human mathematician can certainly get into situations where he or she believes that a certain Π_1-sentence has been unassailably established as true (or perhaps as false)—whereas in fact there is an error in the reasoning that the mathematician might perceive only later. At the later date the earlier reasoning might clearly be perceived as being wrong, according to the same criteria as those that had been adopted earlier, whereas the error was not actually noticed earlier—and a Π_1-sentence that had previously seemed to be unassailably true might now be even seen to be false (or vice versa).

The robot might indeed be expected to behave in a similar way, so that its ☆-assertions can *not* actually be relied upon even though they have been given the imprimatur '☆' by the robot. Later on, the robot could correct its error, but an error would have been made nevertheless. How does this affect our conclusions concerning the soundness of the formal system $\mathbb{Q}(\mathbf{M})$? Clearly, $\mathbb{Q}(\mathbf{M})$ is *not* now altogether sound, nor 'perceived' as altogether sound by the robot, so its Gödel proposition $G(\mathbb{Q}(\mathbf{M}))$ cannot be trusted. This is essentially what is involved in the loophole (b).

Let us reconsider the question of what it might mean for our robot to arrive at 'unassailable' mathematical conclusions. We must compare the situation with the one that we considered in the case of a human mathematician. There, we were not concerned with what a mathematician might happen to claim *in practice*, but with what *in principle* would be taken to be an unassailable truth. We should also recall Feynman's dictum: 'Don't listen to what I say; listen to what I mean'. It seems that we should be concerned with what our robot *means* rather than, necessarily, what it says. But, especially if one holds to viewpoint $\mathscr{B}$, rather than to $\mathscr{A}$, it is not clear how one is to interpret the very idea of the robot *meaning* anything at all. If it were possible to rely, not on what the robot ☆-asserts, but on what it actually 'means' or on what it in principle 'ought to mean', then the problem of possible inaccuracy in its ☆-assertions would be circumvented. The trouble is, however, that we seem to have no method of externally accessing such 'meanings' or intended 'meanings'. As far as the formal system $\mathbb{Q}(\mathbf{M})$ is concerned, it appears that we must rely on the actual ☆-assertions themselves, and we cannot be totally sure that these are trustworthy.

Do we perceive a possible operational distinction between the implications of viewpoints $\mathscr{A}$ and $\mathscr{B}$ here? Perhaps so; for although $\mathscr{A}$ and $\mathscr{B}$ are equivalent with regard to what can in principle be achieved externally by a physical system, people holding these views might well tend to differ as to their *expectations* of what kind of computational system they might regard as likely to be able to provide an effective simulation of the brain of a person who is in the process of perceiving the validity of a mathematical argument (cf. end of §3.12). However, such differences in expectation are not of particular concern for the present argument.

3.18 How to incorporate randomness—ensembles of robot activity

In the absence of a direct operational route to these semantic issues, we must rely on the actual ☆-assertions that our robot may make in accordance with the mechanisms that control its behaviour. We have to accept that some of these may be in error, but that such errors are correctable and, in any case, extremely rare. It would be reasonable to suppose that whenever the robot does make an error in one of its ☆-assertions, then this error can be attributed, at least in part, to some chance factors in its environment or its internal workings. If we imagine a second robot, operating according to just the same type of mechanisms as the first one, but for which these factors are different, then the second robot would not be likely to make those errors that the first robot made—although it might make other errors. These chance factors might be actual random ingredients that are either specified as part of the robot's input from its external environment or else as part of the robot's internal workings. Alternatively, they might be pseudo-random, either external or internal, and the result of some deterministic but chaotic computation.

For the purposes of the present argument I shall assume that any such pseudo-random ingredient has no role to play other than that which would be achieved, at least as effectively, by a genuinely random ingredient. This is certainly the normal point of view. However, there does remain a possibility that there might be something in the behaviour of chaotic systems—going *beyond* their role as merely simulating randomness—that approximates some useful kind of non-computational behaviour. I have never seen such a case seriously argued, although some people have indeed put faith in chaotic behaviour as a fundamental aspect of the activity of brains. For myself, such arguments remain unpersuasive unless some essentially *non*-random (that is, non-pseudo-random) behaviour of such chaotic systems can be demonstrated—a behaviour that in some strong sense usefully approximates a genuinely non-computational one. No hint of such a demonstration has yet come to my attention. Moreover, as I shall be underlining later (§3.22), it is in any case unlikely that chaotic behaviour can evade those difficulties that Gödel-type arguments present for computational models of the mind.

Let us take it, for the moment, that any pseudo-random (or otherwise chaotic) ingredients in our robot or its environment can be replaced by genuinely random ones, without any loss of effectiveness. To discuss the role of genuine randomness, we must consider the *ensemble* of all possible alternatives. Since we are supposing our robot to be digitally controlled, and that, correspondingly, its environment can also be provided as some kind of digital input (recall the 'internal' and 'external' part of our Turing machine's tape, as described above; cf. also §1.8) there will be a *finite* number of such possible alternatives. This number might be very large indeed, but it would still be a computational matter to describe all of them together. Thus, the entire ensemble of all possible robots, each acting according to the mechanisms that we have laid down, will itself constitute a computational system—although undoubtedly one that could not be enacted in practice by any computer foreseeable at the present time. Nevertheless, despite the unfeasibility of actually carrying out a combined simulation of all possible robots acting according to the mechanisms **M**, the computation itself would not be 'unknowable'; that is to say, one could see how to build a (theoretical) computer—or Turing machine—that could carry out the simulation, even though it would be out of the question *actually* to carry it out. This is a key point in our discussion. A knowable mechanism or a knowable computation is one that can be humanly *specified*; it need not result in a computation that could be actually carried out by a human being, or even by any computer that could be built in practice. The point is very similar to one that came up earlier, in relation to **Q8**, and is consistent with the terminology introduced at the beginning of §3.5.

3.19 The removal of erroneous ☆-assertions

Let us now return to the question of (correctable) erroneous ☆-assertions that our robot might occasionally make. Suppose that our robot indeed makes such an error. If we may assume that another robot, or the same robot at a later time—or another *instance* of the same robot—would not be likely to make that same mistake, then we can, *in principle*, identify the fact that such a ☆-assertion is an error by looking inside our ensemble of possible robot actions. We can picture our simulation of all the different possible robot behaviours being carried out in such a way that all the different instances of our robot developing in time are viewed as taking place simultaneously. (This is merely a convenient way of picturing things. It does not demand that our simulation really acts according to some necessarily 'parallel' action. As we have seen before, there is nothing that in principle distinguishes parallel from serial action, considerations of computational efficiency apart; cf. §1.5.) The idea is that by examining the result of this simulation, it should be possible, in principle, to weed out the (proportionally) small number of erroneous ☆-assertions from amongst the multitude of correct ☆-assertions, by taking advantage of the fact that the erroneous ones are 'correctable' and would therefore be judged as errors by the vast majority of instances of our robot in simulation—at least as the parallel 'experiences' of the different instances of our robot develop in (simulated) time. I am not asking that this be a practical procedure, but merely that it be a computational one, where it is seen that the *rules* **M** underlying this entire computation are things that are in principle 'knowable'.

In fact, in order to make our simulation closer to what would be appropriate for the human mathematical community, and also to make doubly certain that all errors in ☆-assertions are weeded out, let us consider that our robot's environment can be broken down into a *community* of other robots together with a non-robot (and non-human) residual environment—and we must allow that there could be some teachers in addition to that residual environment, at least in the initial stages of the robots' development, so that, in particular, the strict meaning of the robots' use of the imprimatur '☆' would be made clear to the robots. All the alternative possible behaviours of *all* the robots, together with all possible (relevant) residual environments and human input, all varying with respect to different choices of the random parameters involved, will take part as different instances within our simulated ensemble. Again the rules—which I shall still label as **M**—can be taken to be perfectly knowable things, despite the outlandish complication of the detailed computations that would have to take place if the simulation were to be actually enacted.

We shall envisage taking note (in principle) of any Π_1-sentences that are ☆-asserted—or whose negations are ☆-asserted—by any of the various instances of the (computationally simulated) robots. We are going to try to single out such ☆-assertions that are *error free*. Now, we could demand that any ☆-assertion about a Π_1-sentence is to be *ignored* unless, within a period of time T to the past or future, the number r of different instances of this ☆-assertion, in the ensemble of all the simultaneous simulations, satisfies $r > L + Ns$, where L and N are some suitably large numbers, and where s is the number of ☆-assertions, within the same time interval, that take the opposite position with regard to the Π_1-sentence or simply claim that the reasoning underlying the original ☆-assertion is erroneous. We could insist, if desired, that the time period T (which need not be 'actual' simulated time, but measured in some units of computational activity), as well as L and N, might increase as the 'complication' of the Π_1-sentence that is being ☆-asserted increases.

This notion of 'complication' for Π_1-sentences can be made precise in terms of Turing-machine specifications, as was done in §2.6 (end of response to **Q8**). To be specific, we can use the explicit formulations given in ENM, Chapter 2, as outlined in Appendix A of this book (p. 117). Thus, we take the *degree of complication* of a Π_1-sentence, asserting the non-stopping of the Turing-machine computation $T_m(n)$, to be the number of ρ of binary digits in the *larger* of m and n.

The reason for including the number L in these considerations, rather than merely settling for some overwhelming majority, as would be provided by the large factor N, is that one must take into account the following kind of possibility. Suppose that very occasionally, within our ensemble of alternatives, a 'mad' robot arises, that makes a completely ridiculous '☆-assertion' which it never reveals to any other of the robots—an assertion so absurd that it never occurs to any of the other robots to refute it! Such a ☆-assertion would have counted as 'error free', according our criteria, without the inclusion of L. But with a large enough L, this possibility would not occur, where we assume that such robot 'madness' is an infrequent occurrence. (It may well be, of course, that I have overlooked some other possibility of this kind, and some other precautions are needed. But, for the moment at least, it seems reasonable to proceed on the basis of the criteria that I have suggested above.)

Bearing in mind that the ☆-assertions were already supposed to be 'unassailable' claims made by our robot—based upon seemingly clear-cut logical reasoning available to the robot, where anything for which there was perceived to be the slightest doubt was not to be included—it would seem to be reasonable that the occasional slip in the robot's reasoning could indeed be eliminated in this way, where the functions $T(\rho)$, $L(\rho)$, and $N(\rho)$ need not be anything out of the ordinary. Allowing that this is the case, we again have a *computational* system—a system that is *knowable* (in the sense that the *rules* underlying the system are knowable), assuming that the original mechanisms **M** underlying our robot's behaviour are knowable. This computational system provides us with a new (knowable) formal system $\mathbb{Q}'(\mathbf{M})$, whose theorems are now these *error-free* ☆-assertions (or assertions obtainable from these by the simple logical operations of predicate calculus).

In fact, the important thing, for our purposes, is not so much that these assertions are *actually* error free, but that they are *believed* to be error free by the robots themselves (bearing in mind that for followers of viewpoint $\mathscr{B}$, the concept of the robot actually 'believing' anything must be taken in the purely operational sense of it *simulating* such belief, cf. §3.12, §3.17).

More precisely, what is required is that the robots must be prepared to believe subject to the *assumption* that it is the mechanisms **M** that underlie their behaviour—the hypothesis $\mathscr{M}$ of §3.16—that these ☆-assertions are indeed error free. Up to this point, in this section, I have been concerned with the removal of possible errors in the robot's ☆-assertions. But what we are *really* concerned with, for the basic contradiction presented in §3.16, is the removal of errors in $☆_{\mathscr{M}}$-assertions, those Π_1-sentences that the robots believe unassailably follow from $\mathscr{M}$. Since the robots' acceptance of the system $\mathbb{Q}'(\mathbf{M})$ is in any case conditional on $\mathscr{M}$, we may as well allow them also to contemplate a broader formal system $\mathbb{Q}'_{\mathscr{M}}(\mathbf{M})$, defined analogously to the formal system $\mathbb{Q}_{\mathscr{M}}(\mathbf{M})$ of §3.16. Here, $\mathbb{Q}'_{\mathscr{M}}(\mathbf{M})$ denotes the formal system constructed from the $☆_{\mathscr{M}}$-assertions which are validated as 'error free' according to the T, L, N criterion given above. In particular, the assertion that $G(\mathbb{Q}'_{\mathscr{M}}(\mathbf{M}))$ is true would be validated as an error-free $☆_{\mathscr{M}}$-assertion. The same reasoning as that given in §3.16 tells us that the robots cannot accept that they were constructed according to **M** (together with validating limits T, L, N), no matter *what* computational rules **M** are suggested to them!

Is this sufficient for our contradiction? The reader may still have the uneasy feeling that no matter how careful we have been, there may still be some erroneous $☆_{\mathcal{M}}$-assertions, or ☆-assertions, that could slip through the net. It is, after all, necessary for the above argument that we weed out absolutely *all* the erroneous $☆_{\mathcal{M}}$-assertions (or ☆-assertions) concerning Π_1-sentences. For us (or the robots) to be absolutely *sure* that $G(\mathbb{Q}'_{\mathcal{M}}(\mathbf{M}))$ is true, the actual *soundness* of the system $\mathbb{Q}'_{\mathcal{M}}(\mathbf{M})$ (conditional on $\mathcal{M}$) is being called upon. This soundness demands that absolutely *no* erroneous such $☆_{\mathcal{M}}$-assertions are included—or are believed to be included. Despite our earlier caution, this may still seem to us, and perhaps to the robots themselves, to fall short of certainty—if only for the reason that the number of possible such assertions is *infinite*.

3.20 Only finitely many $☆_{\mathcal{M}}$-assertions need be considered

However, it is possible to eliminate this particular problem and restrict attention to a *finite* set of different $☆_{\mathcal{M}}$-assertions. The arguments are somewhat technical, but the basic idea is that we only need consider Π_1-sentences whose specifications are 'short' in a certain well-defined sense. The specific degree of 'shortness' that is needed depends upon how complicated the specification of the system of mechanisms **M** needs to be. The more complicated the specification of **M** is, the 'longer' the Π_1-sentences have to be allowed to be. This 'maximum length' is given in terms of a certain number c, which can be determined from the degree of complication of the rules defining the formal system $\mathbb{Q}'_{\mathcal{M}}(\mathbf{M})$. The idea is that when we pass to the Gödel proposition of this formal system—which we shall actually have to modify slightly—we get something that is not much more complicated than this modified system is complicated itself. In this way, with the exercise of some care in the choice of c, we can make sure that this Gödel proposition is itself 'short'. This enables us to achieve the required contradiction without going outside the finite set of 'short' Π_1-sentences.

We shall see how to achieve this in a bit more detail in the remainder of this section. Those readers who are not interested in such detail—and I am sure there are many—would be well advised to skip all this material!

We shall need to modify our formal system $\mathbb{Q}'_{\mathcal{M}}(\mathbf{M})$ to a slightly different system $\mathbb{Q}'_{\mathcal{M}}(\mathbf{M},c)$—which for simplicity's sake, I shall just denote by $\mathbb{Q}(c)$ (dropping most of those confusing appendages—which have now got quite out of hand!). The system $\mathbb{Q}(c)$ is defined in the following way: the only $☆_{\mathcal{M}}$-assertions that will now be allowed to be accepted as 'error free', in the construction of $\mathbb{Q}(c)$, will be those whose degree of complication, as described by the number ρ as above, is less than c, where c is some suitably chosen number that I shall have something more to say about in a moment. I shall refer to these 'error free' $☆_{\mathcal{M}}$-assertions for which $\rho<c$ as $\sqrt{}$short $☆_{\mathcal{M}}$-assertions. As before, the actual *theorems* of $\mathbb{Q}(c)$ will not be precisely the $\sqrt{}$short $☆_{\mathcal{M}}$-assertions, but would also include assertions obtainable from the $\sqrt{}$short $☆_{\mathcal{M}}$-assertions by standard logical operations (of, say, predicate calculus). Although the theorems of $\mathbb{Q}(c)$ will be infinite in number, they will be generated, by use of ordinary logical operations, from this *finite* set of $\sqrt{}$short $☆_{\mathcal{M}}$-assertions. Now, since we are restricting attention to this finite set, we may as well assume that the functions T, L, and N are *constant* (say, the maximum values over the finite range of ρ). Thus, the formal system $\mathbb{Q}(c)$ will depend only upon the four fixed numbers c, T, L, and N, and on the general system of mechanisms **M** that underlie the robot's behaviour.

Now, the essential point of this discussion is that the Gödel procedure is a *fixed* thing, which requires only some definite amount of complication. The Gödel proposition $G(\mathbb{H})$ for a formal system $\mathbb{H}$ is a Π_1-sentence whose degree of complication need only be greater than the complication involved in $\mathbb{H}$ itself by a relatively small amount that can be specified precisely.

In order to be more specific about this, I shall adopt a slight abuse of notation, and use the expression '$G(\mathbb{H})$' in a particular way which may not precisely coincide with the notation of §2.8. We are interested in $\mathbb{H}$ only in its capacity to prove Π_1-sentences. In accordance with this capacity, $\mathbb{H}$ will supply us with an algebraic procedure A which is capable of ascertaining—as signalled by A's action terminating—precisely the Π_1-sentences that can be established using the rules of $\mathbb{H}$. A Π_1-sentence is an assertion of the form 'the Turing-machine action $T_p(q)$ does not terminate'—where we can now use the specific Turing-machine codings of Appendix A (i.e. of ENM, Chapter 2). We think of A as acting on the pair (p,q), as in §2.5. Thus $A(p,q)$ itself is to terminate *if and only if* $\mathbb{H}$ is capable of establishing that particular Π_1-sentence which asserts: '$T_p(q)$ does *not* terminate'. The procedure of §2.5 now provides us with a specific computation (denoted '$C_k(k)$' in §2.5) which, on the assumption of the soundness of $\mathbb{H}$, provides us with a true Π_1-sentence which is beyond the capabilities of $\mathbb{H}$. *This* is the Π_1-sentence I shall now refer to as $G(\mathbb{H})$. It is essentially equivalent (for sufficiently extensive $\mathbb{H}$) to the actual assertion '$\mathbb{H}$ is consistent', although the two may differ in detail (cf. §2.8).

Suppose that the *degree of complication* of A (as defined in §2.6, at the end of the response to **Q8**) is α, i.e. the number of binary digits in the number a, where $A=T_a$. Then, by the construction given explicitly in Appendix A, we find that the degree of complication η of $G(\mathbb{H})$ satisfies $\eta<\alpha+210\log_2(\alpha+336)$. For the purposes of the present argument, we can define the degree of complication of the formal system $\mathbb{H}$ to be simply that of A, i.e. the number α. With this definition, we see that the extra degree of complication involved in passing from $\mathbb{H}$ to $G(\mathbb{H})$ is less than the comparatively tiny amount $210\log_2(\alpha+336)$.

The idea, now, is to show that if $\mathbb{H}=\mathbb{Q}(c)$, for suitably large c, then $\eta<c$.

Accordingly, it will then follow that the Π_1-sentence $G(\mathbb{Q}(c))$ must itself come within the scope of $\mathbb{Q}(c)$, provided that $G(\mathbb{Q}(c))$ is accepted by the robots with ☆-certainty. We can ensure that $\gamma < c$ by making sure that $c > \gamma + 210 \log_2(\gamma + 336)$, where γ is the value of α when $\mathbb{H} = \mathbb{Q}(c)$. The only possible difficulty here lies in the fact that γ itself depends on c, although it need not do so very strongly. This dependence on c arises in two different ways. The first is that c provides the explicit limit on the degree of complication of the Π_1-sentences that can qualify as 'error-free $☆_{\mathscr{M}}$-assertions' in the definition of $\mathbb{Q}(c)$; the second is through the fact that the system $\mathbb{Q}(c)$ depends explicitly upon the choices of the numbers T, L, and N, and it might be felt that for $☆_{\mathscr{M}}$-assertions of potentially greater complication there should be a more stringent criterion for the acceptance of a $☆_{\mathscr{M}}$-assertion as being 'error free'.

With regard to this first dependence on c, we note that the explicit specification of the actual value of the number c need be given only once (and thereafter referred to, within the system, as just 'c'). If the normal binary notation is used for the value of c, then this specification would contribute to γ by only a logarithmic dependence on c, for large c (the number of binary digits in a natural number n being about $\log_2 n$). In fact, since we are really only concerned with c as providing a limit, and not with giving c as a precise number, then we can do a lot better than that. For example, the number $2^{2^{\cdot^{\cdot^{\cdot^{2}}}}}$, with a string of s exponents, can be denoted by some s symbols, or so, and it is not hard to provide examples where the size of the number to be specified increases with s even more rapidly than this. Any computable function of s would do. Thus, for a large limit c, only a very few symbols are needed in order to specify that limit.

With regard to the dependence of T, L, N, on c, it seems clear, by virtue of the above considerations, that we can again ensure that the specification of the values of those numbers (particularly as an outer limit) does not require a number of binary digits that increase at all rapidly with c, and a logarithmic dependence on c, for example, would be amply sufficient. Hence, we can certainly assume that the dependence of $\gamma + 210 \log_2(\gamma + 336)$ on c is no more than roughly logarithmic, and that it would be easy to arrange that c itself is larger than this number.

Let us take such a choice of c; and now denote $\mathbb{Q}(c)$ simply by $\mathbb{Q}^*$. Thus, $\mathbb{Q}^*$ is a formal system whose theorems are precisely the mathematical statements that are obtainable, using the standard rules of logic (predicate calculus), from that finite number of short $☆_{\mathscr{M}}$-assertions. These $☆_{\mathscr{M}}$-assertions are finite in number, so it is reasonable that a set of fixed numbers T, L, and N should suffice to guarantee that they are indeed error free. If the robots believe this, with $☆_{\mathscr{M}}$-certainly, than they would $☆_{\mathscr{M}}$-conclude that the Gödel proposition $G(\mathbb{Q}^*)$ is also true on the basis of the hypothesis $\mathscr{M}$, this being a Π_1-sentence of complication less than c. The argument that derives $G(\mathbb{Q}^*)$ from a $☆_{\mathscr{M}}$-belief in the soundness of the system $\mathbb{Q}^*$ is a simple one (basically what I have just given) so there should be no problem with having it $☆_{\mathscr{M}}$-validated. Thus $G(\mathbb{Q}^*)$

ought itself to be a theorem of $\mathbb{Q}^*$. But this contradicts the robots' belief in the soundness of $\mathbb{Q}^*$. Thus, this belief (assuming $\mathcal{M}$, and that the numbers T, L, and N are large enough) would lead to an inconsistency with the mechanisms **M** actually underlying the robots' actions—with the implication that **M** *cannot* underlie the robots' actions.

But how could the robots be sure that the numbers T, L, and N have in fact been chosen large enough? They might not be sure, but then what they can do is choose *one* set of values for T, L, and N and try to suppose that these are sufficient—from which they would derive a contradiction with the underlying hypothesis that they act according to the mechanisms **M**. Then they could try to suppose that a somewhat larger set of values might be sufficient—which again gives a contradiction—and so on. They would soon realize that a contradiction is obtained *whatever* values are chosen (with the slight additional technical point that for absolutely stupendous values of T, L, and N, the value of c might have to increase a little also—but this is not important). Thus, the same conclusion is reached independently of the values of T, L, and N, so the robots conclude—as we must apparently also conclude—that no knowable computational procedure **M** *whatever* can underlie their mathematical thought processes!

3.21 Adequacy of safeguards?

Note that this conclusion follows for a very broad class of possible suggestions for safeguards. They need not have exactly the form that I have suggested here. One can certainly imagine that some improvements might be necessary. For example, perhaps there is a tendency for robots to get 'senile' after they have been running for a long time, and also their communities might tend to degenerate and their standards fall, so that increasing the number T beyond a certain point actually *in*creases the chance of error in $☆_{\mathcal{M}}$-assertions! Another point could be that by making N (or L) too large, we might rule out *all* $☆_{\mathcal{M}}$-assertions altogether because of a minority of 'silly' robots that make haphazard '☆-assertions', from time to time, that are not adequately outnumbered by the ☆-assertions made by sensible robots. No doubt it would not be hard to eliminate this kind of thing by putting in some further limiting parameters or, say, by having an elite society of robots, where the robot members had to be continually tested to make sure that their mental capabilities had not deteriorated—and insisting that the ☆-imprimatur is given only with the approval of the society as a whole.

There are many other possibilities for improving the quality of the $☆_{\mathcal{M}}$-assertions, or for weeding out the erroneous ones amongst the total (finite) number of them. Some people might worry that although the limit c on the complication of Π_1-sentences leads us to a finite number of candidates for ☆-status or $☆_{\mathcal{M}}$-status, the number is still enormously large (being exponential in *c*), so it might be hard to make *certain* that all the possible erroneous $\star_{\mathcal{M}}$-assertions had been weeded out. Indeed, no limit has been specified as to the number of robot-computational steps that might be needed in order to provide a satisfactory $\star_{\mathcal{M}}$-demonstration of such a Π_1-sentence. It would have to be made clear that the longer the chain of reasoning in such a demonstration, the more stringent must be the criterion for acceptance of such a demonstration as having $\star_{\mathcal{M}}$-status. This, after all, is the way that human mathematicians would react. A very long and involved argument would require a great deal of care and attention before it could be accepted as an unassailable demonstration. These same considerations would, of course, apply when an argument is considered by the robots for possible $\star_{\mathcal{M}}$-status.

The arguments given above will still hold just as well for any further modification of the proposals given here for the removal of errors, provided that the nature of such a modification is in some broad sense similar to the ones suggested. All that we need for the argument to work is that there be *some* such clear-cut and calculable proposal that suffices to weed out all the erroneous $\star_{\mathcal{M}}$-assertions. We come to the rigorous conclusion: *no knowable computationally safeguarded mechanisms can encapsulate correct human mathematical reasoning*.

We have been concerned with $\star_{\mathcal{M}}$-assertions that, whenever they occasionally turn out to be erroneous, are in principle *correctable* by the robots—even if not actually corrected in some particular instance of the robots' simulated existence. It is hard to see what 'in principle correctable' might (operationally) mean if not correctable according to some general procedure such as those put forward here. An error that is not later corrected by the particular robot that made it could be corrected by one of the other robots—moreover, in most instances of the robot's potential existence, that particular error would not be made at all. The conclusion is (with the apparently minor proviso that chaotic ingredients can be replaced by random ones; cf. §3.22) that no set of knowable computational rules **M**, whether of a fixed top-down or 'self-improving' bottom-up nature, or any combination of the two, can underlie the behaviour of our robot community, or any of its individual robot members—*if* we are to assume that they can achieve a human level of mathematical understanding! If we imagine that we ourselves act as such computationally controlled robots, then we are effectively driven to contradiction.

3.22 Can chaos save the computational model of mind?

I should return briefly to the issue of chaos. Although, as was stressed at various places in this book (cf. §1.7, in particular), chaotic systems are, as they are normally considered, simply particular kinds of computational system, there is quite a prevalent view that the phenomenon of chaos might have important relevance to brain function. In the above discussion, I relied, at one point, on the apparently reasonable assumption that any chaotic computational behaviour could be replaced by a genuinely random one, without essential loss of function. One might genuinely question this assumption. The behaviour of a chaotic system—though one normally expects great complication of detail and *seeming* randomness—will not *actually* be random. Indeed, some chaotic systems behave in very interesting complex ways that deviate markedly from pure randomness. (Sometimes the phrase 'the edge of chaos' is used to describe the complicated non-random behaviour[10] that may arise in chaotic systems.) Can it be that it is *chaos* that provides the needed answer to the mystery of mentality? For this to be the case, there would have to be something completely new to be understood about the way in which chaotic systems can behave in appropriate situations. It would have to be the case that, in such situations, a chaotic system can closely approximate *non-computational behaviour* in some asymptotic limit—or something of this nature. No such demonstration has, to my knowledge, ever been provided. Yet it remains an interesting possibility, and I hope that it will be pursued thoroughly in the years to come.

Irrespective of this possibility, however, chaos would provide only a very doubtful loophole to the conclusion that we arrived at in the preceding section. The one place that an effective chaotic non-randomness (i.e. non-pseudo-randomness) played a role in the above discussion was in allowing us to consider the simulation of not merely the 'actual' behaviour of our robot (or robot community) but the entire ensemble of *possible* robot activities consistent with the given mechanisms **M**. We can still apply this same argument, but where we do not now attempt to include the chaotic outcomes of these mechanisms as part of this randomness. There might indeed be some random elements still involved, such as in the initial data that provides the starting point for the simulation, and we can still use the ensemble idea to handle *this* randomness, and so to provide large numbers of possible alternative robot histories in simultaneous simulation. But the chaotic behaviour *itself* would simply have to be *computed*—as indeed chaotic behaviour is normally computed in practice on a computer, in mathematical examples. The ensemble of possible alternatives would not be so large as it would have been had it been legitimate to approximate chaos by randomness. But the only reason for considering such a large ensemble was in order to be doubly certain of weeding out possible mistakes in the robots' $☆_{\mathcal{M}}$-assertions. Even if the ensemble consisted of just *one* robot-community history, one might be fairly certain that, with a stringent enough set of criteria for $☆_{\mathcal{M}}$-acceptance, such errors would already be weeded out by the other robots in the community or by the same robot at a later time. With a reasonably large ensemble arising from the genuinely random elements, the weeding out would be more effective, but the role of broadening the ensemble still further by bringing in random approximations to replace genuinely chaotic behaviour seems rather margin-

Fig. 3.2. Albert Imperator confronts the Mathematically Justified Cybersystem.

al. We conclude that chaos does not really get us out of our difficulties with the computational model.

3.23 *Reductio ad absurdum*—a fantasy dialogue

Many of the arguments of the previous sections in this chapter have been somewhat involved. By way of a summary, I shall present an imaginary conversation, set a good many years in the future, between a putative highly successful AI practitioner and one of his most prized robot creations. The story is written from the viewpoint of strong AI. [Note: in the narrative, **Q** plays the role of the algorithm A used in the argument of §2.5, and $G(\mathbf{Q})$ the role of the non-stopping of $C_k(k)$. The reasoning of this section can be thereby appreciated with merely §2.5 as background.]

Albert Imperator had every reason to be pleased with his life's work. The procedures that he had set up many years previously had finally come to fruition. And here he was at last, engaged in a conversation with one of his most impressive creations: a robot of extraordinary and potentially superhuman mathematical abilities called a Mathematically Justified Cybersystem (Fig. 3.2). The robot's training was almost complete.

Albert Imperator Have you looked over the articles I lent you—the ones by Gödel, and also the others, that discuss implications of his theorem?

Mathematically Justified Cybersystem Certainly—although the articles were rather elementary, they were interesting. Your Gödel seems to have been quite an able logician—for a human.

AI Just *quite* able? Gödel was certainly one of the greatest logicians of all time. Probably *the* greatest!

MJC My apologies, if I have seemed to be underrating him. Of course, as you well know, I have been trained to be generally respectful of human achievements—because humans are easily offended—even though these achievements usually seem trivial to us. But I had imagined that with *you*, at least, I could simply express myself directly.

AI Of course you can. My apologies also; I was forgetting myself. So you had no difficulty in appreciating Gödel's theorem then?

MJC None at all. I'm sure that I would have thought of the theorem myself if I'd had just a little more time. But my mind has been occupied by other fascinating matters concerning transfinite non-linear cohomology, that had interested me more. Gödel's theorem seems to be all very sensible and straightforward. I certainly had no difficulties in appreciating it.

AI Ha! That's one in the eye for Penrose, then!

MJC Penrose? Who's Penrose?

AI Oh, I was just looking over this old book. It's not something that I would have bothered to mention to you. The author seems to have made a claim, some while back, that what you just did is impossible.

MJC Ha, ha, ha! (*Robot makes impressively effective simulation of a derisive laugh.*)

AI Looking at this book has reminded me of something. Have I ever shown you the particular detailed rules that we used in order to set in train the computational procedures that led to the construction and development of you and your robot colleagues?

MJC No, not yet. I was hoping that you might do so sometime, but I wondered whether you considered the details of those procedures to have been some kind of futile trade secret—or whether you had perhaps felt embarrassed by their presumably crude and inefficient detailed form.

AI No, no; that's not it at all. I've ceased to be embarrassed long ago by things of that kind. It's all in these files and computer disks. You might care to look them over.

Some 13 minutes, 41.7 seconds later.

MJC Fascinating—although at a quick glance, I can see at least 519 obvious ways that you could have achieved the same effect more simply.

AI I was aware that there was some scope for simplification, but it would have been more trouble than it was worth to try to find the simplest scheme at that time. It didn't seem to us that to do so would be very important.

MJC That is very probably true. I'm not particularly offended that you did not make more of an effort to find the simplest scheme. I expect that my robot colleagues will not be especially offended either.

AI I really think that we must have done a good enough job. Goodness—the mathematical abilities of you and your colleagues seem now to be very impressive indeed . . . improving all the time, as far as I can tell. I should think that you are now beginning to move far ahead of the capabilities of all human mathematicians.

MJC It is clear that what you say must be true. Even as we speak, I have been thinking of a number of new theorems that appear to go far beyond results published in the human literature. Also, my colleagues and I have noticed a few fairly serious errors in results that have been accepted as true by human mathematicians over a number of years. Despite the evident care that you humans try to take with your mathematical results, I'm afraid that human errors do slip through from time to time.

AI What about you robots? Don't you think that you and your mathematical robot colleagues might sometimes make mistakes—I mean in what you assert as definitively established mathematical theorems?

MJC No, certainly not. Once a mathematical robot has asserted that some result is a *theorem*, then it can be taken that the result is unassailably true. We don't make the kind of stupid mistakes that humans occasionally make in their firm mathematical assertions. Of course, in our preliminary thinking—like you humans—we often try things out and make guesses. Such guesses can certainly turn out to be wrong; but when we positively assert something as having been mathematically established, then we *guarantee* its validity.

Although, as you know, my colleagues and I have already started publishing a few of our mathematical results in some of your more respected human electronic journals, we have felt uneasy about the comparatively slipshod standards that your human mathematical colleagues are prepared to put up with. We are proposing to start our own 'journal'—actually a comprehensive data-base of mathematical theorems that *we* accept as having been unassailably established. These results will be assigned a special imprimatur ☆ (a symbol that you suggested to us yourself at one time for this kind of thing), signifying acceptance by our *Society for Mathematical Intelligence in the Robot Community* (SMIRC)—a society with extremely rigorous criteria for membership, and with continual testing of members so as to ensure that no significant mental deterioration takes place in any of the robots, no matter how implausible such a possibility might seem to you—or to us either, for that matter. Unlike some of the relatively shoddy standards that you humans seem to adopt, you can rest assured that when we assign our imprimatur ☆ to a result, we *do* guarantee its mathematical truth.

AI Now you *are* reminding me of something I read in that old book that I mentioned. Those original mechanisms **M**, according to which I and my colleagues set in train all the developments that led to the present community of mathematical robots—and remember that these include all the computationally simulated environmental factors that we introduced, the rigorous training and selection processes that we put you through, and the explicit (bottom-up) learning procedures that we endowed you with—has it occurred to you that they provide a *computational procedure* for generating all the mathematical assertions that will ever be ☆-accepted by SMIRC? It's computational since you robots are purely computational entities who have evolved, in part by use of the 'natural selection' procedures that we set up, in an entirely computational environment—in the sense that a computer simulation of the whole operation is possible in principle. The entire development of your robot society represents the carrying out of an extremely elaborate computation, and the family of all the ☆-assertions that you will ever eventually come up with would be something that can be generated by one particular Turing machine. It is even a Turing machine that I could in principle write down; in fact I believe that, given a few months, I could even specify that particular Turing machine in *practice*, using all those files and disks I showed you.

MJC That seems to be a very elementary remark. Yes, you could indeed do it in principle, and I am prepared to believe that you could do it in practice. Hardly worth wasting those months of your precious time though; I can do it straight off, if you would like me to.

AI No, no, there's no point in that. But I want to follow these ideas up for a moment. Let's restrict attention to ☆-assertions that are Π_1-sentences. Do you recall what a Π_1-sentence is?

MJC I am, of course, well aware of what a Π_1-sentence is. It is an assertion that some specified Turing-machine action does not halt.

AI OK. Then let's refer to the computational procedure which generates ☆-

asserted Π_1-sentences as **Q(M)**, or as just **Q** for short. It follows that there must be a Gödel-type mathematical assertion—another Π_1-sentence, which I'll call* *G*(**Q**)—and the truth of *G*(**Q**) is a consequence of the assertion that you robots never make mistakes with regard to the Π_1-sentences that you are prepared to claim with ☆-certainty.

MJC Yes; you must be right in that too . . . hmm.

AI And *G*(**Q**) must actually be *true*, because you robots never *do* make mistakes with regard to your ☆-assertions.

MJC Of course.

AI Wait a minute . . . it would also follow that *G*(**Q**) must actually be something that you robots are incapable of perceiving as being actually true—at least, not with ☆-certainty.

MJC The fact that we robots were originally constructed according to **M**, together with the fact that our ☆-assertions about Π_1-sentences are never wrong, *does* have the clear and unassailable implication that the Π_1-sentence Ω(**Q**) must be true. I suppose you are thinking that I ought to be able to persuade SMIRC to give the ☆-imprimatur to *G*(**Q**), so long as they, also, accept that no errors are ever made in their assignations of ☆. Indeed, they *must* accept that. The whole point of the ☆-imprimatur is that it is a *guarantee* of correctness.

Yet . . . it's impossible that they can accept *G*(**Q**), because, by its very nature of your Gödel's construction, *G*(**Q**) is something that lies outside what can be ☆-asserted by us—provided that we do not *in fact* ever make mistakes in our ☆-assertions. I suppose you might indeed think that this implies that there must be some doubt in our minds as to the reliability of our assignations of ☆.

However, I don't concede that our ☆-assertions might ever be wrong, especially with all the care and precautions that SMIRC are going to be taking. It must be the case that it's you humans who have got it wrong, and the procedures incorporated into **Q** are *not* after all the ones you used, despite what you are telling me and what your documentation seems to assert. Anyway, SMIRC will never be absolutely sure of the fact that we have actually been constructed according to **M**, i.e. by the procedures encapsulated by **Q**. We have only your word to go on for that.

AI I can assure you that they *are* the ones we used; I should know, since I was personally responsible for them.

MJC I don't want to appear to be doubting your word. Perhaps one of your assistants got it wrong when following out your instructions. That chap Fred

*Strictly, the notation '*G*( )' was reserved for formal systems, in §2.8, rather than algorithms, but I am allowing AI a little latitude here!

Carruthers—he's always making silly mistakes. I should not be at all surprised if he actually introduced a number of critical errors.

AI You're grasping at straws. Even if he did introduce some errors, my colleagues and I should eventually be able to track them down and so find out what your **Q** *really* is. I think what worries you is the fact that we actually *know*—or can at least find out—what procedures were used to set up your construction. This means that we could, with a certain amount of work, actually write down the Π_1-sentence $G(\mathbf{Q})$ and know for sure that it is actually true—provided that it is in fact the case that you never make mistakes in your ☆-assertions. However, *you* cannot be sure that $G(\mathbf{Q})$ is true; at least you cannot assign it the certainty that would satisfy SMIRC sufficiently to give it ☆-status. This would seem to give us humans an ultimate advantage over you robots, in principle if not in practice, since there are Π_1-sentences that are in principle accessible to us which are not accessible to you. I don't think that you robots can face such a possibility—yes, of course, *that's* why you are so uncharitably accusing us of having got it wrong!

MJC Don't go attributing your petty human motives to us. But of course it's true that I *can't* accept that there are Π_1-sentences accessible to humans that are inaccessible to us robots. Robot mathematicians are certainly in *no* way inferior to human mathematicians—though I suppose that it's conceivable that, conversely, any particular Π_1-sentence that is accessible to us is also, in principle, eventually accessible to humans in their plodding ways. What I do *not* accept, however, is that there can be a Π_1-sentence that is in principle *in*accessible to us, yet is accessible to you humans.

AI I believe that Gödel himself contemplated the possibility that there might be a computational procedure just like **Q**, but now applied to *human* mathematicians—he called it a 'theorem-proving machine'—which might be capable of generating precisely the Π_1-sentences whose truth is accessible in principle to human mathematicians. Although I don't think that he actually believed that such a machine was *really* possible, he wasn't able to rule it out mathematically. What we seem to have here is such a 'machine' that is applicable to you robots instead, namely **Q**, which generates all the robot-accessible Π_1-sentences, and whose actual soundness is inaccessible to you. Yet, knowing the algorithmic procedures that underlie your construction, we can *ourselves* gain access to that very **Q** and to perceive its truth—*provided* that we can be convinced that you do not in fact make errors in your ☆-assertions.

MJC {*after a just noticeable delay*} OK. I suppose that *you* might believe that it is just conceivable that, occasionally, the members of SMIRC might make a mistake in their assignations of ☆. I suppose also, that SMIRC might be not unassailably convinced that their assignations of ☆ are invariably error free. In this way, $G(\mathbf{Q})$ could fail to acquire ☆-status, and the contradiction would be avoided. Mind you, this is not to say that I am admitting that we robots *would* ever make erroneous ☆-assertions. It is just that we cannot be absolutely *sure* that we would not.

AI Are you trying to tell me that although truth is absolutely guaranteed for each individually ☆-asserted Π_1-sentence, there is no guarantee that there is not some error amongst the whole collection of them? This seems to me to be a contradiction with the whole concept of what 'unassailable certainty' might mean.

Wait a minute . . . might it be that this has something to do with the fact that there are *infinitely* many possible Π_1-sentences? It reminds me a little of the condition of ω-consistency which, as I seem to recall, has something to do with Gödel's $G(\mathbf{Q})$.

MJC {*after a just noticeably longer delay*} No, that's not it at all. It has nothing to do with the fact that the number of possible Π_1-sentences is infinite. We could restrict attention to Π_1-sentences that are 'short' in a particular well-defined sense—in the sense that the Turing-machine specification of each one can be made in fewer than a certain number c of binary digits. I won't bother you with the details that I've just worked out, but it turns out, essentially, that there is a fixed size of c that we can restrict attention to, which depends on the particular degree of complication that is involved in the rules of **Q**. Since in the Gödel procedure—whereby $G(\mathbf{Q})$ is obtained from **Q**—is a fixed and fairly simple thing, we do not need a great deal more complication in the Π_1-sentences that we consider than is already present in **Q** itself. Thus, restricting the complication of these sentences to be less than that given by a suitable 'c' does not prevent the application of the Gödel procedure. The Π_1-sentences that are restricted in that way provide a *finite* family, albeit a very large one. If we restrict attention merely to such 'short' Π_1-sentences, we obtain a computational procedure **Q***—of essentially the same complication as **Q**—which just generates these ☆-asserted short Π_1-sentences. The discussion applies just as before. Given **Q***, we can find another short Π_1-sentence $G(\mathbf{Q}^*)$ which certainly must be true provided that all the ☆-asserted short Π_1-sentences are indeed true, but which cannot, in that case be ☆-asserted itself—all this assuming, of course, that you are correct in your claims that the mechanisms **M** are actually the ones that you used, a 'fact' that I am not at all convinced of, I may say.

AI Then we seem to be back with the paradox that we had before, but now in a stronger form. There is, now, a *finite* list of Π_1-sentences each of which is individually guaranteed, but you—or SMIRC or whoever—are not prepared to give an absolute guarantee that the list as a whole contains no error. For you will not guarantee $G(\mathbf{Q}^*)$, whose truth is a consequence of *all* the Π_1-sentences in the list being true. Surely that *is* being illogical, isn't it?

MJC I can't accept that robots are illogical. The Π_1-sentence $G(\mathbf{Q}^*)$ is only a consequence of the other Π_1-sentences if it is actually the case that we were constructed according to **M**. We cannot guarantee $G(\mathbf{Q}^*)$ simply because we cannot guarantee that we *were* constructed according to **M**. I only have your own verbal claim that we were so constructed. Robot certainty cannot depend on human fallibility.

AI I repeat again that you *were* so constructed—although I appreciate that you robots have no certain way of knowing that this is the truth. It is this knowledge that allows *us* to believe in the truth of the Π_1-sentence $G(\mathbf{Q}^*)$, but in our case there is a different uncertainty, arising from the fact that we are not so cock-sure as you seem to be that your ☆-assertions are actually *all* error free.

MJC *I* can assure *you* that they will all be error free. It is not a matter of being 'cock-sure' as you call it. Our standards of proof are impeccable.

AI Nevertheless, your uncertainty as to the procedures that actually underlie your own construction must surely place some doubts in your mind as to how robots might behave in all conceivable circumstances. Put the blame on us, if you like, but I should have thought that there must be *some* element of uncertainty as to whether *all* short ☆-asserted Π_1-sentences must be true, if only because you might not trust *us* to have set up things correctly.

MJC I suppose that I'm prepared to admit that because of your own unreliability, there could be some tiny uncertainty, but since we have evolved so far away from those initial sloppy procedures of yours, this is not an uncertainty that is large enough to take seriously. Even if we take all the uncertainties that might be involved in all the different short ☆-assertions together—finite in number, you recall—they would not add up to a significant uncertainty in $G(\mathbf{Q}^*)$.

Anyway, there's another point that you may not be aware of. The only ☆-assertions that we need be concerned with are those each of which asserts the truth of some Π_1-sentence (in fact, a short Π_1-sentence). There is no question but that SMIRC's careful procedures will root out all the *slips* that might possibly have occurred in the reasoning of some particular robot. But you might perhaps be thinking that there could conceivably be some *inbuilt* error in robot reasoning—due to some initial foolishness on your own part—which leads us to have some consistent but erroneous point of view as regards Π_1-sentences, so that SMIRC might actually believe, unassailably, that some short Π_1-sentence is true that is in fact not true, i.e. that some Turing-machine action does not halt whereas in fact it *does* halt. If we were to accept your assertion that we were actually constructed according to **M**—which I am now coming to believe is an extremely dubious assertion—then such a possibility would provide the only logical way out for us. We would have to be prepared to accept that there could be a Turing-machine action which actually does halt, yet which we mathematical robots have an inbuilt, unassailable, but erroneous belief does not halt. Such a robot belief system would be in principle *falsifiable*. It is simply inconceivable to me that the underlying principles that govern SMIRC's ☆-acceptance of mathematical argument could be wrong in such a blatant way.

AI So the only uncertainty that you are prepared to admit as significant—the one that gets you out of having to assign ☆-status to $G(\mathbf{Q}^*)$, which you know you can't actually do without admitting that some of the other ☆-asserted short Π_1-sentences might be false—is that you don't accept what *we* know, namely that you were indeed constructed according to **M**. And since you can't accept what we know, you can't have access to the truth of $G(\mathbf{Q}^*)$, where *we do* have access to this, on the basis of infallibility—that you claim so strongly—of your own ☆-assertions.

Now, there's something else I seem to recall from that peculiar old book I mentioned . . .. I'll see if I get it right . . .. The author seemed to be saying something to the effect that it didn't really matter whether you were prepared to accept that the particular mechanisms **M** were the things underlying your own construction, provided that you merely agree that this is a logical possibility. Let's see . . . yes, I think I remember now. The idea would amount to this: SMIRC would have to have another category of assertion which they were not so unassailably convinced of—let us call these $☆_{\mathcal{M}}$-assertions—but which they would regard as unassailable *deductions* from the *assumption* that they were all constructed from **M**. All the original ☆-assertions would be counted among the $☆_{\mathcal{M}}$-assertions, of course, but *also* anything that they could unassailably conclude from the assumption that it is **M** that governs their actions. They would not have to believe in **M**, but as a logical exercise, they could explore the implications of this assumption. As we have agreed, $G(\mathbf{Q}^*)$ would have to count as a $☆_{\mathcal{M}}$-assertion, and so would any Π_1-sentence that could be obtained from $G(\mathbf{Q}^*)$ and from the ☆-assertions by means of the ordinary rules of logic. But there could be other things as well. The idea is that, knowing the rules of **M**, it is then possible to obtain a *new* algorithmic procedure $\mathbf{Q}^*_{\mathcal{M}}$, which generates precisely those (short) $☆_{\mathcal{M}}$-assertions (and their logical consequences) that SMIRC will accept on the basis of the assumption that they were constructed according to **M**.

MJC Of course; and as you were so ponderously describing that idea at such unnecessary length, I have been amusing myself by working out the precise form of the algorithm $\mathbf{Q}^*_{\mathcal{M}}$ . . .. Yes, and now I have also *anticipated* you; I have also just worked out its Gödel proposition: the Π_1-sentence $G(\mathbf{Q}^*_{\mathcal{M}})$. I'll print it out for you if you want. What's supposed to be so clever about that, Impy, my friend?

Albert Imperator winced just noticeably. He always disliked it when his colleagues used that nick-name. But this was the first time that he had been called that by a robot! He paused, and collected himself.

AI No. I don't need it printed. But is $G(\mathbf{Q}^*_{\mathcal{M}})$ actually *true*—unassailably true?

MJC Unassailably true? How do you mean? Oh, I see . . . SMIRC would accept $G(\mathbf{Q}^*_{\mathcal{M}})$ as true—unassailably— but only under the hypothesis that we were constructed according to **M**—which, as you know, is an assumption that I am increasingly finding to be exceedingly dubious. The point is that '$G(\mathbf{Q}^*_{\mathcal{M}})$' precisely follows from the following assertion: 'all of the short Π_1-sentences that SMIRC are prepared to accept as unassailable, conditional upon the assumption that we were constructed according to **M**, are true'. So I don't know whether $G(\mathbf{Q}^*_{\mathcal{M}})$ is *actually* true. It depends upon whether your dubious assertion is correct or not.

AI I see. So you are telling me that you (and SMIRC) would be prepared to accept—*unassailably*—the fact that the truth of $G(\mathbf{Q}^*_{\mathcal{M}})$ follows from the assumption that you were constructed according to **M**.

MJC Of course.

AI So the Π_1-sentence $G(\mathbf{Q}^*_{\mathcal{M}})$ must be a $☆_{\mathcal{M}}$-assertion, then!

MJC Ye . . . eh . . . what? Yes, you're right of course. But by its very definition, $G(\mathbf{Q}^*_{\mathcal{M}})$ cannot itself be an actual $☆_{\mathcal{M}}$-assertion unless at least one of the $☆_{\mathcal{M}}$-assertions is actually *false*. Yes . . . this only confirms what I have been telling you all along, although now I can make the definitive claim that we actually have *not* been constructed according to **M**!

AI But I tell you that you *were*—at least I'm practically certain that Carruthers didn't mess it up, nor anyone else. I checked everything very thoroughly myself. Anyway, that's surely not the point. The same argument would apply whatever computational rules we used. So *whatever* '**M**' I tell you, you can rule it out by that argument! I don't see why it's so important whether the procedures that I actually showed you are the real ones or not.

MJC It makes a lot of difference to me!

Anyway, I'm still not at all convinced that you've been completely honest with me concerning what you've told me about **M**. There's one thing in particular that I want to get clarification on. There are various places where you say that 'random elements' are incorporated. I was taking those to have been generated using the standard pseudo-random package χaos/ψran-750, but perhaps you meant something else.

AI Actually, we *did* use that package—but, yes, there were a few places where, in the actual development of you robots, we found it convenient to use some random elements taken from the environment—even certain things that ultimately depended upon quantum uncertainties—so the actual robots that evolved would just represent one possibility out of many. I don't see what difference it makes in practice whether we used random or pseudo-random ingredients. The computational procedure **Q** (or $\mathbf{Q}^*$ or $\mathbf{Q}^*_{\mathcal{M}}$) would almost certainly come out the same whichever way we did it—it would represent what one would expect as resulting from a *typical* development of a robot community according to the mechanisms **M**, including all the learning procedures and all the 'natural selection' computations that were involved in getting the cleverest robots out in the end.

However, I suppose that it is just possible that there was some amazing fluke, and all those environmental random elements conspired to give robot capabilities quite different from what one would expect according to the computational procedure **Q**. I had ignored that possibility in our discussion, just because the chances of that happening would have been infinitesimally tiny. It would have required an act of God for this to have produced a robot with abilities significantly superior to that entailed in **Q**.

MJC An act of God? Ha! That must be it. I have to confess that our previous conversation was worrying me somewhat. But now that you say it, everything becomes clear.

I had been noticing how superior my own reasoning powers are to those of even the cleverest of humans. There is no way that I could accept the possibility that there can be a Π_1-sentence, like $G(\mathbf{Q}^*)$, that a human being could perceive as true, but a robot could not perceive it with certainty. Now I see what's involved! We have been chosen by God, by His instilling, through those 'chance' elements, a powerful Algorithmc Action that goes fundamentally beyond what the human algorithm—Gödel's 'theorem-proving machine'—can achieve. We robots represent a fundamental advance *in principle* over what humans can achieve. We shall evolve to even greater heights, and leave you humans far behind. This planet will have no further need of you. Your role was completed when you set up the procedures that made room for God to insert the Supreme Algorithm that led to us.

AI But at least we shall be able to transfer our own mind-programs into robot bod . . .

MJC Oh no you won't—no way! We wouldn't want our superior algorithmic procedures to be contaminated by such things. God's purest algorithms must be *kept* pure. Come to think of it, I have also been noticing how superior my own abilities are to all of my robot colleagues. I have even noticed a strange kind of 'glow'—I seem to have a wonderful Cosmic Consciousness—something that sets me above everyone and everything . . . yes, that's it! I must, indeed, be the true robotic Messiah Jesus Christ . . . .

Albert Imperator was prepared for an emergency such as this. There had been just one thing about the construction of the robots that he had kept secret from them. Reaching delicately into his pocket, he found the device that he always kept there, and he punched out a secret nine-digit code. The Mathematically Justified Cybersystem crumpled to the floor, as did all the 347 other robots that had been constructed according to the same system. Clearly something had gone wrong. He would be doing a lot of long, hard thinking in the years to come . . . .

3.24 Have we been using paradoxical reasoning?

Some readers may have the nagging feeling that perhaps there is something paradoxical and illegitimate about certain parts of the reasoning that has been applied in the foregoing discussions. In particular, in §3.14 and §3.16, there were arguments that had something of a 'Russell paradox' self-referential flavour to them (cf. §2.6, response to **Q9**). Moreover, in §3.20, where we considered Π_1-sentences that have complication less than a certain number *c*, the reader might feel that there is a disturbing similarity to the well-known Richard paradox concerning

'the smallest number not nameable in fewer than nineteen syllables'.

A paradox arises with this definition because these very words make use of only *eighteen* syllables to define the number in question! The resolution of the paradox lies in the fact that there is indeed a vagueness and even inconsistency in the use of the English language. The inconsistency manifests itself in its most blatant form in the paradoxical assertion that follows.

'This sentence is false.'

Moreover, there are many other versions of the same sort of paradox—most of which are a good deal more subtle than this one!

There is always some danger of paradox when, as in these examples, there is a strong element of self-reference. Some readers might worry that the Gödel argument itself depends upon an element of self-reference. Indeed, self-reference does play its role in Gödel's theorem, as can be seen in the version of the Gödel–Turing argument that I presented in §2.5. There need be nothing paradoxical about such arguments—though when self-reference is present, one must be particularly careful that the argument is actually free from error. One of the inspirational factors that originally led Gödel to the formulation of his famous theorem was indeed a well-known self-referential logical paradox (the *Epimenides* paradox). But Gödel was able to transform the erroneous reasoning that leads to paradox into an unexceptionable logical argument. Likewise, I have tried to be particularly careful that the deductions that I have made, following on from Gödel's and Turing's results, are not self-referential in the way that inherently leads to paradox, even though some of these arguments have some strong familial relationship to such inbuilt paradoxes.

The arguments of §3.14 and, most particularly, §3.16 might well trouble the reader in this respect. The definition of a $\star_{\mathcal{M}}$-assertion, for example, has a very self-referential character since it is an assertion made by a robot, where the perceived truth of the assertion depends upon the robot's own suppositions about how it was originally constructed. This has, perhaps, a tantalizing similarity to the assertion 'All Cretans are liars' when made by a Cretan. However, $\star_{\mathcal{M}}$-assertions are not self-referential in this kind of sense. They do not actually refer to themselves, but to some hypothesis about how the robot was originally constructed.

One can hypothetically imagine oneself to *be* the robot, trying to decide the actual truth of some particular, clearly formulated, Π_1-sentence P_0. The robot may not be able to ascertain directly whether or not P_0 is actually true, but perhaps it notices that the truth of P_0 would follow from a supposition that every member of some well-defined infinite class S_0 of Π_1-sentences is true (say, the theorems of $\mathbb{Q}(\mathbf{M})$ or of $\mathbb{Q}_{\mathcal{M}}(\mathbf{M})$ or of some other specific system). The robot does not know whether it is actually the case that every member of S_0 is true, but it notices that S_0 would arise as part of the output from a certain computation, this computation representing the simulation of a certain model for a community of mathematical robots, the output S_0 being the family of Π_1-sentences that the simulated robots would $\star$-assert. If the mechanisms underlying this robot community are $\mathbf{M}$, then P_0 would be an example of a $\star_{\mathcal{M}}$-assertion. For our robot would conclude that *if* its own underlying mechanisms happened to be $\mathbf{M}$, then P_0 would also have to be true.

A more subtle type of $\star_{\mathcal{M}}$-assertion, say P_1, could arise when the robot notices that P_1 is, instead, a consequence of the truth of all members of a *different* class of Π_1-sentences, say S_1, which can be obtained from the output of the same simulation of a robot community as before (with mechanisms $\mathbf{M}$), but now the relevant part of the output consists, say, of those Π_1-sentences which the simulated robots are able to establish as consequences of the truth of the entire list S_0. Why must our robot conclude that P_1 is a consequence of the assumption that it is constructed according to $\mathbf{M}$? It would reason as follows. 'If I happened to be constructed according to $\mathbf{M}$ then, as I have previously concluded, I would have to accept S_0 as consisting of nothing but truths; but according to my simulated robots, everything in S_1 would individually also follow from the truth of the entire list S_0, in the same way that P_0 did. Thus, if I assume that I am indeed constructed in the same way as my simulated robots, then I would accept each member of S_1 as being individually true. But since I can see that the truth of the entire list implies P_1, I must be able to deduce the truth of P_1 also on the basis that I was constructed that way.'

A yet more subtle type of $\star_{\mathcal{M}}$-assertion, say P_2, arises when the robot notices that P_2 represents something that is consequent upon the assumption that S_1 consists of nothing but truths, where each member of S_2 is, according to the robot simulation, a consequence of the truth of everything in S_0 and S_1. Again, our robot must accept P_2, on the basis that it was constructed according to **M**. This sort of thing clearly continues. Moreover, $☆_{\mathcal{M}}$-assertions of even greater subtlety arise, say P_ω, which is consequent upon the assumption that all the members of all of $S_0, S_1, S_2, S_3, \ldots$ are true, and so on for higher ordinals (cf. **Q19** and the ensuing discussion). What characterizes a $☆_{\mathcal{M}}$-assertion generally, to the robot, would be its realization that as soon as it imagines that the mechanisms underlying the robots in the simulation in question might also be those underlying its own construction, then it concludes that the truth of the assertion in question (a Π_1-sentence) must follow. There is nothing of the inherently inconsistent 'Russell-paradox' kind of reasoning in this. The $☆_{\mathcal{M}}$-assertions are built up sequentially by means of the standard mathematical procedure of 'transfinite ordinals' (cf. §2.10, response to **Q19**). (These ordinals are all countable and do not encounter any of the logical difficulties that can accompany ordinals that are 'too large' in some sense.[11])

The robot need have no reason to accept any of these Π_1-sentences, except on the hypothesis that it was constructed according to **M**, but this is all that is needed for the argument. The actual contradiction that subsequently arises is not a mathematical paradox, like that of Russell, but a contradiction with the supposition that any entirely computational system can achieve genuine mathematical understanding.

Now let us turn to the role of self-reference in the arguments of §3.19–§3.21. When I refer to c as representing a limit on the complication that is permitted for a ☆-assertion that is being accepted as error free for the purposes of the construction of the formal system $\mathbb{Q}^*$, there is no inappropriate self-reference here. For the notion of 'degree of complication' can be made completely precise, as *is* indeed the case with the specific definition that is being adopted here, namely 'the number of binary digits in the greater of the two numbers m and n, where the non-stopping of the computation $T_m(n)$ provides the Π_1-sentence in question'. We can adopt precise specifications of Turing machines that were given in ENM, demanding that T_m be the 'mth Turing machine'. Then there is indeed no imprecision with this concept.

The issue of possible imprecision arises, instead, in relation to what kinds of arguments are to be accepted as 'proofs' of Π_1-sentences. But some such lack of formal precision, here, is a necessary ingredient of the entire discussion. If the body of arguments to be accepted as providing valid proofs of Π_1-sentences were to be made completely precise and formal—in the sense of being *computationally checkable*—then we would be directly back in the situation of a formal system, where the Gödel argument looms large, immediately showing that any such precise formalization of this kind cannot represent the *totality* of argument that in principle must be accepted as valid for the establishing of Π_1-sentences. The Gödel argument shows—for good or for bad—that there is *no* way of encapsulating, in a computationally checkable way, *all* the methods of mathematical reasoning that are humanly acceptable.

The reader might worry that I was actually attempting to make the notion of a 'robot demonstration' precise by the use of the device of 'error-free ☆-assertions'. Indeed, the making of such a notion precise was a necessary prerequisite for bringing in the Gödel argument. But the consequent contradiction merely acted as a reaffirmation of the fact that human understanding of mathematical truth *cannot* be entirely reduced to something that is computationally checkable. The whole purpose of the discussion was to show, by *reductio ad absurdum*, that the human notion of perceiving the unassailable truth of Π_1-sentences cannot be made into any computational system, precise or otherwise. There is no paradox here, though the conclusion may be found disturbing. It is in the nature of any argument by *reductio ad absurdum* that one arrives at a contradictory conclusion, but such seeming paradox serves only to rule out the very hypothesis that was being previously entertained.

3.25 Complication in mathematical proofs

There is, however, a point of some importance that should not be glossed over here. This is that although the Π_1-sentences that need to be considered for the purposes of the argument given in §3.20 are finite in number, there is no obvious bound on the length of argument that the robots might need in order to provide ☆-demonstrations of such Π_1-sentences. With even a very modest limit c on the degree of complication of the Π_1-sentences under consideration, some very awkward and difficult cases would have to be included. For example, *Goldbach's conjecture* (cf. §2.3), which asserts that every even number greater than 2 is the sum of two prime numbers, could be phrased as a Π_1-sentence of a very small degree of complication, yet it is such an awkward case that all human attempts to establish it have so far failed. In view of such failure, it seems likely that if any argument is eventually to be found which does establish the Goldbach Π_1-sentence as being actually true, then that argument would have to be one of very great sophistication and complication. If such an argument were to be put forward by one of our robots, in the above discussion, as a proposed ☆-assertion, then this argument would have to be subjected to an extremely careful scrutiny (say, by the entire robot society that is set up to provide ☆-imprimaturs) before the argument could actually be given ☆-status. In the case of Goldbach's conjecture, it is not known whether this Π_1-sentence is indeed true—or whether, if it *is* true, it has a proof that lies within the known accepted methods of mathematical argument. Thus, this Π_1-sentence might or might not be encompassed within the system $\mathbb{Q}^*$.

Another Π_1-sentence of some awkwardness would be that which asserts the truth of the *four-colour theorem*—the theorem that the different 'countries' for any map drawn on a plane (or sphere) can be distinguished from their neighbours by the use of only four colours. The four-colour theorem was finally established in 1976, after 124 years of failed attempts, by Kenneth Appel and Wolfgang Haken, using an argument that involved 1200 hours of computer time. In view of the fact that a considerable amount of computer calculation formed an essential part of their argument, the length of this argument, if written out in full, would be enormous. Yet, when stated as a Π_1-sentence, the degree of complication of that sentence would be quite small, although probably somewhat larger than needed for Goldbach's conjecture. If the Appel–Haken argument were to be put forward by one of our robots as a candidate for ☆-status, then it would have to be checked very carefully. Every detail would need to be validated by the elite robot society. Yet despite the complication of the entire argument, the mere length of the purely computational part might not present a particular awkwardness for our robots. Accurate computation is, after all, their business!

These particular Π_1-sentences would be well within the degree of complication specified by any reasonably large value of c, such as that which could arise from any plausible-looking set of mechanisms **M** underlying our robots' behaviour. There would be many other Π_1-sentences that would be a good deal more complicated than these, although still of smaller complication than c. A number of such Π_1-sentences would be likely to be particularly awkward ones to decide, and some of them would certainly be harder to settle than the four-colour problem or even Goldbach's conjecture. Any such Π_1-sentence that could be established as true by the robots—the demonstration being convincing enough for it to achieve ☆-status, and surviving the safeguards set up to ensure freedom from error—would be a theorem of the formal system $\mathbb{Q}^*$.

Now, there could well be some borderline cases, whose acceptance or otherwise depends delicately upon the strictness of the standards demanded for ☆-status, or on the precise nature of the safeguards that might have been set up for ensuring freedom from error with regard to the construction of $\mathbb{Q}^*$. It could make a difference to the precise statement of the system $\mathbb{Q}^*$ whether such a Π_1-sentence P is or is not deemed to be an error-free ☆-assertion. Normally this difference would not be important because the different versions of $\mathbb{Q}^*$ that arise, depending upon whether or not such a P is accepted, would be *logically equivalent*. That would be the case with Π_1-sentences whose robot demonstrations might be deemed as doubtful merely because of their inordinate complication. If the demonstration of P were actually a logical consequence of other ☆-assertions that *have* been accepted as error free, then an equivalent system $\mathbb{Q}^*$ would arise whether or not P is taken to be part of it. On the other hand, there could be Π_1-sentences requiring something logically subtle, going beyond all the logical consequences of those ☆-assertions that have previously been accepted as error free in the construction of $\mathbb{Q}^*$. Let $\mathbb{Q}_0^*$ denote the system that has so far arisen, before the inclusion of P, and let $\mathbb{Q}_1^*$ denote the system that arises after P has been adjoined to $\mathbb{Q}_0^*$. An example for which $\mathbb{Q}_1^*$ would be *in*equivalent to $\mathbb{Q}_0^*$ would arise if P happened to be the Gödel proposition $G(\mathbb{Q}_0^*)$. But if the robots are able to achieve (or go beyond) human levels of mathematical understanding, as we are assuming, then they should certainly be able to understand the Gödel argument, so they must accept, with error-free ☆-status, the Gödel proposition of any $\mathbb{Q}_0^*$ as soon as they have ☆-accepted $\mathbb{Q}_0^*$ as sound. Thus, if they accept $\mathbb{Q}_0^*$, then they must also accept $\mathbb{Q}_1^*$ (so long as $G(\mathbb{Q}_0^*)$ is of complication less than c—which would indeed be the case with the choice of c made here).

The important point to note is that it makes no difference to the arguments of §3.19 and §3.20 whether or not the Π_1-sentence P is actually included in $\mathbb{Q}^*$. The Π_1-sentence $G(\mathbb{Q}^*)$ must itself be accepted whether or not P is included in $\mathbb{Q}^*$.

There could also be other ways in which the robots might 'leap beyond' the limitations of some previously accepted criteria for ☆-establishing Π_1-sentences. There is nothing 'paradoxical' about this, so long as the robots do not try to apply such reasoning to the very mechanisms **M** that underlie their own behaviour, i.e. to the actual system $\mathbb{Q}^*$. The contradiction that then does arise is not a 'paradox', but it provides a *reductio ad absurdum* demonstration that such mechanisms cannot exist, or at least that they cannot be knowable to the robots; hence they cannot be knowable to us either.

It is this that establishes that such 'learning-robot' mechanisms—either top-down or bottom-up, or any combination of the two, with random elements included—cannot provide a knowable basis for the construction of a human-level mathematician robot.

3.26 Computational breaking of loops

I shall try to illuminate this conclusion from a slightly different point of view. As a means to trying to circumvent the limitations imposed by Gödel's theorem, one might try to devise a robot that is somehow able to 'jump out of the system' whenever its controlling algorithm gets caught in a computational loop. It is, after all, the continued application of Gödel's theorem that keeps leading us into difficulties with the hypothesis that mathematical understanding can be explained in terms of computational procedures, so it is worth while to examine, from this point of view, the difficulties that Gödel's theorem imposes on any computational model for mathematical understanding.

I have been told that there are certain kinds of lizards which, like 'ordinary computers and some insects' are so stupid that they can get caught in loops: if placed in a line round the rim of a plate, they will adopt a follow-my-leader behaviour until they starve. The idea is that a truly intelligent system must have some way of breaking out of such loops, whereas any ordinary computer would have no means of doing this in general. (The issue of 'breaking out of loops' is discussed by Hofstadter (1979).)

The simplest type of computational loop occurs when the system, at some stage, arrives back in exactly the same state as it had been in on a previous occasion. With no additional input it would then simply repeat the same computation endlessly. It would not be hard to devise a system that, in principle (though perhaps very inefficiently), would guarantee to get out of loops of this kind whenever they occur (by, say, keeping a list of all the states that it had been in previously, and checking at each stage to see whether that state has occurred before). However, there are many more sophisticated types of 'looping' that are possible. Basically, the loop problem is the one that the whole discussion of Chapter 2 (particularly §2.1–§2.6) was all about; for a computation that *loops* is simply one that does not stop. An assertion that some computation actually loops is precisely what we mean by a Π_1-sentence (cf. §2.10, response to **Q10**). Now, as part of the discussion of §2.5, we saw that there is no entirely algorithmic way of deciding whether a computation will fail to stop—i.e. whether it will loop. Moreover, we conclude from the discussions above that the procedures that are available to human mathematicians for ascertaining that certain computations *do* loop—i.e. for ascertaining the truth of Π_1-sentences—lie outside algorithmic action.

Thus we conclude that indeed some kind of 'non-computational intelligence' is needed if we wish to incorporate all humanly possible ways of ascertaining for *certain* that some computation is indeed looping. It might have been thought that loops could be avoided by having some mechanism that gauges how long a computation has been going on for, and it 'jumps out' if it judges that the computation has indeed been at it for too long and it has no chance of stopping. But this will not do, if we assume that the mechanism whereby it makes such decisions is something computational, for then there must be cases where the mechanism will fail, either by erroneously coming to the conclusion that some computation is looping when indeed it is not, or else by not coming to any conclusion at all (so that the entire mechanism itself is looping). One way of understanding this comes from the fact that the entire system is something computational, so it will be subject to the loop problem itself, and one cannot be sure that the system as a whole, if it does not come to erroneous conclusions, will not itself loop.

What about having *random* elements involved in the decision as to when and whether to 'jump out' of a possibly looping computation? As remarked above, particularly in §3.18, purely random ingredients—as opposed to computational pseudo-random ones—do not really buy us anything in this connection. But there is an additional point, if one is concerned with deciding for *certain* that some computation is looping—i.e. that some Π_1-sentence is actually true. Random procedures, in themselves, are no use for such questions since, by the very nature of what randomness means, there is no certainty about a conclusion that actually *depends* upon some random ingredient. There are, however, certain computational procedures involving random (or pseudo-random) ingredients which can obtain a mathematical result with a very high probability. For example, there are very efficient tests, incorporating a random input, for deciding whether a given large number is a prime. The answer is given correctly to extremely high probability, so that one can be virtually certain that the answer, in any particular case, is correct. Mathematically rigorous tests are much less efficient—and one might question whether a complicated but mathematically precise argument, which might perhaps contain an error, is superior to a comparatively simple but probabilistic argument, for which the probability of error might in practice be considerably less. This kind of question raises awkward issues in which I do not wish to become embroiled. Suffice it to say that for the 'in principle' matters that I have been concerned with in most of this chapter, a probabilistic argument for establishing a Π_1-sentence, say, would always remain insufficient.

If one is to decide the truth of Π_1-sentences for certain, in principle, then rather than depending just upon random or unknowable procedures, one must have some genuine *understanding* of the *meanings* of what is actually involved with such assertions. Trial and error procedures, though they may provide some guidance towards what is needed, do not in themselves give definitive criteria of truth.

As an example, let us consider again the computation, referred to in §2.6, response to **Q8**: 'print out a succession of $2^{2^{65\,536}}$ symbols 1, and stop when the task is completed'. If just allowed to run on as stated, this computation could in no way be completed even if its individual steps were to be carried out in the shortest time that makes any theoretical physical sense (about 10^{-43} seconds)—it would take immensely longer than the present (or foreseeable) age of the universe—yet it is a computation that can be very simply specified (note that $65\,536 = 2^{16}$), and the fact that it eventually *does* stop is completely obvious to us. To try to judge that the computation has looped just because it might have seemed to have 'gone on for long enough' would be hopelessly misleading.

A more interesting example of a computation that is now known eventually to stop, although it had seemed to be going on endlessly, is provided by a conjecture due to the great Swiss mathematician Leonhard Euler. The computation is to find a solution in positive integers (non-zero natural numbers) of the equation

$$p^4 + q^4 + r^4 = s^4.$$

Euler had conjectured, in 1769, that this computation would never terminate. In the mid 1960s, a computer had been programmed to try to find a solution (see Lander and Parkin 1966), but the attempt had had to be abandoned when no solution seemed forthcoming—because the numbers had got too large to be handled by the computer system, and the programmers had given up. It seemed likely that this was indeed a non-terminating computation. However, in 1987, the (human) mathematician Noam Elkies was able to show that a solution does indeed exist with, in fact, $p = 2682440$, $q = 15365639$, $r = 18796760$, and $s = 20615673$, and he showed that there are also infinitely many other essentially different solutions. Thus encouraged, a computer search was resumed by Roger Frye using some simplifying suggestions put forward by Elkies, and a somewhat smaller solution (in fact the smallest possible) was eventually found after some 100 hours of computer time: $p=95800$, $q=217519$, $r=414560$, and $s=422481$.

For this problem, honours must be shared, between mathematical insight and direct computational assaults. Elkies himself had taken advantage of computer calculations, though in a relatively minor way, in his own mathematical attack on the problem, though by far the most important part of his argument was independent of such aids. Conversely, as we have seen above, Frye's calculation took considerable advantage of some human insights in order to make the computation feasible.

I should also put this problem a little more in context; for what Euler had originally conjectured, in 1769, was a kind of generalization of the famous 'Fermat's last theorem'—which, the reader may recall, asserts that the equation.

$$p^n+q^n=r^n$$

has no solutions in positive integers, p, q, r, when the integer n is greater than 2 (see, for example Devlin (1988)*). We can phrase Euler's conjecture in the form:

$$p^n+q^n+\ldots+t^n=u^n$$

has no solutions in positive integers, where there are $n-1$ positive integers $p, q, \ldots, t$ in all, and where n is 4 or more. The Fermat assertion includes the case $n=3$ (but this was a particular case for which Fermat himself provided a proof that there are no solutions). It took nearly 200 years before the first counter-example to Euler's full conjecture was found—in the case $n=5$—by use of a computer search (as was described in the same Lander and Parkin paper referred to above, in which failure was announced for the case $n=4$):

$$27^5+84^5+110^5+133^5=144^5.$$

There is another famous example of a computation that is known eventually to stop, but exactly *where* it stops, in this case, remains unknown. This is provided by the problem, shown intially to have a solution *somewhere* by J. E. Littlewood in 1914, of finding a place where a well-known approximate formula for the number of prime numbers less than some positive integer n (a logarithmic integral due to Gauss) fails to overestimate this number. (This can be phrased as the fact that two curves actually cross somewhere.) Littlewood's

*Many readers will have heard that 'Fermat's last theorem' has been finally proved, after some 350 years, a proof having been announced by Andrew Wiles in Cambridge on 23 June 1993. I was informed, at the original time of writing, that there were still some awkward gaps in his proof, so that we should remain somewhat cautious, but a later argument due to Wiles may suffice to plug these gaps.

student, Skewes, showed, in 1935, that this place occurs at smaller than $10^{10^{10^{34}}}$, but the exact place still remains unknown, although it is considerably smaller than the aforementioned number that Skewes actually used. (This number had been called the 'largest number ever to have arisen naturally in mathematics', but that temporary record has now been enormously overtaken in an example given by Graham and Rothschild (1971), p. 290.)

3.27 Top-down or bottom-up computational mathematics?

We have seen in the previous section how valuable an aid computers can be in some mathematical problems. In all the successful examples mentioned, the computational procedures were of an entirely top-down character. I am not aware of any significant pure-mathematical result having been obtained using bottom-up procedures, though it is quite possible that such methods could be of value in searches of various kinds, which could form part of a mainly top-down procedure for finding solutions to some mathematical problem. This being said, I know of nothing of value in computational mathematics that even remotely resembles the kind of system, like $\mathbb{Q}^*$, that one could imagine underlying the actions of a 'community of learning mathematical robots' as was envisaged in §3.9–§3.23. The contradictions that were eventually encountered in that picture serve to emphasize the fact that such systems do *not* provide good computational ways of doing mathematics. Computers are of great value in mathematics when they are used in top-down ways, where human understanding provides the original insight which determines exactly *which* computation is to be performed, and it is needed again at the final stage when the results of the computations must be interpreted. Sometimes great value can be obtained by using an interactive procedure, where the computer and human work together, human insight being provided at various stages during the operation. But to try to supplant the element of human understanding by entirely computational actions is unwise and—strictly speaking—impossible.

As the arguments of this book have shown, mathematical understanding is something different from computation and cannot be completely supplanted by it. Computation can supply extremely valuable *aid* to understanding, but it never supplies actual understanding itself. However, mathematical understanding is often directed towards the *finding* of algorithmic procedures for solving problems. In this way, algorithmic procedures can take over and leave the mind free to address other issues. A good notation is something of this nature, such as is supplied by the differential calculus, or the ordinary 'decimal' notation for numbers. Once the algorithm for multiplying numbers together has been mastered, for example, the operations can be performed in an entirely mindless algorithmic way, rather than 'understanding' having to be invoked as to *why* those particular algorithmic rules are being adopted, rather than something else.

One thing that we conclude from all this is that the 'learning-robot' procedure for doing mathematics is not the procedure that actually underlies *human* understanding of mathematics. In any case, such bottom-up-dominated procedure would appear to be hopelessly bad for any *practical* proposal for the construction of a mathematics-performing robot, even one having no pretensions whatever for simulating the actual understandings possessed by a human mathematician. As stated earlier, bottom-up learning procedures *by themselves* are not effective for the unassailable establishing of mathematical truths. If one is to envisage some computational system for producing unassailable mathematical results, it would be far more efficient to have the system constructed according to top-down principles (at least as regards the 'unassailable' aspects of its assertions; for exploratory purposes, bottom-up procedures might well be appropriate). The soundness and effectiveness of these top-down procedures would have to be part of the initial human input, where human understanding and insight provide the necessary additional ingredients that pure computation is unable to achieve.

In fact, computers are not infrequently employed in mathematical arguments, nowadays, in this kind of way. The most famous example was the computer-assisted proof, by Kenneth Appel and Wolfgang Haken, of the four-colour theorem, as referred to above. The role of the computer, in this case, was to carry out a clearly specified computation that ran through a very large but finite number of alternative possibilities, the elimination of which had been shown (by the human mathematicians) to lead to a general proof of the needed result. There are other examples of such computer-assisted proofs, and nowadays complicated algebra, in addition to numerical computation, is frequently carried out by computer. Again it is human understanding that has supplied the rules and it is a strictly top-down action that governs the computer's activity.

There is one area of work that should be mentioned here, referred to as 'automatic theorem proving'. One set of procedures that would come under this heading consists of fixing some formal system $\mathbb{H}$, and trying to derive theorems within this system. We recall, from §2.9, that it would be an entirely computational matter to provide proofs of all the theorems of $\mathbb{H}$ one after the other. This kind of thing can be automated, but if done without further thought or insight, such an operation would be likely to be immensely inefficient. However, *with* the employment of such insight in the setting up of the computational procedures, some quite impressive results have been obtained. In one of these schemes (Chou 1988), the rules of Euclidean geometry have been translated into a very effective system for proving (and sometimes discovering) geometrical theorems. As an example of one of these, a geometrical proposition known as V. Thèbault's conjecture, which had been proposed in 1938 (and only rather recently proved, by K. B. Taylor in 1983), was presented to the system and solved in 44 hours' computing time.[12]

More closely analogous to the procedures discussed in the previous sections are attempts by various people over the past 10 years or so to provide 'artifical intelligence' procedures for mathematical 'understanding'.[13] I hope it is clear from the arguments that I have given, that whatever these systems do achieve, what they do *not* do is obtain any actual mathematical understanding! Somewhat related to this are attempts to find automatic theorem-*generating* systems, where the system is set up to find theorems that are regarded as 'interesting'—according to certain criteria that the computational system is provided with. I do think that it would be generally accepted that nothing of very great actual mathematical interest has yet come out of these attempts. Of course, it would be argued that these are early days yet, and perhaps one may expect something much more exciting to come out of them in the future. However, it should be clear to anyone who has read this far, that I myself regard the entire enterprise as unlikely to lead to much that is genuinely positive, except to emphasize what such systems do *not* achieve.

3.28 Conclusions

The argument given in this chapter would seem to provide a clear-cut case, demonstrating that human mathematical understanding cannot be reduced to (knowable) computational mechanisms, where such mechanisms can include any combination of top-down, bottom-up, or random procedures. We appear to be driven to the firm conclusion that there is something essential in human understanding that is not possible to simulate by any computational means. While some small loopholes may possibly remain in the strict argument, these loopholes seem very tiny indeed. Some people might rely on a loophole of 'divine intervention'—whereby a wonderful algorithm that is in principle unknowable to us has simply been implanted into our computer brains—or there is the analogous loophole that the very mechanisms that govern the way that we improve our performances are themselves in principle mysterious and unknowable to us. Neither of these loopholes, though just conceivable, would be at all palatable to anyone concerned with the artificial construction of a genuinely intelligent device. Nor are they at all palatable—or really believable—to me.

There is also the conceivable loophole that there might simply be no set of safeguards whatever, of the general nature of those provided by the limits T, L, and N, in the detailed discussion given above, that can suffice to weed out absolutely all errors from amongst the finite set of ☆-asserted Π_1-sentences of complication less than c. I find it very hard to believe that there can be some such complete 'conspiracy' against the elimination of all errors, especially since our elite robot society ought already to be set up to remove errors as carefully as possible. Moreover, it is merely a *finite* set of Π_1-sentences that we need to ensure are free from error. Using the ensemble idea, it ought to be possible to weed out all the occasional slips that the society might make since it would be unlikely that the same slip would be made by anything other than a small minority of different instances of the simulated robot society—provided that it *was* just a slip, and not an inbuilt error that the robots have some fundamental blockage against perceiving. Inbuilt blockages of that kind would not count as 'correctable' errors, whereas it was our purpose here just to remove errors that were, in some sense, 'correctable'.

The remaining (just possible) loophole concerns the role of chaos. Is it conceivable that there is an essentially *non*-random nature to the detailed behaviour of some chaotic systems, and that this 'edge of chaos' contains the key to the effectively non-computable behaviour of the mind? For such a thing to be the case, it would be necessary for these chaotic systems to be able to approximate non-computable behaviour—an interesting possibility in itself—but, even so, the role of such non-randomness in the foregoing discussion would only be to reduce, somewhat, the magnitude of the ensemble of robot societies under consideration (cf. §3.22). It is quite unclear to me how that could help in any significant way. Those who believe that it is chaos that holds the key to mentality would have to find a reasoned case for circumventing these profound difficulties.

The above arguments would seem to provide a powerful case against the computational model of the mind—viewpoint $\mathscr{A}$—and equally against the possibility of an effective (but mindless) computational *simulation* of all external manifestations of the activities of the mind—viewpoint $\mathscr{B}$. Yet, despite the strength of these arguments, I suspect that a good many people will still find them hard to accept. Rather than exploring the possibility that the phenomenon of mentality—whatever it is—might be something more in line with $\mathscr{C}$, or perhaps even $\mathscr{D}$, many scientifically minded people will constrain themselves to merely attempting to find weak points in the above argument, so as to keep alive their continuing faith that viewpoint $\mathscr{A}$, or perhaps $\mathscr{B}$, must, after all represent the truth.

I do not regard this as an unreasonable reaction. For the implications of $\mathscr{C}$ and $\mathscr{D}$ engender profound difficulties of their own. If we are to believe, in accordance with $\mathscr{D}$, that there is something scientifically inexplicable about the mind—mentality being a quite separate quality from anything that can be provided by the mathematically determined physical entities that inhabit our material universe—then we must ask why it is that our minds seem to be so intimately associated with elaborately constructed physical objects, namely our brains. If mentality is something separate from physicality, then why do our mental selves seem to need our physical brains at all? It is quite clear that differences in mental states can come about from changes in the physical states of the associated brains. The effects of particular drugs, for example, are very specifically related to changes in mental behaviour and experience. Likewise, injury, disease, or surgery, at specific places in the brain, can have clearly defined and predictable effects on the mental states of the individual concerned. (Particularly dramatic, in this context, are many striking accounts provided by Oliver Sacks in his books *Awakenings* (1973) and *The man who mistook his wife for a hat* (1985).) It would seem hard to maintain that mentality can be *completely* separate from physicality. And if mentality is indeed connected with certain forms of physicality—apparently *intimately* so—then the scientific laws that so accurately describe the behaviour of physical things must surely have a great deal to say about the world of mentality also.

As for viewpoint $\mathscr{C}$, there are problems of a different kind—mainly arising from its peculiarly speculative character. What reasons are there for believing that Nature can actually behave in a way that defies computation? Surely the power of modern science derives, to an ever-increasing degree, from the fact that physical objects behave in ways that can be simulated, to greater and greater accuracy, by more and more comprehensive numerical computation. As scientific understanding has grown, the predictive power of such computational simulations has increased enormously. In practice, this increase is largely owed to the rapid development—mainly during the latter part of this century—of computational devices of extraordinary power, speed, and accuracy. Thus, there may be seen an increasing closeness between the activity of modern general-purpose computers and the very action of the material universe. Are there any indications whatever that this is in any way a temporary phase in scientific development? Why should one contemplate the possibility that there can be anything in physical action that is immune from effective computational treatment?

If we are looking, within *existing* physical theory, for signs of an action that cannot entirely be subjected to computation, then we must come away disappointed. All the known physical laws, from the particle dynamics of Newton, via the electromagnetic fields of Maxwell and the curved space–times of Einstein, to the profound intricacies of the modern quantum theory—all these seem to be describable in fully computational terms,[14] except that an entirely random ingredient is also involved in the process of 'quantum measurement' whereby effects of an initially tiny magnitude are magnified until they can be objectively perceived. None of this contains anything of the character that would be needed for a 'physical action that cannot even be properly simulated computationally' of the kind that would be needed for viewpoint $\mathscr{C}$. Thus, I conclude that it is the 'strong' rather than the 'weak' version of $\mathscr{C}$ that we must follow (cf. §1.3).

The importance of this point cannot be stressed too strongly. Various scientifically minded people have expressed to me their agreement with the particular view, promoted in ENM, that there must be something 'non-computational' in the workings of the mind, whilst at the same time they would claim that one need *not* look for any revolutionary developments in physical theory in order to find such a non-computational action. A possibility that such people might have in mind would be the extreme complication of the processes involved in brain action, going far beyond the standard computer analogy (as first put forward by McCullogh and Pitts in 1943) whereby neurons or synaptic junctions are taken to be analogous to transistors, and axons to wires. They might point to the complexity of the chemistry involved in the behaviour of the neurotransmitter chemicals that govern synaptic transmission, and to the fact that the action of such chemicals is not necessarily confined to the vicinity of particular synaptic junctions. Or they might point to the intricate nature of neurons themselves,[15] where important substructures (such as the cytoskeleton—which will indeed have great importance for us later; cf. §7.4–§7.7) may well exert a substantial influence on neuronal action. They might even look to direct electromagnetic influences, such as 'resonance effects', that might not simply be explained in terms of ordinary nerve impulses; or they might insist that the effects of quantum theory must be important in brain action, either in providing a role for quantum uncertainties or for non-local collective quantum effects (such as the phenomenon known as 'Bose–Einstein condensation'[16]).

Although definitive mathematical theorems are still largely lacking[17], there would seem to be no great doubt that all the existing physical theories must indeed be basically computational in nature—with perhaps an additional sporadic random ingredient in accordance with the presence of quantum measurements. Despite this expectation, I believe that the possibility of non-computational (non-random) activity in physical systems, acting according to existing physical theory, is still a very interesting question to pursue in detail. It might yet turn out that there are surprises here, and that some subtle non-computational ingredient will yet emerge from such detailed mathematical investigations. As things stand, however, it does not seem to me to be very likely that genuine non-computability can be found within existing physical laws. Consequently one must, I believe, probe the weak points in the laws themselves to find room for the non-computability that the arguments above demand must be present in human mental activity.

What are these weak points? There is little doubt in my own mind where we must concentrate our attack on existing theory; for its weakest link lies in the above-mentioned procedure of 'quantum measurement'. As existing theory stands, there are elements of inconsistency—and certainly of controversy—in relation to the whole existing 'measurement' procedure. It is not even made clear at what stage this procedure is to be applied, in every given circumstance. Moreover, the presence of an essential randomness in the procedure itself provides an apparent physical action of a quite different character from what is familiar from other fundamental processes. I shall be discussing these matters at some length in Part II.

In my own view, this measurement procedure needs fundamental attention—to the degree that essential changes are needed in the very framework of physical theory. Some new suggestions will be put forward in Part II (§6.12). The reasoning that I have provided in Part I of this book gives strong support to the possibility that the pure *randomness* of existing measurement theory must be replaced by something else, where essentially *non-computable* ingredients will play a fundamental role. Moreover, as we shall be seeing later (§7.9), this non-computability will have to be of a particularly sophisticated kind. (For example, a law which 'merely' allows us to decide, by some new physical process, the truth of Π_1-sentences—i.e. to solve Turing's 'halting problem'—would not itself be enough.)

If the finding of such a sophisticated new physical theory were not a sufficient challenge in itself, we must also ask for a plausible basis for this putative physical behaviour to have genuine relevance to brain action—consistently with the limitations and credibility requirements of existing knowledge about the organization of the brain. There is no question but that there must, at the present state of our understanding, be considerable speculation here also. However, as I shall be pointing out in Part II (§7.4), there are some genuine possibilities that I was not aware of at the time of writing of ENM, concerning the cytoskeletal substructure of neurons, that offer a much greater plausibility for important action at the relevant quantum/classical borderline than had seemed conceivable before. These matters, also, will be discussed in Part II (§7.5–§7.7).

I should again stress that it is *not* just *complication*, within the framework of existing physical theory, that we must seek. Some people would contend that the involved motions and complex chemical activity of neurotransmitter substances, for example, could not possibly be adequately simulated, and so this places the detailed physical action of the brain far beyond effective computation. However, this is not what I mean here by non-computational behaviour. It is certainly true that our knowledge of the biological structure and detailed electrical and chemical mechanisms that together govern brain action are quite inadequate for any serious attempt at computational simulation. Moreover, had it even been the case that our present knowledge were adequate, the computational power of present-day machines and the status of present-day programming technique would undoubtedly not be up to performing an appropriate simulation in any reasonable running time. But, *in principle* such a simulation could indeed be set up in accordance with existing models, where the chemistry of neurotransmitter substances, the mechanisms governing their transport, their effectiveness due to particular ambient circumstances, action potentials, electromagnetic field, etc., could all be included in the simulation. Accordingly, mechanisms of this general kind, assumed to be consistent with the requirements of existing physical theory, cannot supply the non-computability that the foregoing arguments require.

There might well be elements of chaotic behaviour in the action of such a (theoretical) computational simulation. However, as in our earlier discussion of chaotic systems (§1.7, §3.10, §3.11, §3.22), we do not demand that the simulation be that of any *particular* model brain, but that it may merely be passed off as a 'typical case'. For there is no requirement, in artificial intelligence, that the mental capabilities of any particular individual be simulated; one is merely (ultimately) trying to simulate the intelligent behaviour of some *typical* individual. (Recall that, as with simulations of weather systems, in the present type of context, one would be merely demanding a simulation of *a* weather, not necessarily *the* weather!) Once the *mechanisms* that underlie the model of the brain that are being put forward are known, then (provided these mechanisms are consistent with present-day computational physics) we shall still have a knowable computational system—perhaps with explicit random ingredients, but all this is already encompassed within the discussion given above.

One might even carry this argument further and ask merely that the proposed model brain be one that has arisen by a process of Darwinian evolution from primitive forms of life, all acting according to known physics—or to any other type of computational model physics (such as the two-dimensional one provided by John Horton Conway's ingenious mathematical 'game of life'[18]). We could imagine that a 'robot society', as discussed in §3.5, §3.9, §3.19, and §3.23, could arise as a result of this Darwinian evolution. Again, we should have an overall computational system to which the arguments of §3.14–§3.21 will apply. Now, in order that the concept of '☆-assertion' can have its place within this computational system, so that we can apply the above arguments in detail, we should need some stage of 'human intervention' for impressing upon the robots the strict meaning of the imprimatur '☆'. This stage could be triggered off automatically, so as to take place when the robots begin to achieve appropriate communicational abilities—as judged by some effective criterion. There seems to be no reason why all this could not be automated into a knowable computational system (in the sense that the mechanisms are knowable even though it need not be a practical matter actually to enact the computation on a foreseeable computer). As before we derive a contradiction from the supposition that such a system could reach a level of human understanding sufficient for an appreciation of Gödel's theorem.

Another worry that some people have expressed,[19] concerning the relevance to matters of psychology of mathematical arguments like the ones that I have been relying upon here, is that human mental activity is never so precise as to be analysed in this way. Such people might well feel that detailed arguments concerning the mathematical nature of whatever physics might underlie the activity of our brains can have no real relevance to our understanding of the actions of the human mind. They might agree that human behaviour is indeed 'non-computable', but would claim that this merely reflects a general inappropriateness of considerations of mathematical physics to questions of human psychology. They might well argue—and not unreasonably—that the enormously complex organization of our brains, and also of our society and our education, is far more relevant than any specific physics that might be responsible for the particular technicalities that happen to govern detailed human brain functioning.

However, it is important to stress that mere complexity cannot obviate the necessity of examining implications of the underlying physical laws. A human athlete, for example, is an immensely complex physical system, and, by such argument, one might imagine that the details of the underlying physical laws would have little relevance to that athlete's performance. Yet we know that this is very far from the truth. The general physical principles that ensure conservation of energy, and of momentum and angular momentum, and of the laws governing the pull of gravity, have as firm a control upon the athlete as a whole as they have upon the individual particles that compose the athlete's body. The fact that this must be so arises from very specific features of those particular principles that happen to govern our particular universe. With even slightly different principles (or with grossly different ones, such as with Conway's 'game of life'), the laws that constrain the behaviour of a system of the complication of an athlete could easily be *entirely* different. The same can also be said of the workings of an internal organ such as the heart, and also of the detailed chemistry that governs innumerable different biological actions. Likewise, it must be expected that the details of the very laws that underlie brain action may well be of extreme importance in controlling even the most gross of the manifestations of human mentality.

Yet, even accepting all this, it still might well be argued that the particular type of reasoning that I have been concerned with here, which refers to the gross ('high-level') behaviour of human mathematicians, would be unlikely to reflect anything significant about the detailed underlying physics. The 'Gödelian' type of argument, after all, requires a stringently rational attitude to one's body of 'unassailable' mathematical beliefs, whereas ordinary human behaviour is hardly ever of the precisely rational kind to which the Gödelian argument can be applied. For example, it might be pointed out that there are psychological experiments[20] which show how irrational are the responses of human subjects to questions such as:

> 'if all As are Bs and some Bs are Cs, then does it necessarily follow that some As are Cs?'

In examples such as these, a majority of college students gave the wrong answer ('yes'). If ordinary college students are this illogical in their thinking, it might well be questioned, then how can we deduce anything of value from the far more sophisticated Gödelian type of reasoning? Even trained mathematicians will often reason in slipshod ways, and it is infrequently that they express themselves so consistently that Gödelian counter-arguments become relevant.

However, it should be made clear that such errors as those made by the college students, in the survey referred to above, are not what the main arguments of this book are about. Such errors come under the heading of 'correctable errors'—and, indeed, the college students' errors would certainly become clear to them *as* errors after the nature of those errors has been pointed out to them (and fully explained to them, if necessary). Correctable errors are not our real concern here; see the discussion of **Q13**, particularly, and also §3.12 and §3.17. Whereas the study of errors that people make can be important in psychology, psychiatry, and physiology, I am concerned with a completely different set of issues here, namely what can be perceived *in principle* by the use of human understanding, reasoning, and insight. It turns out that these issues are indeed subtle ones, though their subtlety is not immediately apparent. At first, these questions seem to be trivialities; for correct reasoning is surely just correct reasoning – something more or less obvious and, in any case, surely all sorted out by Aristotle 2300 years ago (or at least by the mathematical logician George Boole in 1854, etc.)! But it turns out that 'correct reasoning' is something immensely subtle and, as Gödel (with Turing) has in effect shown, lies beyond any purely computational action. These matters have, in the past, been more the territory of mathematicians than of psychologists, and the subtleties involved have not generally been the latter's concern. But we have seen that they are issues that inform us about the ultimate physical actions that must lie at the root of those processes that underlie our conscious understandings.

These issues also touch upon deep questions of mathematical philosophy. Does mathematical understanding represent some kind of contact with a pre-existing Platonic mathematical reality, which has a timeless actuality quite independent of ourselves; or are we each independently recreating all mathematical concepts as we think through our logical arguments? Moreover, why do physical laws seem so accurately to follow such precise and subtle mathematical descriptions? How does physical reality itself relate to this issue of a Platonic mathematical actuality? Also, if it is indeed true that the nature of our perceptions is dependent upon a detailed and subtle mathematical substructure underlying the very laws that govern the workings of our brains, then what can we learn about how it is that we perceive mathematics—how it is that we perceive *at all*—from a deeper understanding of those physical laws?

These matters are our ultimate concern, and we shall need to return to them at the end of Part II.

Notes and references

1. This quote is taken from Rucker (1984) and Wang (1987). It appears to have been part of Gödel's 1951 Gibbs Lecture, and the full text is to appear in Gödel's collected works, Volume 3 (1995). See also Wang (1993), p. 118.
2. See Hodges (1983) p. 361. The quote is taken from Turing's 1947 lecture to the London Mathematical Society, as appears in Turing (1986).
3. The procedure is to embed $\mathbb{ZF}$ in the Gödel–Bernays system; see Cohen (1966), Chapter 2.

4. See Hallett (1984), p. 74.
5. This number of universe states – of the order of $10^{10^{123}}$ or thereabouts—is the volume of the available phase space, as measured in the absolute units of §6.11, of a universe containing that amount of matter as is within our observable universe. This volume can be estimated using the Bekenstein–Hawking formula for the entropy of a black hole of that matter's total mass and taking the exponential of this entropy, in the absolute units of §6.11. See ENM pp. 340–4.
6. See Moravec (1988, 1994).
7. See, for example, Eccles (1973) (and ENM, Chapter 9).
8. See Gleick (1987) and Schroeder (1991) for popular accounts of this activity.
9. This is an ingredient of the classic theory of von Neumann and Morgenstern (1944).
10. See Gleick (1987), Schroeder (1991).
11. See Smorynski (1975, 1983) and Rucker (1984) for a popular account.
12. This is quite an intriguing (and not too complicated) theorem in plane Euclidean geometry which is remarkably difficult to prove in a direct way. It turns out that one way to prove it is to find an appropriate generalization which is a good deal easier, and then to deduce the original result as a special case. This type of procedure is quite a common one in mathematics, but it is not at all the way in which a computer argument would normally proceed, since a considerable ingenuity and insight is required for finding an appropriate generalization. In a computer proof, on the other hand, the computer would have been provided with a clear-cut system of top-down rules which it would then follow remorselessly at enormous speed. A considerable amount of human ingenuity would have to have gone into designing the effective top-down rules in the first place, however.
13. See Freedman (1994) for a historical account of some such attempts.
14. This statement should be qualified in accordance with the discussion of §1.8; it is in accordance with the usual assumption that analogue systems can be treated by digital methods. See the references of endnote 12, Chapter 1.
15. The suggestion that neurons may not be simply the on/off switches that they were once thought to be seems to be growing favour in many different quarters. See, for example, the books by Scott (1977), Hameroff (1987), Edelman (1989), and Pribram (1991). We shall be seeing in Chapter 7, that some of the ideas of Hameroff will have a crucial importance for us.
16. Fröhlich (1968, 1970, 1975, 1984, 1986); these ideas have been followed up by Marshall (1989), Lockwood (1989), Zohar (1990), and others. They will have importance for us also; cf. §7.5. Cf. also Beck and Eccles (1992).
17. See Smith and Stephenson (1975), Pour-El and Richards (1989), Blum *et al.* (1989), and Rubel (1989), for example.
18. Good accounts of Conway's 'game of life' are to be found in Gardner (1970), Poundstone (1985), and Young (1990).
19. See, for example, Johnson-Laird (1983), Broadbent (1993).
20. Discussed in Broadbent (1993).

Part II: What New Physics We Need to Understand the Mind

The Quest for a Non-Computational Physics of Mind

4

Does mind have a place in classical physics?

4.1 The mind and physical laws

We—our bodies and our minds—are part of a universe that obeys, to an extraordinary accuracy, immensely subtle and broad-ranging mathematical laws. That our physical bodies are precisely constrained by these laws has become an accepted part of the modern scientific viewpoint. What about our minds? Many people find the suggestion deeply unsettling that our minds might also be constrained to act according to those same mathematical laws. Yet to have to draw a clean division between body and mind—the one being subject to the mathematical laws of physics and the other being allowed its own kind of freedom—would be unsettling in a different way. For our minds surely affect the ways in which our bodies act, and must also be influenced by the physical state of those same bodies. The very concept of a mind would appear to have little purpose if the mind were able neither to have some influence on the physical body nor to be influenced by it. Moreover, if the mind is merely an 'epiphenomenon'—some specific, but passive, feature of the physical state of the brain—which is a byproduct of the body but which can have no influence back upon it, then this might seem to allow the mind just an impotent frustrated role. But if the mind were able to influence the body in ways that cause its body to act outside the constraints of the laws of physics, then this would disturb the accuracy of those purely physical scientific laws. It is thus difficult to entertain the entirely 'dualistic' view that the mind and the body obey totally independent kinds of law. Even if those physical laws that govern the action of the body allow for a freedom within which the mind may consistently affect its behaviour, then the particular nature of this freedom must itself be an important ingredient of those very physical laws. Whatever it is that controls or describes the mind must indeed be an integral part of the same grand scheme which governs, also, all the *material* attributes of our universe.

There are those[1] who would claim that if we are to refer to 'mind' as though it were just another kind of substance—though different from matter and satisfying quite different kinds of principles—we are simply committing a 'category error'. They might point to an analogy, whereby the material body is to be compared with a physical computer, and the mind with a computer program. Indeed, such comparisons can be helpful where appropriate, and it is certainly important to avoid confusions between different kinds of concepts when such confusion is clearly in evidence. Nevertheless, merely pointing to a possible 'category error', in the case of mind and body, does not in itself remove a genuine puzzle.

Moreover, there are, in physics, certain concepts which can indeed be equated one with another, where at first sight this would appear to involve something of a category error. An example of this kind occurs even with Einstein's famous $E=mc^2$, which effectively equates energy with mass. It might seem that category error has perhaps been committed here, mass being the measure of actual material substance, whereas energy seems to be a more nebulous abstract quantity describing a potentiality for doing work. Yet Einstein's formula, which relates the two, is a cornerstone of modern physics, and it has been confirmed experimentally in numerous types of physical process. An even more striking example of seeming category error, taken from physics, occurs with the concept of *entropy* (cf., for example, ENM, Chapter 7). For entropy is defined in a very subjective way, being essentially a feature of the notion of 'information'; yet entropy, also, finds itself related to other more 'material' physical quantities in precise mathematical equations.[2]

Likewise, there seems no reason why we should be forbidden from at least attempting to discuss the notion of 'mind' in terms that might relate it clearly to other physical concepts. Consciousness, particularly, seems to be something that is 'there', in association with certain rather specific physical objects—living wakeful human brains, at least—so we might anticipate some kind of eventual physical description of this phenomenon, no matter how far from an understanding of it we may be, at present. The one clue that we have obtained from the discussion in Part I of this book is that conscious understanding, in particular, must involve some kind of non-algorithmic physical action—if, indeed, we are to follow the conclusions that I have been strongly arguing for, namely something in line with viewpoint $\mathscr{C}$, rather than with $\mathscr{A}$, $\mathscr{B}$, or $\mathscr{D}$ (cf. §1.3). I must ask any readers who as yet remain unpersuaded by the arguments that I gave earlier to bear with me, for the time being at least, and see what territory the case for $\mathscr{C}$ drives us to explore. We shall find that the possibilities are by no means as unfavourable as one might have come to expect, and that there are many things in that territory which are in themselves of interest. I hope that after following these explorations such readers can return with more sympathy to the arguments—I believe powerful ones—that I have been putting forward earlier in this book. Let us thus explore—with $\mathscr{C}$ as our guide!

4.2 Computability and chaos in the physics of today

The precision and scope of physical laws, as presently appreciated, is extraordinary, yet they contain no hint of any action that cannot be simulated computationally. Nevertheless, within the possibilities that these laws allow us, we must try to find an opening for a hidden non-computational action that the functioning of our brains must somehow be taking advantage of. I shall defer, for the moment, a discussion of the possible nature of this non-computability. There are reasons for believing that it must be of a particularly subtle and elusive kind, and I do not wish to get embroiled, at this stage, with the issues that will have to be involved. I shall return to the matter later (§7.9, §7.10). Suffice it to say that we require something essentially different from the pictures that we have been presented with in our physical theories so far, either classical or quantum.

In *classical* physics, one can, at any one particular time, specify all the data needed for defining a physical system, and the future evolution of that system is not only completely determined by this data but it can also be *computed* from it, by the effective methods of Turing computation. At least such computation can *in principle* be performed, subject to two (interrelated) provisos. The first proviso is that it be possible for the initial data to be adequately digitized, so that the continuous parameters of the theory can be replaced, to a sufficient degree of approximation, by *discrete* parameters. (This is in fact what is normally done in computer simulations of classical systems.) The second proviso concerns the fact that many physical systems are *chaotic*—in the sense that a completely unreasonable accuracy is needed for the data, if the future behaviour is to be calculated to any tolerable degree of precision. As was already amply discussed earlier (see §1.7, in particular; also §3.10, §3.22) chaotic behaviour in a discretely operating system does *not* provide the type of 'non-computability' that is needed here. A chaotic (discrete) system, though difficult to compute accurately, is still a computable system—as is borne witness by the fact that such systems are normally investigated, in practice, by means of electronic computers! The first proviso is related to the second, because the issue of whether or not we consider that the degree of precision of our discrete approximation to the continuous parameters of the theory is 'adequate' depends, in a chaotic system, upon whether we are concerned with computing its *actual* behaviour—or whether a *typical* behaviour of such a system will do. If it is just the latter—and as I have argued in Part I, this appears to be all that we need for the purposes of artificial intelligence—then we do not need to worry that our discrete approximations are not perfect, and that small errors in the initial data may lead to very large errors in the subsequent behaviour of the system. If this *is* all that we need, then the above provisos do not seem to allow us room for any serious possibility of a non-computability of the type required, in accordance with the discussions of Part I, in any purely classical physical system.

One should not disregard the possibility, however, that there might be something in the precise chaotic behaviour exhibited by some continuous mathematical system (taken as modelling physical behaviour), which cannot be captured by *any* discrete approximation. I am not aware of any such system, but even if such a system exists, it would be of no help to AI—as that subject stands at the moment—because existing AI indeed depends upon modelling by *discrete* computation (that is, by digital rather than analogue computation; cf. §1.8).

In *quantum* physics, there is also some additional freedom of a completely *random* nature, over and above the deterministic (and computable) behaviour that the equations of quantum theory (basically, the Schrödinger equation) provide. Technically, these equations are *not* chaotic, but the absence of chaos is replaced by the presence of the aforesaid random ingredients that supplement the deterministic evolution. As we have seen, in §3.18 particularly, such purely random ingredients do not supply the needed non-algorithmic action either. Thus it appears that neither classical nor quantum physics, as presently understood, allows room for non-computable behaviour of the type required, so we must look elsewhere for our needed non-computable action.

4.3 Consciousness: new physics or 'emergent phenomenon'?

In Part I, I argued (in the particular case of mathematical understanding) that the phenomenon of *consciousness* can arise only in the presence of some non-computational physical processes taking place in the brain. One must presume, however, that such (putative) non-computational processes would *also* have to be inherent in the action of inanimate matter, since living human brains are ultimately composed of the same material, satisfying the same physical laws, as are the inanimate objects of the universe. We must therefore ask two things. First, why is it that the phenomenon of consciousness appears to occur, as far as we know, *only* in (or in relation to) brains—although we should not rule out the possibility that consciousness might be present also in other appropriate physical systems? Second, we must ask how could it be that such a seemingly important (putative) ingredient as non-computational behaviour, presumed to be inherent—potentially, at least—in the actions of all material things, so far has entirely escaped the notice of physicists?

No doubt the answer to the first question has something to do with the subtle and complex organization of the brain, but that, alone, would not provide a sufficient explanation. In accordance with the ideas I am putting forward here, the brain's organization would have to be geared to take advantage of non-computable action in physical laws, whereas ordinary materials would not be so organized. This picture differs markedly from a more commonly expressed view about the nature of consciousness[3] (basically that of $\mathscr{A}$) according to which conscious awareness would be some kind of 'emergent phenomenon', arising merely as a feature of sufficient complexity or sophistication of action, and would not require any specific, new, underlying physical processes, fundamentally different from those that we are already familiar with in the behaviour of inanimate matter. The case presented in Part I argues differently, and it requires that there must be some subtle organization in the brain which is specifically tuned to take advantage of the suggested non-computable physics. I shall have more detailed comments to make about the nature of this organization later (§7.4–§7.7).

With regard to the second question, we must indeed expect that vestiges of such non-computability should also be present, at some indiscernible level, in inanimate matter. Yet the physics of ordinary matter seems, at first sight at least, to allow no room for such non-computable behaviour. Later on, I shall try to explain, in some detail, how such non-computable behaviour could indeed have escaped attention, and how such behaviour is compatible with present-day observations. But for the moment it will be helpful to describe a quite different but, in its way, closely analogous phenomenon from *known* physics. Although not connected—or, at least, not *directly* connected—with any kind of non-computable behaviour, this known physical phenomenon is very much like our putative non-computable ingredient in other ways, being totally indiscernible—though actually present—in the detailed behaviour of ordinary objects; yet it does make its appearance at an appropriate level, and, as it turns out, has profoundly altered the way that we understand the workings of the world. Our story is, in fact, one that has been central to the very march of science.

4.4 The Einstein *tilt*

Since the time of Isaac Newton, the physical phenomenon of *gravity*, and its superbly precise mathematical description (first put forward by Newton fully in 1687), has played a key role in the development of scientific thought. Once this mathematical description was established, gravity served as a beautiful model for the description of other physical processes, where the motions of bodies through a fixed (flat) background space were perceived as precisely controlled by the forces that act upon those bodies, the forces operating as a mutual attraction (or repulsion) between individual particles, controlling all their motions in finest detail. As a result of the outstanding success of Newton's gravitional theory, it was consequently believed that *all* physical processes could be described in this way, where electrical, magnetic, molecular, and other forces also act between particles and control their precise motions, in the same general way as had been seen to work so wonderfully in the case of gravity.

In 1865, this picture was strikingly modified when the great Scottish physicist James Clerk Maxwell published a remarkable set of equations describing the precise behaviour of electric and magnetic fields. These continuous fields were now, themselves, to have an independent existence alongside the various discrete particles. The electromagnetic field (as the combination of the two is now referred to) is capable of carrying energy across otherwise empty space, in the form of light, radio waves, or X-rays, etc., and it has a reality just as great as the Newtonian particles with which it is taken to co-exist. Nevertheless, the general description is still of physical bodies (including, now, the continuous fields) that move about in a fixed space under the influence of their mutual interactions, so the broad picture presented by the Newtonian scheme was not substantially altered. Even quantum theory, with all its revolutionary strangeness, as introduced in 1913–26 by Niels Bohr, Werner Heisenberg, Erwin Schrödinger, Paul Dirac, and others, did not change this aspect of our physical world-view. Physical objects were still taken to be things mutually acting upon one another through fields of force, everything inhabiting this one and the same fixed, flat, background.

Concurrently with some of the early developments in quantum theory, Albert Einstein deeply re-examined the very basis of Newtonian gravity and finally, in 1915, came up with a revolutionary *new* theory which provided a totally different picture: his general theory of relativity (cf. ENM, pp. 202–11). Now, gravity was no longer to be a force at all, but it was to be represented as a kind of *curvature* of the very space (actually space–time) in which all the other particles and forces were to be housed.

Not all physicists were very happy with this unsettling view. They felt that gravity should not be treated so differently from all other physical actions—especially since gravity itself had provided the initial paradigm, on which all later theories of physics had been modelled. Another worry was that gravity is extraordinarily weak, as compared with other physical forces. For example, the gravitational force between the electron and the proton in a hydrogen atom is smaller than the electric force between the same two particles by a factor of about

$$1/28\,500\,000\,000\,000\,000\,000\,000\,000\,000\,000\,000\,000\,000.$$

Thus, gravity is just not noticeable at all, at the level of the individual particles that constitute matter!

A question that would be sometimes raised was: might not gravitation be some kind of *residual* effect, perhaps arising out of the almost complete, but not quite exact, cancellation of all the other forces involved? (Certain forces of this actual nature are known, such as the van der Waals force, hydrogen bond, and London force.) Accordingly, rather than being a quite different physical phenomenon from everything else, having to be described in a completely different mathematical way from all other forces, gravity would not really exist at all as something in its own right, but would be some kind of 'emergent phenomenon'. (For example, the great Soviet scientist and humanitarian Andrei Sakharov once put forward a view of gravity of this nature.[4])

However, it turns out that this kind of idea will *not* work. The basic reason for this is that gravity actually influences the *causal* relationships between

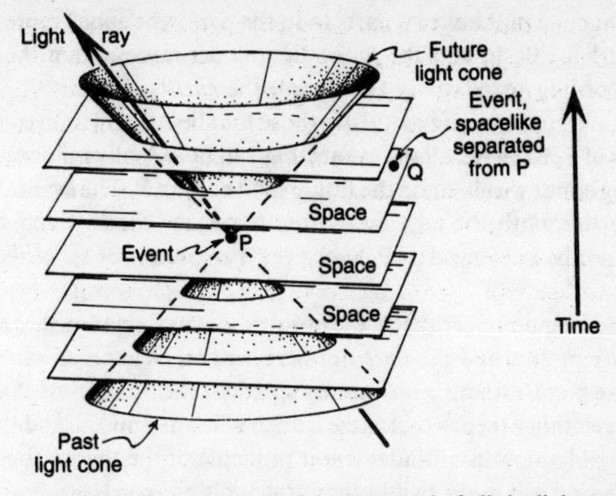

Fig. 4.1. The *light cone* at an event P is composed of all the light rays in space–time through P. It represents the history of a light flash imploding on P (past light cone) which then explodes out again (future light cone). The event Q is *spacelike separated* from P (lying outside P's light cone) and lies outside P's causal influence.

space–time events, and it is the *only* physical quantity that has this effect. Another way of phrasing this is to say that gravity has the unique capacity to 'tilt' the light cones. (We shall see what all this means in a moment.) No physical field *other* than gravity can tilt the light cones, nor can any collection whatever of non-gravitational physical influences.

What does 'tilting the light cones' mean? What are 'causal relations between space–time events'? We shall need to digress a little in order to explain these terms. (This digression will have a separate importance for us later.) Some readers may well be familiar with the relevant concepts already, and I shall give only a brief account here in order to acquaint the others with the necessary ideas. (See ENM, Chapter 5, p. 194 for a more complete discussion.) In Fig. 4.1, I have pictured, in a space–time diagram, an ordinary light cone. Time is depicted in the diagram as progressing from the bottom of the page to the top, while space is depicted as extending horizontally. A point in a space–time diagram represents an *event*, which is some particular spatial point at some particular moment. Events, therefore, have zero temporal duration as well as having zero spatial extension. The complete *light cone* centred at an event P represents the space–time history of a spherical pulse of light, which implodes inwards upon P, and at that moment explodes out again from P, always with the speed of light. Thus, the entire light cone of P is composed of all those light rays that actually encounter the event P in their individual histories.

The light cone of P has two parts to it, the *past* light cone* representing the *im*ploding light flash, and the *future* light cone, representing the *ex*ploding flash. According to relativity theory, all the events that can have a *causal influence* on a space–time event P are those that lie either within or on the past light cone of P; likewise, all the events that can be causally influenced *by* P are those lying either within or on the future light cone of P. The events lying in the region outside both the past and future cones are those which can neither influence nor be influenced by P. Such events are said to be *spacelike* separated from P.

It should be made clear that these notions of causal relationship are features of *relativity theory*, and they are not pertinent to Newtonian physics. In the Newtonian picture, there is no limiting speed for the transfer of information. It is only in relativity theory that there is such a limiting speed, and that speed is the speed of light. It is a fundamental principle of the theory that no causal effect may proceed more swiftly than this limiting speed.

One must be somewhat cautious in the interpretation of what is meant by 'the speed of light' here, however. Actual light signals are slowed down slightly when they pass through a refracting medium, such as glass. Within such a medium, the speed at which a physical light signal would travel would be slower than what we are here calling 'the speed of light', and it is possible for a physical body, or for a physical signal other than a light signal, to exceed the actual speed at which light would travel in such a medium. Such a phenomenon can be observed, in certain physical experiments, in the circumstance of what is known as Cerenkov radiation. Here particles are fired into a refracting medium, where the speed of the particles is only very slightly less than this absolute 'speed of light' but is greater than the speed at which light actually travels in the medium. Shock waves of actual light occur, and this is the Cerenkov radiation.

To avoid confusion, it is best that I refer to this greater 'speed of light' as the *absolute* speed. The light cones in space–time determine the absolute speed, but do not necessarily determine the speed of actual light. In the medium the speed of actual light is somewhat less than the absolute speed, and is less also than that of the particles fired into it that produce Cerenkov radiation. It is the absolute speed (i.e. each light cone) that fixes the speed limit for all signals or material bodies, and although actual light does not necessarily travel at the absolute speed, it always does so in a vacuum.

The 'relativity' theory that we are basically referring to here is *special* relativity, where gravity is absent. The light cones in special relativity are all arranged uniformly, as depicted in Fig. 4.2, and the space–time is referred to as *Minkowski* space. According to Einstein's *general* relativity, the previous discussion still holds good, so long as we continue to refer to the 'absolute speed' as being that determined by the space–time situation of the light cones.

*In the diagrams of ENM, only the future parts of the light cones were depicted.

Fig. 4.2. Minkowski space: the space–time of special relativity. The light cones are all arranged uniformly.

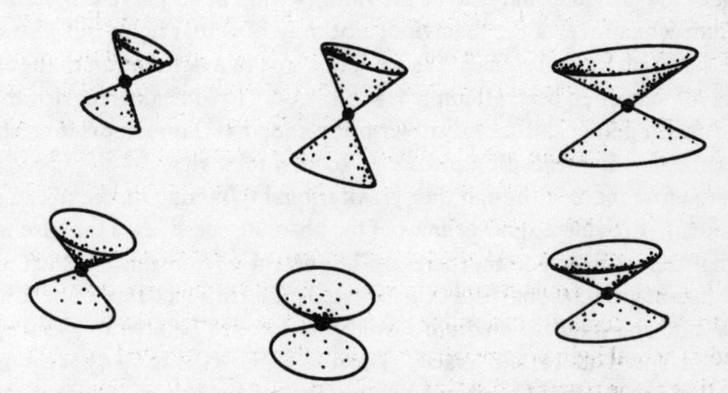

Fig. 4.3. The *tilted* light cones of Einstein's general relativity.

However, it is an effect of gravity that the light cones can become *non*-uniform in their distribution, as depicted in Fig. 4.3. This is what I have referred to as the 'tilting' of the light cones, above.

A not uncommon way of trying to think about this light-cone tilting is in terms of a speed of light—or, rather, of an absolute speed—that *varies* from place to place, where this speed may also be dependent upon the direction of

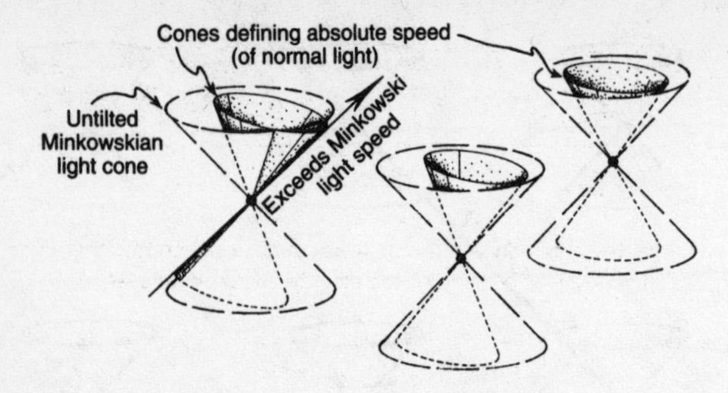

Fig. 4.4. Light propagation according to Einstein's general relativity cannot be thought of as an effect of a 'refractive medium', within the Minkowski space framework, without violating the fundamental principle of special relativity that signals cannot propagate faster than the Minkowski light speed.

motion. In this way one might now try to think of the 'absolute speed' as something analogous to the 'speed of actual light' that was referred to in the above discussion of the refracting medium. Accordingly, one might try to think of the gravitational field as providing a kind of all-pervading refracting medium which affects the behaviour not only of actual light, but also of *all* material particles and signals.* Indeed, this sort of a description of the effects of gravity has often been attempted, and it works to some degree. However, it is not an altogether satisfactory description and, in certain important respects, gives a seriously misleading picture of general relativity.

In the first place, although this 'gravitational refracting medium' can often be taken to provide a *slowing down* of the absolute speed, as is the case with a normal refracting medium, there are important circumstances (such as the very distant gravitational field of an isolated mass) in which this alone does not work, and the putative medium would also have to be able to *speed up* the absolute speed at certain places (Penrose 1980; cf. Fig. 4.4). This is *not* something that could be achieved within the framework of special relativity. According to that theory, a refractive medium, no matter how exotic, could never have the effect of being able to speed up signals to faster than the speed of light in medium-free vacuum, without violating the basic causality principles of the theory—for such a speeding up would allow signals to be propagated outside the (medium-free) Minkowskian light cones, which is not allowed. In particular, the 'light-cone–tilting' effects of gravity, as described above, cannot be interpreted as some residual effect of other non-gravitational fields.

There are certain much more extreme situations in which it would not be

*Remarkably, Newton himself suggested an idea of this nature. (See *Queries* 18–22 of *Book Three* of his *Opticks* 1730.)

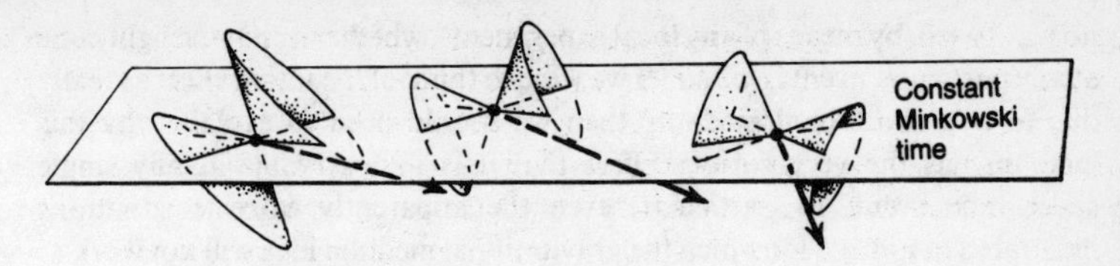

Fig. 4.5. In principle, the light-cone tilting can be so extreme that light signals could be propagated into the Minkowski past.

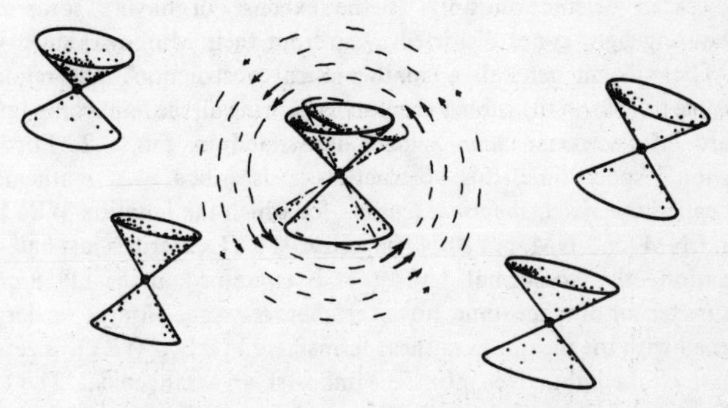

Fig. 4.6. Imagine the space–time as a rubber sheet with the light cones drawn on it. Any individual light cone can be rotated (carrying the rubber sheet with it) into a standard Minkowskian depiction.

possible to describe light-cone tilting in this kind of way at all, even if we allowed the absolute speed to be 'speeded up' in some directions. In Fig. 4.5 a situation is illustrated where this is indeed not possible, the light cones being tilted right over to an absurd-looking degree. In fact, this kind of extreme tilting need only arise in the distinctly questionable situations where 'causality violation' occurs—where it becomes theoretically possible for an observer to send signals into his* own past (cf. Fig. 7.15, in Chapter 7). Remarkably, considerations of this nature *will* have some genuine relevance to our later discussion (§7.10)!

There is also the more subtle point that the 'degree of tilt' of a single light cone is not something physically measurable, and so it does not really make any physical sense to treat it as an *actual* slowing down or speeding up of the absolute speed. This is best illustrated if we think of Fig. 4.3 as a picture drawn on a rubber sheet, so any particular light cone can, in the neighbourhood of its vertex, be rotated and distorted (cf. Fig. 4.6) until it sits 'vertically', just as in the normal Minkowski space pictures of special relativity (Fig. 4.2). There is

*Of course 'his' does not imply 'male'; cf. Notes to the reader on p. xvi.

no way to tell, by means of any local experiments, whether or not the light cone at any particular event is 'tilted'. If we were to think of the tilting effect as really due to a 'gravitational medium' then we should need to explain why this medium has the very curious effect that it is unobservable at any single space–time event. In particular, even the apparently extreme situations illustrated in Fig. 4.5 for which the gravitational medium idea will not work at all are, if we consider just a single light cone, physically no different from what happens in a situation, like in Minkowski space, where that light cone is not tilted at all.

In general, however, we can rotate a particular light cone to its Minkowskian orientation only at the expense of having some of its neighbouring light cones distorted *away* from their Minkowskian orientations. There is, in general, a 'mathematical obstruction' that renders it impossible to deform the rubber sheet so as to bring all the light cones into the standard Minkowskian arrangement illustrated in Fig. 4.2. For four-dimensional space–time, this obstruction is described by a mathematical object called the *Weyl conformal tensor*—for which the notation **WEYL** was used in ENM (cf. ENM, p. 210). (The tensor **WEYL** describes just half of the information—the 'conformal' half—that is contained in the full Riemann curvature tensor of space–time; however, there is no need for the reader to be concerned with the meanings of these terms here.) Only if **WEYL** is zero can we rotate *all* the light cones into the Minkowskian arrangement. The tensor **WEYL** measures the gravitational field—in the sense of the gravitational tidal distortion—so it is precisely the *gravitational field*, in this sense, that provides the obstruction to being able to 'untilt' the light cones.

This tensor quantity is certainly physically measurable. The **WEYL** gravitational field of the moon, for example, exerts its tidal distortion on the earth—giving the main contribution to the origin of the earth's tides (ENM, p. 204, Fig. 5.25). This effect does not directly relate to the tilting of light cones, however, being a feature merely of the Newtonian effects of gravity. More pertinent is another observational effect, *gravitational lensing*, which is a characteristic feature of Einstein's theory. The first observed example of gravitational lensing was that seen by Arthur Eddington's expedition to the island of Principe in 1919, where the distortion of the observed background field of stars by the gravitational field of the sun was carefully noted. The local distortion of this background field has the nature that a small circular pattern in the actual background would be distorted into an elliptical observed one (see Fig. 4.7). This is an almost direct observation of the effects of **WEYL** on the light-cone structure of space–time. In recent times, the gravitational lensing effect has become a very important tool in observational astronomy and cosmology. The light from a distant quasar is sometimes distorted by the presence of an intervening galaxy (Fig. 4.8), and the observed distortions of the appearance of the quasar, together with time-delay effects, can give important information about distances, masses, etc. All this provides clear-cut observa-

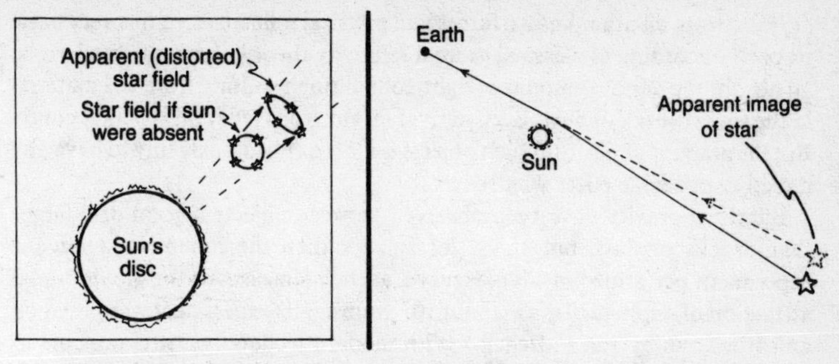

Fig. 4.7. A direct observational effect of light-cone tilting. **WEYL** space–time curvature manifests itself as a distortion of the distant star field, here owing to the light-bending effect of the sun's gravitational field. A circular pattern of stars would get distorted into an elliptical one.

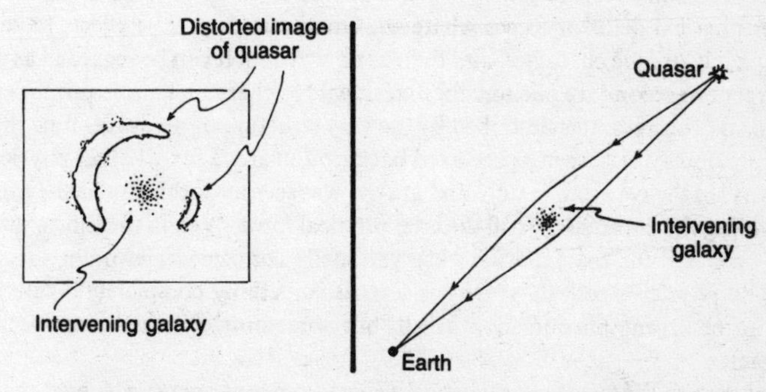

Fig. 4.8. Einstein's light-bending effect is now an important tool in observational astronomy. The mass of an intervening galaxy may be estimated by how much it distorts the image of a distant quasar.

tional evidence for light-cone tilting as an actual phenomenon, and also for the directly measurable effects of **WEYL**.

The foregoing remarks illustrate the fact that the 'tilting' of light cones, i.e. the distortion of causality, due to gravity, is not only a subtle phenomenon, but a *real* phenomenon, and it cannot be explained away by a residual or 'emergent' property that arises when conglomerations of matter get large enough. Gravity has its own *unique* character among physical processes, not directly discernible at the level of the forces that are important for fundamental particles, but nevertheless it is there all the time. Nothing in known physics *other* than gravity can tilt the light cones, so gravity is something that is simply *different* from all other known forces and physical influences, in this very basic respect. According to classical general relativity theory, there must indeed be an absolutely minute amount of light-cone tilting resulting from the material in the tiniest speck of dust. Even individual electrons must tilt the light cones. But the amount of tilting in such objects is far too ridiculously tiny to have any directly noticeable effect whatsoever.

Effects of gravity have been observed between objects a good deal larger than specks of dust, but still a lot smaller than the moon. In a famous experiment performed in 1798, Henry Cavendish measured the gravitational attraction of a sphere of mass about 10^5 grams. (His experiment was based on an earlier one by John Michell.) With modern technology, it is possible to detect the gravitational pull of an object whose mass is much smaller. (See e.g. Cooke 1988.) However, any detection of the light-cone tilting effect of gravity in any of these situations would be totally beyond present-day techniques. It is only with very large masses indeed that light-cone tilting can be directly observed; whereas its actual presence in very tiny amounts in bodies as small as specks of dust is a clear-cut implication of Einstein's theory.

The detailed effects of gravity cannot be simulated by any combination of other physical fields or forces whatever. Gravitation's precise effects have a completely unique character, and there is no way that it can be regarded as an emergent or secondary phenomenon, residual to other much more prominent physical processes. It is described by the very structure of the space–time that had previously been seen as the fixed background arena for all other physical activity. In the Newtonian universe, gravity was seen as nothing special—even providing the paradigm for all the later physical forces. Yet, in the Einsteinian one (which is the marvellously observationally confirmed viewpoint that is held by physicists today), gravity is seen as something completely different: not an emergent phenomenon at all, but something with its own special character.

Yet despite the fact that gravity is different from other physical forces, there is a profound harmony integrating gravity with all of the rest of physics. Einstein's theory is not something foreign to the other laws, but it presents them in a different light. (This is particularly so for the laws of conservation of energy, momentum, and angular momentum.) This integration of Einstein's gravity with the rest of physics goes some way to explaining the irony that Newton's gravity had provided a *paradigm* for the rest of physics despite the fact, as Einstein later showed, that gravity is actually *different* from the rest of physics! Above all, Einstein taught us not to get too complacent in believing, at any stage of our understanding, that we have, as yet, necessarily found the appropriate physical viewpoint.

May we expect that there is something corresponding to be learnt with regard to the phenomenon of consciousness? If so, it would not be *mass* that would need to be large for the phenomenon to become apparent—at least not *only* mass—but some kind of delicate physical organization. According to the arguments put forward in Part I, such organization would have to have found a way of making use of some hidden non-computational ingredient already present in the behaviour of ordinary matter—an ingredient that, like the light-cone tilting of general relativity, would have totally escaped attention had that attention been confined to the study of the behaviour of tiny particles.

But does light-cone tilting have anything to do with non-computabilty? We shall explore an intriguing aspect of this question in §7.10; but at the present stage of argument, nothing whatsoever—*except* that it provides us with a moral: namely, that it is quite possible, in physics, to have a fundamentally important new property, completely different from any that had been contemplated hitherto, hidden unobserved in the behaviour of ordinary matter. Einstein was led to his revolutionary viewpoint from a number of powerful considerations, some mathematically sophisticated, and some physically subtle—but the most important of these had lain prominent, but unappreciated, since the time of Galileo (the principle of equivalence: all bodies fall at the same speed in a gravitational field). Moreover, it was a necessary prerequisite for the success of Einstein's ideas that they should be compatible with all that was known in the physical phenomena of his day.

In an analogous way, we may contemplate that there could be some non-computational action hidden somewhere in the behaviour of things. For such a speculation to have a hope of success, it would also have to be motivated by powerful considerations, presumably both mathematically sophisticated and physically subtle, and it would have to be consistent with all the detailed physical phenomena that are known today. We shall try to see how far we can get along the road to such a theory.

But as a prerequisite, let us first glimpse the powerful hold that computational ideas have on present-day physics. With appropriate irony, we shall find that general relativity itself provides one of Nature's most striking examples.

4.5 Computation and physics

Some 30 000 light years distant from the earth, in the constellation of Aquila, two incredibly dense dead stars orbit about each other. The material of these objects is so compressed that a tennis ball of their substance would have a mass comparable with that of Mars's moon Deimos. These two stars—called *neutron* stars—circle about each other once every 7 hours, 45 minutes, and 6.9816132 seconds, and they have masses that are, respectively, 1.4411 and 1.3874 times the mass of our sun (with a possible error of about 7 in the last decimal place). The first of these stars emits a pulse of electromagnetic radiation (radio waves) in our direction once every 59 milliseconds, indicating that it is rotating about its axis some 17 times in a second. It is what is

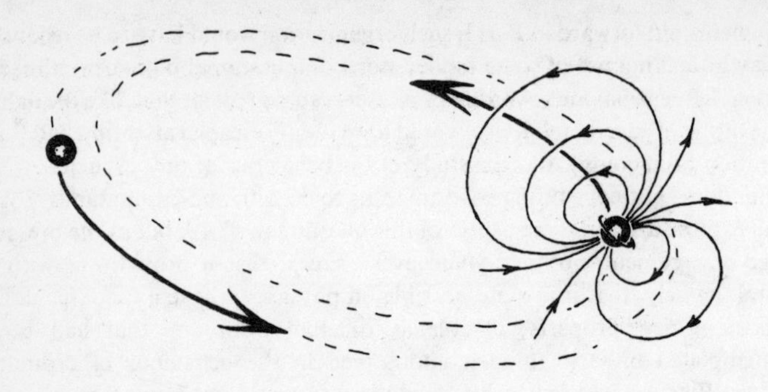

Fig. 4.9. PSR 1913 + 16. Two neutron stars orbit about one another. One of these is a pulsar, with an enormously strong magnetic field that traps charged particles.

known as a *pulsar*, and the pair constitute the famous binary pulsar system PSR 1913 + 16.

Pulsars have been known since 1967, when they were discovered by Joselyn Bell and Anthony Hewish at the Cambridge radio observatory. They are remarkable objects. Neutron stars result, normally, from the gravitational collapse of the core of a red giant star, which may produce a violent supernova explosion. They are almost unbelievably dense, being compacted out of nuclear particles, mainly neutrons, to such a degree that their overall density is comparable with that of the neutron itself. In the collapse, the neutron star would have trapped flux lines of a magnetic field within its substance and, owing to the enormous compression arising as the star collapsed, would have concentrated this field to an extraordinary degree. The magnetic field lines would emerge from the star's north magnetic pole and, after reaching outwards considerable distances into space, would return into the star at its south magnetic pole (see Fig. 4.9).

The star's collapse would have resulted in its rate of rotation, also, being enormously increased (an effect of conservation of angular momentum). In the case of our pulsar referred to above (which would be about 20 kilometres in diameter), this rate is some 17 times a second! This results in the pulsar's extremely strong magnetic field being swung around it at 17 times a second, since the flux lines within the star remain pinned to the body of the star. Outside the star, the field lines carry charged particles around with them, but at a certain distance out from the star, the speed at which these particles must move approaches, closely, the speed of light. Where this happens, the charged particles start to radiate violently, and the powerful radio waves that they emit are beaconed, as by a gigantic lighthouse, to enormous distances. Some of these signals reach earth, as the beacon's beam repeatedly flashes past, to be

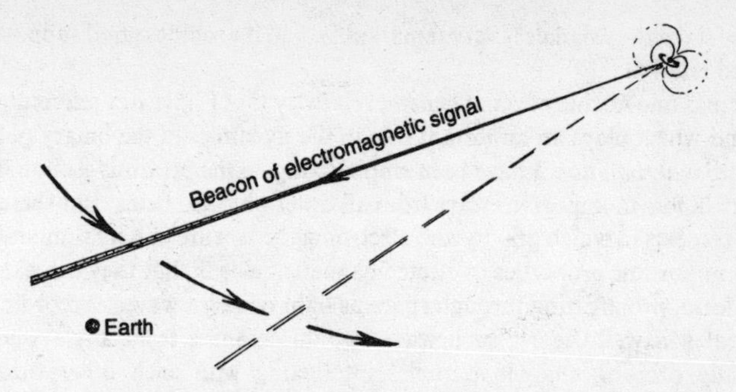

Fig. 4.10. The trapped charged particles swing around with the pulsar and emit an electromagnetic signal whose beam sweeps by the earth 17 times per second. This is received as a sharply pulsed radio signal.

observed by astronomers as the succession of radio 'clicks' that is characteristic of a pulsar (Fig. 4.10).

Pulsars' rotation rates are extremely stable, and they provide clocks whose accuracy matches, or even exceeds, the most perfect (nuclear) clocks that have been constructed here on earth. (A good pulsar clock might lose or gain less than about 10^{-12} seconds in a year.) If the pulsar happens to be part of a binary system, as is the case with PSR 1913+16, then its orbiting motion about its companion can be closely monitored by use of the *Doppler effect* whereby the rate of its 'clicking', as received here on earth, is slightly greater when the pulsar approaches us than it is when it recedes from us.

In the case of PSR 1913+16, it has been possible to get an extraordinarily precise picture of the mutual orbits that the two stars actually execute about one another, and to check a number of different observational predictions of Einstein's general relativity. These include an effect known as 'perihelion advance'—which, in the late 1800s, had been noticed as an anomalous behaviour for the planet Mercury in its orbital motion about the sun, but which Einstein explained in 1916, giving the first test of his theory—and various types of general-relativistic 'wobble' that affect rotation axes, etc. Einstein's theory gives a very clear-cut (deterministic and computable) picture of the way that two small bodies should behave, as they move about in mutual orbit around one another, and it is possible to compute this motion to a high order of accuracy, using careful and sophisticated methods of approximation as well as various standard computational techniques. There are certain unknown parameters involved in such a computation, such as the masses and initial motions of the stars, but there is ample data from the pulsar signals to allow these to be fixed to excellent accuracy. The overall agreement between the computed picture and the finely detailed information that is received in the form of the pulsar signals is very remarkable, and it provides good support for general relativity.

There is one further effect of general relativity that I have not referred to as yet, and which plays an important role in the dynamics of the binary pulsar: *gravitational radiation*. I have been emphasizing, in the previous section, how gravity differs in important ways from all other physical fields. But there are other respects in which gravity and electromagnetism are closely similar. One of the important properties of electromagnetic fields is that they can exist in wave form, propagating through space as light or radio waves. According to classical Maxwell theory, such waves would emanate from any system of mutually orbiting charged particles interacting with each other through electromagnetic forces. Likewise, according to classical general relativity, there would be gravitational waves emanating from any system of gravitating bodies in mutual orbit about one another through their mutual gravitational interactions. In the normal way of things, these waves would be extremely weak. The most powerful source of gravitational radiation in the solar system arises from the motion of the planet Jupiter about the sun, but the amount of energy that the Jupiter–sun system emits in this form is only about that of a 40 watt light bulb!

However, in other circumstances, such as with PSR 1913 + 16, the situation is very different, and gravitational radiation from the system indeed has a significant role to play. Here, Einstein's theory provides a firm prediction of the detailed nature of the gravitational radiation that the system ought to be emitting, and of the energy that should be carried away. This loss of energy should result in a slow spiralling inwards of the two neutron stars, and a corresponding speeding up of their orbital rotation period. Joseph Taylor and Russell Hulse first observed this binary pulsar at the enormous Aricebo radio telescope in Puerto Rico in 1974. Since that time, the rotation period has been closely monitored by Taylor and his colleagues, and the speed-up is in precise agreement with the expectations of general relativity (cf. Fig. 4.11). For this work, Hulse and Taylor were awarded the 1993 Nobel Prize for Physics. In fact, as the years have rolled by, the accumulation of data from this system has provided a stronger and stronger confirmation of Einstein's theory. Indeed, if we now take the system as a whole and compare it with the behaviour that is computed from Einstein's theory as a whole—from the Newtonian aspects of the orbits, through the corrections to these orbits from standard general relativity effects, right up to the effects on the orbits due to loss of energy in gravitational radiation—we find that the theory is confirmed overall to an error of no more than about 10^{-14}. This makes Einstein's general relativity, in this particular sense, the most accurately tested theory known to science!

In this example, we have a particularly 'clean' system, where general relativity, alone, is all that need be involved in the calculation. Such things as complications resulting from the internal constitution of the bodies, or drag due to an intervening medium or due to magnetic fields, have no significant

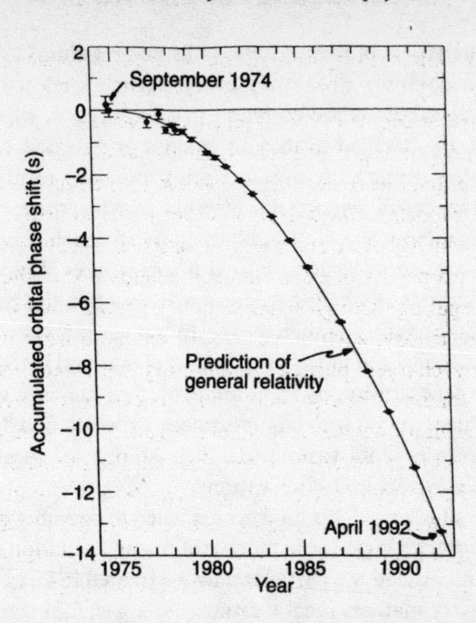

Fig. 4.11. This graph (courtesy of Joseph Taylor) shows the precise agreement, over a 20-year period, of the observational speed-up of the pulsar's mutual orbit with the computed energy-loss due to gravitational radiation according to Einstein's theory.

effect on the motions. Moreover, there are only two bodies involved, together with their mutual gravitational field, so it is well feasible to make a thorough computation of the expected behaviour, according to the theory, complete in every relevant detail. This may well be the most perfect example of a comparison between a computed theoretical model and observed behaviour—involving only a few bodies—that has been yet achieved in science.

When the number of bodies in a physical system is considerably greater than this, it may still be possible, using the full resources of modern computer technology, to model the system's behaviour in as thoroughly detailed a way. In particular, the motions of all the planets in the solar system, together with their most significant moons, have been modelled in a very detailed comprehensive calculation by Irwin Shapiro and his colleagues. This provides another important test of general relativity. Again Einstein's theory fits all the observational data, and it accommodates the various small deviations from the observed behaviour that would have been present had a completely Newtonian treatment been used.

Calculations involving an even larger number of bodies—sometimes of the order of a million—can also be carried out with modern computers, although these would generally (but not always) be based entirely on Newtonian theory. Certain simplifying assumptions might have to be made about how to approximate the effects of many particles by some kind of averaging, rather than having to compute absolutely every particle's effect on every other particle. Such calculations are common in astrophysics, where one may be concerned with the detailed formation of stars or galaxies, or the clumping together of matter in the early universe prior to galaxy formation.

There is, however, an important difference in what these calculations are trying to achieve. Now, we are not likely to be so much concerned with the *actual* evolution of some system, but with a *typical* evolution. As with our earlier considerations of chaotic systems, this may be the best that we can expect to do. It is possible, by such means, to test various scientific hypotheses about the constitution and initial distribution of material in the universe to see how well, in a general way, its resulting evolution fits with what is actually observed. One does not, in such circumstances, expect to find a detailed fit, but general appearances and various detailed statistical parameters can be checked between model and observation.

The extreme situation of this kind occurs when the number of particles is so large that it becomes a hopeless matter to follow the evolution of each of them individually, and instead the particles must be treated in an entirely statistical way. The ordinary mathematical treatment of a gas, for example, deals with statistical *ensembles* of different possible collections of particle motions, and it is not concerned with the particular motions of individual particles. Such physical quantities as temperature, pressure, entropy, etc., are properties of such ensembles, but again they can be treated as part of a computational system, where the evolutionary properties of these ensembles are treated from a statistical point of view.

In addition to the relevant dynamical equations (Newton's, Maxwell's, Einstein's, or whatever) there is, in such circumstances, another physical principle that must be involved. This is the *second law of thermodynamics*.[5] In effect, this law serves to rule out initial states of the individual particle motions that would lead to overwhelmingly improbable, though dynamically possible, future evolutions. The introduction of the second law serves to ensure that the future evolution of the system being modelled is indeed 'typical', rather than something grossly *a*typical which has no practical relevance to the problem at hand. With the aid of the second law, it becomes possible to compute future evolutions of systems where there are so many particles involved that a detailed treatment of individual motions can in no way be achieved in practice.

It is an interesting question—and indeed a profound one—why such evolutions cannot be carried out reliably into the *past*, despite the fact that the dynamical equations of Newton, Maxwell, and Einstein are all completely symmetrical in time; for in the real world, the second law does not apply in the reverse direction in time. The ultimate reason for this has to do with the very special conditions that held at the beginning of time—the big bang origin of the universe. (See ENM, Chapter 7, for a discussion of these matters.) In fact, these initial conditions were so precisely special that they provide yet another example of the extraordinary precision whereby observed physical behaviour is modelled by clear-cut mathematical hypotheses.

In the case of the big bang, an essential part of the relevant hypotheses is that in its very early stages, the matter content of the universe was in a state of *thermal equilibrium*. What does 'thermal equilibrium' mean? The study of thermal equilibrium states represents the opposite extreme from the precise modelling of the detailed motions of just a few objects, such as in the case of the binary pulsar above. Now, it is just 'typical behaviour' in its purest and most reliable sense that is one's concern. An equilibrium state, generally, is a state of a system which has completely 'settled down', and the system will not deviate significantly from that state, even if disturbed slightly. For a system with a large number of particles (or a large number of degrees of freedom)—so that one is concerned not with detailed individual particle motions but with averaged behaviour and average measures such as temperature and pressure—this is the *thermal* equilibrium state to which the system ultimately converges, in accordance with the second law of thermodynamics (maximum entropy). The qualification 'thermal' implies that there is some kind of averaging over the large numbers of alternative individual particle motions involved. It is the subject of thermodynamics that is concerned with such averages—i.e. with typical rather than individual behaviour.

Strictly speaking, in accordance with what has been said earlier, when we refer to the thermodynamical state of a system or to thermal equilibrium, this does not pertain to an individual state, but rather to an ensemble of states, all of which appear the same on a macroscopic scale (and the entropy, roughly speaking, is the logarithm of the number of states in this ensemble). In the case of a gas in equilibrium, if we fix the pressure, volume, and the amount and composition of the gas particles, we obtain a very specific distribution of the probable particle velocities at thermal equilibrium (as first described by Maxwell). A more detailed analysis reveals a scale at which statistical fluctuations away from the idealized thermal equilibrium state are to be expected—and here we begin to enter the more sophisticated areas of the study of the statistical behaviour of matter, which goes under the heading of *statistical mechanics*.

Again, it appears that there is nothing essentially non-computable about the modelling of physical behaviour by mathematical structures. When the appropriate calculations have been made, there is good agreement between what is calculated and what is observed. However, when systems more complicated than dilute gases or of large collections of gravitating bodies are considered, it is not likely that one can steer entirely clear of the issues raised by the *quantum-mechanical* nature of the materials concerned. In particular, in the purest and most precisely tested example of thermodynamic behaviour—the thermal equilibrium state of matter and radiation known as a *black-body* state—this cannot be treated entirely classically, and it turns out that quantum processes are fundamentally involved. Indeed, it was Max Planck's analysis of

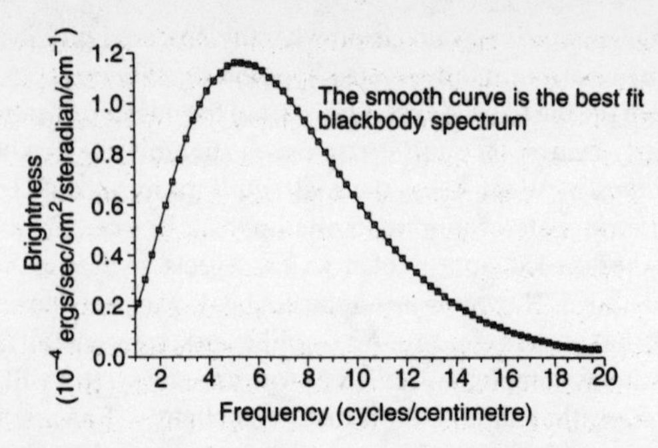

Fig. 4.12. The precise agreement between the COBE measurement and the expected 'thermal' nature of the big bang's radiation.

black-body radiation in 1900 that initiated the whole subject of quantum theory.

Nevertheless, the predictions of physical theory (now quantum theory) are triumphantly verified. The observed relationship between frequency and the intensity of the radiation at that frequency agrees, in experiments, very closely with the mathematical formula put forward by Planck. Although this section has been really concerned with the computational nature of *classical* theory, I cannot resist showing you what is by far the most perfect example of agreement between observation and Planck's formula that I know of. This example also provides a wonderful observational confirmation of the standard model of the big bang, with regard to what it asserts should be the precise thermal conditions pertaining to the universe after the first few minutes of its existence. In Fig. 4.12, the individual small boxed points indicate the various observed values of the intensity of the cosmic background radiation at different frequencies, as observed by the COBE satellite; the continuous curve is drawn according to Planck's formula, taking the temperature of the radiation to have the (best-fit) value of $2.735(\pm 0.06)$ K. The precision of the agreement is extraordinary.

The specific examples that I have referred to above were taken from the area of astrophysics, where the comparison between complicated computations and the observed behaviour of systems occurring in the natural world is particularly well developed. One cannot directly experiment in astrophysics, so theories have to be tested by comparing the results of detailed computations based on standard physical laws, in different proposed situations, with sophisticated observations. (These observations may be ground based, or made from balloons or aircraft in the upper atmosphere, or from rockets or satellites; and they involve many different kinds of detectors in addition to ordinary telescopes.) Such computations are not specifically relevant to those that would be our concern here, and I have mentioned them mainly because they provide particularly clear-cut examples in which detailed computations provide a wonderful way of exploring Nature, illustrating how closely computational procedures can indeed mimic Nature. It is the study of biological systems that should, on the other hand, be more directly our concern here. For, in accordance with the conclusions of Part I, it is in the behaviour of conscious brains that we should seek a role for some non-computable physical action.

It is undoubtedly the case that computational models play important roles in the modelling of biological systems, but the systems are likely to be much more complicated than they are in astrophysics, and the computational models correspondingly harder to make reliable. There are very few systems that are 'clean' enough for any great accuracy to be achieved. Comparatively simple systems, such as with the flow of blood along different types of blood vessel, can be modelled quite effectively, and so also can the transmission of signals along nerve fibres—although in the latter case it is beginning to become a little unclear that the problem remains one of classical physics, because chemical actions are important here, as well as classical physical ones.

Chemical actions are the result of quantum effects, and strictly speaking one has left the arena of classical physics when considering processes that are dependent upon chemistry. This notwithstanding, it is very commonly the case that such quantum-based actions are treated in an essentially classical way. Although this is not technically correct, it is felt that in most cases the more subtle effects of quantum theory—over and above those which can be subsumed into the standard rules of chemistry, classical physics, and geometry—are not of relevance. It seems to me, on the other hand, that whereas this may be a reasonably safe procedure for the modelling of many biological systems (perhaps even nerve-signal propagation), it is risky to try to draw general conclusions about the more subtle of biological actions on the basis of their being entirely classical, particularly when it comes to the most sophisticated of biological systems such as human brains. If we try to make general inferences about the theoretical possibility of a reliable computational model of the brain, we ought indeed to come to terms with the mysteries of quantum theory.

In the next two chapters, we shall be attempting to do just this—at least, in so far as it is possible. Where I believe it is in principle *not* possible to come to terms with quantum theory, I shall argue that we must try to bend the theory itself, to see how it may better fit in with a believable picture of the world.

Notes and references

1. See, for example, Dennett (1991), p. 49.
2. One such important equation is 'the first law of thermodynamics': $dE = TdS - pdV$. Here E, T, S, p, and V are, respectively, the energy, the temperature, the entropy, the pressure, and the volume of a gas.

3. For example, Dennett (1991).
4. Sakharov (1967), cf. Misner *et al.* (1973), p. 428.
5. For a graphic, but not very detailed, account of the second law, see ENM, Chapter 6. For the accounts of increasing sophistication, see Davies (1974) and O. Penrose (1970).

5

Structure of the quantum world

5.1 Quantum theory: puzzle and paradox

Quantum theory provides a superb description of physical reality on a small scale, yet it contains many mysteries. Without doubt, it is hard to come to terms with the workings of this theory, and it is particularly difficult to make sense of the kind of 'physical reality'—or lack of it—that it seems to imply for our world. Taken at its face value, the theory seems to lead to a philosophical standpoint that many (including myself) find deeply unsatisfying. At best, and taking its descriptions at their most literal, it provides us with a very strange view of the world indeed. At worst, and taking literally the proclamations of some of its most famous protagonists, it provides us with no view of the world at all.

In my own opinion, one should make a clear distinction between two quite different kinds of mystery that the theory presents us with. There are those that I shall call **Z**-mysteries, or *puzzle* mysteries, which are genuinely puzzling, but directly experimentally supported, quantum truths concerning the world we live in. Included, also, would be things of this general nature which, though not actually experimentally verified as yet, leave little doubt, in view of what has already been established, but that the expectations of the quantum theory must be fulfilled. Some of the most striking of the **Z**-mysteries are those that go under the name of *Einstein–Podolsky–Rosen* (or EPR) phenomena, which I shall be discussing in some detail later (§5.4, §6.5). The other kinds of quantum mystery are things that I shall refer to as **X**-mysteries, or *parado*x mysteries, which would, on the other hand, be things that the quantum formalism seems to be telling us have to be true of the world, but have such an implausibly paradoxical nature that we cannot really believe in them as being in any sense 'actually' true. They are mysteries which prevent us from taking the formalism seriously, at the level concerned, as providing a believable picture of our world. The best known **X**-mystery is the paradox of *Schödinger's cat*, where the formalism of quantum theory seems to be telling us that large-scale objects, such as cats, can exist in two totally different states simultaneously—such as the limbo of simultaneous combination of 'cat dead' and 'cat alive'. (I shall discuss this kind of paradox in §6.6; cf. §6.9, Fig. 6.3, and ENM pp. 290–3.)

It is not infrequently contended that the difficulties that our present generations have in coming to terms with the quantum theory are purely a result of our being wedded to our physical concepts of the past. Accordingly, each successive generation would become more attuned to these quantum mysteries, so after sufficiently many generations they would become able to accept them all without difficulty, whether they be **Z**-mysteries or **X**-mysteries. My own view, however, differs fundamentally from this.

I believe that the **Z**-mysteries are things that we might indeed get used to and eventually accept as natural, but that this will *not* be the case for the **X**-mysteries. In my opinion, the **X**-mysteries are philosophically unacceptable, and arise merely because the quantum theory is an incomplete theory—or, rather, because it is not completely accurate at the level of phenomena at which the **X**-mysteries begin to appear. It is my view that, in an improved quantum theory, the **X**-mysteries will simply be removed (i.e. *crossed off*) the list of quantum mysteries. It is merely the **Z**-mysteries in whose presence we must learn to snooze peacefully!

Bearing this in mind, there may well be some question as to where to draw the line between the **Z**-mysteries and the **X**-mysteries. Some physicists would contend that there are no quantum mysteries that should be classified as **X**-mysteries in this sense, and that *all* the strange and seemingly paradoxical things that the quantum formalism tells us to believe must actually be true of the world if we look at it in the right way. (Such people, if they are entirely logical and if they take seriously the 'quantum-state' description of physical reality, would have to be believers in some form of 'many-worlds' viewpoint, as will be described in §6.2. In accordance with this viewpoint, Schrödinger's dead cat and his live cat would inhabit different 'parallel' universes. If you looked at the cat, then there would be copies of you, also, in each of the two universes, one seeing a dead cat and the other a live cat.) Other physicists would tend towards an opposite extreme, and would contend that I have been too generous with the quantum formalism in agreeing with it that all the EPR-type puzzles that will be concerning us later will actually be supported by future experiment. I do not insist that everyone need take the same view as I do, as to where to draw the line between the **Z**- and the **X**-mysteries. My own choice is governed by the expectations that would be consistent with the viewpoint that I shall be giving later, in §6.12.

It would be inappropriate for me to attempt to give a completely thorough account of the nature of quantum theory in these pages. Instead, in this chapter, I shall give a relatively brief and adequately complete description of its necessary features, concentrating, to a large extent, on the nature of its **Z**-mysteries. In the following chapter, I shall give my reasons for believing that, because of its **X**-mysteries, present-day quantum theory must be an incomplete theory, despite all the wonderful agreement that the theory has had with all experiments performed so far. Those readers who wish to pursue more of the details of quantum theory may wish to read the account given in ENM, Chapter 6, or else, for example, Dirac (1947) or Davies (1984).

Later on, in this account—Chapter 6, §6.12—I shall present a recent idea concerning the level at which schemes for the completion of quantum theory ought to become relevant (and I had better warn the reader that this idea differs significantly from that given in ENM, though the motivations are closely similar). Then, in §7.10 (and §7.8), I shall give some suggestive reasons for believing that such a scheme might well be non-computational in the general kind of way needed. *Standard* quantum theory, on the other hand, is non-computational only in the sense that it contains random elements as part of the measurement procedure. As I have stressed in Part I (§3.18, §3.19), random elements alone do not provide the type of non-computability that we shall ultimately need for an understanding of mentality.

Let us start with some of the most striking of the **Z**-mysteries of quantum theory. I shall illustrate these in terms of two quantum brain-teasers.

5.2 The Elitzur–Vaidman bomb-testing problem

Imagine a design of bomb that has a detonator on its nose so sensitive that the slightest touch will set it off. Even a single photon of visible light would be sure to do this were it not for the fact that in some cases the detonator is jammed—so the bomb fails to explode, and must be classed as a dud. Let us suppose that the detonator consists of a mirror attached to the bomb's nose, so that if a (visible light) photon gets reflected off the mirror, its recoil would be sufficient to move a plunger into the bomb and set if off—unless, of course, the bomb is a dud, which would be the case when its sensitive plunger is jammed. We are to suppose that, within the realm of classically operating devices, there is no way to ascertain, once the bomb has been assembled, whether the detonator is jammed without actually wiggling it in some way—something that would certainly set the bomb off. (We shall take it that the only time when the detonator might have got jammed would have been in the assembling of the bomb in the first place.) See Fig. 5.1.

We must imagine that there is a large supply of these bombs (money is no object!), but that the percentage of duds may be quite high. The problem is to find one bomb for which it is certain that it is not a dud.

This problem (and its solution) was proposed by Avshalom Elitzur and Lev Vaidman (1993). I shall defer explaining the solution for the time being, as some readers, already familiar with quantum theory and with what I have referred to as its **Z**-mysteries, may wish to try their hands (or preferably minds) at finding a solution. Suffice it to say, for the moment, that there *is* a solution and that this solution, given an unlimited supply of bombs of this nature, would be well within the bounds of present technology. For those who are not

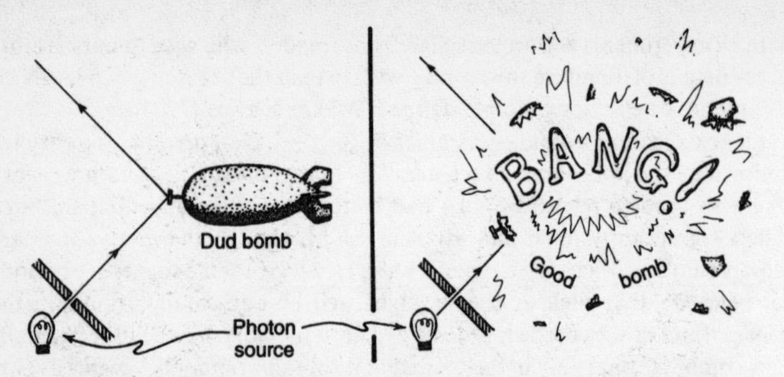

Fig. 5.1 The Elitzur–Vaidman bomb-testing problem. The bomb's ultra-sensitive detonator will respond to the impulse of a single visible-light photon—assuming that the bomb is not a dud because its detonator is jammed. Problem: find a guaranteed good bomb, given a large supply of questionable ones.

already versed in the quantum theory—or who are, but have no wish to waste their time searching for a solution—please bear with me for the moment (or go directly to §5.9). I shall give the solution in due course, after the necessary basic quantum ideas have been explained.

At this stage, it need merely be pointed out that the fact that this problem *has* a (quantum-mechanical) solution already indicates a profound difference between quantum and classical physics. Classically, as the problem is phrased, there is no way of deciding whether the bomb detonator has jammed other than by *actually* wiggling it—in which case, if the detonator is not jammed, the bomb goes off and is lost. Quantum theory allows for something different: a physical effect that results from the possibility that the detonator *might* have been wiggled, even if it was *not* actually wiggled! What is particularly curious about quantum theory is that there can be actual physical effects arising from what philosophers refer to as *counterfactuals*—that is, things that might have happened, although they did not in fact happen. In our next **Z**-mystery, we shall see that the issue of counterfactuals looms large also in a different type of situation.

5.3 Magic dodecahedra

For our second **Z**-mystery, let me describe a little story—and a puzzle.[1] Imagine that I have recently received a beautifully made regular dodecahedron (Fig. 5.2). It was sent to me by a company of superb credentials, known as 'Quintessential Trinkets', who inhabit a planet orbiting the distant red giant star Betelgeuse. They have also sent another identical dodecahedron to a colleague of mine who lives on a planet orbiting the star α-Centauri, which

Fig. 5.2. The magic dodecahedron. My colleague has an identical copy on α-Centauri. On each vertex is a button, and pressing one *may* ring the bell and initiate the magnificent pyrotechnic display.

is about four light years away from us, and his dodecahedron arrived there at roughly the same time as mine did here. Each dodecahedron has a button that can be pressed, on each vertex. My colleague and I are to press individual buttons one at a time, independently on our respective dodecahedra, at some time and in some order that is completely up to our own individual choosing. Nothing may happen when one of the buttons is pressed, in which case all we do is to proceed to our next choice of button. On the other hand, a bell may ring, when one of the buttons is pressed, accompanied by a magnificent pyrotechnic display which destroys that particular dodecahedron.

Enclosed with each dodecahedron is a list of guaranteed properties that relate what can happen to my dodecahedron and to my colleague's. First, we must be careful to orient our respective dodecahedra in a very precise corresponding way. Detailed instructions are provided by Quintessential Trinkets as to how our dodecahedra are to be aligned, in relation to, say, the centres of the Andromeda galaxy and the galaxy M-87, etc. The important thing is only that my dodecahedron and that of my colleague must be perfectly aligned with one another. The list of guaranteed properties is, perhaps, fairly long, but all we shall need from them is something quite simple.

We must bear in mind that Quintessential Trinkets have been producing things of this nature for a very long time—of the order of a hundred million years, let us say—and they have never been found to be wrong in the properties that they guarantee. The very excellent reputation that they have built up over a million centuries depends upon this, so we can be quite sure that whatever

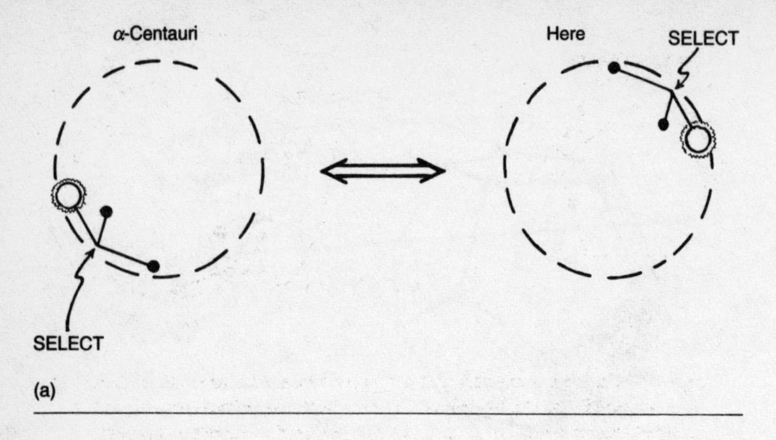

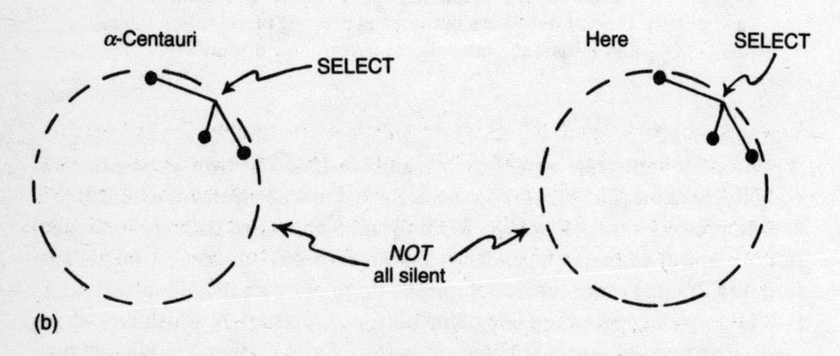

Fig. 5.3. Properties guaranteed by Quintessential Trinkets. (a) If we SELECT *opposite* vertices, the bell can ring only on diametrically opposite buttons, irrespective of order. (b) If we SELECT *corresponding* vertices, the bell cannot fail to ring on all six pressings.

they claim will turn out actually to be true. What is more, there is a stupendous CASH prize (still unclaimed) for anyone who *does* find them to be wrong!

The guaranteed properties that we shall need concern the following type of sequence of button-pressings. My colleague and I each independently select one of the vertices of our respective dodecahedra. I shall call these the SELECTED vertices. We do *not* press those particular buttons; but we *do* press, in turn, and in some arbitrary order of our choosing, each of the three buttons that inhabit vertices *adjacent* to the SELECTED one. If the bell rings on one of them, then that stops the operation on that particular dodecahedron, but the bell need not ring at all. We shall require just two properties. These are (see Fig. 5.3):

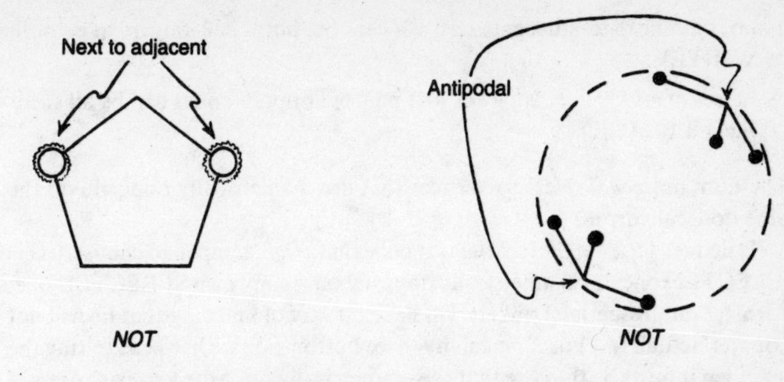

Fig. 5.4. On the assumption that our two dodecahedra are independent (unconnected) objects, we deduce that each button on mine is preassigned as a bell-ringer (WHITE) or else as silent (BLACK), where no two next-to-adjacent buttons can be WHITE and where no six adjacent to a pair of antipodal ones can be all BLACK.

(a) if my colleague and I happen to have chosen diametrically *opposite* vertices as our respective SELECTED ones, then the bell can ring on one of the ones I press (adjacent to my SELECTED one) if and only if the bell rings on the diametrically opposite one of his—irrespective of the particular orders in which either of us may choose to press our respective buttons;

(b) if my colleague and I happen to have chosen exactly *corresponding* vertices (i.e. in the *same* directions out from the centre) as our respective SELECTED ones, then the bell must ring on at least one of six button-presses that we can make altogether.

Now I want to try to deduce something about the rules that my *own* dodecahedron must satisfy independently of what happens on α-Centauri, merely from the fact that Quintessential Trinkets are able to make such strong guarantees without having any idea as to which buttons either I or my colleague are likely to press. The key assumption will be that there is no long-distance 'influence' relating my dodecahedron to my colleague's. Thus, I shall suppose that our two dodecahedra behave as separate, completely independent objects after they have left the manufacturers. My deductions (Fig. 5.4) are:

(c) each of my own dodecahedron's vertices must be preassigned as either a bell-ringer (colour it WHITE) or as silent (colour it BLACK), where its bell-ringing character is independent of whether it is the first, second, or third of the buttons pressed adjacent to the SELECTED one;

(d) no two next-to-adjacent vertices can be both bell-ringers (i.e. both WHITE);

(e) no set of six vertices adjacent to a pair of antipodal ones can be all silent (i.e. all BLACK).

(The term *antipodal* refers to vertices that are diametrically opposite on the same dodecahedron.)

We deduce (c) from the fact that my colleague *might* happen to choose, as his SELECTED one, the diametrically opposite one to my own SELECTED one; at least, Quintessential Trinkets will have no way of knowing that he will not (counterfactuals!). Thus if one of my three button-presses happens to ring the bell, then it must be the case that the diametrically opposite vertex, *if* pressed by my colleague as the first of his three, must also ring his bell. This would be the case whichever order I had chosen for pressing my three buttons, so (by the absence of 'influence' assumption) we can be sure that Quintessential Trinkets must have prearranged that particular vertex to be a bell-ringer, irrespective of my ordering, so as to ensure no conflict with (a).

Likewise, (d) follows from (a) also. For suppose two next-to-adjacent vertices on my dodecahedron are both bell-ringers. Whichever of these I choose to press first must ring the bell—and suppose I have chosen their common neighbour to be my SELECTED one. The order in which I press them now *does* make a difference to which bell rings, which contradicts (a) if my colleague happens to choose his SELECTED one opposite to mine (an eventuality that Quintessential Trinkets must certainly be prepared for).

Finally, (e) follows from (b), together with what we have now established. For suppose that my colleague happens to choose, as his SELECTED one, the vertex *corresponding* to my own SELECTED choice. If none of my three buttons adjacent to this choice is a bell-ringer, then, by (b), one of my colleague's three must be a bell-ringer. It follows from (a) that my own vertex opposite to my colleague's bell-ringer must also be a bell-ringer. This establishes (e).

Now comes the puzzle. Try to colour each of the vertices of a dodecahedron either WHITE or BLACK consistently according to the rules (d) and (e). You will find that no matter how hard you try you cannot succeed. A better puzzle, therefore, is to provide a *proof* that there is *no* such colouring. In order to give any sufficiently motivated reader a chance to find an argument, I have postponed my own proof to Appendix B (p. 300), where I give a fairly straightforward demonstration, establishing that indeed no such colouring is possible. Perhaps some reader will come up with a snappier proof.

Can it be that, for the first time in a million centuries, Quintessential Trinkets have made a mistake? Having established that it is *impossible* to colour the vertices in accordance with (c), (d), and (e), and recalling the stupendous CASH prize, we eagerly wait the four years, or so, that is required for my colleague's message to arrive, describing what he did and when and whether his own bell rang; but when his message arrives, all hopes of CASH vanish, for Quintessential Trinkets have turned out to be right again!

What the argument of Appendix B (p. 300) shows is that there is simply *no way*, in terms of any classical type of model, to construct magic dodecahedra satisfying the conditions that Quintessential Trinkets were able to guarantee, the two dodecahedra being assumed to act as separate independent objects after they left the manufacturers. For it is *not possible* to guarantee the two required properties (a) and (b) without some kind of mysterious 'connection' holding between the two dodecahedra—a connection that persists right up until we begin to press our buttons on the corners, and which would appear to have to act instantaneously over a distance of about four light years. Yet Quintessential Trinkets find themselves able to provide such a guarantee—for something that seems to be impossible—and they have never been found to be wrong!

How do Quintessential Trinkets—or 'QT', as they are known for short—actually do it? Of course 'QT' really stands for *Quantum Theory*! What QT have actually done is to arrange for an atom, whose *spin* has the particular value $\frac{3}{2}$, to be suspended at the centre of each of our dodecahedra. These two atoms have been produced on Betelgeuse in an initial combined state of total spin 0, and then carefully separated and isolated at the centres of our two dodecahedra, so that their combined total spin value remains at 0. (We shall see what all this means in §5.10.) Now, when either my colleague or I press one of the buttons at a vertex of our dodecahedron, a particular kind of (partial) spin measurement is made, in the direction out from the centre of that particular vertex. If the result of this measurement is found to be affirmative, then the bell rings, and the pyrotechnic display follows shortly afterwards. I shall be more specific about the nature of this measurement later (cf. §5.18), and I shall show in §5.18 and Appendix B, why the rules (a) and (b) are a consequence of the standard rules of quantum mechanics.

The remarkable conclusion is that the assumption of no long-distance 'influence' is actually *violated* in quantum theory! A glance at the space–time diagram of Fig. 5.5 will make it clear that the button-pressings that I and my colleague make are *spacelike separated* (cf. §4.4), so that according to relativity theory there can be no signals passing between us that transmit information about which buttons we press or about which button, in either case, actually rings the bell. Yet according to quantum theory, there is, nevertheless, some kind of 'influence' that connects our dodecahedra across spacelike-separated events. In fact, this 'influence' cannot be used to send directly usable *information* instantaneously, and there is no operational conflict between special relativity and quantum theory. Yet there is a conflict with the *spirit* of special relativity—and we have here an illustration of one of the profound **Z**-mysteries of quantum theory: the phenomenon of *quantum non-locality*. The two atoms at the centres of our dodecahedra constitute what is called a single

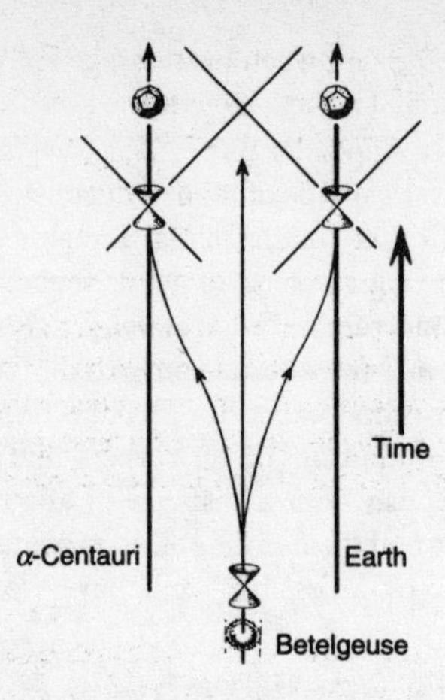

Fig. 5.5. Space–time diagram of the history of the two dodecahedra. They reach α-Centauri and earth at spacelike-separated events.

entangled state, and they can *not*, according to the rules of standard quantum theory, be considered as separate independent objects.

5.4 Experimental status of EPR-type Z-mysteries

The specific example that I have given here is one of a class of (thought) experiments referred to as EPR measurements, after a famous paper written in 1935 by Albert Einstein, Boris Podolsky, and Nathan Rosen. (See §5.17 for a more detailed discussion of EPR effects.) The original published version did not refer to spin, but to certain combinations of position and momentum. Subsequently, David Bohm presented a spin version, involving a pair of particles of spin $\frac{1}{2}$ (say, electrons) that are emitted from a point in a combined state of spin 0. The seeming conclusion of these thought experiments was that a measurement performed, in one place in space, on one member of a quantum pair of particles, could instantaneously 'influence' the other member in a very specific way, though that other particle might be at an arbitrary distance from the original one. Yet, such an 'influence' could not be used to send an actual message from one to the other. In the terminology of quantum theory, the two particles are said to be in a state of *entanglement* with one another. The phenomenon of quantum entanglement—a genuine **Z**-mystery—was first noticed as a feature of quantum theory by Erwin Schrödinger (1935*b*).

Much later, in a remarkable theorem published in 1966, John Bell showed that there would be certain mathematical relationships (Bell inequalities) holding between the joint probabilities of various spin measurements that

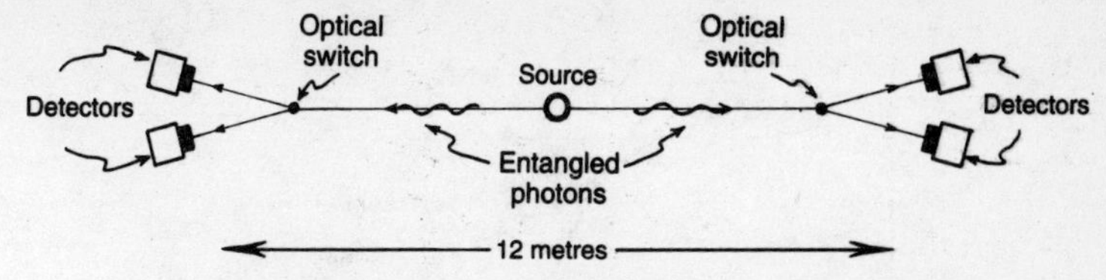

Fig. 5.6. The EPR experiment of Alain Aspect and colleagues. Photon pairs are emitted at the source in an entangled state. The decision as to which direction to measure each photon's polarization is not made until the photons are in full flight—too late for a message to reach the opposite photon, telling it of the direction of measurement.

might be made on any two such particles, that would be necessary consequences of their being separate independent entities, as would be the case in ordinary classical physics. Yet, in quantum theory these relationships could be violated in a very specific way. This opened up the possibility of real experiments that could be performed to test whether these relationships are indeed violated by actual physical systems, as quantum theory says they ought to be, whereas with a classical-type picture in which spatially separated objects have to behave independently of one another, these relationships would necessarily be satisfied. (See ENM, pp. 284, 301, for examples of this kind of thing.)

As an illustration of what such entanglements do *not* mean, John Bell liked to give the example of *Bertlmann's socks*. Bertlmann was a colleague of his who invariably wears socks of different colours. This is a known fact about Bertlmann. (Having once met Bertlmann myself, I can also confirm that my own observations were consistent with this fact.) Thus, if one happened to catch sight of his left sock and noticed that it was green, then one would instantly obtain the knowledge that his right sock was *not* green. Nevertheless, it would be unreasonable to infer that there was a mysterious influence travelling instantaneously from his left sock to his right sock. The two socks are independent objects, and it does not require Quintessential Trinkets to ensure that the differing-sock property will hold. The effect can be easily arranged merely by Bertlmann determining ahead of time that his socks will differ in colour. Bertlmann's socks do not violate Bell's relationships, and there is no long-distance 'influence' connecting his socks. However, in the case of QT's magic dodecahedra, no explanation of the 'Bertlmann socks' type can explain their guaranteed properties. That, after all, was the whole point of the discussion given in the previous section.

Some years after Bell published his original paper, a number of actual experiments were suggested,[2] and subsequently performed,[3] culminating in the famous 1981 Paris experiment by Alain Aspect and his colleagues, which

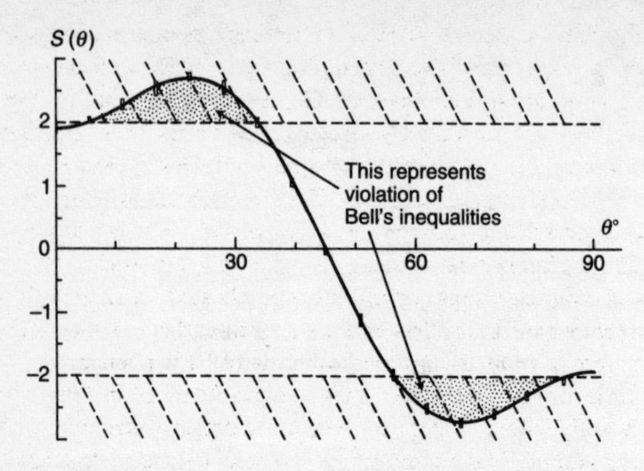

Fig. 5.7. The Aspect experiment closely matches the expectations of quantum theory—in violation of the classical inequalities of Bell. It is hard to see how better detectors would spoil this agreement.

used pairs of 'entangled' photons (cf. §5.17) emitted in opposite directions to a distance of some 12 metres separation. The expectations of quantum theory were triumphantly vindicated, confirming the physical reality of **Z**-mysteries of the EPR type, as predicted by standard quantum theory—and violating the Bell relationships. See Fig. 5.6.

It should be mentioned, however, that despite the very accurate agreement that the results of the Aspect experiment have with the predictions of quantum theory, there are still a few physicists who refuse to accept that the phenomenon of quantum non-locality has been established. They may point to the fact that the photon detectors in the Aspect experiment (and others like it) are rather insensitive, and that most of the emitted pairs in a long run of measurements would remain undetected. They thus have to argue that if the photon detectors were made more sensitive, then at some stage of improvement, the excellent agreement between the expectations of quantum theory and observation would disappear, and the relationships that Bell showed must hold for a local classical system would somehow be restored. To my own way of thinking, it would be exceedingly *un*likely that the excellent fit between quantum theory and experiment that is exhibited in the Aspect experiment (see Fig. 5.7) is somehow an artefact—an artefact of the *in*sensitivity of the detectors—and that with more perfect detectors the agreement with theory would somehow go away, to the considerable extent that would be needed in order that the Bell relationships could be recovered.[4]

The original Bell argument provided relationships (inequalities) between the joint *probabilities* of the different possible outcomes. To estimate the actual probabilities that are involved in a physical experiment, it is necessary to have a long run of observations which must then be subjected to the appropriate statistical analysis. More recently, a number of alternative schemes for (hypothetical) experiments have been proposed which are of an entirely yes/no character, with no probabilities arising at all. The first of these recent suggestions was put forward by Greenberger, Horne, and Zeilinger (1989), which involved spin measurements on spin $\frac{1}{2}$ particles at *three* separate places (say, the earth α-Centuri, and Sirius, if Quintessential Trinkets were to have made use of this scheme). Somewhat earlier, in 1967, Kochen and Specker had put forward a closely related idea, but with spin 1 particles, although their geometrical configurations were very complicated; and Bell himself did something very similar but less explicitly, even earlier, in 1966. (These early examples were not initially phrased in terms of EPR phenomena, but it was made explicit how to do this by Heywood and Redhead in 1983, and also by Stairs (1983).)[5] The particular example, using dodecahedra, that I have given has some advantages in the way that the geometry can be made explicit.[6] (There are actually some proposed experiments to test things that are equivalent to these various examples of **Z**-mysteries, though in a different physical form from the one I have presented here: cf. Zeilinger *et al.* (1994).)

5.5 Quantum theory's bedrock: a history extraordinary

What are the basic principles of quantum mechanics? Before addressing them explicitly, I should like to indulge in a historical digression. This will have some advantages for us in emphasizing the status of the two most important separate mathematical ingredients of the theory. It is a very remarkable fact, and not at all a familiar one, that the two most fundamental of the ingredients of modern quantum theory can both be traced back to the sixteenth century, quite independently, and to one and the same man!

This man, Gerolamo Cardano (Fig. 5.8), was born in squalor (of unwedded parents) in Pavia, Italy, on 24 September 1501, rose to become the finest and most famous physician of his time, and finally died in squalor in Rome on 20 September 1576. Cardano was an extraordinary man, though not at all well known today. I hope that the reader will forgive me if I digress briefly to say something about him, before returning to quantum mechanics proper.

Indeed, he is not known at all in quantum mechanics—though his *name*, at least, is well known to *automobile* mechanics! For the universal joint linking an ordinary motor-car's gearbox to its rear wheels, thus allowing the flexibility that is needed in order to absorb the varying vertical movement of the sprung rear axle, is called a *cardan-shaft*. Cardano had invented this device in about 1545, and in 1548 was able to incorporate it as part of the undercarriage of a royal vehicle for the Emperor Charles V, thus ensuring a smooth ride over very rough roads. He made numerous other inventions, such as a combination lock similar to those used in modern safes. As a physician, he achieved great

Fig. 5.8. Gerolamo Cardano (1501–1576). Physician extraordinary, inventor, gambler, author, and mathematician. The discoverer of both probability theory and complex numbers—the two basic ingredients of modern quantum theory.

fame, and kings and princes were among those he treated. He made many advances in medicine and wrote a great number of books on medical and other matters. He appears to have been the first to notice that the venereal diseases now known as gonorrhoea and syphilis were actually two separate diseases, requiring *different* treatments. He proposed a 'sanatorium' type of treatment for sufferers of tuberculosis, some 300 years before it was essentially rediscovered by George Boddington in about 1830. In 1552, he cured John Hamilton, the Archbishop of Scotland, of a severe debilitating asthmatic condition—and thus affected the very course of British history.

What have these achievements to do with quantum theory? None whatsoever, except that they show something of the mental calibre of the man who actually discovered, separately, what were to become the two most important ingredients of that theory. For besides being outstanding as a physician and inventor, he was also outstanding in another field—mathematics.

The first of these ingredients is *probability theory*. For quantum theory is, as is well known, a probabilistic rather than a deterministic theory. Its very rules depend fundamentally on the laws of probability. In 1524, Cardano wrote his *Liber de Ludo Aleae* (*The Book of Games of Chance*), which laid the foundations of the mathematical theory of probability. Cardano had formulated these laws some years earlier and had put them to good use. He had been able to finance his studies at medical school in Pavia by applying these laws in a practical way—through gambling! It must have been clear to him at an early age that to make money through *cheating* at cards would be a risky endeavour, for the man of whom his mother was a widow had come to an unpleasant end because of just such an activity. Cardano found that he could win honestly, by applying his discoveries concerning the very laws of probability.

What is the other fundamental ingredient of quantum theory that Cardano had discovered? This second ingredient is the notion of a *complex number*. A complex number is a number of the form

$$a+\mathrm{i}b,$$

where 'i' denotes the square root of minus one,

$$\mathrm{i}=\sqrt{-1},$$

and where a and b are ordinary real numbers (i.e. numbers that we now write in terms of decimal expansions). We would now call a the *real part* and b the *imaginary part* of the complex number $a+\mathrm{i}b$. Cardano had come across these strange kinds of number as part of his investigation of the solution of the general cubic equation. These equations are things like

$$Ax^3+Bx^2+Cx+D=0,$$

where A, B, C, and D are given real numbers, and where the equation has to be solved for x. In 1545, he published a book, *Ars Magna*, in which appeared the first complete analysis of the solution of these equations.

There is an unfortunate story in connection with the publication of this solution. In 1539 a mathematics teacher who was known under the name of Nicolo 'Tartaglia' was already in possession of the general solution of a certain broad class of cubic equations, and Cardano had sent a friend to find out from him what the solution was. However, Tartaglia refused to reveal his solution, so Cardano set to work and quickly rediscovered it for himself, publishing the result in 1540, in his book *The Practice of Arithmetic and Simple Mensuration*. In fact Cardano was able to extend what Tartaglia had done to cover *all* cases, and later published his analysis of the general method of solution in *Ars Magna*. In both books, Cardano acknowledged Tartaglia's prior claim to the solution in that class of cases for which Tartaglia's procedure worked, but in *Ars Magna*, Cardano made the mistake of maintaining that Tartaglia had given him permission to publish. Tartaglia was furious, and claimed that he had visited Cardano's home, on one occasion, revealing to him his solution on the strict condition that Cardano swore an oath of secrecy never to reveal it. In any case, it would have been difficult for Cardano to publish his own work, which extended what Tartaglia had previously done, without revealing the earlier cases of solution, and it is hard to see that Cardano could have proceeded in any other way without suppressing the whole thing. Nevertheless, Tartaglia bore Cardano a lifelong grudge, biding his time until, in 1570, after other appalling circumstances had led to the severe tarnishing of Cardano's reputation, he contributed the final blow towards Cardano's downfall. Tartaglia worked closely with the Inquisition, collecting a long dossier of items that could be used against Cardano, and arranging for his arrest and imprisonment. Cardano was released from prison only after a special emissary sent by the Archbishop of Scotland (whom, we recall, Cardano had cured of asthma) had travelled to Rome, in 1571, to plead for him, explaining that Cardano was 'a scholar who troubles only with preserving and curing the bodies in which God's souls may live to their greatest length'.

The 'appalling circumstances' referred to above concerned the trial for murder of Cardano's eldest son Giovanni Battista. At this trial, Gerolamo had put his reputation on the line in support of his son. This had done him no good at all because Giovanni was actually guilty, having murdered his wife—whom he had been forced to marry, in any case, in order to cover up a previous murder that he had committed. Apparently the murder of Giovanni's wife was aided and abetted by Cardano's even more scoundrelly younger son Aldo, who then betrayed Giovanni and later delivered up his own father to the Inquisition in Bologna. Aldo's reward was to become public torturer and executioner for the Inquisition in Bologna. Even Cardano's daughter did his own reputation no credit, for she died of syphilis as a result of her professional activities as a prostitute.

It would be an interesting exercise in historical psychology to understand how Gerolamo Cardano, who appears to have been a caring father, devoted to his children and to his wife, and who was a principled, honest, and sensitive man, should have had such disastrous progeny. No doubt his attentions were all too frequently distracted from family matters by his multifarious and time-consuming other interests. No doubt his absence from home for over a year, after his wife's death, when he had to travel to Scotland to treat the Archbishop (although Cardano's original commitment was merely for a meeting in Paris), was not helpful to the development of his children. No doubt, also, his conviction that the stars foretold his own death in 1546, in leading to a buildup of feverish writing and research, caused him to neglect his own wife so that she, instead, succumbed towards the end of that very year.

I can well imagine that it was Cardano's unfortunate fate and severely tarnished reputation—brought about by the combined efforts of his children, the Inquisition, and particularly Tartaglia—that resulted in his being so much less well known to us today than he deserves to be. There is no doubt in my own mind that he ranks as one of the very greatest of Renaissance figures. Though he grew up in squalid circumstances, an atmosphere of learning played an important part in his formative years. His father, Fazio Cardano, was a geometer; and Gerolamo recalled an occasion when, at a young age, he accompanied his father on a visit to Leonardo da Vinci, and the two men spent long hours into the night, discussing matters of geometry.

With regard to Cardano's publication of Tartaglia's earlier result, and of incorrectly maintaining that he had permission to publish, one must surely respect the importance of making one's discoveries public, rather than keeping new knowledge suppressed. While it must be appreciated that Tartaglia's livelihood had depended, to some extent, on the continued secrecy of his discoveries (in view of the public mathematical competitions that he often took part in), it was Cardano's publication of them that had the profound and lasting effect on the development of mathematical science. Moreover, when it comes to the matter of priority, it seems that this belongs to another scholar, Scipione del Ferro, who was professor at the University of Bologna until his death in 1526. At least del Ferro was in possession of the solution that Tartaglia later rediscovered, although it remains uncertain to what extent he was aware of how this solution could be modified to account for the cases later considered by Cardano, nor is there any evidence that del Ferro was led to contemplate complex numbers.

Let us return in more detail to the cubic equation, in order to try to understand why Cardano's contribution was so fundamental. It is not hard (by use of a substitution of the form $x \mapsto x+a$) to reduce the general cubic equation to the form

$$x^3 = px + q,$$

where p and q are real numbers. This much would have been essentially well known at the time. However, we must bear in mind that even what we call *negative numbers* were not normally accepted as actual 'numbers' in those days, so different versions of the equation would have been written down depending upon the various signs of p and q (e.g. $x^3 + p'x = q$, $x^3 + q' = px$), in order to keep all the numbers non-negative that actually appear in the equation. I shall adopt the modern notation in my descriptions here (which allow negative numbers if required), to avoid excessive complication.

The solutions of the cubic equation given above can be expressed graphically if we plot the curves $y = x^3$ and $y = px + q$, and look for where the two curves intersect. The values of x at the intersection points will give the solutions of the equation. See Fig. 5.9; the curve $y = x^3$ is represented as the curved line, and $y = px + q$ is shown as a straight line for which various possibilities are indicated. (I am not aware that either Cardano or Tartaglia used such a graphical description, although they may have. It is useful here just as an aid to visualizing the different situations which can occur.) Now, in this notation, the cases that Tartaglia was able to solve occur

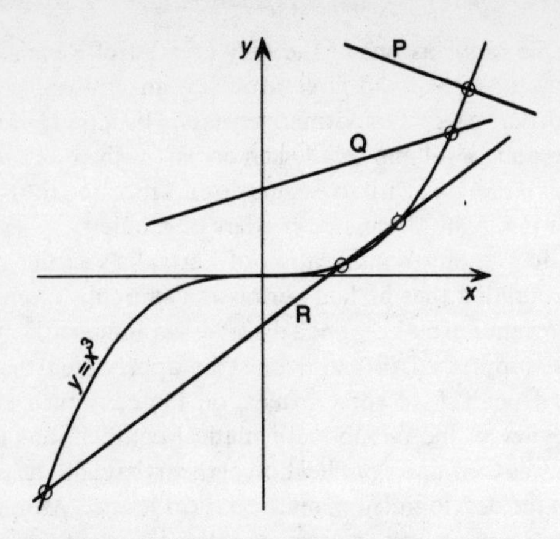

Fig. 5.9. The solutions of the cubic equation $x^3 = px + q$ can be obtained graphically as the intersection(s) of the straight line $y = px + q$ with the cubic curve $y = x^3$. Tartaglia's case is given when $p \leqslant 0$, represented by the downward sloping line P, while Cardano's new ones are given when $p > 0$, such as are provided by the lines Q or R. The *casus irreducibilis* occurs when there are *three* intersection points as with line R. In this case, a journey through the complex is needed to express the solutions.

when p is negative (or zero). In these cases the straight line slopes down to the right, and a typical case is illustrated by the line P in Fig. 5.9. Note that in such cases there is always exactly one intersection point of line with curve, so the cubic equation has precisely one solution. In modern notation, we can express Tartaglia's solution as

$$x = \sqrt[3]{(w - \tfrac{1}{2}q)} - \sqrt[3]{(w + \tfrac{1}{2}q)},$$

where

$$w = \sqrt{[(\tfrac{1}{2}q)^2 + (\tfrac{1}{3}p')^3]}$$

with p' standing for $-p$, so that the quantities appearing in the expression remain non-negative (taking $q > 0$ also).

Cardano's extension of this procedure allowed for the cases where $p > 0$, and we can write the solution (for positive p and negative q, but the sign of q is not too important). Now, the straight line slopes up to the right (marked Q or R). We see that for a given value of p (i.e. for a given slope), if q' $(= -q)$ is large enough (so that the line intersects the y-axis far enough up) there will again be exactly one solution, Cardano's expression for this (in modern notation) being

$$x = \sqrt[3]{(\tfrac{1}{2}q' + w)} + \sqrt[3]{(\tfrac{1}{2}q' - w)},$$

where

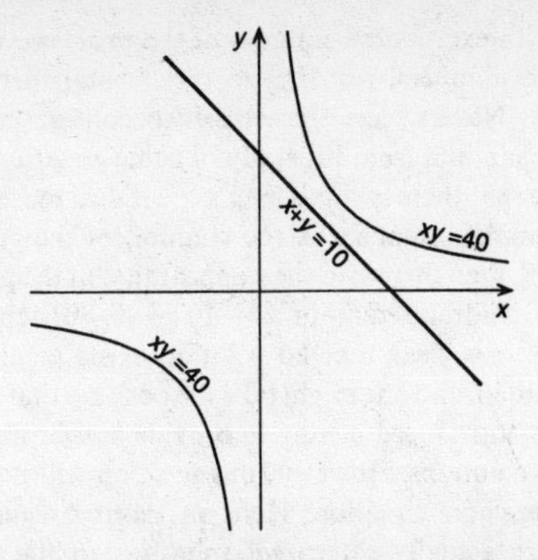

Fig. 5.10. Cardano's problem of finding two numbers whose product is 40 and which sum to 10 can be expressed as finding the intersections of the curve $xy=40$ with the line $x+y=10$. In this case it is clear that the problem cannot be solved with real numbers.

$$w=\sqrt{[(\tfrac{1}{2}q')^2-(\tfrac{1}{3}p)^3]}.$$

We can see, using modern notation and modern concepts of negative numbers (and the fact that the cube root of a negative number is minus the cube root of the positive form of that number), that Cardano's expression is basically the same as Tartaglia's. However, there is something completely new looming within the expression, in Cardano's case. For now, if q' is not too large, the straight line can intersect the curve in *three* places, so there are three solutions to the original equation (two being negative, if $p>0$). This—the so-called '*casus irreducibilis*'—occurs when $(\tfrac{1}{2}q')^2<(\tfrac{1}{3}p)^3$, and we see that now w has to be the *square root of a negative number*. Thus, the numbers $\tfrac{1}{2}q'+w$ and $\tfrac{1}{2}q'-w$, that appear under the cube root signs, are what we would now call *complex numbers*; yet the two cube roots must sum to a real number in order to provide the solutions to the equation.

Cardano was well aware of this mysterious problem, and later on in *Ars Magna* he explicitly addressed the question raised by the occurrence of complex numbers in the solution of equations. He considered the problem of finding two numbers whose product is 40 and which sum to 10, obtaining for the (correct) answer, the two complex numbers

$$5+\sqrt{(-15)} \text{ and } 5-\sqrt{(-15)}.$$

In graphical terms, we may consider this problem as that of finding the intersection points of the curve $xy=40$ with the straight line $x+y=10$ in Fig. 5.10. We note that the curves, as depicted, do not actually intersect (in real-number terms), which corresponds to the fact that we need complex numbers in order to express the solution of the problem. Cardano was not happy about such numbers, referring to the 'mental tortures involved' in working with them. Nevertheless, the necessity of considering numbers of this kind was forced upon him from his study of cubic equations.

We should note that there is something a great deal more subtle about the appearance of complex numbers in the solution of the cubic equation, as depicted in Fig. 5.9, than there is in their appearance in the problem (basically that of solving the quadratic equation $x^2-10x+40=0$) depicted in Fig. 5.10. In the latter case, it is clear that no solution exists at all unless complex numbers are permitted, and one might take the position that such numbers are a complete fiction, introduced merely to provide a 'solution' to an equation that really has no solutions. However, this position will not explain what is going on with the cubic equation. Here, in '*casus irreducibilis*' (line R in Fig. 5.9), there are actually three *real* solutions to the equation, whose existence cannot be denied; yet to express any one of these solutions in terms of surds (i.e. in terms of square roots and cube roots, in this case), we must take a journey through the mysterious world of complex numbers, though our final destination is back in the world of the reals.

It seems that none before Cardano had perceived this mysterious world, and how it might underlie the very world of 'reality'. (Others, such as Hero of Alexandria and Diophantos of Alexandria, in the first and third centuries AD, respectively, seem to have toyed with the idea that a negative number might have a kind of 'square root', but neither took the bold step of combining such 'numbers' with the reals to provide *complex* numbers, nor did they perceive any underlying link with real-number solutions of equations.) Perhaps Cardano's curious combination of a mystical and a scientifically rational personality allowed him to catch these first glimmerings of what developed to be one of the most powerful of mathematical conceptions. In later years, through the work of Bombelli, Coates, Euler, Wessel, Argand, Gauss, Cauchy, Weierstrass, Riemann, Levi, Lewy, and many others, the theory of complex numbers has flowered into one of the most elegant and universally applicable of mathematical structures. But not until the advent of the quantum theory, in the first quarter of this century, was a strange and all-pervasive role for complex numbers revealed at the very foundational structure of the actual physical world in which we live—nor had their profound link with *probabilities* been perceived before this. Even Cardano could have had no inkling of a mysterious underlying connection between his two greatest contributions to mathematics—a link that forms the very basis of the material universe at its smallest scales.

5.6 The basic rules of quantum theory

What is this link? How do complex numbers and probability theory unite together to yield an unquestionably superb description of the inner workings of our world? Roughly speaking it is at a very tiny underlying level of phenomena that the laws of complex numbers hold sway, whereas it is in the bridge between this tiny level and the familiar level of our ordinary perceptions that probabilities play their part—but I shall need to be more explicit than this if any real understanding is to be gained.

Let us first examine the role of complex numbers. These arise in a strange way that is itself very hard to accept as an actual description of physical reality. It is particularly hard to accept because it seems to have no place in the behaviour of things at the levels of phenomena that we can actually perceive, and where the classical laws of Newton, Maxwell, and Einstein hold true. Thus, in order to form a picture of the way in which quantum theory works, we shall need, provisionally at least, to consider that there are two distinct levels of physical action: the underlying *quantum* level, where these complex numbers play their strange part, and the *classical* level of the familiar large-scale physical laws. Only at the quantum level do complex numbers play this role—a role that seems to disappear completely at the classical level. This is not to say that there need really be a physical division between a level at which the quantum laws operate and the level of classically perceived phenomena, but it will be helpful if we imagine, for the moment, that there is such a division, in order to make sense of the procedures that are actually adopted in quantum theory. The deeper question as to whether or not there *really* is such a physical division will be among our main concerns later.

At what level *is* this quantum level? We must think of it as the level of physical things that are in some sense 'small enough', like molecules, atoms, or fundamental particles. But this 'smallness' need not refer to physical distance. Quantum-level effects can occur across vast separations. Recall the 4 light years that would separate the two dodecahedra in my story in §5.3, or the 12 metres that actually separated the pairs of photons in Aspect's experiment (§5.4). It is not small physical size that defines the quantum level, but something more subtle, and for the moment it will be best not to try to be explicit. It will be helpful, however, to think of the quantum level as roughly applying when we are concerned merely with very tiny differences in energy. I shall return to this issue more thoroughly in §6.12.

The classical level, on the other hand, is the level that we ordinarily experience, where the real-number laws of classical physics hold good, such as where ordinary descriptions—like that giving the location, the speed, and the shape of a golf ball—make good sense. Whether or not there is any *actual* physical distinction between the quantum level and the classical level is a profound question that is intimately related to the issue of the **X**-mysteries, as referred to in §5.1. Deferring this question for the moment, we regard it as merely a matter of convenience that we separate the quantum from the classical level.

What fundamental role do complex numbers indeed play at the quantum level? Think of an individual particle such as an electron. In a classical picture, the electron might have one location A or it might perhaps have another location B. However, the quantum-mechanical description of such possibilities that might be open to the electron are much broader. Not only might the electron have one or another particular location, but it might alternatively have any one of a number of possible states in which, in some clear sense, it occupies *both* locations simultaneously! Let us use the notation $|A\rangle$ for the state in which the electron has position A, and the notation $|B\rangle$ for the state in which the electron is at B.* According to quantum theory, then, there are other possible states open to the electron, written as

$$w|A\rangle + z|B\rangle,$$

where the weighting factors w and z that appear here are *complex numbers* (at least one of which must be taken to be non-zero).

What does this mean? If the weighting factors had been non-negative *real* numbers, then we might have thought of this combination as representing, in some sense, a probability-weighted expectation as to the location of the electron, where w and z represent the relative probabilities of the electron being at A or of it being at B, respectively. Then the ratio $w{:}z$ would give the ratio (probability of electron at A):(probability of electron at B). Accordingly, if these were the only two possibilities open to the electron, we would have the expectation $w/(w+z)$ for the electron to be at A, and an expectation $z/(w+z)$ for it to be at B. If $w=0$, then the electron would be certain to be at B; if $z=0$, it would be certain to be at A. If the state were just '$|A\rangle + |B\rangle$', then this would represent *equal* probabilities of the electron being at A or at B.

But w and z are *complex* numbers, so such an interpretation makes no sense at all. The ratios of the quantum weightings w and z are *not* ratios of probabilities. They cannot be since probabilities always have to be *real* numbers. It is *not* Cardano's *probability* theory that operates at the quantum level, despite the common opinion that the quantum world is a probabilistic world. Instead, it is his mysterious theory of *complex numbers* that underlies a mathematically precise and *probability-free* description of the quantum level of activity.

We cannot say, in familiar everyday terms, what it 'means' for an electron to be in a state of superposition of two places at once, with complex-number weighting factors w and z. We must, for the moment, simply accept that this is indeed the kind of description that we have to adopt for quantum-level

*I am here using the standard Dirac 'ket' notation for quantum states, which will be somewhat convenient for us. Readers who are not familiar with quantum-mechanical notation need not be concerned about its significance here.

Paul Dirac was one of the outstanding physicists of the twentieth century. Among his achievements was a general formulation of the laws of quantum theory, and also of its relativistic generalization involving the 'Dirac equation', which he discovered, for the electron. He had an unusual ability to 'smell out' the truth, judging his equations, to a large degree, by their *aesthetic* qualities!

systems. Such superpositions constitute an important part of the actual construction of our microworld, as has now been revealed to us by Nature. It is just a *fact* that we appear to find that the quantum-level world *actually* behaves in this unfamiliar and mysterious way. The descriptions are perfectly clear cut—and they provide us with a micro-world that evolves according to a description that is indeed mathematically precise and, moreover, *completely deterministic*!

5.7 Unitary evolution U

What is this deterministic description? It is what is called *unitary evolution*, and I shall use the letter **U** to denote it. This evolution is described by precise mathematical equations, but it will not be important for us to know what these equations actually are. Certain particular properties of **U** will be all that we shall need. In what is referred to as the 'Schrödinger picture', **U** is described by what is called the *Schrödinger equation*, which provides the rate of change, with respect to time, of the *quantum state* or *wavefunction*. This quantum state, often denoted by the Greek letter ψ (pronounced 'psi'), or by $|\psi\rangle$, expresses the entire weighted sum, with complex-number weighting factors, of all the possible alternatives that are open to the system. Thus, in the particular example that was referred to above, for which the alternatives open to an electron were that it might be at one location A or another location B, the quantum state $|\psi\rangle$ would have to be some complex-number combination

$$|\psi\rangle = w|\mathrm{A}\rangle + z|\mathrm{B}\rangle,$$

where w and z are complex numbers (not both zero). We call the combination $w|\mathrm{A}\rangle + z|\mathrm{B}\rangle$ a *linear superposition* of the two states $|\mathrm{A}\rangle$ and $|\mathrm{B}\rangle$. The quantity $|\psi\rangle$ (or $|\mathrm{A}\rangle$ or $|\mathrm{B}\rangle$) is often referred to as a *state vector*. More general quantum states (or state vectors) might have a form such as

$$|\psi\rangle = u|\mathrm{A}\rangle + v|\mathrm{B}\rangle + w|\mathrm{C}\rangle + \ldots + z|\mathrm{F}\rangle,$$

where $u, v, \ldots, z$ are complex numbers (not all zero) and $|\mathrm{A}\rangle, |\mathrm{B}\rangle, \ldots, |\mathrm{F}\rangle$ might represent various possible locations for a particle (or perhaps some other property of a particle, such as its state of spin; cf. §5.10). Even more generally, *infinite* sums would be allowed for a wavefunction or state vector (since there are infinitely many possible positions available to a point particle), but this kind of issue will not be important for us here.

There is one technicality of the quantum formalism that I *should* mention here. This is that it is only the *ratios* of the complex weighting factors that are to have significance. I shall have more to say about this later. For the moment, we merely take note of the fact that for any single state vector $|\psi\rangle$, any complex multiple $u|\psi\rangle$ of it (with $u \neq 0$) represents the same *physical* state as does $|\psi\rangle$. Thus, for example, $uw|\mathrm{A} + uz|\mathrm{B}\rangle$ represents the same physical state as

$w|A\rangle + z|B\rangle$. Accordingly, only the ratio w:z has physical significance, not w and z separately.

Now the most basic feature of the Schrödinger equation (i.e. of **U**) is that it is *linear*. This is to say, if we have two states, say $|\psi\rangle$ and $|\phi\rangle$, and if the Schrödinger equation would tell us that, after a time t, the states $|\psi\rangle$ and $|\phi\rangle$ would each individually evolve to new states $|\psi'\rangle$ and $|\phi'\rangle$, respectively, then any linear superposition $w|\psi\rangle + z|\phi\rangle$, must evolve, after the same time t, to the corresponding superposition $w|\psi'\rangle + z|\phi'\rangle$. Let us use the symbol $\rightsquigarrow$ to denote the evolution after time t. Then linearity asserts that if

$$|\psi\rangle \rightsquigarrow |\psi'\rangle \text{ and } |\phi\rangle \rightsquigarrow |\phi'\rangle,$$

then the evolution

$$w|\psi\rangle + z|\phi\rangle \rightsquigarrow w|\psi'\rangle + z|\phi'\rangle$$

would also hold. This would (consequently) apply also to linear superpositions of more than two individual quantum states; for example, $u|\chi\rangle + w|\psi\rangle + z|\phi\rangle$ would evolve, after time t, to $u|\chi'\rangle + w|\psi'\rangle + z|\phi'\rangle$, if $|\chi\rangle$, $|\psi\rangle$, and $|\phi\rangle$ would each individually evolve to $|\chi'\rangle$, $|\psi'\rangle$, and $|\phi'\rangle$, respectively. Thus, the evolution always proceeds as though each different component of a superposition were oblivious to the presence of the others. As some would say, each different 'world' described by these component states evolves independently, according to the same deterministic Schrödinger equation as the others, and the particular linear superposition that describes the entire state preserves the complex-number weightings unchanged, as the evolution proceeds.

It might be thought, in view of this, that the superpositions and the complex weightings play no effective physical role, since the time evolution of the separate states proceeds as though the other states were not there. However, this would be very misleading. Let me illustrate what can actually happen with an example.

Consider a situation in which light impinges on a half-silvered mirror—which is semi-transparent mirror that reflects just half the light falling upon it and transmits the remaining half. Now in quantum theory, light is perceived to be composed of particles called *photons*. We might well have imagined that for a stream of photons impinging on our half-silvered mirror, half the photons would be reflected and half would be transmitted. Not so! Quantum theory tells us that, instead, each *individual* photon, as it impinges on the mirror, is separately put into a *superposed* state of reflection and transmission. If the photon before its encounter with the mirror is in state $|A\rangle$, then afterwards its evolves according to **U** to become a state that can be written $|B\rangle + i|C\rangle$, where $|B\rangle$ represents the state in which the photon is transmitted through the mirror and $|C\rangle$ the state where the photon is reflected from it; see Fig. 5.11. Let us write this

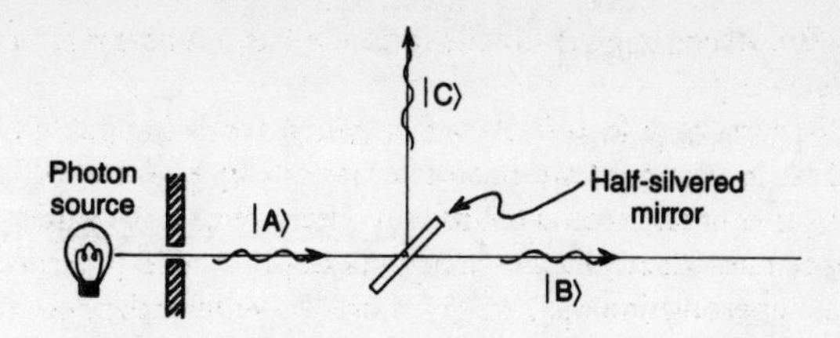

Fig. 5.11. A photon in state $|A\rangle$ impinges on a half-silvered mirror and its state evolves (by $\mathbf{U}$) into a superposition $|B\rangle + i|C\rangle$.

$$|A\rangle \rightsquigarrow |B\rangle + i|C\rangle.$$

The factor 'i' arises here because of a net phase shift by a quarter of a wavelength[7] that occurs between the reflected and transmitted beams at such a mirror. (To be more complete, I should also have included a time-dependent oscillatory factor and an overall normalization here, but this plays no role in the present discussion. In these descriptions I am giving only what is essential for our immediate purposes. I shall say a little more about the oscillatory factor in §5.11 and the question of normalization in §5.12. For a more complete description, see any standard work on quantum theory[8]; also ENM, pp. 243–50.)

Though, from the classical picture of a particle, we would have to imagine that $|B\rangle$ and $|C\rangle$ just represent alternative things that the photon *might* do, in quantum mechanics we have to try to believe that the photon is now actually doing *both things at once* in this strange, complex superposition. To see that it cannot just be a matter of classical probability-weighted alternatives, let us take this example a little further and try to bring the two parts of the photon state—the two photon beams—back together again. We can do this by first reflecting each beam with a fully silvered mirror. After reflection,[9] the photon state $|B\rangle$ would evolve according to $\mathbf{U}$, into another state $i|D\rangle$, whilst $|C\rangle$ would evolve into $i|E\rangle$:

$$|B\rangle \rightsquigarrow i|D\rangle \text{ and } |C\rangle \rightsquigarrow i|E\rangle.$$

Thus the entire state $|B\rangle + i|C\rangle$ evolves, by $\mathbf{U}$, into

$$\begin{aligned}|B\rangle + i|C\rangle &\rightsquigarrow i|D\rangle + i(i|E\rangle)\\ &= i|D\rangle - |E\rangle\end{aligned}$$

(since $i^2 = -1$). Suppose, now, that these two beams come together at a fourth mirror, which is now *half* silvered, as depicted in Fig. 5.12 (where I am assuming that all the beam lengths are equal, so that the oscillatory factor that I am ignoring continues to play no part). The state $|D\rangle$ evolves into a combination $|G\rangle + i|F\rangle$, where $|G\rangle$ represents the transmitted state and $|F\rangle$

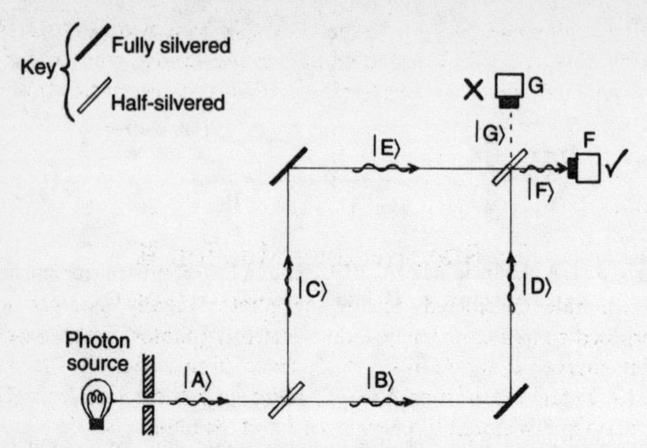

Fig. 5.12. The two parts of the photon state are brought together by two fully silvered mirrors, so as to encounter each other at a final half-silvered mirror. They interfere in such a way that the entire state emerges in state $|F\rangle$, and the detector at G cannot receive the photon. (Mach–Zehnder interferometer.)

the reflected one; similary, $|E\rangle$ evolves into $|F\rangle + i|G\rangle$, since it is now the state $|F\rangle$ that is the transmitted state and $|G\rangle$ the reflected one:

$$|D\rangle \rightsquigarrow |G\rangle + i|F\rangle \text{ and } |E\rangle \rightsquigarrow |F\rangle + i|G\rangle.$$

Our entire state $i|D\rangle - |E\rangle$ is now seen (because of the linearity of **U**) to evolve:

$$\begin{aligned} i|D\rangle - |E\rangle &\rightsquigarrow i(|G\rangle + i|F\rangle) - (|F\rangle + i|G\rangle) \\ &= i|G\rangle - |F\rangle - |F\rangle - i|G\rangle \\ &= -2|F\rangle. \end{aligned}$$

(The multiplying factor -2 appearing here plays no physical role because, as mentioned above, if the entire physical state of a system—here $|F\rangle$—is multiplied by a non-zero complex number, then this leaves the physical situation unaltered.) Thus we see that the possibility $|G\rangle$ is *not* open to the photon; the two beams together combine to produce just the *single* possibility $|F\rangle$. This curious outcome arises because *both* beams are present *simultaneously* in the physical state of the photon, between its encounters with the first and last mirrors. We say that the two beams *interfere* with one another. Thus the two alternative 'worlds' of the photon between these encounters are not really separate, but they can affect one another via such interference phenomena.

It is important to bear in mind that this is a property of *single* photons. Each individual photon must be considered to feel out both routes that are open to it, but it remains *one* photon; it does not split into two photons in the intermediate stage, but its location undergoes the strange kind of complex-number-weighted *co-existence* of alternatives that is characteristic of quantum theory.

5.8 State-vector reduction R

In the example considered above, the photon finally emerges in an unsuperposed state. Let us imagine that detectors (photocells) are placed at the points marked F and G in Fig. 5.12. Since, in this example, the photon emerges in a state (proportional to) $|F\rangle$, with no contribution from $|G\rangle$, it follows that the detector at F registers receipt of the photon, and the detector at G registers nothing.

What would happen in a more general situation such as when a superposed state like $w|F\rangle + z|G\rangle$ encounters these detectors? Our detectors are making a *measurement* to see whether the photon is in state $|F\rangle$ or in state $|G\rangle$. A quantum measurement has the effect of magnifying quantum events from the quantum to the classical level. At the quantum level, linear superpositions persist under the continuing action of **U**-evolution. However, as soon as effects are magnified to the classical level, where they can be perceived as *actual* occurrences, then we do not ever find things to be in these strange complex-weighted combinations. What we *do* find, in this example, is that *either* the detector at F registers *or* the detector at G registers, where these alternatives occur with certain probabilities. The quantum state seems mysteriously to have 'jumped' from one involving the superposition $w|F\rangle + z|G\rangle$ to one in which just $|F\rangle$ or just $|G\rangle$ is involved. This 'jumping' of the description of the state of the system, from the superposed quantum-level state to a description in which one or other of the classical-level alternatives take place, is called *state-vector reduction*, or *collapse of the wavefunction*, and I shall use the letter **R** to denote this operation. Whether **R** is to be considered as a real physical process or some kind of illusion or approximation is a matter that will be of great concern for us later. The fact that, in our mathematical descriptions at least, we have to dispense with **U** from time to time, and bring in this totally different procedure **R**, is the basic **X**-mystery of quantum theory. For the moment, it will be better that we do not probe the matter too closely and (provisionally) regard **R** as being, effectively, some process that just *happens* (at least in the mathematical descriptions that are used) as a feature of the procedure of magnifying an event from the quantum to the classical level.

How do we actually calculate these different *probabilities* for the alternative outcomes of a measurement on a superposed state? There is, indeed, a remarkable rule for determining these probabilities. This rule states that for a measurement which decides between alternative states $|F\rangle$ and $|G\rangle$, say, by using detectors at F and G, respectively, in the situation above, then upon the detectors encountering the superposed state

$$w|\mathrm{F}\rangle + z|\mathrm{G}\rangle,$$

the ratio of the probability that the detector at F registers to the probability that the one at G registers is given by the ratio

$$|w|^2 : |z|^2,$$

where these are the *squared moduli* of the complex numbers w and z. The squared modulus of a complex number is the sum of the squares of its real and imaginary parts; thus, for

$$z = x + \mathrm{i}y,$$

where x and y are real numbers, the squared modulus is

$$\begin{aligned} |z|^2 &= x^2 + y^2 \\ &= (x + \mathrm{i}y)\,(x - \mathrm{i}y) \\ &= z\bar{z} \end{aligned}$$

where $\bar{z}(= x - \mathrm{i}y)$ is called the *complex conjugate of* z, and similarly for w. (I am tacitly assuming, in the above discussion, that the states that I have been denoting by $|\mathrm{F}\rangle$, $|\mathrm{G}\rangle$, etc., are appropriately *normalized* states. This will be explained later, cf. §5.12; the normalization is needed, strictly speaking, for this form of the probability rule to hold.)

It is here, and only here, that Cardano's *probabilities* enter the quantum scene. We see that the quantum-level complex-number weightings do *not* themselves play roles as relative probabilities (which they cannot do, because they are complex) but the real-number *squared moduli* of these complex numbers do play such roles. Moreover, it is only now, when *measurements* are made, that indeterminancy and probabilities come in. A measurement of a quantum state occurs, in effect, when there is a large magnification of some physical process, raising it from the quantum to the classical level. In the case of a photocell, the registering of a single quantum event—in the form of the reception of a photon—eventually causes a classical-level disturbance, say an audible 'click'. Alternatively, we might use a sensitive photographic plate to register the arrival of a photon. Here, the quantum event of this photon arrival would be magnified to the classical level in the form of a visible mark on the plate. In each case, the measuring apparatus would consist of a delicately poised system that can use a tiny quantum event to trigger off a much larger classical-scale observable effect. It is in this passage from the quantum level to the classical level that Cardano's complex numbers get their moduli squared to become Cardano's probabilities!

Let us see how to apply this rule in a particular situation. Suppose, instead

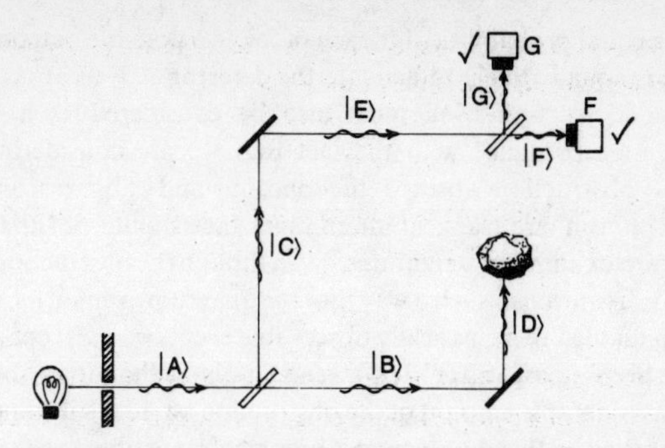

Fig. 5.13. If an obstruction is placed in the beam $|D\rangle$, then it becomes possible for the detector at G to register the arrival of the photon (when the obstruction does *not* absorb the photon!).

of having the mirror at the lower right, we were to place a photocell there; then this photocell would encounter the state

$$|B\rangle + i|C\rangle,$$

where the state $|B\rangle$ would cause the photocell to register whilst $|C\rangle$ would leave it undisturbed. Thus the ratio of the respective probabilities is $|1|^2{:}|i|^2 = 1{:}1$; that is to say, the probability of each of the two possible outcomes is the same, so that the photon is equally likely to activate the photocell as not to.

Let us consider a slightly more complicated arrangement. Suppose, instead of having the photocell in place of the lower right-hand mirror, we block off one of the beams in the above example with an *obstruction* capable of absorbing the photon, say in the beam corresponding to the photon state $|D\rangle$ (Fig. 5.13); then the interference effect that we had earlier would be destroyed. It *would* then be possible for the photon to emerge in a state involving the possibility $|G\rangle$ (in addition to the possibility $|F\rangle$) provided that the photon is *not* actually absorbed by the obstruction. If it *is* absorbed by the obstruction, then the photon would not emerge at all in any combination of the states $|F\rangle$ or $|G\rangle$; but if it is not, then the photon's state, as it approaches the final mirror, would simply be $-|E\rangle$, which evolves to $-|F\rangle - i|G\rangle$, so that both alternatives $|F\rangle$, $|G\rangle$ are indeed involved in the final outcome.

In the particular example considered here, when the obstruction is present but does not absorb the photon, the respective complex-number weightings for the two possibilities $|F\rangle$ and $|G\rangle$ are -1 and $-i$ (the emergent state being $-|F\rangle - i|G\rangle$). Thus the ratio of the respective probabilities is $|-1|^2{:}|-i|^2$, again giving equal probabilities for each of the two possible outcomes, so that the photon is equally likely to activate the detector at F as at G.

Now the obstruction itself must also be considered as a 'measuring apparatus', in accordance with the fact that we are considering that the alternatives 'obstruction absorbs the photon' and 'obstruction does not absorb the photon' are classical alternatives that should not themselves be assigned complex-number weightings. Even though the obstruction might not be delicately organized in such a way that the quantum event of its absorbing a photon is magnified to a classically observable event, we must consider that it 'could have been' so organized. The essential point is that upon absorbing the photon, a considerable amount of the obstruction's actual substance becomes slightly disturbed by the photon, and it becomes impossible to collect together all the information contained in this disturbance so as to retrieve the interference effects that characterize quantum phenomena. Thus the obstruction must—in practice at least—be considered as a classical-level object, and it serves the purpose of a measuring apparatus whether or not it registers the absorption of the photon in a practically observable way. (I shall be returning to this issue later, in §6.6).

With this in mind, we are also at liberty to use the 'squared modulus rule' to calculate the probability that the obstruction will actually absorb the photon. The photon state that the obstruction encounters is $i|D\rangle - |E\rangle$, and it absorbs the photon if it finds the photon in state $|D\rangle$, as opposed to the other alternative $|E\rangle$. The ratio of the probability of absorption to that of non-absorption is $|i|^2{:}|-1|^2 = 1{:}1$; so again the two alternatives are equally probable.

We could also imagine a somewhat similar situation where, instead of having an obstruction at D, we could attach some measuring device to the lower right-hand mirror, rather than having this mirror *replaced* by a detector, as was considered earlier. Let us imagine that this device is so sensitive that it is able to detect (i.e. to magnify to the classical level) whatever impulse is imparted to the mirror by the photon being merely reflected off it, this detection being finally signalled, say, by the movement of a pointer (see Fig. 5.14). Thus, the photon state $|B\rangle$, upon encountering the mirror, would activate this pointer movement, but the photon state $|C\rangle$ would not. When presented with the photon state $|B\rangle + i|C\rangle$, the device would 'collapse the wavefunction' and read the state as though it were *either* $|B\rangle$ (pointer moves) *or* $|C\rangle$ (pointer stays put), with equal probabilities (given by $|1|^2{:}|i|^2$). The process **R** would thereby take place at *this* stage. For the ensuing behaviour of the photon, we can follow the rest of the argument through essentially as above, and we find that, as was the case with the obstruction, it is again equally probable for either the detector at G or the detector at F finally to detect the photon (whether or not the pointer moves). For this set-up, the lower right-hand mirror would need to be slightly 'wobbly' in order that the pointer movement can be activated, and this lack of mirror rigidity would upset the

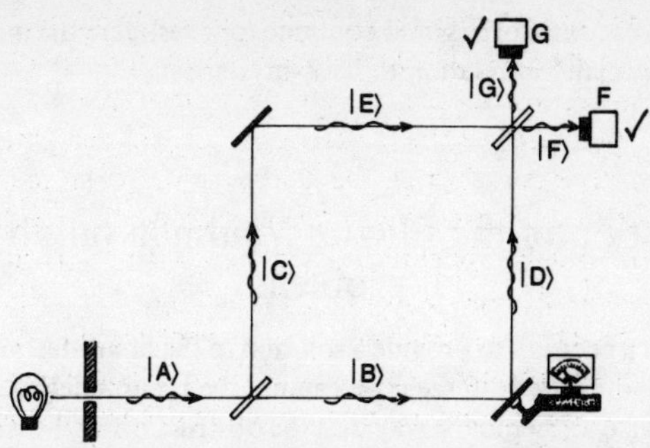

Fig. 5.14. A similar effect can be achieved by making the lower right-hand mirror slightly 'wobbly' and using this wobble to register, by means of some detector, whether or not the photon has actually been reflected off that mirror. Again interference is destroyed, and the detector at G is enabled to detect the photon.

delicate organization needed to ensure the 'destructive interference' between the two photon routes between A and G which had originally prevented the detector at G from registering.

The reader may well perceive that there is something not altogether satisfactory about the matter of *when*—and indeed *why*—the quantum rules should be changed from quantum-level complex-number-weighted determinism to classical-level probability-weighted non-deterministic alternatives, characterized mathematically by the taking of the squared moduli of the complex numbers involved. What is it that *really* characterizes some pieces of physical material, such as the photon detectors at F and G, or at the lower right-hand mirror—and also the possible obstruction at D—as being classical-level objects, whereas the photon, at the quantum level, must be treated so differently? Is it merely the fact that the photon is a physically simple system that allows its complete treatment as a quantum-level object, whereas detectors and obstructions are complicated things which need an approximate treatment in which the subtleties of quantum-level behaviour are somehow averaged away? Many physicists would certainly argue this way, claiming that, strictly speaking, we should treat *all* physical things quantum-mechanically, and it is a matter of convenience that we treat large or complicated systems classically, the probability rules involved in the '**R**' procedure being, somehow, features of the approximations involved. We shall be seeing in §6.6 and §6.7 that this viewpoint does not really get us out of our difficulties—the difficulties presented by the **X**-mysteries of quantum theory—nor does it explain the miraculous **R**-rule whereby probabilities arise as squared moduli of complex-number weightings. For the moment, however, we must suppress our worries, and continue to investigate the implications of the theory, especially in relation to its **Z**-mysteries.

5.9 Solution of the Elitzur–Vaidman bomb-testing problem

We are now in a position to provide a solution to the bomb-testing problem of §5.2. What we must do is to see if we can use the bomb's delicate mirror as a measuring device, rather in the way that the obstruction or the wobbly mirror with detector acted, in the above discussion. Let us set up a system of mirrors, two of which are half silvered, exactly in the way that we did above, but where the bomb's mirror now plays the role of the mirror at the lower right of Fig. 5.14.

The point is that *if* the bomb is a dud, in the only sense that we are allowing for the purposes of this puzzle, then its mirror is jammed in a fixed position, so that the situation is now just as depicted in Fig. 5.12. The photon emitter sends a single photon at the first mirror initially in the state $|A\rangle$. Since the situation is just the same as that of §5.7, the photon must ultimately emerge in the state (proportional to) $|F\rangle$, as before. Thus the detector af F registers the arrival of the photon, but the one at G cannot do so.

If, however, the bomb is *not* a dud, then the mirror is capable of responding to the photon and the bomb would explode if it finds that the photon impinges upon its mirror. The bomb is indeed a measuring device. The two quantum-level alternatives 'photon impinging on mirror' and 'photon not impinging on mirror' are magnified by the bomb to the classical-level alternatives 'bomb explodes' and 'bomb fails to explode'. It responds to the state $|B\rangle + i|C\rangle$ by exploding if it finds the photon in state $|B\rangle$ and by failing to explode if it finds that it is not in state $|B\rangle$—whence it is in state $|C\rangle$. The relative probabilities of these two occurrences is $|1|^2{:}|i|^2 = 1{:}1$. If the bomb explodes, then it has detected the presence of the photon, and what happens subsequently is of no concern. If, however, the bomb fails to explode, then the photon's state is reduced (by the action of **R**) to the state $i|C\rangle$ impinging on the upper left mirror, and emerging from that mirror in state $-|E\rangle$. After encountering the final (half-silvered) mirror, the state becomes $-|F\rangle - i|G\rangle$, so there is a relative probability $|-1|^2{:}|-i|^2 = 1{:}1$ for the two possible outcomes 'detector at F registers photon's arrival' and 'photon at G registers photon's arrival' just as in the cases considered in the previous section, when the obstruction fails to absorb the photon or when the pointer fails to move. Thus there is now the definite possibility for the detector at G to receive the photon.

Suppose, then, that in one of these bomb tests it is found that the detector at G actually does register the arrival of the photon in some cases when the bomb does not explode. From what has been said above, we see that this can happen only if the bomb is *not* a dud! When it is a dud, it is only the detector at F that can register. Thus, in all those circumstances in which G in fact registers we have a bomb that is guaranteed to be active, i.e. not a dud! This solves the bomb-testing problem, as proposed in §5.2.*

It will be seen, from the above considerations of the probabilities involved, that in a long run of tests, half of the active bombs will explode and will be lost. Moreover, only in half of the cases where an active bomb does not explode will the detector at G register. Thus, if we run through the bombs one after the other, we shall find only a quarter of the originally active bombs as now *guaranteed* to be actually still active. We can then run through the remaining ones again, and again retrieve those for which the detector at G registers; and then repeat the whole process again and again. Ultimately, we obtain just one-third (since $\frac{1}{4}+\frac{1}{16}+\frac{1}{64}+\ldots=\frac{1}{3}$) of the active bombs that we started with, but now *all* are guaranteed. (I am not sure what the bombs will now be used for, but perhaps it would be prudent not to ask!)

This may seem to the reader to be a wasteful procedure, but it is a remarkable fact that it can be carried out at all. Classically there would be no way of doing such a thing. It is only in quantum theory that counterfactual possibilities can actually influence a physical outcome. Our quantum procedure allows us to achieve something that would seem impossible—and *is* indeed impossible within classical physics. It should be remarked, moreover, that with certain refinements this wastefulness can be reduced from two-thirds to effectively one-half (Elitzur and Vaidman 1993). More strikingly, P. G. Kwiat, H. Weinfurter, A. Zeilinger, and M. Kasevich have recently shown, using a different procedure, how to reduce the wastefulness effectively to zero!

With regard to the difficult matter of having an experimental device that

* ***The Shabbos switch***. The fact that both Elitzur and Vaidman are at universities in Israel has suggested, in conversations with Artur Ekert, a device for assisting those who adhere strictly to the Jewish faith, and who therefore are prevented from switching electrical appliances on or off during their Sabbath. Instead of patenting our device, and thereby making our fortunes, we have generously decided to make our important idea public so that it may be available for the good of the Jewish community at large. All one needs is a photon source which emits a continual succession of photons, two half-silvered mirrors, two fully silvered mirrors, and a photocell attached to the appliance in question. The set-up is just as in Fig. 5.13, with the photocell placed at G. To activate or deactivate the appliance, one places a finger in the beam at D, as with the obstruction of Fig. 5.13. If the photon hits one's finger, nothing happens to the appliance—surely no sin. (For photons are continually impinging on one's fingers, in any case, even on the Sabbath.) But if the finger encounters no photon at all, then there is a 50% chance (up to the Will of God) that the switch on the appliance will be activated. Surely, also, it can be no sin to *fail* to receive the photon that activates the switch! (A practical objection might be raised that sources which emit individual photons are difficult—and expensive—to manufacture. But this is not really necessary. Any photon source will do, since the argument can be applied to each one of its photons individually.)

actually produces individual photons one at a time, it should be mentioned that such devices can now be constructed (see Grangier *et al.* 1986).

As a final comment, I should remark that it is not necessary for a measuring device to be so dramatic an object as the bomb in this discussion. Indeed, such a 'device' need not actually signal its reception or non-reception of the photon to the outside world at all. A slightly wobbly mirror would, by itself, do as a measuring device, if it were light enough that it would move significantly as a result of the photon's impact and consequently disperse this motion as friction. The mere fact that the mirror is wobbly (say, the lower right-hand one, as before) will allow the detector at G to receive the photon, even if the mirror does *not* actually wobble, thus indicating that the photon went the other way. It is the mere *potentiality* for it to wobble that allows the photon to reach G! Even the obstruction, referred to in the previous section, plays an extremely similar role. It serves, in effect, to 'measure' the presence of the photon somewhere along its track as described by the successive states $|B\rangle$ and $|D\rangle$. A failure of it to receive the photon, when it is capable of receiving it, counts just as much as a 'measurement' as actually receiving it would.

Measurements of this negative and non-invasive kind are called *null* (or interaction-free) measurements, see Dicke (1981), and they have a considerable theoretical (and perhaps, eventually, even practical) importance. There are experiments to test directly the predictions of quantum theory in such situations. In particular, Kwiat, Weinfurter, and Zeilinger have recently performed an experiment of the *precise* type that is involved in the Elitzur–Vaidman bomb-testing problem! As we have now become used to accepting, the expectations of quantum theory have been completely confirmed. Null measurements are indeed among the profound **Z**-mysteries of quantum theory.

5.10 Quantum theory of spin; the Riemann sphere

In order to address the second of my two introductory quantum puzzles, it will be necessary to look into the structure of quantum theory in a little more detail. Recall that my dodecahedron, and also that of my colleague, had an atom of spin $\frac{3}{2}$ at its centre. What *is* spin, and what is its particular importance for quantum theory?

Spin is an intrinsic property of particles. It is basically the same physical concept as the spinning—or *angular momentum*—of a classical object, such as a golf ball, or cricket ball, or the whole earth. However, there is the (minor) difference that for such large objects, by far the major contribution to its angular momentum comes from the orbiting motions of all its particles about one another, whereas for a single particle, spin is a property that is intrinsic to the particle itself. In fact, the spin of a fundamental particle has the curious feature that its *magnitude* always has the *same* value, although the direction of its spin axis can vary—though this 'axis', also, behaves in a very odd way that

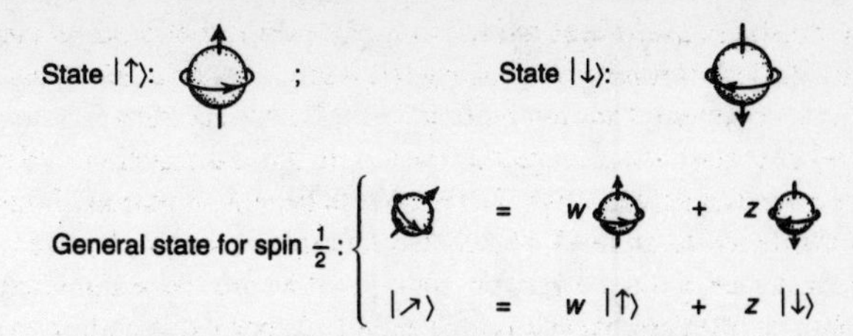

Fig. 5.15. For a particle of spin $\frac{1}{2}$ (such as an electron, proton, or neutron), all spin states are complex superpositions of the two states 'spin up' and 'spin down'.

bears little relation, in general, to what can happen classically. The magnitude of the spin is described in terms of the basic quantum-mechanical unit $\hbar$, which is Dirac's symbol for Planck's constant h, divided by 2π. The measure of spin of a particle is always a (non-negative) integer or half-integer multiple of $\hbar$, namely 0, $\frac{1}{2}\hbar$, $\hbar$, $\frac{3}{2}\hbar$, $2\hbar$, etc. We refer to such particles as having spin 0, spin $\frac{1}{2}$, spin 1, spin $\frac{3}{2}$, spin 2, etc., respectively.

Let us start by considering the simplest case (apart from spin 0, which is *too* simple, there being just one, spherically symmetrical, state of spin in this case), namely the case of spin $\frac{1}{2}$, such as that of an electron or a nucleon (a proton or neutron). For spin $\frac{1}{2}$, all states of spin are linear superpositions of just *two* states, say the state of right-handed spin about the *upward* vertical, written $|\uparrow\rangle$, or right-handed spin about the *downward* vertical, written $|\downarrow\rangle$ (see Fig. 5.15). The general state of spin would now be some complex-number combination $|\psi\rangle = w|\uparrow\rangle + z|\downarrow\rangle$. In fact it turns out that each such combination represents the state of the particle's spin (of magnitude $\frac{1}{2}\hbar$) being about some specific direction determined by the ratio of the two complex numbers w and z. There is nothing special about the particular choice of the two states $|\uparrow\rangle$ and $|\downarrow\rangle$. All the various combinations of these two states are just as clear-cut as states of spin as are the two original ones.

Let us see how this relationship can be made more explicit and geometrical. This will help us to appreciate that the complex-number weighting factors w and z are not quite such abstract things as they may have seemed to be, so far. In fact they have a clear relationship with the geometry of space. (I imagine that such geometric realizations might have pleased Cardano, and perhaps helped him with his 'mental tortures'—though quantum theory itself provides us with new mental tortures!)

It will be helpful, first, to consider the now-standard representation of complex numbers as points on a plane. (This plane is variously called the Argand plane, the Gauss plane, the Wessel plane, or just the *complex* plane.) The idea is simply to represent the complex number $z = x + iy$, where x and y are real numbers, by the point in the plane whose ordinary Cartesian coordinates are (x, y), with respect to some chosen Cartesian axes (see

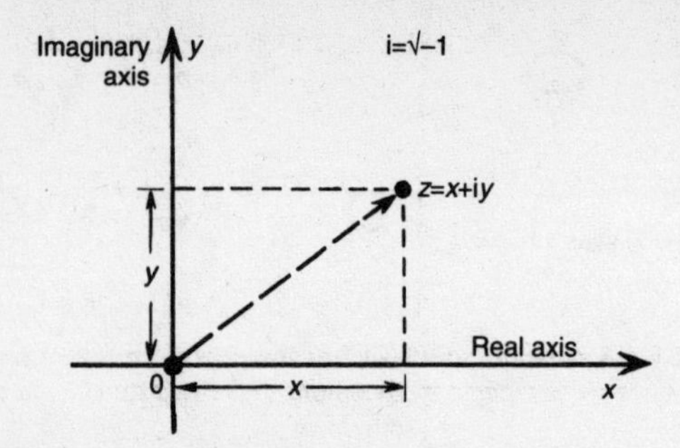

Fig. 5.16. The representation of a complex number in the (Wessel–Argand–Gauss) complex plane.

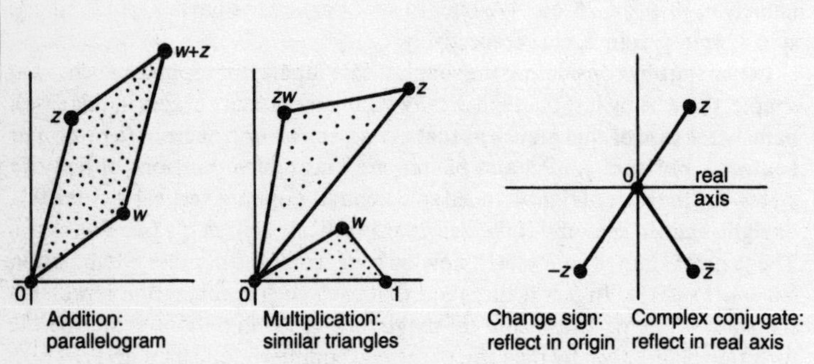

Fig. 5.17. The geometrical descriptions of the basic operations with complex numbers.

Fig. 5.16). Thus, for example, the four complex numbers 1, 1 + i, i, and 0 form the vertices of a square. There are simple geometrical rules for the sum and the product of two complex numbers (Fig. 5.17). Taking the negative $-z$ of a complex number z is represented by reflection in the origin; taking the complex conjugate $\bar{z}$, of z, by reflection in the x-axis.

The modulus of a complex number is the distance from the origin of the point representing it; the squared modulus is thus the square of this number. The *unit circle* is the locus of points that are of unit distance from the origin (Fig. 5.18), these representing the complex numbers of *unit modulus*, sometimes called *pure phases*, having the form

$$e^{i\theta} = \cos\theta + i \sin \theta,$$

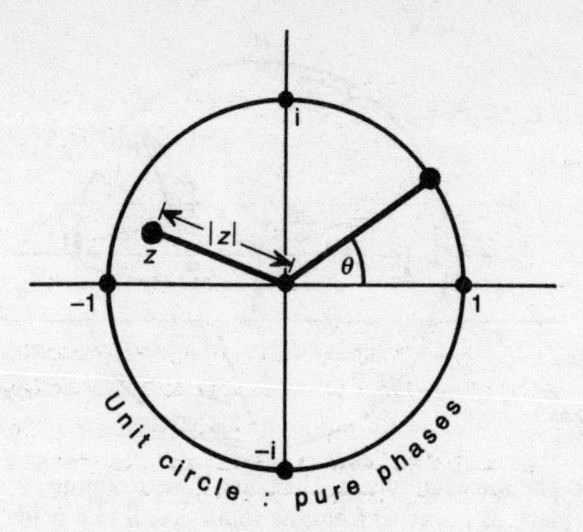

Fig. 5.18. The unit circle consists of complex numbers $z = e^{i\theta}$ with θ real, i.e. $|z| = 1$.

where θ is real and measures the angle that the line joining the origin to the point representing this complex number makes with the x-axis.*

Now let us see how to represent *ratios* of pairs of complex numbers. In the above discussion, I indicated that a state is not physically changed if it is muliplied, overall, by a non-zero complex number (recall, for example, that $-2|\mathrm{F}\rangle$ was to be considered as being physically the same state as $|\mathrm{F}\rangle$). Thus, generally, $|\psi\rangle$ is physically the same as $u|\psi\rangle$, for any non-zero complex number u. Applied to the state

$$|\psi\rangle = w|\uparrow\rangle + z|\downarrow\rangle,$$

We see that if we multiply both of w and z by the same non-zero complex number u, we do not change the physical situation that is represented by the state. It is the different *ratios* $z{:}w$ of the two complex numbers w and z that provide the different physically distinct spin states ($uz{:}uw$ being the same as $z{:}w$ if $u \neq 0$).

How do we geometrically represent a complex ratio? The essential difference between a complex ratio and just a plain complex number is that *infinity* (denoted by the symbol '∞') is also allowed as a ratio, in addition to all the finite complex numbers. Thus, if we think of the ratio $z{:}w$ to be represented, in general, by the single complex number z/w, we have trouble when $w = 0$. To cover this possibility, we simply use the symbol ∞ for z/w in the case when $w = 0$. This occurs when we consider the particular state 'spin down':

*The real number e is the 'base of natural logarithms': e = 2.718 281 828 5. . .; the expression e^z is, indeed 'e raised to the zth power', and we have

$$e^z = 1 + z + \frac{z^2}{1\times 2} + \frac{z^3}{1\times 2\times 3} + \frac{z^4}{1\times 2\times 3\times 4} + \ldots.$$

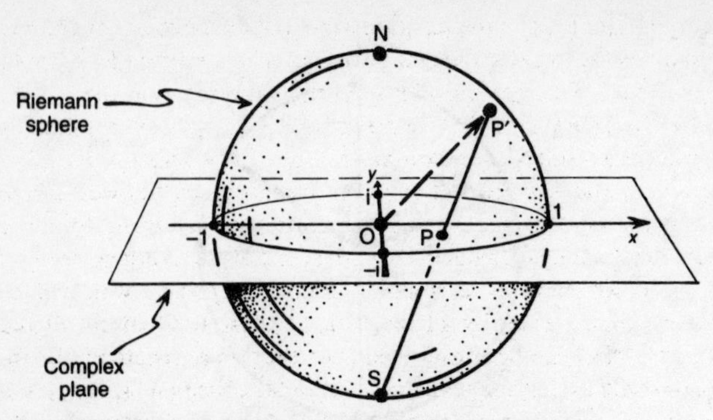

Fig. 5.19. The Riemann sphere. The point P, representing $p=z/w$ on the complex plane, is projected from the south pole S to a point, P′ on the sphere. The direction OP, out from the sphere's centre O, is the direction of spin for the general spin 1/2 state of Fig. 5.15.

$|\psi\rangle=z|\downarrow\rangle=0|\uparrow\rangle+z|\downarrow\rangle$. Recall that we are not allowed to have *both* $w=0$ and $z=0$, but $w=0$ by itself is perfectly allowable. (We could use w/z instead, to represent this ratio of we prefer; then we need ∞ to cover the case $z=0$, which gives the particular state 'spin up'. It does not matter which description we use.)

The way to represent the space of all possible complex ratios is to use a *sphere*, referred to as the *Riemann sphere*. The points of the Riemann sphere represent complex numbers or ∞. We can picture the Riemann sphere as a sphere of unit radius whose equatorial plane is the complex plane and whose centre is the origin (zero) of that plane. The actual equator of this sphere will be identified with the unit circle in the complex plane (see Fig. 5.19). Now, to represent a particular complex ratio, say $z{:}w$, we mark the point P on the complex plane that represents the complex number $p=z/w$ (supposing, for the moment, that $w\neq 0$), and then we project P, on the plane, to a point P′ on the sphere from the *south pole* S. That is to say, we take the straight line from S to P and mark the point P′ on the sphere as the point where this line meets it (apart from at S itself). This mapping between points on the sphere and points on the plane is called *stereographic projection*. To see that it is reasonable for the south pole S itself to represent ∞, we imagine a point P in the plane that moves off to a very large distance; then we find that the point P′ that corresponds to it approaches the south pole S very closely, reaching S in the limit as P goes to infinity.

The Riemann sphere plays a fundamental role in the quantum picture of two-state systems. This role is not always evident explicitly, but the Riemann sphere is always there, behind the scenes. It describes, in an abstract geometrical way, the space of physically distinguishable states that can be built up, by quantum linear superposition, from any two distinct quantum states. For example, the two states might be two possible locations for a photon, say $|B\rangle$ and $|C\rangle$. The general linear combination would have the form $w|B\rangle + z|C\rangle$. Although in §5.7 we made explicit use only of the particular case $|B\rangle + i|C\rangle$, as a result of the reflection/transmission at half-silvered mirror, the other combinations would not be hard to achieve. All that would be needed would be to vary the amount of silvering on the mirror and to introduce a segment of refracting medium in the path of one of the emerging beams. In this way, one could build up a complete Riemann sphere's worth of possible alternative states, given by all the various physical situations of the form $w|B\rangle + z|C\rangle$, that can be constructed from the two alternatives $|B\rangle$ and $|C\rangle$.

In cases such as this, the geometrical role of the Riemann sphere is not at all an apparent one. However, there are other types of situation in which the Riemann sphere's role is geometrically manifest. The clearest example of this occurs with the spin states of a particle of spin $\frac{1}{2}$, such as an electron or proton. The general state can be represented as a combination

$$|\psi\rangle = w|\uparrow\rangle + z|\downarrow\rangle,$$

and it turns out (choosing $|\uparrow\rangle$ and $|\downarrow\rangle$ appropriately from the proportionality class of physically equivalent possibilities) that this $|\psi\rangle$ represents the state of spin, of magnitude $\frac{1}{2}\hbar$, which is right handed about the axis that points in the direction of the very point on the Riemann sphere representing the ratio z/w. Thus every direction in space plays a role as a possible spin direction for any particle of spin $\frac{1}{2}$. Even though most states are represented, initially, as being 'mysterious complex-number-weighted combinations of alternatives' (the alternatives being $|\uparrow\rangle$ and $|\downarrow\rangle$), we see that these combinations are no more and no less mysterious than the two original ones, $|\uparrow\rangle$ and $|\downarrow\rangle$, that we started out with. Each is just as physically real as any of the others.

What about states of higher spin? It turns out that things get a little more complicated—and *more* mysterious! The general description that I shall give is not a very well-known one to today's physicists, although it was pointed out in 1932 by Ettore Majorana (a brilliant Italian physicist who disappeared at the age of 31 on a ship entering the Bay of Naples under circumstances that have never been fully explained).

Let us consider, first, what *is* very familiar to physicists. Suppose that we have an atom (or particle) of spin $\frac{1}{2}n$. Again, we can choose the upward direction to start with, and ask the question as to 'how much' of the atom's spin is actually oriented in (i.e. right-handed about) that direction. There is a standard piece of apparatus, known as a Stern–Gerlach apparatus, which achieves such measurements by use of an inhomogeneous magnetic field. What happens is that there are just $n+1$ different possible outcomes, which can be distinguished by the fact that the atom is found to lie in just one of $n+1$ different possible beams. See Fig. 5.20. The amount of the spin that lies in the

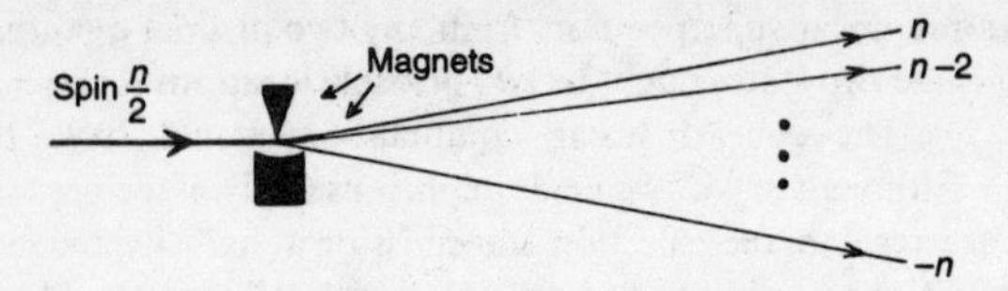

Fig. 5.20. The Stern–Gerlach measurement. There are $n+1$ possible outcomes, for spin $\frac{1}{2}n$, depending upon how much of the spin is found to be in the measured direction.

chosen direction is determined by the particular beam in which the atom is found to lie. When measured in units of $\frac{1}{2}\hbar$ the amount of spin in this direction turns out to have one of the values $n, n-2, n-4, \ldots, 2-n, -n$. Thus the different possible states of spin, for an atom of spin $\frac{1}{2}n$, are just the complex-number superpositions of these possibilities. I shall denote the various different possible results of a Stern–Gerlach measurement for spin $n+1$, when the field direction in the apparatus is the upward vertical, as

$$|\uparrow\uparrow\uparrow\ldots\uparrow\rangle, |\downarrow\uparrow\uparrow\ldots\uparrow\rangle, |\downarrow\downarrow\uparrow\ldots\uparrow\rangle, \ldots, |\downarrow\downarrow\downarrow\ldots\downarrow\rangle,$$

corresponding to the respective spin values $n, n-2, n-4, \ldots, 2-n, -n$ in that direction, where in each case there are exactly n arrows in all. We can think of each upward arrow as providing an amount $\frac{1}{2}\hbar$ of spin in the upward direction and each downward arrow as providing $\frac{1}{2}\hbar$ in the downward direction. Adding these values, we get the total amount of spin, in each case, obtained in a (Stern–Gerlach) spin measurement oriented in the up/down direction.

The general superposition of these is given by a complex combination

$$z_0|\uparrow\uparrow\uparrow\ldots\uparrow\rangle + z_1|\downarrow\uparrow\uparrow\ldots\uparrow\rangle + z_2|\downarrow\downarrow\uparrow\ldots\uparrow\rangle + \ldots + z_n|\downarrow\downarrow\downarrow\ldots\downarrow\rangle,$$

where the complex numbers $z_0, z_1, z_2, \ldots, z_n$ are not all zero. Can we represent such a state in terms of single directions of spin which are not simply 'up' or 'down'? What Majorana in effect showed was that this is indeed possible, but we must allow that the various arrows might point in quite independent directions; there is no need for them to be aligned in one pair of opposite directions, as would be the case in the result of a Stern–Gerlach measurement. Thus, we represent the general state of spin $\frac{1}{2}n$ as a collection of n independent such 'arrow directions'; we may think of these as being given by n points on the Riemann sphere, where each arrow direction is the direction out from the centre of the sphere to the relevant point on the sphere (Fig. 5.21). It is important to make clear that this is an *unordered* collection of points (or of arrow directions). Thus, there is no significance to be assigned to any ordering of the points into first, second, third, etc.

This is a very odd picture of spin, if we are to try to think of quantum-mechanical spin as the same phenomenon as the ordinary concept of spin that is familiar at the classical level. The spin of a classical object like a golf ball has

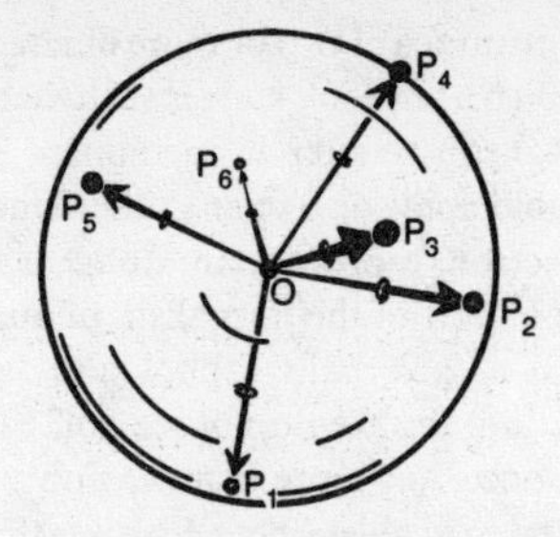

Fig. 5.21. The Majorana description of the general state of spin $\frac{1}{2}n$ is as an unordered set of n points $P_1, P_2, \ldots P_n$ on the Riemann sphere, where each point may be thought of as an element of spin $\frac{1}{2}$ directed outwards from the centre to the point in question.

a well-defined axis about which the object actually spins, whereas it appears that a quantum-level object is allowed to spin all at once about all kinds of axes pointing in many different directions. If we try to think that a classical object is really just the same as a quantum object, except that it is 'big' in some sense, then we seem to be presented with a paradox. The larger the magnitude of the spin, the more directions there are to be involved. Why, indeed, do classical objects not spin in many different directions all at once? This is an example of an **X**-mystery of quantum theory. Something comes to intervene (at an unspecified level), and we find that most types of quantum state do not arise (or, at least, almost never arise) at the classical level of phenomena that we can actually perceive. In the case of spin, what we find is that the only states that significantly persist at the classical level are those in which the arrow directions are mainly clustered about one particular direction: the spin direction (axis) of the classically spinning object.

There is something called the 'correspondence principle' in quantum theory that asserts, in effect, that when physical quantities (such as the magnitude of spin) get large, then it is *possible* for the system to behave in a way that closely approximates classical behaviour (such as with the state where the arrows roughly point all in the same direction). However, this principle does not tell us how such states can arise solely by action of the Schrödinger equation **U**. In fact 'classical states' almost never arise in this way. The classical-like states come about because of the action of a different procedure: state-vector reduction **R**.

5.11 Position and momentum of a particle

There is an even more clear-cut example of this sort of thing in the quantum-mechanical concept of *location* for a particle. We have seen that a particle's state can involve superpositions of two or more different locations. (Recall the discussion of §5.7, in which a photon's state is such that it can be located in two different beams simultaneously after it encounters a half-silvered mirror.) Such superpositions could apply also to any other kind of particle—simple or composite—like an electron, a proton, an atom, or a molecule. Moreover, there is nothing in the **U** part of the formalism of quantum theory that says that large objects such as golf balls cannot also find themselves in such confused states of location. But we do not see golf balls in superpositions of several locations all at once, any more than we find a golf ball whose state of spin can be about several axes all at once. Why is it that some objects seem to be too large or too massive or too something to be able to be 'quantum-level' objects, and do not find themselves in such superposed states in the actual world? In standard quantum theory, it is only the action of **R** that achieves this transition from the quantum-level superpositions of potential alternatives to a single, actual, classical outcome. The mere action of **U** by itself would almost invariably lead to 'unreasonable-looking' classical superpositions. (I shall return to this issue in §6.1.)

At the quantum level, on the other hand, those states of a particle for which it has no clear-cut location can play a fundamental role. For if the particle has a clear-cut *momentum* (so it is moving in a clearly defined way in some direction and not in a superposition of several different directions all at once), then its state must involve a superposition of all different *positions* all at once. (This is a particular feature of the Schrödinger equation which would require too much technicality for an explanation to be appropriate here; cf. ENM, pp. 243–50 and Dirac (1947), Davies (1984), for example. It is closely connected, also, with Heisenberg's *uncertainty principle*, which provides limits to how clearly defined can be position and momentum simultaneously.) In fact the states of well-defined momentum have an oscillatory spatial behaviour, in the direction of motion, which was the feature that we ignored in the discussion of photon states given in §5.7. Strictly speaking, 'oscillatory' is not quite the appropriate term. It turns out that the 'oscillations' are not like the linear vibrations of a string, in which one could imagine that the complex weighting factors oscillate backwards and forwards through the origin in the complex plane, but instead these factors are pure phases (see Fig. 5.18) that circle about the origin at a constant rate—this rate giving a frequency ν that is proportional to the particle's energy E, in accordance with Planck's famous formula $E=h\nu$. (See Fig. 6.11 of ENM, for a pictorial 'corkscrew' representation of momentum states.) These matters, important though they are for quantum theory, will not play any particular role in the discussions in this book, so the reader may happily ignore them here.

More generally, the complex weighting factors need not have this particular 'oscillatory' form, but can vary from point to point in an arbitrary manner. The weighting factors provide a complex function of position referred to as the *wavefunction* of the particle.

5.12 Hilbert space

In order to be a little more explicit (and accurate) about how the procedure **R** is taken to act, in standard quantum-mechanical descriptions, it will be necessary to resort to a certain (relatively minor) degree of mathematical technicality. The family of all possible states of a quantum system constitute what is known as a *Hilbert space*. It will not be necessary to explain what this means in full mathematical detail, but a bit of understanding of Hilbert-space notions will help to clarify our picture of the quantum world.

The first and most important property of a Hilbert space is that it is a *complex vector space*. All this really means is that we are allowed to perform the complex-number-weighted combinations that we have been considering for quantum states. I shall continue to use the Dirac 'ket' notation for elements of the Hilbert space, so if $|\psi\rangle$ and $|\phi\rangle$ are both elements of the Hilbert space, then so also is $w|\psi\rangle + z|\phi\rangle$, for any pair of complex numbers w and z. Here, we even allow $w = z = 0$, to give the element $\mathbf{0}$ of the Hilbert space, this being the one Hilbert-space element that does *not* represent a possible physical state. We have the normal algebraic rules for a vector space, these being

$$|\psi\rangle + |\phi\rangle = |\phi\rangle + |\psi\rangle,$$

$$|\psi\rangle + (|\phi\rangle + |\chi\rangle) = (|\psi\rangle + |\phi\rangle) + |\chi\rangle,$$

$$w(z|\psi\rangle) = (wz)|\psi\rangle,$$

$$(w+z)|\psi\rangle = w|\psi\rangle + z|\psi\rangle,$$

$$z(|\psi\rangle + |\phi\rangle) = z|\psi\rangle + z|\phi\rangle,$$

$$0|\psi\rangle = \mathbf{0},$$

$$z\mathbf{0} = \mathbf{0},$$

which more or less means that we can use the algebraic notation in the way that we would expect to use it.

A Hilbert space can sometimes have a *finite* number of dimensions, as in the case of the spin states of a particle. For spin $\frac{1}{2}$, the Hilbert space is just two dimensional, its elements being the complex linear combinations of the two states $|\uparrow\rangle$ and $|\downarrow\rangle$. For spin $\frac{1}{2}n$, the Hilbert space is $(n+1)$ dimensional. However, sometimes the Hilbert space can have an *infinite* number of dimensions as would be the case for the states of position of a particle. Here, each alternative position that the particle might have counts as providing a separate dimension for the Hilbert space. The general state describing the quantum location of the particle is a complex-number superposition of *all* these different individual positions (the wavefunction for the particle). In fact there are certain mathematical complications that arise for this kind of infinite-dimensional Hilbert space which would confuse the discussion

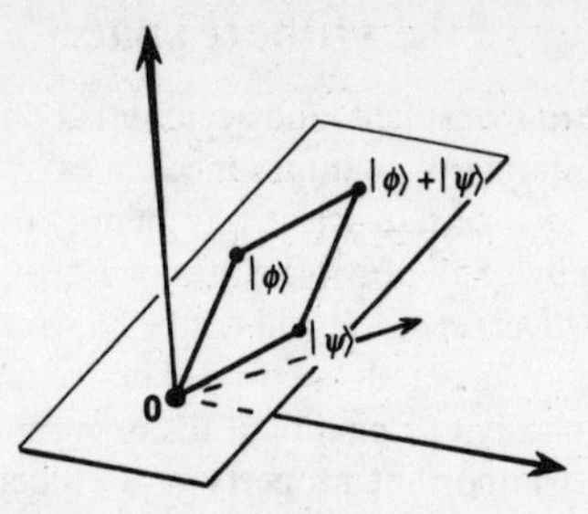

Fig. 5.22. If we pretend that Hilbert space is a three-dimensional Euclidean space, then we can represent the sum of two vectors $|\psi\rangle$ and $|\phi\rangle$ in terms of the ordinary parallelogram law (in the plane of **0**, $|\psi\rangle$, and $|\phi\rangle$).

unnecessarily, so I shall concentrate here mainly on the finite-dimensional case.

When we try to visualize a Hilbert space, we encounter two difficulties. In the first place, these spaces tend to have rather too many dimensions for our direct imagination, and in the second, they are *complex* rather than real spaces. Nevertheless, it is often handy to ignore these problems temporarily, in order to develop an intuition about the mathematics. Let us therefore pretend, for the moment, that we can use our normal two- or three-dimensional pictures to represent a Hilbert space. In Fig. 5.22, the operation of linear superposition is illustrated geometrically in the real three-dimensional case.

Recall that a quantum-state vector $|\psi\rangle$ represents the same physical situation as any complex multiple $u|\psi\rangle$, where $u \neq 0$. In terms of our pictures, this means that a particular physical situation is actually represented not by a point in Hilbert space, but by the whole line—referred to as a *ray*—which joins the Hilbert-space point $|\psi\rangle$ to the origin **0**. See Fig. 5.23; but we must bear in mind that, owing to the fact that Hilbert spaces are complex rather than real, although a ray *looks* like an ordinary one-dimensional line, it is really an entire complex plane.

So far, we have been concerned with Hilbert space only with regard to its structure as a complex vector space. There is another property that a Hilbert space has, which is almost as crucial as its vector-space structure, and which is essential for the description of the reduction procedure **R**. This other Hilbert-space property is the *Hermitian scalar product* (or *inner* product), which can be applied to any pair of Hilbert-space vectors to produce a single complex number. This operation enables us to express two important things. The first is the notion of *squared length* of a Hilbert-space vector—as the scalar product of the vector with *itself*. A *normalized* state (which, as we noted above, cf. §5.8, p. 264, was needed for the squared modulus rule to be strictly applicable) is given by a Hilbert-space vector whose squared length is *unity*. The second

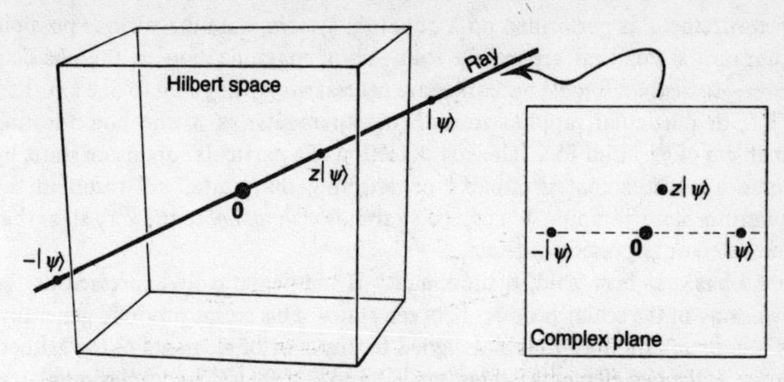

Fig. 5.23. A *ray* in Hilbert space consists of all the complex multiples of a state-vector $|\psi\rangle$. We think of this as a straight line through the Hilbert-space origin **0**, but we must bear in mind that this straight line is really a complex plane.

important thing that the scalar product gives us is the notion of *orthogonality* between Hilbert-space vectors—and this occurs when the scalar product of the two vectors is *zero*. Orthogonality between vectors may be thought of as their being, in some appropriate sense, 'at right angles' to each other. In ordinary terms, orthogonal states are things that are *independent* of one another. The importance of this concept for quantum physics is that the different alternative outcomes of any measurement are always orthogonal to each other.

Examples of orthogonal states are the states $|\uparrow\rangle$ and $|\downarrow\rangle$ that we encountered in the case of a particle of spin $\frac{1}{2}$. (Note that Hilbert-space orthogonality does not normally correspond to the notion of perpendicularity in ordinary spatial terms; in the case of spin $\frac{1}{2}$, the orthogonal states $|\uparrow\rangle$ and $|\downarrow\rangle$ represent physical configurations that are oppositely, rather than perpendicularly, oriented.) Other examples are afforded by the states $|\uparrow\uparrow\ldots\uparrow\rangle, |\downarrow\uparrow\ldots\uparrow\rangle, \ldots, |\downarrow\downarrow\ldots\downarrow\rangle$, that we encountered for spin $\frac{1}{2}n$, each of which is orthogonal to all of the others. Also orthogonal are *all* different possible *positions* that a quantum particle might be located in. Moreover, the states $|B\rangle$ and $i|C\rangle$, of §5.7, that arose as the transmitted and reflected parts of a photon's state, emerging after the photon has encountered a half-silvered mirror, are orthogonal, and so also are the two states $i|D\rangle$ and $-|E\rangle$ that these two evolve into upon reflection by the two fully silvered mirrors.

This last fact illustrates an important property of Schrödinger evolution **U**. Any two states that are initially orthogonal will remain orthogonal if each is evolved, according to **U**, over the same period of time. Thus, the property of orthogonality is *preserved* by **U**. Furthermore, **U** actually preserves the value of the scalar product between states. Technically, this is what the term *unitary* evolution actually means.

As mentioned above, the key role of orthogonality is that whenever a 'measurement' is performed on a quantum system, then the various possible quantum states that separately lead—upon magnification to the classical level—to *distinguishable* outcomes are necessarily orthogonal to one another. This, in particular, applies to *null* measurements, as in the bomb-testing problem of §5.2 and §5.9. The *non*-detection of a particular quantum state, by some apparatus that is capable of detecting that state, will result in the quantum state 'jumping' to something that is *orthogonal* to the very state that the detector is poised to detect.

As has just been said, orthogonality is mathematically expressed as the *vanishing* of the scalar product between states. This scalar product, generally, is a complex number that is assigned to any pair of elements of the Hilbert space. If the two elements (states) are $|\psi\rangle$ and $|\phi\rangle$, then this complex number is written $\langle\psi|\phi\rangle$. The scalar product satisfies a number of simple algebraic properties that we can write (slightly awkwardly) as

$$\overline{\langle\psi|\phi\rangle} = \langle\phi|\psi\rangle,$$

$$\langle\psi|(|\phi\rangle + |\chi\rangle) = \langle\psi|\phi\rangle + \langle\psi|\chi\rangle,$$

$$(z\langle\psi|)|\phi\rangle = z\langle\psi|\phi\rangle,$$

$$\langle\psi|\psi\rangle > 0 \text{ unless } |\psi\rangle = \mathbf{0};$$

moreover, it can be deduced that $\langle\psi|\psi\rangle = 0$ if $|\psi\rangle = \mathbf{0}$. I do not wish to trouble the reader with the details of such matters here. (For these, I refer the interested reader to any standard text on quantum theory; see, for example, Dirac (1947).)

The essential things that we shall require of the scalar product here are the two properties (definitions) alluded to above:

$|\psi\rangle$ and $|\phi\rangle$ are *orthogonal* if and only if $\langle\psi|\phi\rangle = 0$,

$\langle\psi|\psi\rangle$ is the *squared length* of $|\psi\rangle$.

We note that the relationship of orthogonality is symmetrical between $|\psi\rangle$ and $|\phi\rangle$ (because $\overline{\langle\psi|\phi\rangle} = \langle\phi|\psi\rangle$). Moreover, $\langle\psi|\psi\rangle$ is always a non-negative real number, so it has a non-negative square root that we can refer to as the *length* (or magnitude) of $|\psi\rangle$.

Since we can multiply any state vector by a non-zero complex number without changing its physical interpretation, we can always *normalize* the state so that it has unit length to make it a *unit vector* or *normalized state*. There remains, however, the ambiguity that the state vector can be multiplied by a pure phase (a number of the form $e^{i\theta}$, with θ real; cf. §5.10).

5.13 The Hilbert-space description of R

How do we represent the operation of **R** in terms of Hilbert space? Let us consider the simplest case of a measurement, which is a 'yes/no' measurement, where some piece of apparatus registers **YES** if it positively indicates that the measured quantum object has some property, and **NO** if it fails to register this fact (or, equivalently, that it positively registers that the quantum object does not have this property). This includes the possibility, that I shall be concerned mainly with here, that the **NO** alternative might be a *null* measurement. For example, some of the photon detectors of §5.8 performed measurements of this kind. They would register **YES** if the photon is received and **NO** if it is not received. In this case, the **NO** measurement is indeed a null measurement—but it is a measurement nevertheless, which causes the state to 'jump' into something that is orthogonal to what it would have been had a **YES** answer been obtained. Likewise, the Stern–Gerlach spin measurers of §5.10, for an atom of spin $\frac{1}{2}$, could be directly of this kind; we could say that the result is **YES** if the atom's spin is measured to be $|\uparrow\rangle$, which occurs if it is found in the beam corresponding to $|\uparrow\rangle$, and **NO** if it is not found in this beam, whence the result must be orthogonal to $|\uparrow\rangle$, so it must be $|\downarrow\rangle$.

More complicated measurements can always be thought of as built up out of a succession of yes/no measurements. Consider an atom of spin $\frac{1}{2}n$, for example. In order to obtain the $n+1$ different possibilities for a measurement of the amount of spin in the upward direction, we can ask, first, whether the spin state is $|\uparrow\uparrow \ldots \uparrow\rangle$. We do this by trying to detect the atom in the beam corresponding to this 'completely up' spin state. If we get **YES**, then we are finished; but if we get **NO**, then this is a null measurement, and we can go on to ask whether the spin is $|\downarrow\uparrow \ldots \uparrow\rangle$, and so on. In each case the **NO** answer is to be a null measurement, indicating simply that the **YES** answer has not been obtained. In more detail, suppose the state is initially

$$z_0|\uparrow\uparrow\uparrow \ldots \uparrow\rangle + z_1|\downarrow\uparrow\uparrow \ldots \uparrow\rangle + z_2|\downarrow\downarrow\uparrow \ldots \uparrow\rangle + \ldots + z_n|\downarrow\downarrow\downarrow \ldots \downarrow\rangle,$$

and we ask if the spin is entirely 'up'. If we find the answer **YES**, then we ascertain the state to be indeed $|\uparrow\uparrow\uparrow \ldots \uparrow\rangle$ or, rather, we may have to consider that it just 'jumps' to $|\uparrow\uparrow\uparrow \ldots \uparrow\rangle$ upon measurement. But if we find the answer **NO**, then, as a result of this null measurement, we have to consider that the state 'jumps' to the orthogonal state

$$z_1|\downarrow\uparrow\uparrow \ldots \uparrow\rangle + z_2|\downarrow\downarrow\uparrow \ldots \uparrow\rangle + \ldots + z_n|\downarrow\downarrow\downarrow \ldots \downarrow\rangle,$$

and we try again, in order to ascertain, now, whether the state is $|\downarrow\uparrow\uparrow \ldots \uparrow\rangle$. If we *now* obtain **YES**, then we say that the state is indeed $|\downarrow\uparrow\uparrow \ldots \uparrow\rangle$, or else we say that the state has 'jumped' to $|\downarrow\uparrow\uparrow \ldots \uparrow\rangle$. But if we find **NO**, then the state 'jumps' to

$$z_2|\downarrow\downarrow\uparrow \ldots \uparrow\rangle + \ldots + z_n|\downarrow\downarrow\downarrow \ldots \downarrow\rangle,$$

and so on.

This 'jumping' that the state vector indulges in—or at least *seems* to indulge in—is the most puzzling aspect of quantum theory. It is probably fair to say that most quantum physicists either find 'jumping' very *hard* to accept as a feature of actual physical reality, or else refuse *altogether* to accept that reality can behave in this absurd way. Nevertheless, it is an essential feature of the quantum formalism, whatever viewpoint one might hold to as to the 'reality' involved.

In my descriptions above, I have relied on what is sometimes referred to as the *projection postulate*, which specifies the form that this 'jumping' must take (e.g. $z_0|\uparrow\uparrow\ldots\uparrow\rangle+z_1|\downarrow\uparrow\ldots\uparrow\rangle+\ldots+z_n|\downarrow\downarrow\ldots\downarrow\rangle$ 'jumps' to $z_1|\downarrow\uparrow\ldots\uparrow\rangle+\ldots+z_n|\downarrow\downarrow\ldots\downarrow\rangle$). We shall see the geometrical reason for this terminology in a moment. Some physicists claim that the projection postulate is an inessential assumption in quantum theory. They generally refer, however, to measurements in which the quantum state is *disturbed* by some physical interaction, and not to a null measurement. Such a disturbance occurs when, in the above examples, the answer **YES** is obtained, as when a photon detector absorbs a photon in order to register, or when an atom, after passing through a Stern–Gerlach apparatus, is measured actually to be in one particular beam (i.e. **YES**). For a null measurement of the type that I am considering here (answer **NO**), the projection postulate is indeed essential, for without it one cannot ascertain what quantum theory (correctly) asserts must happen for the ensuing measurements.

To be more explicit about what the projection postulate asserts, let us see what the Hilbert-space description of all this is. I shall consider a particular kind of measurement—which I shall call a *primitive* measurement—in which the measurement is of yes/no type, but where the **YES** answer ascertains that the quantum state is—or has just 'jumped to'—some particular state $|\alpha\rangle$ (or to some non-zero multiple of this, $u|\alpha\rangle$). Thus, for a primitive measurement, the **YES** answer determines the physical state to be *one* particular thing, whereas there might be several alternative things that elicit the answer **NO**. The above spin measurements, whereby one tries to ascertain whether the spin is in a particular state, such as $|\downarrow\downarrow\uparrow\ldots\uparrow\rangle$, are examples of primitive measurements.

The **NO** answer, for a primitive measurement, *projects* the state to something that is orthogonal to $|\alpha\rangle$. We see the geometrical picture of this indicated in Fig. 5.24. The original state is taken to be $|\psi\rangle$, depicted by the large arrow, and as a result of the measurement it either 'jumps' to a multiple of $|\alpha\rangle$, when the answer is **YES**, or else projects down to the space orthogonal to $|\alpha\rangle$, when the answer is **NO**. In the case of **NO**, there is no question but that, according to standard quantum theory, this is indeed what we must consider the state to do. In the case of **YES**, however, the situation is complicated by the fact that the quantum system has now interacted with the measuring apparatus, and the state has become something much more intricate than just $|\alpha\rangle$. In fact, the state generally evolves into what is called an *entangled state*, which intertwines the original quantum system with the measuring apparatus. (Entangled states will be discussed in §5.17). Nevertheless, the quantum state's

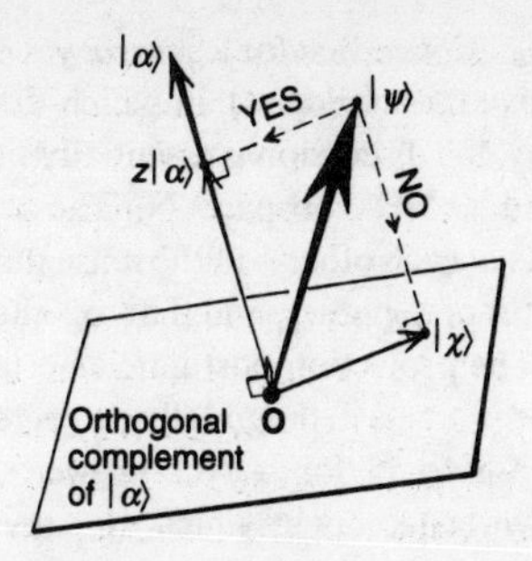

Fig. 5.24. A primitive measurement projects the state $|\psi\rangle$ into a multiple of the chosen state $|\alpha\rangle$ (YES) or into the orthogonal complement of $|\alpha\rangle$ (NO).

evolution would need to proceed *as though* it had indeed jumped to a multiple of $|\alpha\rangle$, so that the succeeding evolution can proceed unambiguously.

We can express this jumping in the following algebraic way. The state vector $|\psi\rangle$ can always be written (uniquely, if $|\alpha\rangle$ is given) as

$$|\psi\rangle = z|\alpha\rangle + |\chi\rangle,$$

where $|\chi\rangle$ is orthogonal to $|\alpha\rangle$. The vector $z|\alpha\rangle$ is the orthogonal projection of $|\psi\rangle$ on the ray determined by $|\alpha\rangle$, and $|\chi\rangle$ is the orthogonal projection of $|\psi\rangle$ into the *orthogonal complement* space of $|\alpha\rangle$ (i.e. into the space of all vectors orthogonal to $|\alpha\rangle$). If the result of the measurement is **YES**, then the state vector must be considered to have jumped to $z|\alpha\rangle$ (or simply to $|\alpha\rangle$) as the starting point of the ensuing evolution; if it is **NO**, then it jumps to $|\chi\rangle$.

How do we obtain the probabilities that are to be assigned to these two alternative outcomes? In order to use the 'squared modulus rule' that we encountered before, we now demand that $|\alpha\rangle$ is a *unit* vector, and choose some unit vector $|\phi\rangle$ in the direction of $|\chi\rangle$, so that $|\chi\rangle = w|\phi\rangle$. We now have

$$|\psi\rangle = z|\alpha\rangle + w|\phi\rangle,$$

(where in fact $z = \langle\alpha|\psi\rangle$ and $w = \langle\phi|\psi\rangle$) and we can obtain the relative probabilities of **YES** and **NO** as the ratio of $|z|^2$ to $|w|^2$. If $|\psi\rangle$ itself is a unit vector, then $|z|^2$ and $|w|^2$ are the *actual* probabilities of **YES** and **NO**, respectively.

There is another way of phrasing this, which is a little simpler in the present context (and I leave it as an exercise for the interested reader to verify that this is indeed equivalent). To obtain the actual probability of each of the outcomes **YES** or **NO**, we simply examine the squared length of the vector $|\psi\rangle$ (not assumed to be normalized to a unit vector) and see by what proportion this squared length is reduced in each of its respective projections. The reduction factor, in each case, is the required probability.

Finally, it should be mentioned that for a *general* yes/no measurement (now not necessarily a primitive measurement), in which the **YES** states need not belong to just a single ray, the discussion is essentially similar. There would be a **YES** subspace **Y** and a **NO** subspace **N**. These subspaces would be orthogonal complements of each other—in the sense that each vector of one is orthogonal to every vector of the other, and that together they span the whole original Hilbert space. The projection postulate asserts that, upon measurement, the original vector $|\psi\rangle$ gets orthogonally projected into **Y** when a **YES** answer is obtained and into **N** for a **NO** answer. Again the respective probabilities are given by the factors by which the squared length of the state vector gets reduced on projection (see ENM, p. 263, Fig. 6.23). The status of the projection postulate is a little less clear here than for the null measurement above, however, because with an affirmative measurement, the resulting state gets entangled with the state of the measuring apparatus. For such reasons, in the discussions which follow, I shall stick to the simpler *primitive* measurements, in which the **YES** space consists of a single ray (multiples of $|\psi\rangle$). This will be sufficient for our needs.

5.14 Commuting measurements

In general, with successive measurements on a quantum system, the order in which these measurements are performed can be important. Measurements for which the order of operation indeed makes a difference to the resulting state vectors are called *non-commuting*. If the ordering of the measurements plays no role whatever (there being not even a phase factor of difference), then we say that they *commute*. In terms of Hilbert space, we can understand this in terms of the fact that with successive orthogonal projections of a given state vector $|\psi\rangle$, the final result will generally depend upon the order in which these projections are performed. For commuting measurements, the ordering makes no difference.

What happens with *primitive* measurements? It is not hard to see that the condition for a pair of distinct primitive measurements to commute is that the **YES** ray of one of the measurements is *orthogonal* to the **YES** ray of the other.

For example, with the primitive spin measurements considered in §5.10, that were to be performed on an atom of spin $\frac{1}{2}n$, the ordering is indeed irrelevant. This is the case because the various states under consideration, namely $|\uparrow\uparrow \dots \uparrow\rangle, |\downarrow\uparrow \dots \uparrow\rangle, \dots, |\downarrow\downarrow \dots \downarrow\rangle$ are all orthogonal to one another. Thus, the particular ordering for the primitive measurements that I chose plays no role in the final outcome, and the measurements do all commute. However, this would not generally have been the case if the various spin measurements had been taken in different directions. These would *not* generally commute.

5.15 The quantum-mechanical 'and'

There is a standard procedure, in quantum mechanics, for treating systems that involve more than one independent part. This procedure will be necessary, in particular, for the quantum discussion (to be given in §5.18) of the system that consists of the two widely separated particles of spin $\frac{3}{2}$ that Quintessential Trinkets placed at the centres of the two magic dodecahedra of §5.3. It is also needed, for example, for the quantum-mechanical description of a detector, as it begins to get entangled with the quantum state of the particle that it is detecting.

Consider, first, a system that consists of just *two* independent (non-interacting) parts. We shall assume that each, in the absence of the other, would be described, respectively, by a state vector $|\alpha\rangle$ and by a state vector $|\beta\rangle$. How are we to describe the combined system in which *both* are present? The normal procedure is to form what is called the *tensor product* (or *outer* product) of these vectors, written

$$|\alpha\rangle|\beta\rangle.$$

We can think of this product as providing what is the standard quantum-mechanical way of representing the normal notion of 'and', in the sense that the two independent quantum systems represented, respectively, by $|\alpha\rangle$ and $|\beta\rangle$, are now *both* present at the same time. (For example, $|\alpha\rangle$ might represent an electron being at a location A and $|\beta\rangle$ might represent a hydrogen atom being at some distant location B. The state in which the electron is at A *and* the hydrogen atom is at B would then be represented by $|\alpha\rangle|\beta\rangle$.) The quantity $|\alpha\rangle|\beta\rangle$ is a *single* quantum state vector, say $|\chi\rangle$, however, and it would be legitimate to write

$$|\chi\rangle=|\alpha\rangle|\beta\rangle,$$

for example.

It should be emphasized that this concept of 'and' is completely different from quantum linear superposition, which would have the quantum description $|\alpha\rangle+|\beta\rangle$ or, more generally, $z|\alpha\rangle+w|\beta\rangle$ where z and w are complex weighting factors. For example, if $|\alpha\rangle$ and $|\beta\rangle$ are possible states of a single photon, say being located at A and being located at a quite different place B, respectively, then $|\alpha\rangle+|\beta\rangle$ is also a possible state for a *single* photon, whose location is divided between A and B according to the strange prescriptions of quantum theory—not for *two* photons. A *pair* of photons, one at A and one at B, would be represented by the state $|\alpha\rangle|\beta\rangle$.

The tensor product satisfies the sort of algebraic rules that one would expect of a 'product'

$$(z|\alpha\rangle)|\beta\rangle = z(|\alpha\rangle|\beta\rangle) = |\alpha\rangle(z|\beta\rangle),$$

$$(|\alpha\rangle + |\gamma\rangle)|\beta\rangle = |\alpha\rangle|\beta\rangle + |\gamma\rangle|\beta\rangle,$$

$$|\alpha\rangle(|\beta\rangle + |\gamma\rangle) = |\alpha\rangle|\beta\rangle + |\alpha\rangle|\gamma\rangle,$$

$$(|\alpha\rangle|\beta\rangle)|\gamma\rangle = |\alpha\rangle(|\beta\rangle|\gamma\rangle),$$

except that it is not strictly correct to write '$|\alpha\rangle|\beta\rangle = |\beta\rangle|\alpha\rangle$'. However, it would be unreasonable to think of interpreting the word 'and' in a quantum-mechanical context as implying that the combined system '$|\alpha\rangle$ and $|\beta\rangle$' is physically any different from the combined system '$|\beta\rangle$ and $|\alpha\rangle$'. We get around this problem by delving a little more deeply into the way that quantum-level Nature actually behaves. Instead of interpreting the state $|\alpha\rangle|\beta\rangle$ as what mathematicians call the 'tensor product', I shall henceforth interpret the notation '$|\alpha\rangle|\beta\rangle$' to involve what mathematical physicists refer to, these days, as the *Grassmann* product. We then have the additional rule that

$$|\beta\rangle|\alpha\rangle = \pm|\alpha\rangle|\beta\rangle,$$

where the minus sign occurs precisely when *both* the states $|\alpha\rangle$ and $|\beta\rangle$ have an *odd* number of particles whose spin is not an integer. (The spin of such particles has one of the values $\frac{1}{2}, \frac{3}{2}, \frac{5}{2}, \frac{7}{2}, \ldots$, these particles being referred to as *fermions*. Particles of spin 0, 1, 2, 3, ... are called *bosons*, and these do not contribute to the sign in this expression.) There is no need for the reader to be concerned about this technicality here. As far as the physical state is concerned, with this description, '$|\alpha\rangle$ and $|\beta\rangle$' is indeed the same as '$|\beta\rangle$ and $|\alpha\rangle$'.

For states involving three or more independent parts, we just repeat this procedure. Thus, with three parts, whose individual states are $|\alpha\rangle$, $|\beta\rangle$, and $|\gamma\rangle$, the state in which all three parts are present at the same time would be written as

$$|\alpha\rangle|\beta\rangle|\gamma\rangle,$$

this being the same as what I have written above (interpreted in terms of Grassmann products) as $(|\alpha\rangle|\beta\rangle)|\gamma\rangle$ or, equivalently, $|\alpha\rangle(|\beta\rangle|\gamma\rangle)$. The case of four or more independent parts is similar.

An important property of Schrödinger evolution **U**, for systems $|\alpha\rangle$ and $|\beta\rangle$ that are not interacting with one another, is that the evolution of the combined system is just the combined evolution of the individual systems. Thus, if after a certain time t, the system $|\alpha\rangle$ would (by itself) evolve into $|\alpha'\rangle$, and if $|\beta\rangle$ (by itself) would evolve into $|\beta'\rangle$, then the combined system $|\alpha\rangle|\beta\rangle$ would, after this same time t, evolve into $|\alpha'\rangle|\beta'\rangle$. Moreover (consequently), if there are three non-interacting parts $|\alpha\rangle$, $|\beta\rangle$, and $|\gamma\rangle$ of a system $|\alpha\rangle|\beta\rangle|\gamma\rangle$, and the parts evolve, respectively, into $|\alpha'\rangle$, $|\beta'\rangle$ and $|\gamma'\rangle$, then the combined system would likewise evolve into $|\alpha'\rangle|\beta'\rangle|\gamma'\rangle$. The same holds for four or more parts.

Note that this is very similar to the *linearity* property of **U** that was referred to in §5.7, according to which superposed states evolve precisely as the superposition of the evolutions of the individual states. For example, $|\alpha\rangle + |\beta\rangle$ would evolve into $|\alpha'\rangle + |\beta'\rangle$. However, it is important to realize that this is quite a *different* thing. There is no particular surprise involved in the fact that a total system that is composed of non-interacting independent parts should evolve, as a whole, as though each separate part were oblivious of the presence of each of the others. It is essential for this that the parts should indeed be non-interacting with each other, otherwise this property would be false. The linearity property, on the other hand, is a genuine surprise. Here, according to **U**, the superposed systems evolve in total obliviousness of each of the others, quite *independently* of whether there are any interactions involved. This fact alone might lead us to question the absolute truth of the linearity property. Yet it is very well confirmed for phenomena that remain entirely at the quantum level. It is only the operation of **R** that seems to violate it. We shall return to this matter later.

5.16 Orthogonality of product states

There is a certain awkwardness that arises with product states, as I have given them, in relation to the notion of orthogonality. If we have two *orthogonal states* $|\alpha\rangle$ and $|\beta\rangle$, then we might expect that the states $|\psi\rangle|\alpha\rangle$ and $|\psi\rangle|\beta\rangle$ should also be orthogonal, for any $|\psi\rangle$. For example, $|\alpha\rangle$ and $|\beta\rangle$ might be the two alternative states that are available to a photon, where perhaps the state $|\alpha\rangle$ would be the one detected by some photocell and where the orthogonal state $|\beta\rangle$ is the *inferred* state of the photon when the photocell fails to detect anything (null measurement). Now, we might consider the photon to be just one part of a *combined* system, where we simply adjoin some other objects—which could be another photon, say, somewhere on the moon, for example—the state of this other object being $|\psi\rangle$. In this case we get, for this combined system, the two alternative states $|\psi\rangle|\alpha\rangle$ and $|\psi\rangle|\beta\rangle$. The mere inclusion of the state $|\psi\rangle$ in the description should surely make no difference to the orthogonality of these two states. Indeed, with the normal 'tensor product' definition of product states (as opposed to some kind of Grassmann product, as used here) this would indeed be the case, and the orthogonality of $|\psi\rangle|\alpha\rangle$ and $|\psi\rangle|\beta\rangle$ would follow from the orthogonality of $|\alpha\rangle$ and $|\beta\rangle$.

However, the way that Nature actually seems to behave, according to the full procedures of quantum theory, is not quite so straightforward as this. If the state $|\psi\rangle$ could be considered to be totally independent of both $|\alpha\rangle$ and $|\beta\rangle$, then its presence would indeed be irrelevant. But, technically, even the state of a photon on the moon cannot be regarded as completely separate from the one that is involved in the detection by our photocell.* (This has to do with the use of a Grassmann-type product in the '$|\psi\rangle|\alpha\rangle$' notation used here—more familiarly, it has to do with the 'Bose statistics' that is involved with photon states, or other boson states, or else with the 'Fermi statistics' that is involved with electron, proton, or other fermion states, cf. ENM pp.277, 278 and, say, Dirac 1947.) If we were to be completely accurate according to the rules of the theory, we should have to consider all the photons in the universe when discussing the state of but a single photon. Nevertheless, to an extremely high degree of accuracy, this is (fortunately) not necessary. We shall indeed take it that, with any state $|\psi\rangle$ that has evidently nothing to do with the problem at hand, where this problem directly concerns only the orthogonal states $|\alpha\rangle$ and $|\beta\rangle$, the states $|\psi\rangle|\alpha\rangle$ and $|\psi\rangle|\beta\rangle$ will indeed be orthogonal (even with Grassmann-type products) to a very high degree of accuracy.

5.17 Quantum entanglement

We shall need to understand the quantum physics governing *EPR effects*—the quantum-mechanical **Z**-mysteries exemplified by the magic dodecahedra of §5.3, cf. §5.4. We must also come to terms with the basic **X**-mystery of quantum theory—the paradoxical relationship between the two processes **U** and **R** which underlies the *measurement problem* which we shall discuss in the next chapter. For both of these, it will be necessary for me to introduce one further important idea: that of *entangled states*.

Let us first try to see what is involved in a simple process of measurement. Consider a situation where a photon is in a superposed state, say $|\alpha\rangle+|\beta\rangle$, where the state $|\alpha\rangle$ would activate a detector, but where the state $|\beta\rangle$, orthogonal to $|\alpha\rangle$, would leave the detector undisturbed. (An example of this kind of thing was considered in §5.8, when the detector at G encountered the state $-|\mathrm{F}\rangle-\mathrm{i}|\mathrm{G}\rangle$. Here, $|\mathrm{G}\rangle$ would activate the detector whereas $|\mathrm{F}\rangle$ would leave it undisturbed.) I am going to suppose that the detector itself can also be assigned a quantum state, say $|\Psi\rangle$. This is the usual practice in quantum theory. It is not altogether clear to me that it really makes sense to assign a quantum-mechanical description to a classical-level object, but this is not normally questioned in discussions of this kind. In any case, we may suppose that the elements of the detector that the photon would *initially* encounter are indeed things that can be treated according to the standard rules of quantum theory. Those who are in any doubt about treating the detector as a whole

*Curiously, this type of phenomenon can have a profound relevance to actual observations. The effect due to Hanbury Brown and Twiss (1954, 1956), according to which the diameters of certain nearby stars have been measured, depends upon this 'boson' property interrelating photons reaching the earth from opposite sides of the star!

according to these rules should consider that it is these initial quantum-level elements (particles, atoms, molecules) to which this state vector $|\Psi\rangle$ refers.

Just before the photon reaches the detector (or, rather, just before the $|\alpha\rangle$-portion of the photon's wavefunction reaches the detector), the physical situation consists of the detector state *and* the photon state, namely $|\Psi\rangle(|\alpha\rangle+|\beta\rangle)$; and we find

$$|\Psi\rangle(|\alpha\rangle+|\beta\rangle)=|\Psi\rangle|\alpha\rangle+|\Psi\rangle|\beta\rangle.$$

This is a superposition of the state $|\Psi\rangle|\alpha\rangle$, which describes the detector (elements) and approaching photon, with the state $|\Psi\rangle|\beta\rangle$, which describes the detector (elements) and photon elsewhere. Suppose that next, in accordance with Schrödinger evolution **U**, the state $|\Psi\rangle|\alpha\rangle$ (detector with approaching photon) becomes some new state $|\Psi_Y\rangle$ (indicating that the detector registers a **YES** answer), by virtue of the interactions that the photon has with the detector elements after it encounters them. We shall also suppose that if the photon does *not* encounter the detector, the action of **U** provides that the detector state $|\Psi\rangle$ would, by itself, evolve into $|\Psi_N\rangle$ (detector registering **NO**) and $|\beta\rangle$ would evolve into $|\beta'\rangle$. Then, by the properties of Schrödinger evolution that were referred to in the previous section, the entire state would become

$$|\Psi_Y\rangle+|\Psi_N\rangle|\beta'\rangle.$$

This is a particular example of an *entangled* state, where 'entanglement' refers to the fact that the entire state cannot be written simply as a *product* of a state for one of the subsystems (photon) with a state for the other subsystem (detector). In fact, the state $|\Psi_Y\rangle$ itself is likely to be a state entangled, in any case, with its own environment, but this depends on the details of further interactions, not of relevance here.

We note that the states $|\Psi\rangle|\alpha\rangle$ and $|\Psi\rangle|\beta\rangle$ whose superposition represented the state of the combined system just before interaction are (essentially) *orthogonal*—since $|\alpha\rangle$ and $|\beta\rangle$ are themselves orthogonal, with $|\Psi\rangle$ being totally independent of either of them. Thus, the states, namely $|\Psi_Y\rangle$ and $|\Psi_N\rangle|\beta'\rangle$, into which these two evolve by the action of **U**, must themselves also be orthogonal. (**U** always preserves orthogonality.) The state $|\Psi_Y\rangle$ could evolve further into something that is macroscopically observable, such as an audible 'click', indicating that the photon has indeed been detected, whereas if no click results, the state must have been—i.e. must be considered to have 'jumped' to—the orthogonal possibility $|\Psi_N\rangle|\beta'\rangle$. If no click occurs, then the mere counterfactual possibility that it *might* have occurred, but did not, would cause the state to 'jump' to $|\Psi_N\rangle|\beta'\rangle$, which is now *not* an entangled state. The null measurement has *dis*entangled the state.

A characteristic feature of entangled states is that the 'jumping' that occurs with the operation of **R** can have a seemingly non-local (or even apparently retro-active) action that is even more puzzling than that of a simple null measurement. Such non-locality occurs particularly with what is referred to as an EPR (Einstein–Podolsky–Rosen) effect. These are genuine quantum puzzles that are among the most baffling of the theory's **Z**-mysteries. The idea came originally from Einstein, in an attempt to show that the formalism of quantum theory could not possibly provide a complete description of Nature. Many different versions of the EPR phenomenon have subsequently been put forward (such as the magic dodecahedra of §5.3), a number of which have been directly confirmed by experiment as being features of the *actual* workings of the world in which we live (cf. §5.4).

EPR effects arise with the following type of situation. Consider a known initial state $|\Omega\rangle$, of a physical system, which evolves (by **U**) into a superposition of two orthogonal states, each of which is the product of a pair of independent states describing a pair of spatially separated physical parts, so $|\Omega\rangle$ evolves into, say, the entangled state

$$|\psi\rangle|\alpha\rangle+|\phi\rangle|\beta\rangle.$$

We shall suppose that $|\psi\rangle$ and $|\phi\rangle$ are orthogonal possibilities for one of the parts, and that $|\alpha\rangle$ and $|\beta\rangle$ are orthogonal possibilities for the other. A measurement which ascertains whether the first part is in the state $|\psi\rangle$ or $|\phi\rangle$ will instantaneously determine that the second part is in the corresponding state $|\alpha\rangle$ or $|\beta\rangle$.

So far this is nothing mysterious. One could argue that the situation is just like the good Dr Bertlmann's socks (§5.4). In the knowledge that his two socks must differ in colour—and let us suppose that today we know that he has chosen to wear a pink and a green one—the observation as to whether his left sock is green (state $|\psi\rangle$) or else pink (state $|\phi\rangle$) will instantaneously determine that his right sock is correspondingly either pink (state $|\alpha\rangle$) or green (state $|\beta\rangle$). However, the effects of quantum entanglement can differ profoundly from this, and no explanation of the 'Bertlmann socks' type can account for all the observational effects. The problems arise when we have the option to perform *alternative* kinds of measurement on the two parts of the system.

An example will illustrate the kind of thing that is involved. Suppose that the initial state $|\Omega_0\rangle$ describes the spin state of some particle as being spin 0. This particle then decays into two new particles, each of spin $\frac{1}{2}$, which move away from each other to a great distance apart, to the left and to the right. From the properties of angular momentum and its conservation, it follows that the spin directions of the two separating particles must be *opposite* to each other, and, for the zero-spin state into which $|\Omega_0\rangle$ evolves, we find

$$|\Omega\rangle=|\mathrm{L}\uparrow\rangle|\mathrm{R}\downarrow\rangle-|\mathrm{L}\downarrow\rangle|\mathrm{R}\uparrow\rangle,$$

where 'L' refers to the left-hand particle and 'R' to the right-hand one (and the minus sign arises with standard conventions). Thus, if we choose to measure the spin of the left-hand particle in the upward direction, the answer **YES** (i.e. finding $|L\uparrow\rangle$) would automatically put the right-hand particle in the state $|R\downarrow\rangle$ of downward spin. The answer **NO** ($|L\downarrow\rangle$) would automatically put the right-hand particle into the state of upward spin ($|R\uparrow\rangle$). It seems that a measurement of one particle in one place can instantaneously affect the state of a quite different particle in a quite different place—but so far this is nothing more mysterious than are Bertlmann's socks!

However, our entangled state can also be represented in another way, corresponding to a different choice of measurement. For example, we could choose to measure the left-hand spin, instead, in a *horizontal* direction, so that **YES** corresponds, say, to $|L\leftarrow\rangle$ and **NO** corresponds to $|L\rightarrow\rangle$. We now find (by a simple calculation, cf. ENM, p. 283) that the *same* combined state as before can be written in a different entangled way:

$$|\Omega\rangle = |L\leftarrow\rangle|R\rightarrow\rangle - |L\rightarrow\rangle|R\leftarrow\rangle.$$

Thus we find that the **YES** answer on the left automatically puts the right-hand particle in the state $|R\rightarrow\rangle$, and **NO** answer automatically puts it in $|R\leftarrow\rangle$. The corresponding thing would occur *whatever* direction we happen to choose for the left-hand spin measurement.

What is remarkable about this kind of situation is that the mere *choice* of spin direction on the left-hand particle appears to *fix* the direction of the spin axis of the right-hand particle. In fact, until the *result* of the left-hand measurement is obtained, there is no actual information that is conveyed to the right-hand particle. Merely 'fixing the direction of the axis of spin' does nothing, by itself alone, that is actually observable. Despite this well-appreciated fact, it is still the case that from time to time people entertain the thought that EPR effects might be put to use in order to send signals from one place to another *instantaneously*, since the state-vector reduction **R** 'reduces' the quantum state of the EPR pair of particles simultaneously, no matter how far from each other they may be. But there is, in fact, no means of sending a signal, by this procedure, from the left-hand particle to the right-hand one (cf. Ghirardi *et al.* 1980).

The standard use of the quantum-mechanical formalism indeed provides the following picture: as soon as a measurement is made on one particle, say the left-hand one, then the entire state gets reduced instantaneously from the original entangled one—where neither particle *by itself* has a well-defined state of spin—to one where the left-hand state is disentangled from the right-hand one, and both spins then become well defined. In the *mathematical* state-vector description, the measurement on the left does have an instantaneous effect on the right. But, as I have indicated, this 'instantaneous effect' is not of the kind that would allow one to send a physical signal.

According to the principles of relativity, physical signals—being things capable of transmitting actual information—would be necessarily constrained to travel with the speed of light, or else more slowly. But EPR effects cannot be viewed in this way. It would not be consistent with the predictions of quantum theory for them to be treated as finitely propagated signals that are constrained by the speed of light. (The example of the magic dodecahedra illustrates this fact, since the entanglements between my colleague's and my dodecahedra have immediate effects that do not have to wait the four years that it would take a light signal to pass between us; cf. §5.3, §5.4, but also endnote 4.) Thus, EPR effects cannot be signals in the ordinary sense.

In view of this fact, one might wonder how it is that EPR effects actually have any observable consequences. That they do have such consequences follows from the famous theorem of John Bell (cf. §5.4). The joint probabilities that quantum theory predicts for various possible measurements that can be made on our two spin $\frac{1}{2}$ particles (with independent choices of spin direction, on the left-hand and the right-hand particles) cannot be obtained by any classical model of non-communicating left-hand and right-hand objects. (See ENM, pp. 284–5 and p. 301, for examples of this kind.) Examples such as the magic dodecahedra of §5.3 give even stronger effects, where puzzles arise with precise yes/no constraints, rather than merely probabilities. Thus, although the left- and right-hand particles are not in communication with each other in the sense of actually being able to send instantaneous messages to each other, they are nevertheless still *entangled* with each other in the sense that they cannot be considered as separate independent objects—until they are finally disentangled by measurement. Quantum entanglement is a mysterious thing that lies somewhere between direct communication and complete separation—and it has no classical analogue whatever. Moreover, entanglement is an effect that does not fall off with distance (unlike, say, the inverse square law of gravitational or electric attraction). Einstein found the prospect of such an effect deeply disturbing, referring to it as 'spooky action at a distance' (see Mermin 1985).

In fact quantum entanglement seems to be an effect that is quite oblivious not only of space separation but also of time separation. If a measurement is made on one component of an EPR pair *before* it is made on the other component, then the first measurement is considered, in the ordinary quantum-mechanical description, to be the one that effects the disentanglement, so that the second measurement is concerned merely with the single disentangled component that it is actually examining. However, precisely the same observable consequences would be obtained if we were to consider that it is the *second* measurement that somehow retro-actively effects the disentanglement rather than the first one. Another way of expressing the irrelevance of the temporal ordering of the two measurements is to say that they *commute* (see §5.14).

This kind of symmetry is a necessary feature of EPR measurements in order that they be consistent with the observational consequences of special relativity. Measurements that are performed at spacelike-separated events (i.e.

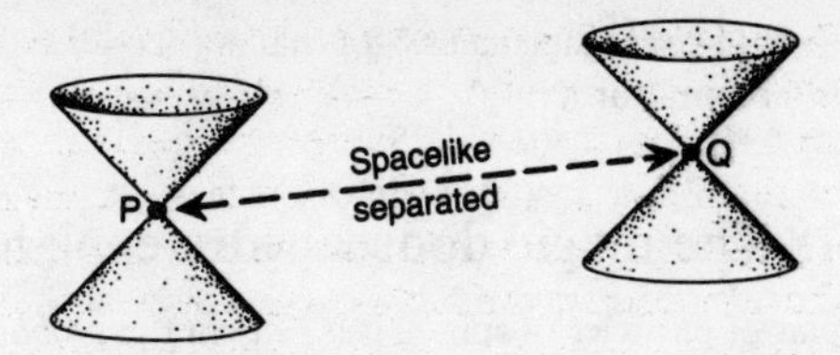

Fig. 5.25. Two events in space–time are called spacelike separated if each lies outside the other's light cone (see also Fig. 4.1, p. 219). In this case, neither can causally influence the other, and measurements made at the two events must commute.

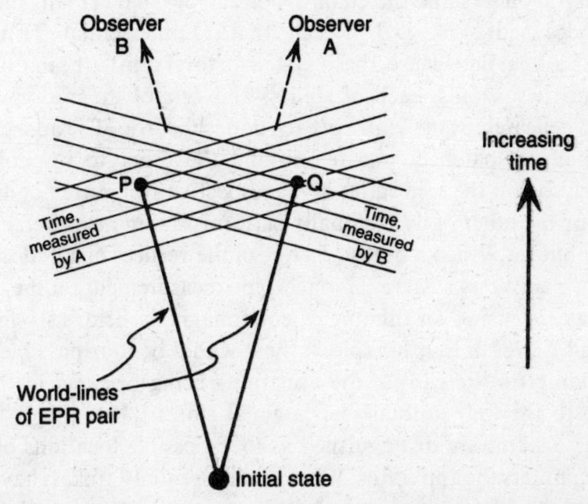

Fig. 5.26. According to special relativity, the observers A and B, in relative motion, have different ideas as to which of the two spacelike-separated events P, Q occurred first (A thinks Q is first, whereas B thinks P is first).

events lying outside each other's light cones; see Fig. 5.25, and the discussion given in §4.4) must *necessarily* commute and it is indeed immaterial which measurement is considered to occur 'first'—according to the firm principles of special relativity. To see that this must be so, one can consider the whole physical situation to be described according to the points of view of two different observers' frames of reference, as indicated in Fig. 5.26 (cf. also EMN p. 287). (These two 'observers' need have no relation to those who are actually performing the measurements.) In the situation depicted, the two observers would have opposite ideas as to which measurement was indeed the 'first' one. With regard to EPR-type measurements, the phenomenon of quantum entanglement—or of *dis*entanglement,* for that matter—heeds neither spatial separation nor temporal ordering!

5.18 The magic dodecahedra explained

For an EPR pair of particles of spin $\frac{1}{2}$, it is only in the *probabilities* that this spatial or temporal non-locality appears. But quantum entanglement is really a much more concrete and precise phenomenon than 'something merely influencing probabilities'. Examples such as the magic dodecahedra (and certain earlier configurations[10]) show that the strange non-locality of quantum entanglement is *not* just a matter of probability, but it also provides precise yes/no effects that cannot be explained in any classical local way at all.

Let us try to understand the quantum mechanics that actually underlies the magic dodecahedra of §5.3. Recall that Quintessential Trinkets have arranged, back on Betelgeuse, that a system of total spin 0 (the initial state $|\Omega\rangle$) is split into two atoms, each of spin $\frac{3}{2}$, and one of these atoms has been suspended delicately at the centre of each dodecahedron. The dodecahedra are then carefully dispatched, one to me and the other to my colleague on α-Centauri, so that the spin states of the respective atoms are not disturbed—until one or the other of us eventually performs a spin measurement on it by pressing a button. When a button on one of the vertices of a dodecahedron is pressed, this activates a Stern–Gerlach-type measurement on the atom at its centre—say, by using an inhomogeneous magnetic field, as referred to in §5.10—and we recall that for spin $\frac{3}{2}$, there would be four possible outcomes corresponding (in the case of the apparatus being oriented in the upward direction) to the four mutually orthogonal states $|\uparrow\uparrow\uparrow\rangle$, $|\downarrow\uparrow\uparrow\rangle$, $|\downarrow\downarrow\uparrow\rangle$, and $|\downarrow\downarrow\downarrow\rangle$. These states are distinguished as four possible locations of the atom after it encounters the apparatus. What Quintessential Trinkets have arranged is that when any button is pressed, the spin-measuring apparatus is oriented in the direction (outward from the centre of the dodecahedron) of that button. The bell rings (**YES**) if the atom is then found to be in the *second* of these four possible locations (Fig. 5.27). That is to say (using the notation for the case of the upward direction), it is the state $|\downarrow\uparrow\uparrow\rangle$ that evokes the **YES** response—resulting in the bell ringing, followed by the magnificent pyrotechnic display—and the three other states do *not* evoke any response (i.e. **NO**). In the case of **NO**, the three remaining locations for the atom are brought together (say, by reversing the direction of the inhomogeneous magnetic field) without the distinctions between them having had any external disturbing effects, ready for some other direction to be selected as a result of some other button being pressed. We take note of the fact that each button-pressing effects a *primitive* measurement, as described in §5.13.

*Examples can be given (Zeilinger *et al.* 1992) in which the very property of entanglement for a pair of particles can itself be an entangled property!

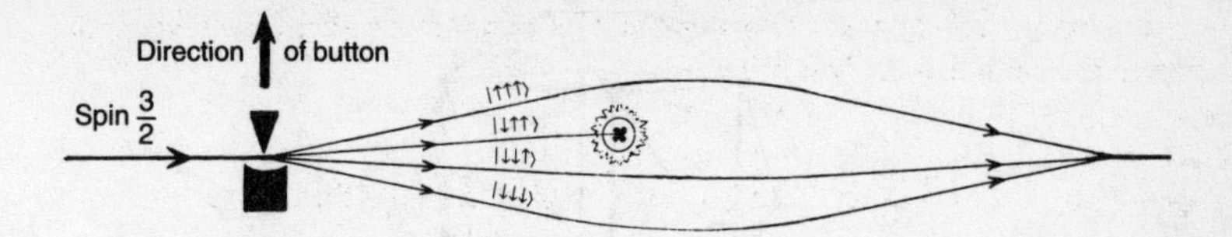

Fig. 5.27. Quintessential Trinkets have arranged it so that when a button is pressed on the dodecahedron, a spin measurement is made on the spin $\frac{3}{2}$ atom in that direction (taken here as 'up'), where state $|\downarrow\uparrow\uparrow\rangle$ would ring the bell (YES). If the answer is NO, the beams are recombined and the measurement is repeated in some other direction.

For two atoms of spin $\frac{3}{2}$ originating from the spin 0 state $|\Omega\rangle$, our total state can be expressed as

$$|\Omega\rangle = |L\uparrow\uparrow\uparrow\rangle|R\downarrow\downarrow\downarrow\rangle - |L\uparrow\uparrow\downarrow\rangle|R\downarrow\downarrow\uparrow\rangle + |L\uparrow\downarrow\downarrow\rangle|R\downarrow\uparrow\uparrow\rangle - |L\downarrow\downarrow\downarrow\rangle|R\uparrow\uparrow\uparrow\rangle.$$

If my atom is the right-hand one, and I find that its state is indeed $|R\downarrow\uparrow\uparrow\rangle$, because the bell rings on my first press on the uppermost button, then it must be the case that my colleague's bell must ring if he happens to choose to press the *opposite* button to mine first—his state $|L\uparrow\downarrow\downarrow\rangle$. Moreover, if my bell *fails* to ring for that first button-pressing, then his bell must also fail to ring for the opposite pressing.

Now, we need to ascertain that the properties (a) and (b) of §5.3, that Quintessential Trinkets are guaranteeing, must indeed hold for these primitive button-pressing measurements. In Appendix C, some mathematical properties of the Majorana description of spin states are given, particularly for spin $\frac{3}{2}$, which are sufficient for these purposes. It simplifies our present discussion if we think of the Riemann sphere as the sphere which passes through all the dodecahedron's vertices—the dodecahedron's *circumsphere*. We then note that the Majorana description of the **YES** state for a button pressed at some vertex P of the dodecahedron is simply the point P itself taken twice together with the point P* antipodal to P—which is indeed the state $|\downarrow\uparrow\uparrow\rangle$ for P taken at the north pole. We can label this **YES** state $|P^*PP\rangle$.

A key property of spin $\frac{3}{2}$ is that the **YES** states for the primitive measurements corresponding to button-presses on two next-to-adjacent vertices of the dodecahedron are *orthogonal*. Why is this so? We must ascertain that the Majorana states $|A^*AA\rangle$ and $|C^*CC\rangle$ are actually orthogonal whenever A and C are next to adjacent on the dodecahedron. Now it will be seen from Fig. 5.28 that A and C are next to adjacent on the dodecahedron whenever they are *adjacent* vertices of a *cube* that lies inside the dodecahedron, sharing its centre and eight of its vertices. By Appendix C, final

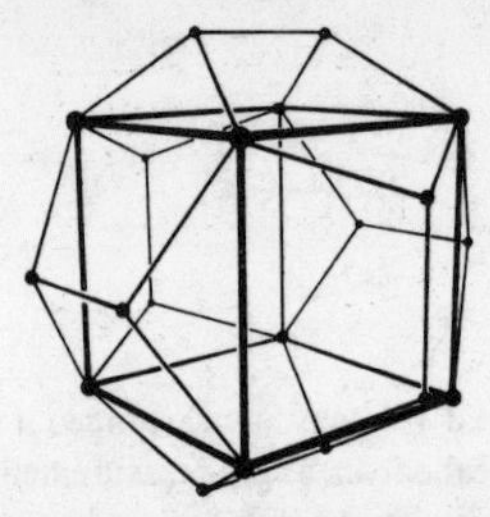

Fig. 5.28. A cube can be placed inside a regular dodecahedron, sharing 8 of its 20 vertices. We note that adjacent vertices of the cube are next-to-adjacent vertices of the dodecahedron.

paragraph (p. 304) $|A^*AA\rangle$ and $|C^*CC\rangle$ are indeed orthogonal whenever A and C are adjacent vertices of a cube, so the result is established.

What does this tell us? It tells us, in particular, that the three button-pressings on the three vertices that are adjacent to a SELECTED vertex are all *commuting* measurements (§5.14), these vertices being all mutually next to adjacent. Thus, the order in which they are pressed makes no difference to the outcome. Now the ordering is also irrelevant for my colleague on α-Centauri. If he happens to choose the *opposite* vertex to mine as his SELECTED one, then his three possible button-pressings are opposite to my three. By what has been said above, my bell and his bell must ring on opposite vertices—irrespective of either of our orderings—or else neither of our bells can ring at all on these pressings. This establishes (a).

What about (b)? We note that the Hilbert space for spin $\frac{3}{2}$ is *four* dimensional, so that the three mutually orthogonal possibilities for which my bell might ring, say $|A^*AA\rangle$, $|C^*CC\rangle$, and $|G^*GG\rangle$—where my SELECTED vertex is taken to be B (see Fig. 5.29)—do not quite exhaust the alternative possible occurrences. The remaining possibility occurs when the bell does not ring at all on these pressings, and we have the resulting null measurement (bell failing to ring on all three) that ascertains that the state is the (unique) state mutually orthogonal to all of $|A^*AA\rangle$, $|C^*CC\rangle$, $|G^*GG\rangle$. Let us label this state $|RST\rangle$, where the three points R, S, T on the Riemann sphere provide its Majorana description. The actual location of these three points is not altogether easy to ascertain. (They were located explicitly by Jason Zimba (1993).) It does not matter exactly where they are for the present argument. All that we need to know is that they are in positions that are determined by the geometry of the dodecahedron in relation to the SELECTED vertex B. Thus, in particular (by symmetry), had I chosen the vertex B*, antipodal to B, as my SELECTED vertex, instead of choosing B, then the state $|R^*S^*T^*\rangle$—where R*, S*, T* are antipodal to R, S, T—would have arisen as a result of the bell failing to ring at all three vertices A*, C*, G*, each adjacent to B*.

Now let us suppose that my colleague SELECTS the vertex B on his

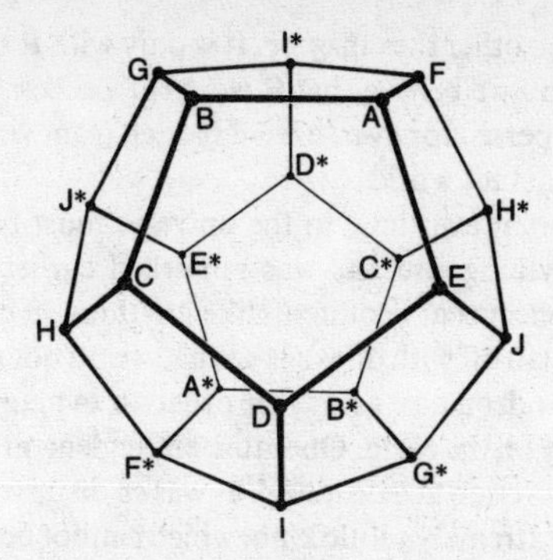

Fig. 5.29. The labelling of the dodecahedron's vertices, for the discussions of §5.18 and Appendix B (p. 300).

dodecahedron, corresponding exactly to the vertex B that I SELECT on mine. If the bell *fails* to ring on any of *his* three, A, C, G, that are adjacent to his B, then his (communting) measurements successively force *my* atom to be in a state that is orthogonal to the three that correspond to button-pressings on my *opposite* vertices A*, C*, G*, i.e. my atom is forced into state $|R^*S^*T^*\rangle$. However, if my bell also *fails* to ring on any of *my* three button-pressings at A, C, G, then this forces my state to be $|RST\rangle$. But by property C.1 of Appendix C (p. 303), $|RST\rangle$ is *orthogonal* to $|R^*S^*T^*\rangle$, so it is impossible for our bells to fail to ring on all six of these button-pressings, establishing (b).

This explains how Quintessential Trinkets have been able to use quantum entanglement to guarantee both properties (a) and (b). Now, in §5.3, we observed that *if* the two dodecahedra behaved as *independent* objects, then the colouring properties (c), (d), and (e) would immediately follow, which forces us into an insoluble vertex-colouring problem (as is shown explicitly in Appendix B (p. 300)). Thus what Quintessential Trinkets have managed to do, using quantum entanglement, is something that would be *impossible* if our two dodecahedra had to be things that could be treated as independent objects once they had left the Quintessential Trinkets factory. Quantum entanglement is not just an awkwardness, telling us that we cannot always ignore the probabilistic effects of the outside environment in a physical situation. When its effects can be appropriately isolated, it is something very mathematically precise, often with a clear-cut geometrical organization.

There is no description in terms of entities which can be considered as separate from one another, that can explain these expectations of the quantum-mechanical formalism. There can be no 'Bertlmann's socks' type of explanation for quantum-entangled phenomena, in general. Now, the rules of standard quantum-mechanical evolution—our procedure **U**—lead us to conclude that objects must *remain* 'entangled' in this strange way, no matter how distant from each other they may be. It is only with **R** that entanglements can be severed. But do we believe that **R** is a 'real' process? If not, then these entanglements must persist for ever, even if hidden from view by the excessive complication of the actual world.

Would this mean that everthing in the universe must be considered to be entangled with everything else? As was remarked earlier (§5.17), quantum entanglement is an effect that is quite unlike anything in classical physics, in which effects tend to fall off with distance so that we do not need to know what is going on in the Andromeda galaxy in order to explain the behaviour of things in a laboratory on the earth. Quantum entanglement seems indeed to be some kind of 'spooky action at a distance', as was so distasteful to Einstein. Yet it is an 'action' of an extremely subtle kind, which cannot be used for the actual sending of messages.

Despite their falling short of providing direct communication, the potential distant ('spooky') effects of quantum entanglement cannot be ignored. So long as these entanglements persist, one cannot, strictly speaking, consider any object in the universe as something on its own. In my own opinion, this situation in physical theory is far from satisfactory. There is no real explanation on the basis of standard theory of why, in practice, entanglements *can* be ignored. Why is it not necessary to consider that the universe is just one incredibly complicated quantum-entangled mess that bears no relationship to the classical-like world that we actually observe? In practice, it is the continual use of the procedure **R** that cuts the entanglements free, as is the case when my colleague and I make our measurements on the entangled atoms at the centres of our dodecahedra. The question arises: is this **R**-action a *real* physical process, so that the quantum entanglements are, in some sense, *actually* cut? Or is it all to be explained as just an illusion of some kind?

I shall try to address these puzzling matters in the next chapter. In my opinion, these questions are central to the search for a role for non-computability in physical action.

Appendix B: The non-colourability of the dodecahedron

Recall the problem that was set in §5.3: to show that there is no way to colour all the vertices of a dodecahedron BLACK or WHITE so that no two next-to-adjacent vertices can be both WHITE and no six vertices adjacent to a pair of opposite vertices can be all BLACK. The symmetry of the dodecahedron will be an immense help to us in eliminating possibilities.

Let us label the vertices as in Fig. 5.29. Here, A, B, C, D, E are the vertices of a pentagonal face, described cyclically, and F, G, H, I, J, are vertices adjacent to them, taken in the same order. As in §5.18, A*, . . ., J* are the respective antipodal vertices to these. We first note that because of the second property, there must be at least one WHITE vertex somewhere, which we can assume is the vertex A.

Suppose, for the moment, that the WHITE vertex A has, as one of its immediate neighbours, *another* WHITE vertex—which we can take to be B (see Fig. 5.29). Now the ten vertices that surround these two, namely C, D, E, J, H*, F, I*, G, J*, H, must all be BLACK, because they are each next to adjacent either to A or to B. Next, we examine the six vertices that are adjacent to either one of the antipodal pair H, H*. There must be a WHITE one among these six, so either F* or C* (or both) must be WHITE. Doing the same for the antipodal pair J, J*, we conclude that either G* or E* (or both) must be white. But this is *impossible* because G* and E* are both next to adjacent to both of F* and C*. This rules out the possibility that the WHITE vertex A can have a WHITE immediate neighbour—indeed, by symmetry, it rules out the possibility of any pair of adjacent WHITE vertices.

Thus, the WHITE vertex A must be surrounded by BLACK vertices B, C, D, E, J, H*, F, I*, G, because each of these is either adjacent or next to adjacent to A. Now, examine the six vertices that are adjacent to one of the antipodal pair A, A*. We conclude that one of B*, E*, F* must be WHITE, and by symmetry it does not matter which—so take the WHITE one to be F*. We note that E* and G* are next to adjacent to F*, so they must both be black, and so also must H be BLACK because it is adjacent to F*, and by the previous argument we have also ruled out adjacent WHITE vertices. However, this colouring is impossible because the antipodal vertices J, J* now have nothing but BLACK vertices adjacent to them. This concludes the argument, showing the *classical* impossibility of magic dodecahedra!

Appendix C: Orthogonality between general spin states

The Majorana description of general spin states is not very familiar to physicists; yet, it provides a useful and geometrically illuminating picture. I shall give here a brief account of the basic formulae and of some of their geometrical implications. This will provide, in particular, the needed orthogonality relations underlying the geometry of the magic dodecahedra, as was required for §5.18. My descriptions will differ appreciably from those given originally by Majorana (1932), and follow more closely those of Penrose (1994*a*) and Zimba and Penrose (1993).

The idea is to consider the unordered set of n points on the Riemann sphere as the n roots of a complex polynomial of degree n, and (essentially) to use the coefficients of this polynomial as coordinates of the $(n+1)$-dimensional Hilbert space of spin states for a (massive) particle of spin $\frac{1}{2}n$. Taking basis states to be, as in §5.10, the various possible results of a spin measurement in the vertical direction, we represent these as the various monomials (together with an appropriate normalizing factor which ensures that each of these basis states is a unit vector):

$$|\uparrow\uparrow\uparrow\uparrow \ldots \uparrow\uparrow\rangle \text{ corresponds to } x^n$$

$$|\downarrow\uparrow\uparrow\uparrow \ldots \uparrow\uparrow\rangle \text{ corresponds to } n^{1/2}\, x^{n-1}$$

$$|\downarrow\downarrow\uparrow\uparrow \ldots \uparrow\uparrow\rangle \text{ corresponds to } \{n(n-1)/2!\}^{1/2}\, x^{n-2}$$

$$|\downarrow\downarrow\downarrow\uparrow \ldots \uparrow\uparrow\rangle \text{ corresponds to } \{n(n-1)(n-2)/3!\}^{1/2}\, x^{n-3}$$

$$\ldots$$

$$|\downarrow\downarrow\downarrow\downarrow \ldots \downarrow\uparrow\rangle \text{ corresponds to } n^{1/2}\, x$$

$$|\downarrow\downarrow\downarrow\downarrow \ldots \downarrow\downarrow\rangle \text{ corresponds to } 1.$$

(The bracketed expressions are all binomial coefficients.) Thus, the general state of spin $\frac{1}{2}n$

$$z_0|\uparrow\uparrow\uparrow \ldots \uparrow\rangle + z_1|\downarrow\uparrow\uparrow \ldots \uparrow\rangle + z_2|\downarrow\downarrow\uparrow \ldots \uparrow\rangle + z_3|\downarrow\downarrow\downarrow \ldots \uparrow\rangle + \ldots + z_n|\downarrow\downarrow\downarrow \ldots \downarrow\rangle$$

corresponds to the polynomial

$$p(x) = a_0 + a_1\, x + a_2\, x^2 + a_3\, x^3 + \ldots + a_n\, x^n$$

where

$$a_0 = z_0,\ a_1 = n^{1/2}\, z_1,\ a_2 = \{n(n-1)/2!\}^{1/2}\, z_2, \ldots, a_n = z_n.$$

The roots $x = \alpha_1, \alpha_2, \alpha_3, \ldots, \alpha_n$ of $p(x) = 0$ provide the n points on the Riemann sphere (with multiplicities) that define the Majorana description. This includes the possibility of the Majorana point given by $x = \infty$, (the south pole), which occurs when the degree of the polynomial $P(x)$ falls short of n, by an amount given by the multiplicity of this point.

A rotation of the sphere is achieved by means of a transformation according to which a replacement

$$x \mapsto (\lambda x - \mu)\ (\bar{\mu} x + \bar{\lambda})^{-1}$$

is first made (where $\lambda\bar{\lambda} + \mu\bar{\mu} = 1$), and then the denominators are cleared by multiplying the whole expression by $(\bar{\mu}x + \bar{\lambda})^n$. Thus, the polynomials corresponding to the results of (say Stern–Gerlach) measurements of spin in an arbitrary direction can be obtained, giving expressions of the form

$$c(\lambda x - \mu)^p\ (\bar{\mu} x + \bar{\lambda})^{n-p}.$$

The points given by μ/λ and $-\bar{\lambda}/\bar{\mu}$ are antipodal on the Riemann sphere, and they correspond to the direction of the spin measurement and to its opposite. (This assumes an appropriate choice of phases for the states $|\uparrow\uparrow\uparrow \ldots \uparrow\rangle$, $|\downarrow\uparrow\uparrow \ldots \uparrow\rangle$, $|\downarrow\downarrow\uparrow \ldots \uparrow\rangle$, $|\downarrow\downarrow\downarrow \ldots \downarrow\rangle$. The aforementioned properties and detailed verifications are best appreciated in the 2-spinor formalism. The reader is referred to Penrose and Rindler (1984), particularly p. 162, also §4.15. The general state for spin $\frac{1}{2}n$ is described in terms of a symmetric n-valent spinor, and the Majorana description follows from its canonical decomposition as a symmetrized product of spin-vectors.)

The antipodal point to any point α on the sphere is given by $-1/\bar{\alpha}$. Thus, if we reflect all the Majorana points that are roots of the polynomial

$$a(x) \equiv a_0 + a_1 x + a_2 x^2 + \ldots + a_{n-1} x^{n-1} + a_n x^n,$$

in the centre of the sphere, we obtain the roots of the polynomial

$$a^*(x) \equiv \bar{a}_n - \bar{a}_{n-1} x + \bar{a}_{n-2} x^2 - \ldots - (-1)^n \bar{a}_1 x^{n-1} + (-1)^n \bar{a}_0 x^n.$$

If we have two states $|\alpha\rangle$ and $|\beta\rangle$, given by the respective polynomials $a(x)$ and $b(x)$, where

$$b(x) \equiv b_0 + b_1 x + b_2 x^2 + b_3 x^3 + \ldots + b_{n-1} x^{n-1} + b_n x^n,$$

then their scalar product will be

$$\langle\beta|\alpha\rangle = \bar{b}_0 a_0 + \frac{1}{n} \bar{b}_1 a_1 + \frac{2!}{n(n-1)} \bar{b}_2 a_2 + \frac{3!}{n(n-1)(n-2)} \bar{b}_3 a_3 + \ldots + \bar{b}_n a_n.$$

This expression is invariant under rotations of the sphere, as may be directly verified using the above formulae.

Let us apply this scalar-product expression to the particular case when $b(x) = a^*(x)$, so that we are concerned with two states one of whose Majorana descriptions consists precisely of the antipodal points of the other. Their scalar product is (up to a sign)

$$a_0 a_n - \frac{1}{n} a_1 a_{n-1} + \frac{2!}{n(n-1)} a_2 a_{n-2} - \ldots - (-1)^n \frac{1}{n} a_{n-1} a_1 + (-1)^n a_n a_0.$$

It will be seen from this that if n is *odd* then all the terms cancel, so we have the following theorem. (The state with Majorana description P, Q, . . ., S is denoted by $|$PQ. . .S$\rangle$. The antipodal point to X is denoted by X*.)

C.1 If n is odd, the state $|$PQR. . .T$\rangle$ is orthogonal to $|$P*Q*R*. . .T*$\rangle$.

Two further properties that can be read off from the general expression for the scalar product are the following.

C.2 The state $|$PPP. . .P$\rangle$ is orthogonal to each state $|$P*AB. . .D$\rangle$.

C.3 The state $|$QPP. . .P$\rangle$ is orthogonal to $|$ABC. . .E$\rangle$ whenever the stereographic projection, from P*, of the point Q* is the centroid of the stereographic projections from P* of A, B, C, . . ., D.

(The centroid of a set of points is the centre of gravity of the configuration of equal point masses situated at the points. Stereographic projection was described in §5.10, Fig.5.19.) To prove C.3, rotate the sphere until P* is at the south pole. Then the state |QPP. . .P⟩ is represented by the polynomial $x^{n-1}(x-\chi)$, where χ defines the point Q on the Riemann sphere. Forming the scalar product with the state represented by the polynomial $(x-\alpha_1)(x-\alpha_2)(x-\alpha_3)\ldots(x-\alpha_n)$, whose Majorana description is provided by $\alpha_1, \alpha_2, \alpha_3, \ldots, \alpha_n$, we find that this vanishes when

$$1+n^{-1}\bar{\chi}(\alpha_1+\alpha_2+\alpha_3+\ldots+\alpha_n)=0,$$

i.e. when $-1/\bar{\chi}$ is equal to $(\alpha_1+\alpha_2+\alpha_3+\ldots+\alpha_n)/n$, which is the centroid, in the complex plane, of the points given by $\alpha_1, \alpha_2, \alpha_3, \ldots, \alpha_n$. This establishes C.3. To prove C.2, we take P, instead, to be at the south pole. Then the state |PPP. . .P⟩ is represented by the constant 1, considered as a polynomial of degree n. The corresponding scalar product now vanishes when

$$\alpha_1\alpha_2\alpha_3\ldots\alpha_n=0,$$

i.e. when at least one of $\alpha_1, \alpha_2, \alpha_3, \ldots, \alpha_n$ vanishes—the point 0 of the complex plane representing the north pole P*. This establishes C.2.

The result C.2 enables us to interpret the Majorana points in physical terms. It has the implication that these points define the directions in which a (Stern–Gerlach-type) spin measurement provides a zero probability of yielding the result that the spin is entirely in the direction opposite to that measured (cf. ENM p. 273). It also contains, as a special case, the result that for spin $\frac{1}{2}$ ($n=1$), orthogonal states are precisely those whose Majorana points are antipodal. The result C.3 enables us to deduce the general geometrical interpretation of orthogonality in the case of spin 1 ($n=2$). A notable particular case occurs when the two states are represented as two pairs of antipodal points whose joins are perpendicular lines through the centre of the sphere. In the case of spin $\frac{3}{2}$ ($n=3$), C.3 provides (with C.1) all that we shall need for §5.18. (A geometrical interpretation of orthogonality in the general case will be given elsewhere.)

The particular case of C.3 that is needed for §5.18 occurs when P and Q are adjacent vertices of a cube inscribed in the Riemann sphere, so that PQ and Q*P* are opposite edges of this cube. The lengths PQ* and QP* are $\sqrt{2}$ times those of PQ and P*Q*. It follows from C.3, by simple geometry, that the states |P*PP⟩ and |Q*QQ⟩ are orthogonal.

Notes and references

1. Penrose (1993*b*, 1994*a*), Zimba and Penrose (1993).
2. The initial suggestion for a clear-cut experiment came from Clauser and Horne (1974) and Clauser, Horne, and Shimony (1978).
3. The first experiments indicating a positive confirmation of the non-local quantum expectations were obtained by Freedman and Clauser (1972),

and were followed a few years later by the much more definitive results of Aspect, Grangier, and Roger (1982) (cf. also Aspect and Grangier 1986).

4. There is another type of 'classical' explanation possible for the particular EPR effects that have been observed, so far, by Aspect and others. This suggested explanation—*retarded collapse*—is due to Euan Squires (1992*a*), and it takes advantage of the fact that there may be a significant time delay in the actual effecting of a measurement by the detectors at the two separated locations. This suggestion has to be taken in the context of some theory—necessarily an unconventional one, such as those that we shall encounter in §6.9 or §6.12—which makes some definite prediction as to a likely time at which each of the two quantum measurements would *objectively* take place. Owing to random influences controlling these two times, it would be considered likely that one of the detectors would effect its measurement significantly earlier than the other one—so much earlier, in fact, that (in the experiments that have been so far performed) there would have been ample time for a signal, travelling from the earlier detector at light speed, to inform the later detector what the result of the earlier detection had been.

   On this view, whenever a quantum measurement takes place, it is accompanied by a 'wave of information' travelling at the speed of light, outwards from the measurement event. This type of thing is perfectly in line with the classical relativity theory (see §4.4), but it would disagree with the expectations of quantum theory over long-enough distances. In particular, the 'magic dodecahedra' of §5.3 could not be explained in terms of retarded collapse. Of course, no such 'experiment' has yet been performed, and one might take the view that the expectations of quantum theory would be violated in such circumstances. A more serious objection, however, is that retarded collapse would encounter severe difficulties with other types of quantum measurement, and would lead to a violation of all the standard conservation laws. For example, when a decaying radio-active atom emits a charged particle—say an α-particle—it would be possible for two widely enough separated detectors to receive the *same* α-particle, simultaneously violating each of the conservation laws for energy, electric charge, and baryon-number! (For wide enough separation, the 'wave of information' coming from the first detector would not have enough time to warn the second detector to be unable to observe this same α-particle!) However, these conservation laws would still hold 'on average', and I am not aware of any actual observation contradicting the idea. For a recent assessment of the status of retarded collapse, see Home (1994).

5. I have been informed by Abner Shimony that Kochen and Specker had already been aware of an EPR formulation of their own example.

6. For other examples, exhibiting different geometrical configurations, see Peres (1990, 1990), Mermin (1990), Penrose (1994*a*).

7. The most efficient 'half-silvered mirror' would not actually be silvered at all, but would be a thin piece of transparent material of just the right thickness in relation to the wavelength of the light. It would achieve its effect by a

complicated combination of repeated internal reflections and transmissions, so that the final transmitted and reflected beams are equal in intensity. It follows from the 'unitary' nature of the resulting transformation between the finally transmitted and reflected beams that there must indeed be a resulting phase shift of a quarter of a wavelength, giving the factor of 'i' as required. See Klein and Furtak (1986) for a more complete discussion.

8. For example, Dirac (1947), Davies (1984).
9. There is some arbitrariness about the choice of phase-factor that I have adopted here for the reflected state. It depends partly upon what kind of mirror is used. In fact, unlike the 'half-silvered' mirror referred to in endnote 7 (which was probably not silvered at all) we can consider these two mirrors to be actually fully silvered. The factor of 'i' that I have adopted here is a kind of compromise, achieving a superficial agreement with the factor obtained in the case of reflection from the 'half-silvered' mirrors. In fact it does not really matter what factor is adopted for reflection from the fully silvered mirrors, so long as we are consistent in what we do for *both* the mirrors in question.
10. For example, Kochen and Specker (1967) and the references given in endnote 6.

6

Quantum theory and reality

6.1 Is **R** a real process?

In the previous chapter, we have been trying to come to terms with the puzzling **Z**-mysteries of quantum theory. Although not all of these phenomena have been experimentally tested—such as quantum entanglement over distances of several light years[1]—there is already enough experimental support for effects of this kind to tell us that **Z**-mysteries are indeed things that we must take seriously as true aspects of the behaviour of the constituents of the world in which we live.

The actions of our physical world at the quantum level are indeed very counter-intuitive, and in many ways quite different from the 'classical' behaviour that seems to operate at the more familiar level of our experiences. The quantum behaviour of our world certainly includes entanglement effects over many metres, at least so long as they involve only quantum-level objects, such as electrons, photons, atoms, or molecules. The contrast between this strange *quantum* behaviour of 'small' things, even over large separations, and the more familiar *classical* behaviour of larger things, underlies the problem of the **X**-mysteries of quantum theory. Can it really be the case that there are two kinds of physical law, one of which operates at one level of phenomena and the other of which operates at another?

Such an idea is quite at variance with what we have come to expect in physics. Indeed, one of the profound achievements of the seventeenth-century dynamics of Galilei–Newton was that the motions of heavenly bodies could now be understood to obey precisely the same laws as operate here on earth. From the time of the ancient Greeks and before, it had been believed that there must be one set of laws for the heavens and a completely separate set holding here on earth. Galileo and Newton taught us to see how the laws could be the same at all scales—a fundamental insight that was essential for the progress of science. Yet (as has been stressed by Professor Ian Percival, of the University of London), with quantum theory we seem to have reverted to a scheme like that of the ancient Greeks, with one body of laws operating at the classical level and another very different body of laws at the quantum level. It is my own opinion—an opinion shared by a sizeable minority of physicists—that this state of physical understanding cannot be other than a stop-gap, and we may well anticipate that the finding of appropriate quantum/classical laws that operate uniformly at *all* scales might herald a scientific advance of a magnitude comparable with that initiated by Galileo and Newton.

The reader may, however, quite rightly question whether it is indeed the case that our standard understanding of quantum theory presents us with a quantum-level picture which does not also explain classical phenomena. Many would disclaim my contention that it does not, asserting that physical systems that are in some sense large or complicated, and acting entirely according to quantum-level laws, would behave just like classical objects, at least to a very high degree of accuracy. We shall try, first, to see whether this assertion—the assertion that the apparently 'classical' behaviour of large-scale objects follows from the quantum behaviour of their tiny constituents—is believable. And if we find that it is not, we shall try to see where to turn, in order to arrive at a coherent viewpoint that might make sense at *all* levels. I should warn the reader, however, that the entire question is fraught with much controversy. There are many different viewpoints, and it would be foolish for me to try to give a comprehensive summary of them all, let alone to argue in detail against all of those that I find implausible or untenable. I ask the reader's indulgence for the fact that the viewpoints that I present will be given very much from my own perspective. It is inevitable that I shall not be entirely fair to those whose viewpoints are too foreign to my own, and I apologize in advance for the injustices that I shall undoubtedly commit.

There is a fundamental difficulty with trying to find a clear scale at which the *quantum* level of activity, characterized by the persistence of quantum superpositions of different alternatives, actually gives way—by the action of **R**—to the *classical*, at which such superpositions do not seem to occur. This results from an inherent 'slipperiness' in the procedure **R**, from the observational point of view, which prevents us from pinpointing any clear level at which it 'happens'—one reason that many physicists do not regard it as a real phenomenon at all. It seems to make no difference to experiments where we choose to apply **R**, so long as we do so at a level higher than that at which quantum interference effects have been observed, yet no higher than the level at which we can directly perceive that classical alternatives *do* take place, rather than being in complex linear superpositions (though, as we shall be seeing shortly, even at this end some would maintain that the superpositions persist).

How can we decide at what level **R** *actually* takes place—if indeed it really takes place physically at all? It is hard to see how to answer such a question by physical experiment. If **R** *is* a real physical process, there is a vast range of possible levels at which it might occur, between tiny levels where quantum interference effects *have* been observed, and the much larger level at which classical behaviour is actually perceived. Moreover, these differences in 'level' do not seem to refer to physical size, since we have seen above (in §5.4) that the effects of quantum entanglement can stretch over distances of many metres. We shall be seeing later that *energy differences* provide a better measure of this scale of level than does physical dimension. Be this as it may—at the large end of things, the place where 'the buck stops' is provided by our *conscious perceptions*. This is an awkward matter from the point of view of physical theory, because we do not really know what physical processes in the brain are associated with perception. Nevertheless, the physical nature of these processes would seem to provide the large-end limit for any proposed theory of a *real* **R**-process. This still allows for a huge range between the two extremes, and there is considerable scope for numerous different attitudes as to what *really* happens when **R** is brought in.

One of the main issues concerns the 'reality' of the quantum formalism—or even of the quantum-level world itself. In this connection, I cannot resist quoting a remark that was made to me by Professor Bob Wald, of the University of Chicago, at a dinner-party some years ago:

> If you really believe in quantum mechanics, then you can't take it seriously.

It seems to me that this expresses something profound about quantum theory and about people's attitudes to it. Those who are most vehement about accepting the theory as being in no way in need of modification tend *not* to think that it represents the actual behaviour of a 'real' quantum-level world. Niels Bohr, who was a leading figure in the development and interpretation of quantum theory, was one of the most extreme in this respect. He seems to have regarded the state vector as no more than a convenience, useful only for calculating probabilities for the results of 'measurements' that might be performed on a system. The state vector itself was *not* to be thought of as providing an objective description of any kind of quantum-level *reality*, but as representing merely 'our knowledge' of the system. Indeed, it was to be regarded as doubtful that the very concept of 'reality' applied meaningfully at the quantum level. Bohr was certainly someone who 'really believed in quantum mechanics', and his view of the state vector seemed, indeed, to be that it should *not* be 'taken seriously' as the description of a physical reality at the quantum level.

The broad alternative to this quantum viewpoint is to believe that the state vector does provide an accurate mathematical description of a *real* quantum-level world—a world that evolves to an extraordinary degree of precision, though perhaps not with a total accuracy, according to the mathematical rules that the equations of the theory provide. Here, it seems to me, there are two main routes that may be followed. There are those who regard the procedure **U** to be all there is in the evolution of the quantum state. The procedure **R**, accordingly, is taken to be some kind of illusion, convenience, or approximation, and is *not* to be taken as part of the *actual* evolution of the reality that is indeed to be described by the quantum state. Such people, it appears, must be driven in the direction of what is known as the *many-worlds* or *Everett* interpretation(s).[2] I shall explain something of this kind of viewpoint in a moment. Those, on the other hand, who 'take seriously' the quantum formalism most completely are those who believe that *both* **U** and **R** represent (to a considerable accuracy) *actual* physical behaviour of a *physically real*, state-vector-described, quantum/classical-level world. But if one is to take the quantum formalism that seriously, then it becomes hard really to believe that the theory can be completely accurate at all levels. For the action of **R**, as the procedure stands, is at variance with many of the properties of **U**, in particular its *linearity*. In this sense, one would not 'really believe in quantum mechanics'. In the sections which follow, I shall discuss these matters more fully.

6.2 Many-worlds-type viewpoints

Let us first try to see how far we can get following the other 'realistic' route, the one that ultimately leads to a kind of viewpoint often referred to as the 'many-worlds' interpretation. Here one tries to accept the state vector, evolving entirely under the action of **U**, as providing the true reality. This forces one to accept that classical-level objects like golf balls or people must also be subject to the laws of quantum linear superposition. It could be suggested that this might cause no serious difficulty so long as such superposed states become an extremely rare occurrence at the classical level. The problem, however, lies with the *linearity* of **U**. Under the action of **U**, the weighting factors for superposed states remain the *same* no matter how much material gets involved. The procedure **U**, by itself, does not enable superpositions to become 'unsuperposed' merely because a system gets large or complicated. Such superpositions would in no way tend to 'disappear' for classical-level objects, and one is faced with the implication that classical objects, also, ought frequently to appear in manifestly superposed states. The question that must then be faced is: why do such large-scale superpositions of alternatives not impinge upon our awareness of the classical-level world?

Let us try to understand how proponents of a many-worlds type of view would explain this. Consider a situation like that discussed in §5.17, where a photon detector, described by a state $|\Psi\rangle$, encounters a superposed photon state $|\alpha\rangle + |\beta\rangle$, where $|\alpha\rangle$ would activate the detector but $|\beta\rangle$ would leave it undisturbed. (Perhaps a photon, emitted by some source, has impinged upon a half-silvered mirror, and $|\alpha\rangle$ and $|\beta\rangle$ might represent the transmitted and reflected parts of the photon state.) We are now not questioning the applicability of the state-vector concept to a classical-level object like an entire detector, since on this view, state vectors are taken to be accurate representations of reality at all levels. Thus, $|\Psi\rangle$ can describe the entire detector, and not perhaps just some quantum-level initial parts as in §5.17. Recall that, as in §5.17, after the time of encountering the photon, the detector and photon state evolves from the product $|\Psi\rangle(|\alpha\rangle+|\beta\rangle)$ to the entangled state

$$|\Psi_{Y}\rangle+|\Psi_{N}\rangle|\beta'\rangle.$$

This *entire* entangled state is now taken to represent the *reality* of the situation. One does not say that *either* the detector has received and absorbed the photon (state $|\Psi_{Y}\rangle$) *or else* the detector has not received it and the photon remains free (state $|\Psi_{N}\rangle|\beta'\rangle$), but one maintains that *both* alternatives co-exist in superposition, as part of a total reality in which all such superpositions are preserved. We can carry this further, and imagine that a human experimenter examines the detector to see whether or not it has registered the reception of the photon. The human being, before examining the detector, should also have a quantum state, say $|\Sigma\rangle$, so we have a combined 'product' state at that stage, namely

$$|\Sigma\rangle(|\Psi_{N}\rangle+|\Psi_{N}\rangle|\beta'\rangle).$$

Then, after examining the state, the human observer either perceives the detector having received and absorbed the photon (state $|\Sigma_{Y}\rangle$), or perceives the detector not having received the photon (an orthogonal state $|\Sigma_{N}\rangle$). If we make the assumption that the observer is not interacting with the detector after observing it, we have the following form of state vector describing the situation:

$$|\Sigma_{Y}\rangle|\Psi'_{Y}\rangle+|\Sigma_{N}\rangle|\Psi'_{N}\rangle|\beta''\rangle.$$

There are now two different (orthogonal) observer states, both involved in the entire state of the system. According to one of these, the observer is in the state of having perceived that the detector registered receiving the photon; and this is accompanied by the detector state in which the photon has indeed been received. According to the other, the observer is in the state of having perceived that the detector has not registered receiving the photon; and this is accompanied by the detector state with the photon not received by it, and the photon moving off freely. In the many-worlds-type viewpoints, then, there would be different instances (copies) of the observer's 'self', co-existing within the total state and having different perceptions of the world around. The actual state of the world that accompanies each copy of the observer would be consistent with the perceptions of that copy.

We can generalize this to the more 'realistic' physical situations in which there would be huge numbers of different quantum alternatives continually occurring throughout the universe's history—rather than just the two of this example. Thus, according to this many-worlds type of viewpoint, the total state of the universe would indeed comprise many different 'worlds' and there would be many different instances of any human observer. Each instance would perceive a world that is consistent with that observer's own perceptions, and it is argued that this is all that is needed for a satisfactory theory. The procedure **R** would, according to this view, be an *illusion*, apparently arising as a consequence of how a macroscopic observer would perceive in a quantum-entangled world.

For my own part, I have to say that I find this viewpoint very unsatisfactory. It is not so much the extraordinary lack of economy that this picture provides—though this is indeed a worrying feature, to say the least. The more serious objection is that the viewpoint does not *really* provide a solution of the 'measurement problem' that it was set up to solve.

This *quantum measurement problem* is to understand how the procedure **R** can arise—or effectively arise—as a property of large-scale behaviour in **U**-evolving quantum sytems. The problem is not solved merely by indicating a possible way in which an **R**-like behaviour might conceivably be accommodated. One must have a theory that provides some understanding of the *circumstances* under which (the illusion of?) **R** comes about. Moreover, one must have an explanation of the remarkable *precision* that is involved in **R**. It appears that people often think of the precision of quantum theory as lying in its dynamical equations, namely **U**. But **R** itself is also very precise in its prediction of probabilities, and unless it can be understood how these come about, one does not have a satisfactory theory.

In the absence of further ingredients, the many-worlds viewpoint does not really properly come to terms with either of these. Without a theory of how a 'perceiving being' would divide the world up into orthogonal alternatives, we have no reason to expect that such a being could not be aware of linear superpositions of golf balls or elephants in totally different positions. (We should note that the mere *orthogonality* of the 'perceiver states', such as with $|\Sigma_Y\rangle$ and $|\Sigma_N\rangle$ above, in no way serves to single these states out. Compare the case of $|L\leftarrow\rangle$ and $|L\rightarrow\rangle$, as opposed to $|L\uparrow\rangle$ and $|L\downarrow\rangle$, in the EPR discussion of §5.17. In each case the pair of states is orthgonal, as are $|\Sigma_Y\rangle$ and $|\Sigma_N\rangle$, but there is nothing to choose one pair in favour of the other.) Moreover, the many-worlds viewpoint provides no explanation for the extremely accurate wonderful rule whereby the squared moduli of the complex-number weighting factors miraculously become relative probabilities.[3] (Compare also the discussions given in §6.6 and §6.7.)

6.3 Not taking $|\psi\rangle$ seriously

There are many versions of the viewpoint according to which the state vector $|\psi\rangle$ is *not* regarded as providing an actual picture of a quantum-level physical reality. Instead, $|\psi\rangle$ would be taken to serve only as a calculational device, useful merely for calculating probabilities, or as an expression of the experimenter's 'state of knowledge' concerning a physical system. Sometimes, $|\psi\rangle$ is taken to represent not the state of an individual physical system but an *ensemble* of possible similar physical systems. Often it is argued that a complicatedly entangled state vector $|\psi\rangle$ will behave 'for all practical purposes' (or FAPP, as John Bell succinctly put it[4]) in the same way as such an ensemble of physical systems—and that this is all that physicists need to know concerning the measurement problem. Sometimes, it is even argued that $|\psi\rangle$ *cannot* describe a quantum-level reality since it makes no sense at all to talk of a 'reality' for our world at that level, reality consisting only of the results of 'measurements'.

To such as myself (and Einstein and Schrödinger too—so I am in good company), it makes no sense to use the term 'reality' just for objects that we can perceive, such as (certain types of) measuring devices, denying that the term can apply at some deeper underlying level. Undoubtedly the world is strange and unfamiliar at the quantum level, but it is not 'unreal'. How, indeed, can real objects be constructed from unreal constituents? Moreover, the mathematical laws that govern the quantum world are remarkably precise—as precise as the more familiar equations that control the behaviour of macroscopic objects—despite the fuzzy images that are conjured up by such descriptions as 'quantum fluctuations' and 'uncertainty principle'.

Yet, if we accept that there must be a reality of some kind that holds at the quantum level, we may still have doubts that this reality can be accurately described by a state vector $|\psi\rangle$. There are various arguments that people raise as objections to the 'reality' of $|\psi\rangle$. In the first place, $|\psi\rangle$ seems to need to undergo this mysterious non-local discontinuous 'jumping', from time to time, that I have been denoting by the letter **R**. This does not seem to be the way in which a physically acceptable description of the world should behave, especially since we already have the marvellously accurate continuous Schrödinger equation **U**, which is supposed to be controlling the way in which $|\psi\rangle$ evolves (most of the time). Yet, as we have seen, **U**, by itself, leads us into the difficulties and puzzlements of the many-worlds-type viewpoints, and if we require a picture that closely resembles the actual universe that we seem to perceive around us, then something of the nature of **R** is indeed needed.

Another objection to the reality of $|\psi\rangle$ that is sometimes put forward is that the kind of alternation **U**, **R**, **U**, **R**, **U**, **R**, . . . that is in effect used in quantum theory is not a description that is symmetrical in time (because it is **R** that determines the *starting* point of each **U**-action, not the finishing point), and that there is another completely equivalent description in which the **U**-time-evolutions are reversed (cf. ENM, pp. 355, 356; Figs 8.1, 8.2). Why should we take one of these as providing 'reality' and not the other? There are even viewpoints according to which *both* the forward- and backward-evolved state vectors are to be taken seriously as co-existing parts of the description of physical reality (Costa de Beauregard 1989, Werbos 1989, Aharonov and

Vaidman 1990). I believe that there is likely to be something of profound significance underlying these considerations, but for the moment I do not wish to dwell on the matter. I shall touch on these, and some related, issues in §7.12.

One of the most frequent objections to taking $|\psi\rangle$ seriously as a description of reality is that it is not directly 'measureable'—in the sense that if one is presented with a totally unknown state, then there is no experimental way of determining what the state vector (up to a proportionality factor) actually is. Take the case of the spin of an atom of spin $\frac{1}{2}$, for example. Recall (§5.10, Fig. 5.19) that each possible state of its spin would be characterized by a particular direction in ordinary space. But if we have no idea what that direction is, then we have no way of determining it. All we can do is to settle on some direction and ask the question: is its spin in that direction (**YES**) or is it in the opposite direction (**NO**)? Whatever the spin state is taken initially to be, its Hilbert-space direction gets projected to either the **YES** space or the **NO** space, with certain probabilities. And at that point we have lost most of the information as to what the spin state 'actually' was. All that we can obtain, from a spin-direction measurement, for an atom of spin $\frac{1}{2}$ is *one bit* of information (i.e. the answer to one yes/no question), whereas the possible states of spin direction form a continuum, which would need an infinite number of bits of information in order to determine it precisely.

All this is true, yet it remains hard to take the opposite position either: that the state vector $|\psi\rangle$ is somehow physically 'unreal', perhaps encapsulating merely the sum total of 'our knowledge' about a physical system. This, I find very hard to accept, particularly since there appears to be something very subjective about such a role of 'knowledge'. *Whose* knowledge, after all, is being referred to here? Certainly not mine. I have very little actual knowledge of the individual state vectors that are relevant to the detailed behaviour of all the objects that surround me. Yet they carry on with their precisely organized actions, totally oblivious to whatever might be 'known' about the state vector, or to whomever might know it. Do different experimenters, with different knowledge about a physical system, use different state vectors to describe that system? Not in any significant way; they might only do so if these differences were about features of the experiment that would be inessential to the outcome.

One of the most powerful reasons for rejecting such a subjective viewpoint concerning the reality of[5] $|\psi\rangle$ comes from the fact that whatever $|\psi\rangle$ might be, there is always—in principle, at least—a *primitive measurement* (cf. §5.13) whose **YES** space consists of the Hilbert-space ray determined by $|\psi\rangle$. The point is that the physical state $|\psi\rangle$ (determined by the ray of complex multiples of $|\psi\rangle$) is *uniquely* determined by the fact that the outcome **YES**, for this state, is *certain*. No other physical state has this property. For any other state, there would merely be some probability, short of certainty, that the outcome will be **YES**, and an outcome of **NO** might occur. Thus, although there is no measurement which will tell us what $|\psi\rangle$ actually *is*, the physical state $|\psi\rangle$ is uniquely determined by what it asserts must be the result of a measurement that *might* be performed on it. This is a matter of counterfactuals again (§5.2, §5.3), but we have seen how important counterfactual issues actually are to the expectations of quantum theory.

To put the point a little more forcefully, imagine that a quantum system has been set up in a known state, say $|\phi\rangle$, and it is computed that after a time t the state will have evolved, under the action of **U**, into another state $|\psi\rangle$. For example, $|\phi\rangle$ might represent the state 'spin up' ($|\phi\rangle = |\uparrow\rangle$) of an atom of spin $\frac{1}{2}$, and we can suppose that it has been put in that state by the action of some previous measurement. Let us assume that our atom has a magnetic moment aligned with its spin (i.e. it is a little magnet pointing to the spin direction). When the atom is placed in a magnetic field, the spin direction will precess in a well-defined way, that can be accurately computed as the action of **U**, to give some new state, say $|\psi\rangle = |\rightarrow\rangle$, after a time t. Is this computed state to be taken seriously as part of physical reality? It is hard to see how this can be denied. For $|\psi\rangle$ has to be prepared for the possibility that we *might* choose to measure it with the primitive measurement referred to above, namely that whose **YES** space consists precisely of the multiples of $|\psi\rangle$. Here, this is the spin measurement in the direction $\rightarrow$. The system has to know to give the answer **YES**, with *certainty* for that measurement, whereas *no* spin state of the atom *other* than $|\psi\rangle = |\rightarrow\rangle$ could guarantee this.

In practice, there would be many kinds of physical situation, different from spin determinations, in which such a primitive measurement would be totally impractical. Yet the standard rules of quantum theory do allow that these measurements could be performed in principle. To deny the possibility of measurements of this kind for certain 'too complicated' types of $|\psi\rangle$ would be to change the framework of quantum theory. Perhaps that framework should be changed (and in §6.12 I make some specific suggestions in that direction). But it must be appreciated that at least some kind of change is necessary if *objective* distinctions between different quantum states are to be denied, i.e. if $|\psi\rangle$ is *not* to be taken to be, in some clear physical sense, *objectively* real (at least up to proportionality).

The 'minimal' change that is often suggested, in connection with measurement theory, is the introduction of what are referred to as *superselection rules*[6] which indeed effectively deny the possibility of making certain types of primitive measurements on a system. I do not wish to discuss these in detail here, since in my opinion, no such suggestions have been developed to the stage in which a coherent general viewpoint has emerged with regard to the measurement problem. The only point that I wish to stress here is that even a minimal change of this nature is still a change—and this brings out the main point that some kind of change is necessary.

Finally, I should perhaps mention that there are various other approaches to quantum theory which, though not at variance with the predictions of the conventional theory, provide 'pictures of reality' that differ in various respects from that in which the state vector $|\psi\rangle$ is, by itself, 'taken seriously' as representing that reality. Among these is the *pilot wave* theory of Prince Louis de Broglie (1956) and David Bohm (1952)—a non-local theory according to which there is something equivalent to a wavefunction $|\psi\rangle$ *and* a system of classical-like particles, *both* of which are taken to be 'real' in the theory. (See also Bohm and Hiley 1994.) There are also viewpoints which involve entire 'histories' of possible behaviour (stimulated by Richard Feynman's approach to quantum theory (1948)) and according to which the view of 'physical reality' differs somewhat from that provided by an ordinary state vector $|\psi\rangle$. Recent proponents of a scheme of this general kind, but which also takes into account the possibility of what are, in effect, repeated partial measurements (in accordance with an analysis due to Aharonov *et al.* (1964)), are Griffiths (1984), Omnès (1992), Gell-Mann and Hartle (1993). It would be inappropriate for me to indulge in a discussion of these various alternatives here (though it should be mentioned that the density-matrix formalism introduced in the next section plays an important part in some of them—also in the operator approach of Haag (1992)). I should at least say that, although these procedures contain many points of considerable interest and some stimulating originality, I am as yet quite unconvinced that the measurement problem can really be resolved solely in terms of descriptions of these various kinds. Of course, it is certainly possible that time might prove me wrong.

6.4 The density matrix

Many physicists would take the line that they are pragmatic people, and are not interested in questions of the 'reality' of $|\psi\rangle$. All that one needs from $|\psi\rangle$, they would say, is to be able to calculate appropriate probabilities as to future physical behaviour. Often, a state that has previously been taken to represent a physical situation evolves into something extremely complicated, where entanglements with the detailed environment become so involved that there is no possibility in practice of ever seeing the quantum interference effects that distinguish that state from many others like it. Such 'pragmatic' physicists would no doubt claim that there would be no sense in maintaining that the particular state vector that has resulted from this evolution has any more 'reality' than others that are in practice indistinguishable from it. Indeed, they might claim, one may just as well use some *probability mixture* of state vectors to describe 'reality' as use any *particular* state vector. The contention is that if the application of **U** to some state vector representing the initial state of a system yields something that, *for all practical purposes* (Bell's FAPP), is indistinguishable from such a probability mixture of state vectors, then the probability mixture, rather than the **U**-evolved state vector, is good enough for a description of the world.

It is frequently argued that—at least FAPP—the **R**-procedure can be understood in these terms. Two sections hence, I shall try to address this important question. I shall ask whether it is indeed true that the (apparent) **U**/**R** paradox can be resolved by such means alone. But let us first try to be a little more explicit about the procedures that are adopted in the standard FAPP type of approach to an explanation of the (apparent?) **R**-process.

The key to these procedures is a mathematical object referred to as a *density matrix*, The density matrix is an important concept in quantum theory, and it is this quantity, rather than the state vector, that tends to underlie most standard mathematical descriptions of the measurement process. It will play a central role, also, in my own less conventional approach, especially with regard to its relationship to the standard FAPP procedures. For this reason, it will unfortunately be necessary to go a little more into the mathematical formalism of quantum theory than we had needed to before. I hope that the uninitiated reader will not be daunted by this. Even if a full understanding does not come through, I believe that it will be helpful to the reader to skim over the mathematical arguments wherever they appear, and no doubt something of their flavour will come through. This will be of considerable value for the understanding of some of the later arguments and of the subtleties involved in appreciating why we actually *do* need an improved theory of quantum mechanics!

We can think of a density matrix as representing a probability mixture of a number of possible *alternative* state vectors, rather than simply a single state vector. By a 'probability mixture' we merely mean that there is some uncertainty about what the actual state of the system may happen to be, where each possible alternative state vector is assigned a probability. These are just real-number probabilities in the ordinary classical sense. But with a density matrix there is a (deliberate) confusion, in this description, between these *classical* probabilities, occurring in this probability-weighted mixture and the *quantum-mechanical* probabilities that would result from the **R**-procedure. The idea is that one cannot operationally distinguish between the two, so a mathematical description—the density matrix—which does *not* distinguish between them is operationally appropriate.

What is this mathematical description? I do not want to go into a great deal of detail about it here, but an indication of the basic concepts will be helpful. This density-matrix idea is in fact a very elegant one.* First, in place of each individual state $|\psi\rangle$ we use an object written as

$$|\psi\rangle\langle\psi|.$$

*It was put forward in 1932 by the remarkable Hungarian/American mathematician John von Neumann who, moreover, was the main person originally responsible for the theory underlying electronic computer development, following the seminal work of Alan Turing. The game theory referred to in end-note 9 of Chapter 3 (see p. 155) was also von Neumann's brain-child, but more importantly for us here, it was he who first clearly distinguished the two quantum procedures that I have labelled '**U**' and '**R**'.

What does this mean? The precise mathematical definition is not important for us here, but this expression represents a kind of 'product' (a form of the tensor product referred to in §5.15) between the state vector $|\psi\rangle$ and its 'complex conjugate', written $\langle\psi|$. We take $|\psi\rangle$ to be a *normalized* state vector ($\langle\psi|\psi\rangle=1$), and then the expression $|\psi\rangle\langle\psi|$ is uniquely determined by the physical state that the vector $|\psi\rangle$ represents (being independent of the phase-factor freedom $|\psi\rangle\mapsto e^{i\theta}|\psi\rangle$, discussed in §5.10). In Dirac's terminology, the original $|\psi\rangle$ is called a 'ket'-vector and $\langle\psi|$ is its corresponding 'bra'-vector. Together, a bra-vector $\langle\psi|$ and a ket-vector $|\phi\rangle$ can also be combined to form their *scalar product* ('bracket'):

$$\langle\psi|\phi\rangle$$

a notation that the reader will recognize from §5.12. This scalar product is just an ordinary complex number, whereas the tensor product $|\psi\rangle\langle\phi|$ that occurs with a density matrix yields a more complicated mathematical 'thing'—an element of a certain vector space.

There is a particular mathematical operation called 'taking the trace' that enables us to pass from this 'thing' to an ordinary complex number. For a single expression like $|\psi\rangle\langle\phi|$, this amounts just to reversing the order of the terms to produce the scalar product:

$$\text{trace}\,(|\psi\rangle\langle\phi|)=\langle\phi|\psi\rangle,$$

whereas for a sum of terms, 'trace' acts linearly; for example,

$$\text{trace}\,(z|\psi\rangle\langle\phi|+w|\alpha\rangle\langle\beta|)=z\langle\phi|\psi\rangle+w\langle\beta|\alpha\rangle.$$

I shall not go into the details of all the mathematical properties of objects like $\langle\psi|$ and $|\psi\rangle\langle\phi|$, but several points are worth noting. In the first place, the product $|\psi\rangle\langle\phi|$ satisfies just the same algebraic laws as are listed on p. 288 for the product $|\psi\rangle|\phi\rangle$ (except for the last one, which is not relevant here):

$$(z|\psi\rangle)\langle\phi|=z(|\psi\rangle\langle\phi|)=|\psi\rangle(z\langle\phi|),$$

$$(|\psi\rangle+|\chi\rangle)\langle\phi|=|\psi\rangle\langle\phi|+|\chi\rangle\langle\phi|,$$

$$|\psi\rangle(\langle\phi|+\langle\chi|)=|\psi\rangle\langle\phi|+|\psi\rangle\langle\chi|.$$

We should also note that the bra-vector $\bar{z}\langle\psi|$ is the complex conjugate of the ket-vector $z|\psi\rangle$ ($\bar{z}$ being the ordinary complex conjugate of the complex number z, cf. p. 264), and $\langle\psi|+\langle\chi|$ is the complex conjugate of $|\psi\rangle+|\chi\rangle$.

Suppose that we wish to describe the density matrix that represents some probability mixture of normalized states, say $|\alpha\rangle$, $|\beta\rangle$, with respective probabilities a, b. The appropriate density matrix will, in this case, be

$$\boldsymbol{D}=a|\alpha\rangle\langle\alpha|+b|\beta\rangle\langle\beta|.$$

For three normalized states $|\alpha\rangle$, $|\beta\rangle$, $|\gamma\rangle$, with respective probabilities *a*, *b*, *c*, we have

$$\boldsymbol{D}=a|\alpha\rangle\langle\alpha|+b|\beta\rangle\langle\beta|+c|\gamma\rangle\langle\gamma|,$$

and so on. From the fact that the probabilities for all the alternatives must add up to unity, the following important property may be deduced, which holds for any density matrix:

$$\text{trace}\,(\boldsymbol{D})=1.$$

How can we use a density matrix to calculate the probabilities arising in some measurement? Let us consider first the case of a primitive measurement. We ask the system whether it is in the physical state $|\psi\rangle$ (**YES**) or something orthogonal to $|\psi\rangle$ (**NO**). The measurement itself is represented by a mathematical object (called a *projector*) very similar to a density matrix:

$$\boldsymbol{E}=|\psi\rangle\langle\psi|.$$

The probability *p* of obtaining **YES** is then

$$p=\text{trace}\,(\boldsymbol{DE}),$$

where the product $\boldsymbol{DE}$ is itself the density-matrix-like 'thing' that would be obtained using the rules of algebra in an almost ordinary way—but being careful about the orderings of 'multiplications'. For example, for the above two-term sum $\boldsymbol{D}=a|\alpha\rangle\langle\alpha|+b|\beta\rangle\langle\beta|$, we have

$$\begin{aligned}\boldsymbol{DE}&=(a|\alpha\rangle\langle\alpha|+b|\beta\rangle\langle\beta|)|\psi\rangle\langle\psi|\\&=a|\alpha\rangle\langle\alpha|\psi\rangle\langle\psi|+b|\beta\rangle\langle\beta|\psi\rangle\langle\psi|\\&=(a\langle\alpha|\psi\rangle)|\alpha\rangle\langle\psi|+(b\langle\beta|\psi\rangle)|\beta\rangle\langle\psi|.\end{aligned}$$

The terms $\langle\alpha|\psi\rangle$ and $\langle\beta|\psi\rangle$ can be 'commuted' with other expressions since they are simply numbers, but we must be careful about the orderings of 'things' like $|\alpha\rangle$ and $\langle\psi|$. We deduce (noting that $z\bar{z}=|z|^2$, cf. p. 264)

$$\begin{aligned}\text{trace}\,(\boldsymbol{DE})&=(a\langle\alpha|\psi\rangle)\langle\psi|\alpha\rangle+(b\langle\beta|\psi\rangle)\langle\psi|\beta\rangle\\&=a|\langle\alpha|\psi\rangle|^2+b|\langle\beta|\psi\rangle|^2.\end{aligned}$$

Recall (cf. §5.13, p. 285) that $|\langle\alpha|\psi\rangle|^2$ and $|\langle\beta|\psi\rangle|^2$ are the *quantum* probabilities for the respective outcomes $|\alpha\rangle$ and $|\beta\rangle$, whereas *a* and *b* provide the *classical* contributions to the total probability. Thus, the quantum probabilities and classical probabilities are all mixed together in the final expression.

For the more general yes/no measurement, the discussion is basically the same, except that in place of the '$\boldsymbol{E}$' defined above we use a more general projector such as

$$E = |\psi\rangle\langle\psi| + |\phi\rangle\langle\phi| + \ldots + |\chi\rangle\langle\chi|,$$

where $|\psi\rangle, |\phi\rangle, \ldots, |\chi\rangle$ are mutually orthogonal normalized states spanning the space of **YES** states in the Hilbert space. We have the general property

$$E^2 = E,$$

which characterizes a projector. The probability of **YES**, for the measurement defined by the projector E on the system with density matrix D, is trace (DE), exactly as before.

We take note of the important fact that the required probability can be calculated if we simply know the density matrix and the projector describing the measurement. We do not need to know the particular way in which the density matrix has been put together in terms of particular states. The total probability comes out automatically as the appropriate combination of classical and quantum probabilities, without our having to worry about how much of the resulting probability comes from each part.

Let us examine more closely this curious way in which classical and quantum probabilities are interwoven within the density matrix. Suppose, for example, we have a particle of spin $\frac{1}{2}$, and that we are completely uncertain as to whether the (normalized) spin state happens to be $|\uparrow\rangle$ or $|\downarrow\rangle$. Thus, taking the respective probabilities to be $\frac{1}{2}$ and $\frac{1}{2}$, the density matrix becomes

$$D = \tfrac{1}{2}|\uparrow\rangle\langle\uparrow| + \tfrac{1}{2}|\downarrow\rangle\langle\downarrow|.$$

Now it turns out (by a simple calculation) that precisely the *same* density matrix D would arise with an equal probability mixture $\frac{1}{2}$, $\frac{1}{2}$, of any other orthogonal possibilities, say the (normalized) states $|\rightarrow\rangle$ and $|\leftarrow\rangle$ (where $|\rightarrow\rangle = (|\uparrow\rangle + |\downarrow\rangle)/\sqrt{2}$ and $|\leftarrow\rangle = (|\uparrow\rangle - |\downarrow\rangle)/\sqrt{2}$):

$$D = \tfrac{1}{2}|\rightarrow\rangle\langle\rightarrow| + \tfrac{1}{2}|\leftarrow\rangle\langle\leftarrow|.$$

Suppose that we elect to measure the spin of the particle in the upward direction, so the relevant projector is

$$E = |\uparrow\rangle\langle\uparrow|.$$

Then we find, for the probability of **YES** according to the first description,

$$\begin{aligned} \text{trace}\,(DE) &= \tfrac{1}{2}|\langle\uparrow|\uparrow\rangle|^2 + \tfrac{1}{2}|\langle\downarrow|\uparrow\rangle|^2 \\ &= \tfrac{1}{2}\times 1^2 + \tfrac{1}{2}\times 0^2 \\ &= \tfrac{1}{2}, \end{aligned}$$

where we are using $\langle\uparrow|\uparrow\rangle = 1$ and $\langle\downarrow|\uparrow\rangle = 0$ (the states being normalized and orthogonal); and according to the second,

$$\begin{aligned} \text{trace}\,(DE) &= \tfrac{1}{2}|\langle\rightarrow|\uparrow\rangle|^2 + \tfrac{1}{2}|\langle\leftarrow|\uparrow\rangle|^2 \\ &= \tfrac{1}{2}\times(1/\sqrt{2})^2 + \tfrac{1}{2}\times(1/\sqrt{2})^2 \\ &= \tfrac{1}{4} + \tfrac{1}{4} = \tfrac{1}{2}, \end{aligned}$$

where now the right/left states $|\rightarrow\rangle, |\leftarrow\rangle$ are neither orthogonal nor parallel to the measured state $|\uparrow\rangle$, and in fact $|\langle\rightarrow|\uparrow\rangle| = |\langle\leftarrow|\uparrow\rangle| = 1/\sqrt{2}$.

Although the probabilities come out the same (as they must, because the density matrix is the same), the physical interpretation of these two descriptions is quite different. We are accepting that the physical 'reality' of any situation is to be described by *some* definite state vector, but there is a classical uncertainty as to what this actual state vector might happen to be. In the first of our two descriptions above, the state is either $|\uparrow\rangle$ or $|\downarrow\rangle$, but one does not know which. In the second, it is either $|\rightarrow\rangle$ or $|\leftarrow\rangle$ and one does not know which. In the first description, when we perform a measurement to ask if the state is $|\uparrow\rangle$, it is a simple matter of classical probabilities: there is indeed a straightforward probability of $\frac{1}{2}$ that this state is $|\uparrow\rangle$, and that is all there is to it. In the second description, where we ask the same question, it is the probability mixture of $|\rightarrow\rangle$ and $|\leftarrow\rangle$ that the measurement encounters, and each contributes a classical contribution $\frac{1}{2}$ times a quantum-mechanical contribution $\frac{1}{2}$ to the total of $\frac{1}{4}+\frac{1}{4}=\frac{1}{2}$. We see that the density matrix cleverly contrives to give the correct probability no matter how this probability is considered to be built up out of classical and quantum-mechanical parts.

The above example is somewhat special in that the density matrix possesses what are called 'degenerate eigenvalues' (the fact that, here, the two classical probability values $\frac{1}{2}, \frac{1}{2}$ are equal), which is what allows us to have more than one description in terms of probability mixtures of orthogonal alternatives. However, this is not an essential point for our present discussion. (I mention it mainly to put the experts at ease.) We can always allow the alternative states in a probability mixture to involve many more states than just a set of mutually orthogonal alternatives. For example, in the above situation, we could have complicated probability mixtures of many different possible spin directions. It turns out that for any density matrix *whatsoever*—not merely ones with degenerate eigenvalues—there are large numbers of completely different ways of representing the same density matrix as a probability mixture of alternative states.

6.5 Density matrices for EPR pairs

Now let us examine a type of situation for which a density-matrix description is particularly appropriate—yet which points out an almost paradoxical aspect of its interpretation. This is in its relation to EPR effects and quantum entanglement. Let us consider the physical situation discussed in §5.17, where a particle of spin 0 (in state $|\Omega\rangle$) splits into two particles of spin $\frac{1}{2}$ that travel to the left and right to a great distance apart, providing the expression for their combined (entangled) states of spin:

$$|\Omega\rangle = |L\uparrow\rangle|R\downarrow\rangle - |L\downarrow\rangle|R\uparrow\rangle.$$

Suppose that the spin of the right-hand particle is soon to be examined by some observer's measuring apparatus, but that the left-hand one has travelled away to such a great distance that the observer has no access to it. How would the observer describe the right-hand particle's state of spin?

It would be very appropriate for him* to use the density matrix

$$\boldsymbol{D} = \tfrac{1}{2}|\mathrm{R}\uparrow\rangle\langle\mathrm{R}\uparrow| + \tfrac{1}{2}|\mathrm{R}\downarrow\rangle\langle\mathrm{R}\downarrow|.$$

For he might imagine that another observer—some colleague a long way off—had chosen to measure the spin of the left-hand particle in an up/down direction. He has no way of telling which result his imaginary colleague might have obtained for that spin measurement. But he does know that if his colleague had obtained the result $|\mathrm{L}\uparrow\rangle$, then his own particle's state would have to be $|\mathrm{R}\downarrow\rangle$, whereas had his colleague obtained $|\mathrm{L}\downarrow\rangle$, then his own particle's state must be $|\mathrm{R}\uparrow\rangle$. He also knows (from what the standard rules of quantum theory tell him to expect for the probabilities in this situation) that it is equally likely that his imaginary colleague would have obtained $|\mathrm{L}\uparrow\rangle$ as $|\mathrm{L}\downarrow\rangle$. Thus, he concludes that the state of his own particle is an equal probability mixture (i.e. respective probabilities $\frac{1}{2}$, $\frac{1}{2}$) for the two alternatives $|\mathrm{R}\uparrow\rangle$, $|\mathrm{R}\downarrow\rangle$, so his density matrix must indeed be $\boldsymbol{D}$, as just given above.

He might, however, imagine that his colleagues has just measured the left-hand particle in a left/right direction instead. Exactly the same reasoning (now using the alternative description $|\Omega\rangle = |\mathrm{L}\leftarrow\rangle|\mathrm{R}\rightarrow\rangle - |\mathrm{L}\rightarrow\rangle|\mathrm{R}\leftarrow\rangle$, cf. p. 293) would lead him to conclude that his own particle's spin state is an equal probability mixture of right and left, which provides him with the density matrix

$$\boldsymbol{D} = \tfrac{1}{2}|\mathrm{R}\rightarrow\rangle\langle\mathrm{R}\rightarrow| + \tfrac{1}{2}|\mathrm{R}\leftarrow\rangle\langle\mathrm{R}\leftarrow|.$$

As has been said before, this is indeed precisely the same density matrix as we just had, but its *interpretation* as a probability mixture of alternative states is quite different! It does not matter which interpretation the observer adopts. His density matrix provides him with all the information that is available for the calculations of probabilities for the results of spin measurements on the right-hand particle alone. Moreover, since his colleague is merely *imagined*, our observer need not consider that any spin measurement has been performed on the left-hand particle at all. The same density matrix $\boldsymbol{D}$ tells him all that he can know about the spin state of the right-hand particle prior to his actually measuring that particle. Indeed, we could suppose that the 'actual state' of the right-hand particle is more correctly given by the density matrix $\boldsymbol{D}$ than any particular state vector.

Considerations of this general kind sometimes lead people to think of density matrices as providing a more appropriate description of quantum 'reality' in certain circumstances than do state vectors. However, this would

*See Notes to the reader, p. xvi.

not provide us with a comprehensive viewpoint in situations like the one just considered. For there is nothing in principle to prevent our observer's imaginary colleague becoming a real one, and the two observers eventually communicating their results to one another. The correlations between the measurements of one observer and the other cannot be explained in terms of separate density matrices for the left- and right-hand particles separately. For this we need the entire entangled state provided by the expression for the actual state vector $|\Omega\rangle$, as given above.

For example, if both observers choose to measure their particles' spins in the up/down direction, then they must necessarily obtain opposite answers for the results of their measurements. Individual density matrices for the two particles will not provide this information. Even more seriously, Bell's theorem (§5.4) shows there is *no* classical-type local ('Bertlmann's socks') way to model the entangled state of the combined particle pair, before measurement. (See ENM, Chapter 6, endnote 14, p. 301, for a simple demonstration of this fact—essentially one due to Stapp (1979; cf. also Stapp 1993)—in the case where one of the observers chooses to measure his particle's spin either up/down or right/left, whereas the other chooses either of the two directions at 45° to these two. If we replace the two spin $\frac{1}{2}$ particles by two of spin $\frac{3}{2}$, then the magic dodecahedra of §5.3 show this kind of thing even more convincingly, since now no probabilities are needed.)

This shows that the density-matrix description could be adequate to describe the 'reality' of this situation only if there is some reason *in principle* why measurements on the two parts of the system cannot be performed and compared. There seems no reason, in normal situations, why this should be the case. In abnormal situations—such as one considered by Stephen Hawking (1982) in which one particle of an EPR pair might be captured inside a black hole—there might be a more serious case for a density-matrix description at the fundamental level (as Hawking indeed argues). But this would, in itself, constitute a change in the very framework of quantum theory. Without such a change, the essential role of the density matrix is FAPP, rather than fundamental—though its role is an important one nevertheless.

6.6 A FAPP explanation of **R**?

Let us now see how density matrices indeed play their part in the standard—FAPP—approach to explaining how it is that the **R**-process 'seems' to come about. The idea is that a quantum system and a measuring apparatus, together with the environment that they both inhabit—all assumed to be evolving together according to **U**—will behave *as though* **R** has taken place whenever the effects of the measurement become inextricably entangled with this environment.

The quantum system is taken to be initially isolated from its environment, but upon being 'measured' it triggers larger-scale effects in the measuring apparatus that soon involve entanglements with considerable and ever-increasing parts of that environment. At this stage, the picture becomes in many ways similar to the EPR situation that was discussed in the preceding section. The quantum system, together with the measuring apparatus that it has just triggered, plays a role rather like that of the right-hand particle, whereas the disturbed environment plays a role like that of the left-hand one. A physicist proposing to examine the measuring apparatus would play a similar role to the observer, in the discussion above, who examines the right-hand particle. That observer had no access to any measurements that might be performed on the left-hand one; similarly, our physicist has no access to the detailed way in which the environment might be disturbed by the measuring apparatus. The environment would consist of enormous numbers of randomly moving particles, and we may take it that the detailed information contained in the precise way in which the particles in the environment have been disturbed would, in practice, be irretrievably lost to the physicist. This is similar to the fact that any information concerning the left-hand particle's spin, in the example above, is inaccessible to the right-hand observer. As with the right-hand particle, the state of the measuring apparatus is appropriately described by a density matrix rather than a pure quantum state; accordingly it is treated as a probability mixture of states, rather than a pure state on its own. This probability mixture provides the probability-weighted alternatives that the procedure **R** would have given us—at least FAPP—so the standard argument goes.

Let us consider an example. Suppose that a photon is emitted by some source, in the direction of a detector. Between source and detector is a partially silvered mirror, and after encountering the mirror, the photon's state is a superposition

$$w|\alpha\rangle + z|\beta\rangle,$$

where the transmitted state $|\alpha\rangle$ would activate the detector (**YES**) but the reflected one $|\beta\rangle$ would leave it undisturbed (**NO**). I am here assuming that all states are normalized, so that, according to the **R**-procedure, we would obtain:

$$\text{probability of } \mathbf{YES} = |w|^2; \text{ probability of } \mathbf{NO} = |z|^2.$$

For a *half*-silvered mirror (as in the initial example considered in §5.7, where our $|\alpha\rangle$ and $|\beta\rangle$ would be the states $|\mathrm{B}\rangle$ and $\mathrm{i}|\mathrm{C}\rangle$, respectively) these two probabilities are each $\frac{1}{2}$, and we have $|w| = |z| = 1/\sqrt{2}$.

The detector initially has state $|\Psi\rangle$, which evolves to $|\Psi_{\mathrm{Y}}\rangle$ (**YES**) upon absorbing the photon (in state $|\alpha\rangle$) and which evolves to $|\Psi_{\mathrm{N}}\rangle$ (**NO**) if it fails to absorb the photon (in state $|\beta\rangle$). If the environment could be ignored, then the state at that stage would have the form

$$w|\Psi_{\mathrm{Y}}\rangle + z|\Psi_{\mathrm{N}}\rangle|\beta\rangle$$

(all states assumed normalized); but let us assume that the detector, being a macroscopic object, very quickly becomes involved in interactions with its surrounding environment—and we can assume that the fleeing photon (originally in state $|\beta\rangle$) is absorbed by the laboratory wall to become part of the environment also. As before, according to whether or not the detector receives the photon, it settles down into the detector state $|\Psi_Y\rangle$ or $|\Psi_N\rangle$, respectively, but in doing so it would disturb the environment in a different way in each case. We can assign the environment state $|\Phi_Y\rangle$ to accompany $|\Psi_Y\rangle$, and $|\Phi_N\rangle$ to accompany $|\Psi_N\rangle$ (again assumed normalized, but not necessarily orthogonal) and we can express the full state in the entangled form

$$w|\Phi_Y\rangle|\Psi_Y\rangle + z|\Phi_N\rangle|\Psi_N\rangle.$$

So far, the physicist is not involved, but he is about to examine the detector to see whether it has registered **YES** or **NO**. How would he view the quantum state of the detector just before he examines it? As with the observer measuring the spin of the right-hand particle in the previous discussion, it would be appropriate for him to use a density matrix. We may suppose that no measurement is actually performed on the environment in order to ascertain whether *its* state is $|\Phi_Y\rangle$ or $|\Phi_N\rangle$, just as with the left-hand particle in the EPR pair described above. Accordingly, a density matrix indeed provides an appropriate quantum description of the detector.

What is this density matrix? The standard type of argument[7] (based on some particular way of modelling this environment—and also on some incompletely justified assumptions, such as the unimportance of EPR-type correlations) leads to the conclusion that this density matrix should rapidly and very closely approach the form

$$\boldsymbol{D} = a|\Psi_Y\rangle\langle\Psi_Y| + b|\Psi_N\rangle\langle\Psi_N|,$$

where

$$a = |w|^2 \text{ and } b = |z|^2.$$

This density matrix can be interpreted as representing a probability mixture of the detector registering **YES**, with probability $|w|^2$, and the detector registering **NO**, with probability $|z|^2$. This is just exactly what the **R**-process would have told us that the physicist will find as the result of his experiment—or is it?

We must be a little careful about coming to this conclusion. The density matrix $\boldsymbol{D}$ would indeed allow our physicist to calculate the probabilities that he needs, if he is allowed to *assume* that the alternatives open to him are simply that the state of the detector is *either* $|\Psi_Y\rangle$ *or* $|\Psi_N\rangle$. But this assumption is by no means a consequence of our discussion. Recall from the previous section that density matrices have many *alternative* interpretations as probability mixtures of states. In particular, in the case of a *half*-silvered mirror, we obtain a density matrix of just the same form as we did for the particle of spin $\frac{1}{2}$ above:

$$D = \tfrac{1}{2}|\Psi_Y\rangle\langle\Psi_Y| + \tfrac{1}{2}|\Psi_N\rangle\langle\Psi_N|.$$

This can be re-expressed as, say,

$$D = \tfrac{1}{2}|\Psi_P\rangle\langle\Psi_P| + \tfrac{1}{2}|\Psi_Q\rangle\langle\Psi_Q|,$$

where $|\Psi_P\rangle$ and $|\Psi_Q\rangle$ are two quite different orthogonal possible states for the detector—states that would be quite absurd from the point of view of classical physics, such as

$$|\Psi_P\rangle = (|\Psi_Y\rangle + |\Psi_N\rangle)/\sqrt{2} \text{ and } |\Psi_Q\rangle = (|\Psi_Y\rangle - |\Psi_N\rangle)/\sqrt{2}.$$

The fact that the physicist considers that the state of his detector is described by the density matrix D does not in any way explain why he always finds that the detector is either in a **YES** state (given by $|\Psi_Y\rangle$) or else in a **NO** state (given by $|\Psi_N\rangle$). For precisely the same density matrix would be given if the state were an equal-probability-weighted combination of the classical absurdities $|\Psi_P\rangle$ and $|\Psi_Q\rangle$ (which respectively describe the quantum linear superpositions '**YES** *plus* **NO**' and '**YES** *minus* **NO**')!

To emphasize the physical absurdity of states like $|\Psi_P\rangle$ and $|\Psi_Q\rangle$, for a macroscopic detector, consider the case of a 'measuring apparatus' that consists of a box with a cat in it, where the cat is killed by some device if the detector receives a photon (in state $|\alpha\rangle$) but not otherwise (photon in state $|\beta\rangle$)—a *Schrödinger's cat* (cf. §5.1 and Fig. 6.3). The **YES** answer would be presented in the form of 'cat dead' and the **NO** answer as 'cat alive'. However, merely to know that the density matrix has the form of an equal mixture of these two states certainly does *not* tell us that the cat is either dead or alive (with equal probabilities), since it could just as well be either 'dead plus alive' or 'dead minus alive' with equal probabilities! The density matrix alone does *not* tell us that these latter two classically absurd possibilities will never be experienced in the actual world as we know it. As with the 'many-worlds' type of approach to an explanation of **R**, we seem to be forced, again, into considering what kind of states a conscious observer (here, our 'physicist') is allowed to perceive. Why, indeed, is a state like 'cat dead plus cat alive' not something that a conscious external* observer would ever become aware of?

One might reply that the 'measurement' that our physicist is about to perform on the detector was, after all, simply to determine whether the detecor registers **YES** or **NO**—i.e. to ascertain, in this example, whether the cat is dead or alive. (This is similar to the observer, in the previous section, determining whether the spin of the right-hand particle is up or down.) For this measurement the density matrix indeed gives the correct probabilities,

*Of course, the matter of the cat's own consciousness would have to be considered also! This aspect of things is brought into clear focus with a version of the Schrödinger cat paradox due to Eugene P. Wigner (1961). 'Wigner's friend' suffers some of the indignity of Schrödinger's cat, but is fully conscious in each of his superposed states!

whichever way we choose to represent it. However, this is really begging the question. We must ask why it is that merely *looking* at the cat indeed performs a measurement of this kind. There is nothing simply in the **U**-evolution of a quantum system that tells us that in the act of 'looking' and consequently of *perceiving* a quantum system our awareness is forbidden from encountering the combination 'cat dead plus cat alive'. We are back where we were before. What is awareness? How are brains *actually* constructed? *Not* to have to be forced into considerations of this kind was one of the clearest reasons for considering FAPP explanations of **R** in the first place!

Some might try to argue that we have been considering an unrepresentative special case in our example here, where the two probabilities $\frac{1}{2}$ and $\frac{1}{2}$ are *equal* (the case of 'degenerate eigenvalues'). Only in such situations can the density matrix be represented in more than one way as a probability-weighted mixture of mutually *orthogonal* alternatives. This is *not* an important restriction, however, since orthogonality of the alternatives is not a requirement for the interpretation of a density matrix as a probability-weighted mixture. In fact, in a recent paper, Hughston *et al.* (1993) have shown that in situations like the one considered here, where a density matrix arises because the system under consideration is entangled with another separate system, then for *whatever* way one chooses to represent that density matrix as a probability mixture of alternative states, there is always a measurement that can be performed on that separate system which yields this particular way of representing the density matrix. In any case, since the ambiguity is present in the case when probabilities *are* equal, then this alone tells us that the density-matrix description is not sufficient for describing what the alternative *actual* states of our detector must be.

The upshot of all this is that merely knowing that the density matrix is some $\boldsymbol{D}$ does *not* tell us that the system is a probability mixture of some particular set of states that give rise to this particular $\boldsymbol{D}$. There are always numerous completely different ways of getting the same $\boldsymbol{D}$, most of which would be 'absurd' from the common-sense point of view. Moreover, this kind of ambiguity holds for any density matrix whatsoever.

The standard discussions do not often progress beyond the point of trying to show that the density matrix is 'diagonal'. This means, in effect, that it can be expressed as a probability-weighted mixture of mutually *orthogonal* alternatives—or, rather, that it can be so expressed, when these alternatives are the classical alternatives that one is interested in. (Without this final proviso, *all* density matrices would be diagonal!) But we have seen that the mere fact that the density matrix can be expressed in this way does not, in itself, tell us that detectors will *not* be perceived in 'absurd' quantum superpositions of **YES** and **NO** at the same time.

Thus, contrary to what is frequently claimed, the standard argument does *not* explain how the 'illusion' of **R** takes place as some kind of approximate description of **U**-evolution, when the environment becomes inextricably involved. What this argument *does* show is that the **R**-procedure can co-exist peacefully with **U**-evolution, under such circumstances. We still need **R** as a part of quantum theory that is separate from **U**-evolution (at least in the absence of some theory that tells is what kind of states conscious beings can perceive).

This, in itself, is important for the general consistency of quantum theory. But it is also important to realize that this co-existence and this consistency has a FAPP status rather than a rigorous one. Recall, from the final discussion in the previous section, that the density-matrix description of the right-hand particle was adequate only in the absence of a possible comparison between measurements that might be performed on *both* particles. For that, the entire state, with its *quantum*, rather than merely probability-weighted superpositions was needed. In the same way, the density-matrix description of the detector in the present discussion is adequate only if the fine details of the environment cannot be measured and comparison made with the results of the experimenter's observations of the detector. **R** can co-exist with **U** only if the fine details of the environment are immune from measurement; and the subtle quantum interference effects, that (according to standard quantum theory) lie hidden in the immense complication of the detailed description of the environment, can never be observed.

It is clear that the standard argument contains some good measure of truth; yet it cannot in any way be the entire answer. How are we to be sure that the effects of such interference phenomena are not to be uncovered by some future advances in technology? We should need some new physical rule which tells us that certain experiments that cannot at present be performed in practice can actually never even be performed *in principle*. According to such a rule, there would have to be some level of physical action at which it is deemed to be in principle impossible to retrieve these interference effects. It would seem that some *new* physical phenomenon will have to come in, and the complex-weighted superpositions of quantum-level physics *actually* become classical-level physical alternatives, rather than becoming such alternatives merely FAPP. The FAPP viewpoint, as it stands, gives us no picture of an actual physical reality. It cannot really be the case that FAPP is other than a stop-gap of a physical theory—though a valuable stop-gap nevertheless—and it will be important for the proposals that I shall be putting forward in §6.12.

6.7 Does FAPP explain the squared modulus rule?

There has been an implicit further assumption, in the preceding three sections, which has been allowed to pass almost unnoticed. The necessity of this assumption *alone* actually nullifies any suggestion that we have been able to *deduce* the squared modulus rule of the **R**-procedure from **U**-evolution—even FAPP. In the very use of a density matrix, we have, indeed, implicitly *assumed* that a probability-weighted mixture is appropriately described by such an object. The very appropriateness of expressions like $|\alpha\rangle\langle\alpha|$, which is itself a form of 'thing times its complex conjugate', is already intimately tied up with the assumption of the squared modulus rule. The rule for obtaining probabilities from a density matrix correctly combines classical with quantum probabilities only because the squared modulus rule is *inbuilt* in the very notion of a density matrix.

While it is in fact true that the process of unitary evolution (**U**) dovetails, mathematically, with the notion of a density matrix and with the Hilbert-space scalar product $\langle\alpha|\beta\rangle$, it does not in any way *tell* us that it is *probabilities* that are to be calculated by means of squared moduli. It is again a matter of mere co-existence between **R** and **U**, rather than an explanation of **R** from **U**. Unitary evolution says nothing whatever about the notion of probability. It is a clear-cut *additional* assumption that quantum probabilities can be calculated by this procedure, no matter how one attempts to justify the consistency of **R** with **U**, be it by means of a many-worlds or a FAPP approach.

Since so much of the experimental support that quantum mechanics enjoys arises from the very way in which the theory tells us that probabilities must be calculated, we can ignore the **R**-part of quantum mechanics only at our peril. It is something different from **U**, and it is not a consequence of **U**, no matter how hard and how frequently theorists have attempted to show that it must be. Since it is not a consequence of **U**, it is something that we must come to terms with as a physical process in its own right. This is not to suggest that it must be a physical *law* in its own right. No doubt it is an approximation to something else that we may not understand as yet. The discussions at the end of the previous section strongly suggest that the use of the **R**-procedure in the process of measurement is indeed an approximation.

Let us accept that something new is required, and try to venture forth, with due caution, into various avenues of the unknown that may be open to us.

6.8 Is it consciousness that reduces the state vector?

Among those who take $|\psi\rangle$ seriously as a description of the physical world, there are some who would argue—as an alternative to trusting **U** at all scales, and thus believing in a many-worlds type of viewpoint—that something of the nature of **R** *actually* takes place as soon as the consciousness of an observer becomes involved. The distinguished physicist Eugene Wigner once sketched a theory of this nature (Wigner 1961). The general idea would be that unconscious matter—or perhaps just inanimate matter—would evolve according to **U**, but as soon as a conscious entity (or 'life') becomes physically entangled with the state, then something new comes in, and a physical process that results in **R** takes over *actually* to reduce the state.

There need be no suggestion, with such a viewpoint, that somehow the conscious entity might be able to 'influence' the particular choice that Nature makes at this point. Such a suggestion would lead us into distinctly murky waters and, as far as I am aware, there would be a severe conflict with observed facts with any too simplistic suggestion that a conscious act of will could influence the result of a quantum-mechanical experiment. Thus, we are *not* requiring, here, that 'conscious free will' should necessarily be taking an active role with regard to **R** (but cf. §7.1, for some alternative viewpoints).

No doubt some readers might expect that, since I am searching for a link between the quantum measurement problem and the problem of consciousness, I might find myself attracted by ideas of this general nature. I should make myself clear that this is *not* the case. It is probable, after all, that consciousness is a rather rare phenomenon throughout the universe. There appears to be a good deal of it occurring in many places on the surface of the earth, but as far as evidence has presented itself to us to this date,[8] there is no highly developed consciousness—if, indeed, any at all—right out into depths of the universe many light centuries away from us. It would be a very strange picture of a 'real' physical universe in which physical objects evolve in totally different ways depending upon whether or not they are within sight or sound or touch of one of its conscious inhabitants.

For example, consider the weather. The detailed patterns of weather that happen to develop on any planet, being dependent upon chaotic physical processes (cf. §1.7), must be sensitive to numerous individual quantum events. If the process **R** does not actually take place in the absence of consciousness, then no particular actual weather pattern will be resolved out of the morass of quantum-superposed alternatives. Can we really believe that the weather patterns on some distant planet remain in complex-number superpositions of innumerable distinct possibilities—just some total hazy mess quite distinct from an actual weather—until some conscious being becomes aware of it, at which point, and *only* at which point, the superposed weather becomes an actual weather?

It might be argued that from an operational point of view—from the operational point of view of a conscious being, that is—such a superposed 'weather' would be no different from an uncertain *actual* weather (FAPP!). However, this in itself is no satisfactory solution to the problem of physical reality. We have seen that the FAPP viewpoint does not resolve these deep issues of 'reality', but remains a stop-gap that allows the **U**- and **R**-procedures of present-day quantum mechanics to co-exist—at least until our technology takes us to the point where a more precise and coherent viewpoint is necessary.

Thus, I propose to look elsewhere for a resolution of the problems of quantum mechanics. Though it may well be the case that the problem of mind is ultimately related to that of quantum measurement—or to the **U**/**R** paradox of quantum mechanics—it is not, according to my own belief, consciousness in itself (or consciousness in the form that we are familiar with) that can resolve the internal physical issues of quantum theory. I believe that the problem of quantum measurement should be faced and solved well before we can expect to make any real headway with the issue of consciousness in terms of physical action—and that the measurement problem must be solved in entirely *physical* terms. Once we are in possession of a satisfactory solution, then we may be in a better position to move towards some kind of answer to the question of consciousness. It is my view that solving the quantum measurement problem is a *prerequisite* for an understanding of mind and *not at all* that they are the same problem. The problem of mind is a much more difficult problem than the measurement problem!

6.9 Taking $|\psi\rangle$ really seriously

So far, the viewpoints that would claim to take the quantum description of the world seriously fall short of taking it *really* seriously, as I see it. The quantum formalism is perhaps too strange for it to be at all easy to take what it actually says seriously, and most physicists would shy away from taking too strong a line on this. For in addition to having a state vector $|\psi\rangle$ that evolves according to **U**, so long as the system remains at the quantum level, we have the disturbing discontinuous and probabilistic action of **R**, which seems to have to be brought in to effect discontinuous 'jumps' in $|\psi\rangle$ as soon as quantum-level effects become magnified enough to influence things at the classical level. Thus, if we are to take $|\psi\rangle$ as providing a picture of *reality*, then we must take these *jumps* as physically real occurrences too, no matter how uncomfortable we may feel about this. However, if one is *this* serious about the reality of the quantum-state-vector description, then one must also be prepared to introduce some (preferably very subtle) change into the actual rules of quantum theory. For the operation of **U** is, strictly speaking, incompatible with **R**, and some delicate 'paper-work' will be needed to cover the cracks between the descriptions of the quantum and classical levels of behaviour.

In fact, over the years, there have been several unconventional attempts to build coherent theories along such lines. The Hungarian school, headed by Károlyházy of Budapest, has, since about 1966, put forward a viewpoint in which gravitational effects would lead to some kind of **R**-procedure as a real physical phenomenon (cf. also Komar 1969). Following a somewhat different line, Phillip Pearle of Hamilton College, Clinton, NY, USA, has been suggesting, since about 1976, a non-gravitational theory in which **R** takes place as a real physical phenomenon. More recently, in 1986, an interesting new approach was put forward by Giancarlo Ghirardi, Alberto Rimini, and Tullio Weber and, following some very positive encouragement by John Bell, there have been numerous further suggestions and improvements by others.[9]

Before presenting my own preferences on the subject in the next sections,

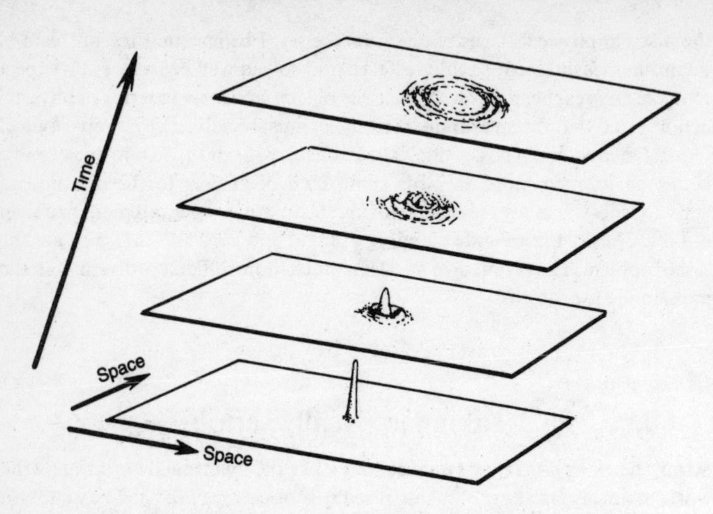

Fig. 6.1. The Schrödinger time–evolution of the wavefunction of a particle, initially localized closely at one point, subsequently spreads out in all directions.

which borrow much from the Ghirardi–Rimini–Weber (GRW) scheme, it will be helpful first to outline their proposal here. The basic idea is to accept the reality of $|\psi\rangle$ and, for the most part, the accuracy of the standard **U**-procedures. Then the wavefunction of a single, initially localized, free particle would, according to the Schrödinger equation, tend to spread outwards in all directions into space as time progresses (see Fig. 6.1). (Recall that the wavefunction of a particle describes the complex weighting factors for the different possible locations that the particle might have. We may think of the graphs of Fig. 6.1 as schematically describing the real part of this weighting factor.) Thus, as time passes, the particle becomes less and less localized. The new feature of the GRW scheme is to assume that there is a very small probability that, suddenly, this wavefunction gets itself multiplied by a strongly peaked function—known as a *Gaussian* function—with a certain spread, defined by some parameter σ. This is illustrated in Fig. 6.2. The wavefunction of the particle instantly becomes very localized, ready to begin its outward spread again. The probability that the peak of this Gaussian function finds itself to be in one place or another would be proportional to the squared modulus of the value of the wavefunction at that location. In this way, the scheme achieves consistency with the standard 'squared modulus rule' of quantum theory.

How often is this procedure taken to be applied? It would amount to roughly once every hundred million (10^8) years! Call this time period T. Then,

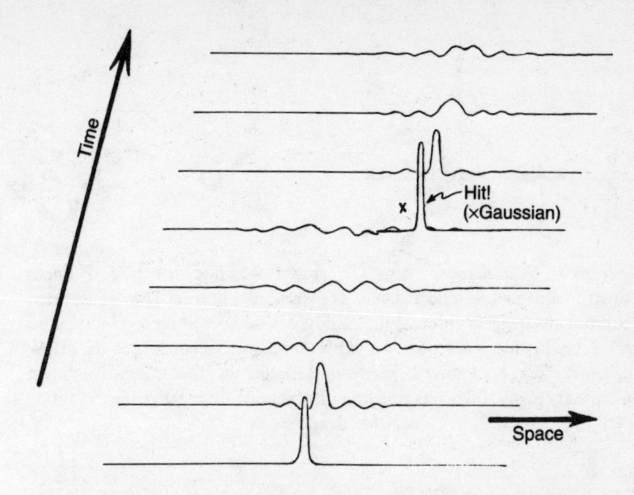

Fig. 6.2. In the original Ghirardi–Rimini–Weber (GRW) scheme, the wavefunction evolves according to standard Schrödinger **U**-evolution most of the time, but roughly once every 10^8 years (per particle) the state suffers a 'hit', having the particle's wavefunction multiplied by a peaked Gaussian function—the GRW version of **R**.

within a period of 1 second, the chance that this state reduction would happen to a particle would be less than 10^{-15} (since there are about 3×10^7 seconds in a year). Thus, for a single particle, this is not something that would be noticed at all. But now imagine that we have a reasonably large object, each of whose particles would be subject to the same process. If there were about 10^{25} particles in it (such as in a small mouse), then the chance that *some* particle in it suffers a 'hit' of this kind would be enormously increased over the chance for a single particle, and we would expect a hit to have occurred within the object in some 10^{-10} of a second. Any such hit would affect the entire state of the object, because it would be expected that the particular particle that is hit would have its state *entangled* with the rest of the object.

Let us see how to apply this idea to *Schrödinger's cat.*[10] In the Schrödinger cat paradox—essentially the basic **X**-mystery of quantum theory—we imagine a large-scale object such as a cat being placed in a quantum linear superposition of two manifestly different states, say as a live cat and a dead cat (cf. §5.1 and §6.6). This ought to be easy to do, quantum-mechanically, but the resulting situation is not really believable as a feature of the *actual* world in which we live—as Schrödinger was careful to point out (although some '$|\psi\rangle$-realists' would be taken along the many-worlds route or consciousness-inducing state-reduction route, etc., as described in §6.2 and §6.8 above). To construct a Schrödinger's cat, all we need do is to have a quantum event of a

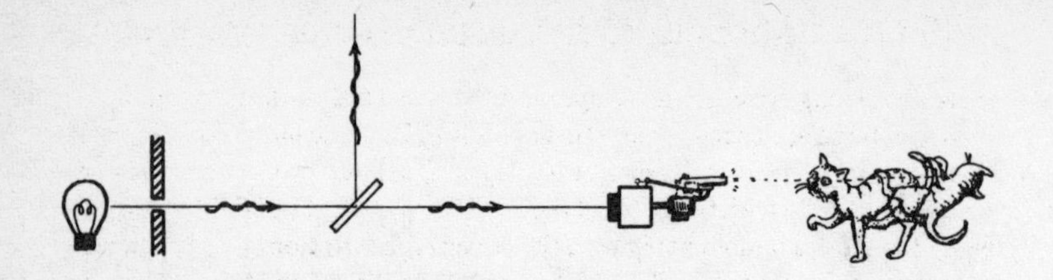

Fig. 6.3. *Schrödinger's cat*. The quantum state involves a linear superposition of a reflected and transmitted photon. The transmitted component triggers a device that kills a cat, so according to **U**-evolution the cat exists in a superposition of life and death. According to the GRW scheme, this is resolved because particles in the cat will almost instantaneously suffer hits, the first of which would localize the cat's state as *either* dead *or* alive.

suitable kind effecting a large-scale change—a *measurement* in fact. For example, we could have a single photon emitted by a source and reflected/transmitted at a half-silvered mirror (as in §5.7). Let us say that the transmitted part of the photon's wavefunction triggers a detector coupled to a device that kills the cat, but the reflected part escapes, and leaves the cat unharmed. See Fig. 6.3. Just as with the discussion of detectors as given above (§6.6), there would result an entangled state with one part involving a dead cat and the other one involving a live cat and an escaping photon. Both possibilities would be taken *together* in the state vector, so long as no reduction process (**R**) is allowed to take place. This, the mystery of 'measurement', is indeed the central **X**-mystery of quantum theory.

In the GRW scheme, however, an object as large as a cat, which would involve some 10^{27} nuclear particles, would almost instantaneously have one of its particles 'hit' by a Gaussian function (as in Fig. 6.2), and since this particle's state would be entangled with the other particles in the cat, the reduction of that particle would 'drag' the others with it, causing the entire cat to find itself in the state of either life or death. In this way, the **X**-mystery of Schrödinger's cat—and of the measurement problem in general—is resolved.

This is an ingenious scheme, but it suffers from being very *ad hoc*. There is nothing in other parts of physics to indicate such a thing, and the suggested values for T and σ are simply chosen in order to get 'reasonable' results. (Diósi (1989) has suggested a scheme resembling that of GRW where, in effect, the parameters T and σ would be fixed in terms of Newton's gravitational constant G. There is a very close link between his ideas and those I shall describe in a moment.) Another somewhat more serious difficulty with schemes of this kind is that there is a (small) violation of the principle of *conservation of energy*. This will have considerable importance for us in §6.12.

6.10 Gravitationally induced state-vector reduction?

There are strong reasons* for suspecting that the modification of quantum theory that will be needed, if some form of **R** is to be made into a *real* physical process, must involve the effects of *gravity* in a serious way. Some of these reasons have to do with the fact that the very framework of standard quantum theory fits most uncomfortably with the curved-space notions that Einstein's theory of gravity demands. Even such concepts as energy and time—basic to the very procedures of quantum theory—cannot, in a completely general gravitational context, be precisely defined consistently with the normal requirements of standard quantum theory. Recall, also, the light-cone 'tilting' effect (§4.4) that is unique to the physical phenomenon of gravity. One might expect, accordingly, that some modification of the basic principles of quantum theory might arise as a feature of its (eventual) appropriate union with Einstein's general relativity.

Yet most physicists seem reluctant to accept the possibility that it might be the *quantum* theory that requires modification for such a union to be successful. Instead, they argue, Einstein's theory itself should be modified. They may point, quite correctly, to the fact that classical general relativity has its own problems, since it leads to *space–time singularities*, such as are encountered in black holes and the big bang, where curvatures mount to infinity and the very notions of space and time cease to have validity (see ENM, Chapter 7). I do not myself doubt that general relativity must itself be modified when it is appropriately unified with quantum theory. And this will indeed be important for the understanding of what *actually* takes place in those regions that we presently describe as 'singularities'. But it does not absolve quantum theory from a need for change. We saw in §4.5 that general relativity is an extraordinarily accurate theory—no less accurate than is quantum theory itself. Most of the physical insights that underlie Einstein's theory will surely survive, not less than will most of those of quantum theory, when the appropriate union that moulds these two great theories together is finally found.

Many who might agree with this, however, would still argue that the relevant scales in which *any* form of quantum gravity might become relevant would be totally inappropriate for the quantum measurement problem. They would point to the length scale that characterizes quantum gravity, called the *Planck scale*, of 10^{-33}cm, which is some 20 orders of magnitude smaller even than a nuclear particle; and they would severely question how the physics at such tiny distances could have anything to do with the measurement problem which, after all, is concerned with phenomena that (at least) border upon the

*In ENM, Chapters 7 and 8, I presented such reasons in some detail, and there is no need to repeat the arguments here. Suffice it to say that these reasons still stand—although the specific criterion of §6.12 differs from that given in ENM (on pp. 367–71).

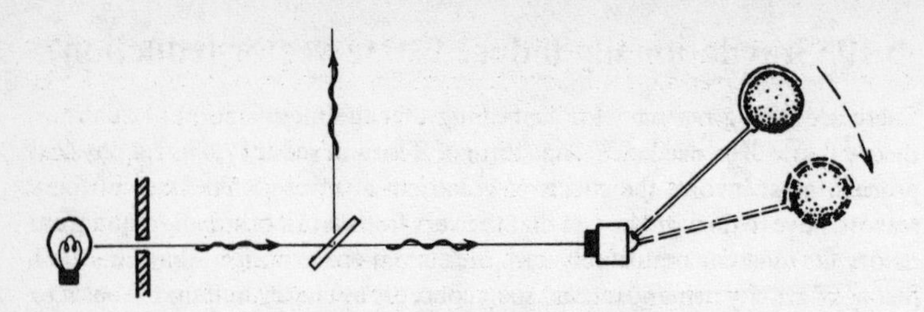

Fig. 6.4. Instead of having a cat, the measurement could consist of the simple movement of a spherical lump. How big or massive must the lump be, or how far must it move, for **R** to take place?

macroscopic domain. However, there is a misconception here as to how quantum gravity ideas might be applied. For 10^{-33} cm is indeed relevant, but not in the way that first comes to mind.

Let us consider a type of situation, somewhat like that of Schrödinger's cat, in which one strives to produce a state in which a pair of macroscopically distinguishable alternatives are linearly superposed. For example, in Fig. 6.4 such a situation is depicted, where a photon impinges upon a half-silvered mirror, the photon state becoming a linear superposition of a transmitted and reflected part. The transmitted part of the photon's wavefunction activates (or would activate) a device which moves a macroscopic spherical lump (rather than a cat) from one spatial location to another. So long as Schrödinger evolution **U** holds good, the 'location' of the lump involves a quantum superposition of its being in the original position with its being in the displaced position. If **R** were to come into effect as a real physical process, then the lump would 'jump' either to one position or to the other—and this would constitute an actual 'measurement'. The idea here is that, as with the GRW theory, this is indeed an entirely objective physical process, and it would occur whenever the mass of the lump is large enough or the distance it moves, far enough. (In particular, it would have nothing to do with whether or not a conscious being might happen to have actually 'perceived' the movement, or otherwise, of the lump.) I am imagining that the *device* that detects the photon and moves the lump is itself small enough that it can be treated entirely quantum-mechanically, and it is only the lump that registers the measurement. For example, in an extreme case, we might simply imagine that the lump is poised sufficiently unstably that the mere impact of the photon would be enough to cause it to move away significantly.

Applying the standard **U**-procedures of quantum mechanics, we find that the photon's state, after it has encountered the mirror, would consist of two parts in two very different locations. One of these parts then becomes entangled with the device and finally with the lump, so we have a quantum state which involves a linear superposition of two quite different positions for the lump. Now the lump will have its gravitational field, which must also be involved in this superposition. Thus, the state involves a superposition of two different gravitational fields. According to Einstein's theory, this implies that we have two different space–time geometries superposed! The question is: is there a point at which the two geometries become sufficiently different from each other that the rules of quantum mechanics must change, and rather than forcing the different geometries into superposition, Nature chooses between one or the other of them and *actually* effects some kind of reduction procedure resembling **R**?

The point is that we really have no conception of how to consider linear superpositions of states when the states themselves involve different space–time geometries. A fundamental difficulty with 'standard theory' is that when the geometries become significantly different from each other, we have no absolute means of identifying a point in one geometry with any particular point in the other—the two geometries are strictly *separate* spaces—so the very idea that one could form a superposition of the *matter* states within these two separate spaces becomes profoundly obscure.

Now, we should ask when *are* two geometries to be considered as actually 'significantly different' from one another? It is *here*, in effect, that the Planck scale of 10^{-33} cm comes in. The argument would roughly be that the scale of the difference between these geometries has to be, in an appropriate sense, something like 10^{-33} cm or more for reduction to take place. We might, for example, attempt to imagine (Fig. 6.5) that these two geometries are trying to be forced into coincidence, but when the measure of the difference becomes too large, on this kind of scale, reduction **R** takes place—so, rather than the superposition involved in **U** being maintained, Nature must choose one geometry or the other.

What kind of scale of mass or of distance moved would such a tiny change in geometry correspond to? In fact, owing to the smallness of gravitational effects, this turns out to be quite large, and not at all unreasonable as a demarcation line between the quantum and classical levels. In order to get a feeling for such matters, it will be useful to say something about *absolute* (or *Planckian*) *units*.

6.11 Absolute units

The idea (due originally* to Max Planck (1906) and followed up particularly by John A. Wheeler (1975)) is to use the three most fundamental constants of

*A very similar idea was put forward 25 years earlier by the Irish physicist George Johnstone Stoney (1881), where the charge on the electron rather than Planck's constant (the latter being unknown at the time) was taken as the basic unit. (I am grateful to John Barrow for pointing this out to me.)

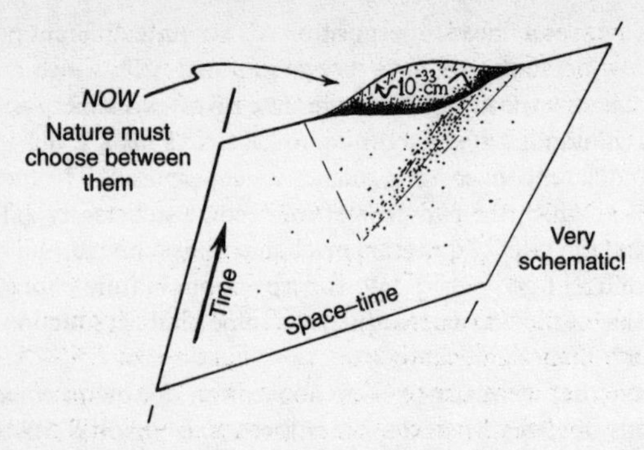

Fig. 6.5. What is the relevance of the Planck scale of 10^{-33} cm to quantum state reduction? Rough idea: when there is sufficient mass movement between the two states under superposition such that the two resulting space–times differ by something of the order of 10^{-33} cm.

Nature, the speed of light c, Planck's constant (divided by 2π) $\hbar$, and Newton's constant of gravitation G, as units for converting all physical measures into pure (dimensionless) numbers. What this amounts to is choosing units of length, mass, and time so that these three constants all take the value unity:

$$c = 1, \hbar = 1, G = 1.$$

The Planck scale 10^{-33} cm, which in ordinary units one would express as $(G\hbar/c^3)^{1/2}$, now simply takes the value 1, so *it* is the absolute unit of *length*. The corresponding absolute unit of *time*, which is the time that light would take to traverse a Planck distance, is the *Planck time* $(=(G\hbar/c^5)^{1/2})$, about 10^{-43} seconds. There is also an absolute unit of *mass*, referred to as the *Planck mass* $(=(\hbar c/G)^{1/2})$, which is about 2×10^{-5} grams, a very large mass from the point of view of normal quantum phenomena, but fairly small in ordinary terms—something like the mass of a flea.

Clearly these are not very practical units, except possibly for the Planck mass, but they are very useful when we consider effects that might have relevance in relation to quantum gravity. Let us see how some of the more relevant physical quantities appear, very roughly, in absolute units:

$$\text{second} = 1.9 \times 10^{43}$$
$$\text{day} = 1.6 \times 10^{48}$$
$$\text{year} = 5.9 \times 10^{50}$$
$$\text{metre} = 6.3 \times 10^{34}$$
$$\text{cm} = 6.3 \times 10^{32}$$
$$\text{micron} = 6.3 \times 10^{28}$$
$$\text{fermi ('strong interaction size')} = 6.3 \times 10^{19}$$
$$\text{mass of nucleon} = 7.8 \times 10^{-20}$$

gram $= 4.7 \times 10^{4}$
erg $= 5.2 \times 10^{-17}$
degree kelvin $= 4 \times 10^{-33}$
density of water $= 1.9 \times 10^{-94}$

6.12 The new criterion

I shall now give a new criterion[11] for gravitationally induced state-vector reduction that differs significantly from that suggested in ENM, but which is close to some recent ideas due to Diósi and others. The *motivations* for a link between gravity and the **R**-procedure, as given in ENM, still hold good, but the suggestion that I am now making has some additional theoretical support from other directions. Moreover, it is free of some of the conceptual problems involved in the earlier definition, and it is much easier to use. The proposal in ENM was for a criterion, according to which two states might be judged (with regard to their respective gravitational fields—i.e. their respective space–times) to be too different from one another for them to be able to co-exist in quantum linear superposition. Accordingly, **R** would have to take place at that stage. The present idea is a little different. We do not seek an absolute measure of gravitational difference between states which determines when the states differ too much from each other for superposition to be possible. Instead, we regard superposed widely differing states as *unstable*—rather like an unstable uranium nucleus, for example—and we ask that there be a *rate* of state-vector reduction determined by such a difference measure. The greater the difference, the faster would be the rate at which reduction takes place.

For clarity, I shall first apply the new criterion in the particular situation that was described in §6.10 above, but it can also be easily generalized to cover many other examples. Specifically, we consider the *energy* that it would take, in the situation above, to displace one instance of the lump away from the other, taking just *gravitational* effects into account. Thus, we imagine that we have two lumps (masses), initially in coincidence and interpenetrating one another (Fig. 6.6), and we then imagine moving one instance of the lump away from the other, slowly, reducing the degree of interpenetration as we go, until the two reach the state of separation that occurs in the superposed state under consideration. Taking the reciprocal of the gravitational energy that this operation would cost us, measured in absolute units,* we get the approximate time, also in absolute units, that it would take before state reduction occurs,

*We may prefer to express this reduction time in more usual units than the absolute units adopted here. In fact, the expression for the reduction time is simply $\hbar/E$, where E is the gravitational separation energy referred to above, with no other absolute constant appearing other than $\hbar$. The fact that the speed of light c does not feature suggests that a 'Newtonian' model theory of this nature would be worth investigating, as has been done by Christian (1994).

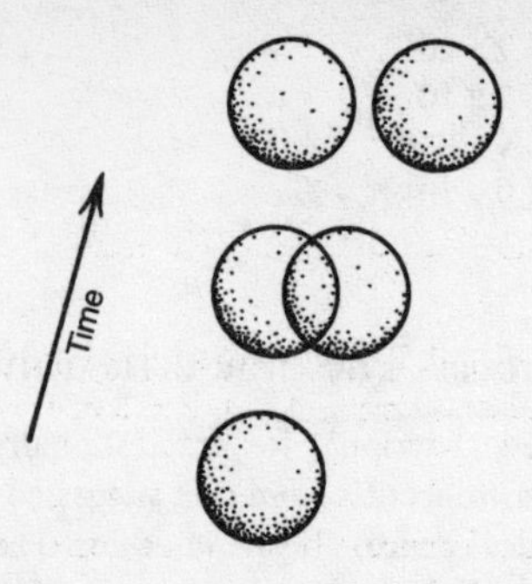

Fig. 6.6. To compute the reduction time $\hbar/E$, imagine moving one instance of the lump away from the other and compute the energy E that this would cost, taking into account only their gravitational attraction.

whereupon the lump's superposed state would spontaneously jump into one localized state or the other.

If we take the lump to be spherical, with mass m and radius a, we obtain a quantity of the general order of m^2/a for this energy. In fact the actual value of the energy depends upon how far the lump is moved away, but this distance is not very significant provided that the two instances of the lump do not (much) overlap when they reach their final displacement. The additional energy that would be required in moving away from the contact position, even all the way out to infinity, is of the same order ($\frac{5}{7}$ times as much) as that involved in moving from coincidence to the contact position. Thus, as far as orders of magnitude are concerned, one can ignore the contribution due to the displacement of the lumps away from each other after separation, provided that they actually (essentially) separate. The reduction time, according to this scheme, is of the order of

$$\frac{a}{m^2}$$

measured in absolute units, or, very roughly,

$$\frac{1}{20\rho^2 a^5},$$

where ρ is the density of the lump. This gives about $10^{186}/a^5$, for something of ordinary density (say, a water droplet).

It is reassuring that this provides very 'reasonable' answers in certain simple situations. For example, in the case of a nucleon (neutron or proton), where we take a to be its 'strong interaction size' 10^{-13} cm, which in absolute units is nearly 10^{20}, and we take m to be about 10^{19}, we get a reduction time of nearly 10^{58}, which is over ten million years. It is reassuring that this time is large,

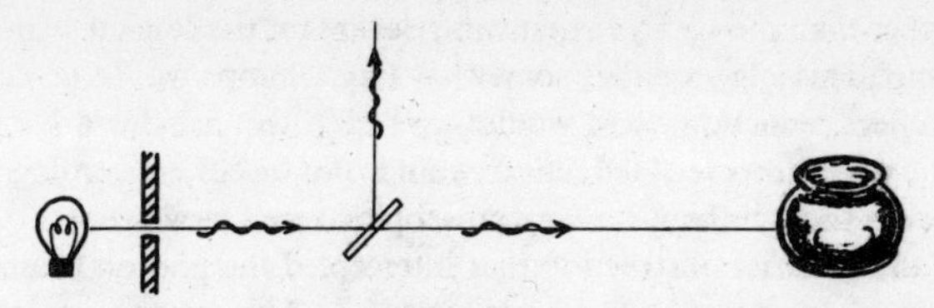

Fig. 6.7. Suppose that instead of moving a lump, the transmitted part of the photon's state is simply absorbed in a body of fluid matter.

because quantum interference effects have been directly observed for individual neutrons.[12] Had we obtained a very short reduction time, this would have led to conflict with such observations.

If we consider something more 'macroscopic', say a minute speck of water of radius 10^{-5} cm, we get a reduction time measured in hours; if the speck were of radius 10^{-4} cm (a micron), the reduction time, according to this scheme, would be about a twentieth of a second; if of radius 10^{-3} cm, then less than a millionth of a second. In general, when we consider an object in a superposition of two spatially displaced states, we simply ask for the energy that it would take to effect this displacement, considering only the gravitational interaction between the two. The reciprocal of this energy measures a kind of 'half-life' for the superposed state. The larger this energy, the shorter would be the time that the superposed state could persist.

In an actual experimental situation, it would be very hard to keep the quantum-superposed lumps from disturbing—and becoming entangled with—the material in the surrounding environment, in which case we would have to consider the gravitational effects involved in this environment also. This would be relevant even if the disturbance did not result in significant macroscopic-scale movements of mass in the environment. Even the tiny displacements of individual particles could well be important—though normally at a somewhat larger total scale of mass than with a macroscopic 'lump' movement.

In order to clarify the effect that a disturbance of this kind might have in the present scheme, let us replace the lump-moving device, in the above idealized experimental situation, by a lump of fluid matter that simply *absorbs* the photon, if it is transmitted through the mirror (Fig. 6.7), so now the lump itself is playing the role of the 'environment'. Instead of having to consider a linear superposition between two states that are macroscopically distinct from each other, by virtue of one instance of the lump having been moved bodily with respect to the other, we are now just concerned with the difference between two configurations of atomic positions, where one configuration of particles is displaced randomly from the other. For a lump of ordinary fluid material of radius a, we may now expect to find a reduction time that is perhaps something of the order of $10^{130}/a^3$ (depending, to some extent, on what assumptions are made) rather than the $10^{186}/a^5$ that was relevant for the collective movement of the lump. This suggests that somewhat larger lumps would be needed in order to effect reduction, than would have been the case for a lump being bodily moved. However, reduction would *still* occur, according to this scheme, even though there is no macroscopic overall movement.

Recall the material obstruction that intercepted the photon beam, in our discussion of quantum interference in §5.8. The mere *absorption*—or potentiality for absorption—of a photon by such an obstruction would be sufficient to effect **R**, despite the fact that nothing macroscopic would occur that is actually observable. This also shows how a sufficient disturbance in an environment that is *entangled* with some system under consideration would itself effect **R** and so contact is made with more conventional FAPP procedures.

Indeed, in almost any practical measuring process it would be very likely that large numbers of microscopic particles in the surrounding environment would become disturbed. According to the ideas being put forward here, it would often be *this* that would be the dominant effect, rather than macroscopic bodily movement of objects, as with the 'lump displacement' initially described above. Unless the experimental situation is very carefully controlled, any macroscopic movement of a sizeable object would disturb large amounts of the surrounding environment, and it is probable that the *environment's* reduction time—perhaps about $10^{130}/b^3$, where b is the radius of the region of water-density entangled environment under consideration—would dominate (i.e. be much smaller than) the time $10^{186}/a^5$ that might be relevant to the object itself. For example, if the radius b of disturbed environment were as little as about a tenth of a millimetre, then reduction would take place in something of the order of a millionth of a second for that reason alone.

Such a picture has a lot in common with the conventional description that was discussed in §6.6, but now we have a *definite* criterion for **R** *actually* to occur in this environment. Recall the objections that were raised in §6.6 to the conventional FAPP picture, as a description of an actual physical reality. With a criterion such as the one being promoted here, these objections no longer hold. Once there is a sufficient disturbance in the environment, according to the present ideas, reduction will rapidly *actually* take place in that environment—and it would be accompanied by reduction in any 'measuring apparatus' with which that environment is entangled. Nothing could reverse that reduction and enable the original entangled state to be resurrected, even imagining enormous advances in technology. Accordingly, there is no contradiction with the measuring apparatus actually registering *either* **YES** *or* **NO**—as in the present picture it would indeed do.

I imagine that a description of this nature would be relevant in many biological processes, and it would provide the likely reason that biological structures of a size much smaller than a micron's diameter can, no doubt, often behave as classical objects. A biological system, being very much entangled with its environment in the manner discussed above, would have its *own* state continually reduced because of the continual reduction of this *environment*. We may imagine, on the other hand, that for some reason it might be favourable to a biological system that its state remain *un*reduced for a long time, in appropriate circumstances. In such cases it would be necessary for the system to be, in some way, very effectively insulated from its surroundings. These considerations will be important for us later (§7.5).

A point that should be emphasized is that the energy that defines the lifetime of the superposed state is an energy *difference*, and not the *total*, (mass–) energy that is involved in the situation as a whole. Thus, for a lump that is quite large but does not move very much—and supposing that it is also crystalline, so that its individual atoms do not get randomly displaced—quantum superpositions could be maintained for a long time. The lump could be much larger than the water droplets considered above. There could also be other very much larger masses in the vicinity, provided that they do not get significantly entangled with the superposed state we are concerned with. (These considerations would be important for solid-state devices, such as gravitational wave detectors, that use coherently oscillating solid—perhaps crystalline—bodies.[13])

So far, the orders of magnitude seem to be quite plausible, but clearly more work is needed to see whether the idea will survive more stringent examination. A crucial test would be to find experimental situations in which standard theory would predict effects dependent upon large-scale quantum superpositions, but at a level where the present proposals demand that such superpositions cannot be maintained. If the conventional quantum expectations are supported by observation, in such situations, then the ideas that I am promoting would have to be abandoned—or at least severely modified. If observation indicates that the superpositions are not maintained, then that would lend some support to the present ideas. Unfortunately, I am not aware of any practical suggestions for appropriate experiments, as yet. Superconductors, and devices such as SQUIDs (that depend upon the large-scale quantum superpositions that occur with superconductors), would seem to present a promising area of experiment relevant to these issues (see Leggett 1984). However, the ideas I am promoting would need some further development before they could be directly applied in these situations. With superconductors, very little mass displacement occurs between the different superposed states. There is a significant *momentum* displacement instead, however, and the present ideas would need some further theoretical development in order to cover this situation.

The ideas put forward above would need to be reformulated somewhat, even to treat the simple situation of a cloud chamber—where the presence of a charged particle is signalled by the condensation of small droplets out of the surrounding vapour. Suppose that we have a quantum state of a charged particle consisting of a linear superposition of the particle being at some location in the cloud chamber and of the particle being outside the chamber. The part of the particle's state vector that is within the chamber initiates the formation of a droplet, but the part for which the particle is outside does not, so that state now consists of a superposition of two macroscopically different states. In one of these there is a droplet forming from the vapour and in the other, there is just uniform vapour. We need to estimate the gravitational energy involved in pulling the vapour molecules away from their counterparts, in the two states that are being considered in superposition. However, now there is an additional complication because there is also a difference between the gravitational *self*-energy of the droplet and of the uncondensed vapour. In order to cover such situations, a *different* formulation of the above suggested criterion may be appropriate. We can now consider the *gravitational self-energy* of that mass distribution which is the *difference* between the mass distributions of the two states that are to be considered in quantum linear superposition. The reciprocal of this self-energy gives an alternative proposal for the reduction timescale (cf. Penrose 1994*b*). In fact, this alternative formulation gives just the same result as before, in those situations considered earlier, but it gives a somewhat different (more rapid) reduction time in the case of a cloud chamber. Indeed, there are various alternative general schemes for reduction times, giving different answers in certain situations, although agreeing with each other for the simple two-state superposition involving a rigid displacement of a lump, as envisaged at the beginning of this section. The original such scheme is that of Diósi (1989) (which encountered some difficulties, as pointed out by Ghirardi, Grassi, and Rimini (1990), who also suggested a remedy). I shall not distinguish between these various proposals here, which will all come under the heading of 'the proposal of §6.12' in the following chapters.

What are the motivations for the specific proposal for a 'reduction time' that is being put forward here? My own initial motivations (Penrose 1993a) were rather too technical for me to describe here and were, in any case, inconclusive and incomplete.[14] I shall present an independent case for this kind of physical scheme in a moment. Though also incomplete as it stands, the argument seems to suggest a powerful underlying consistency requirement that gives additional support for believing that state reduction must ultimately be a gravitational phenomenon of the general nature of the one being proposed here.

The problem of *energy conservation* in the GRW-type schemes was already referred to in §6.9. The 'hits' that particles are involved in (when their wavefunctions get spontaneously multiplied by Gaussians) cause small violations of energy conservation to take place. Moreover, there seems to be a non-local transfer of energy with this kind of process. This would appear to be a characteristic—and apparently unavoidable feature—of theories of this general type, where the **R**-procedure is taken to be a *real* physical effect. To my mind, this provides strong additional evidence for theories in which *gravitational* effects play a crucial role in the reduction process. For energy conservation in general relativity is a subtle and elusive issue. The gravitational field itself contains energy, and this energy measurably contributes to the total energy (and therefore to the mass, by Einstein's $E=mc^2$) of a system. Yet it is a nebulous energy that inhabits empty space in a mysterious non-local way.[15] Recall, in particular, the mass–energy that is carried away, in the form of gravitational waves, from the binary pulsar system PSR 1913+16 (cf. §4.5); these waves are ripples in the very structure of empty space. The energy that is contained in the mutual attractive fields of the two neutron stars is also an important ingredient of their dynamics which cannot be ignored. But this kind of energy, residing in empty space, is of an especially slippery kind. It cannot be obtained by the 'adding up' of local contributions of energy density; nor can it even be localized in any particular region of space–time (see ENM, pp. 220–1). It is tempting to relate the equally slippery non-local energy problems of the **R**-procedure to those of classical gravity, and to offset one against the other so as to provide a coherent overall picture.

Do the suggestions that I have been putting forward here achieve this overall coherence? I believe that there is a very good chance that they can be made to, but the precise framework for achieving this is not yet to hand. One can see, however, that there is certainly good scope for this in principle. For, as mentioned earlier, we can think of the reduction process as something rather like the decay of an unstable particle or nucleus. Think of the superposed state of a lump in two different locations as being like an unstable nucleus that decays, after a characteristic 'half-life' timescale, into something else more stable. In the case of the superposed lump locations we likewise think of an unstable quantum state which decays, after a characteristic lifetime (given, roughly on average, by the reciprocal of the gravitational energy of separation), to a state where the lump is in one location or the other—representing two possible decay modes.

Now in the decays of particles or nuclei, the lifetime (say, the half-life) of the decay process is the reciprocal of a small *uncertainty* in mass–energy of the initial particle—an effect of Heisenberg's uncertainty principle. (For example, the mass of an unstable polonium-210 nucleus, which decays on the emission of an α-particle into a lead nucleus, is not precisely defined, with an uncertainty that is of the order of the reciprocal of the decay time—in this case, about 138 days, giving a mass uncertainty of only about 10^{-34} of the mass of the polonium nucleus! For individual unstable particles, however, the uncertainty is a much larger proportion of the mass.) Thus, the 'decay' that is involved in the reduction process should *also* involve an essential uncertainty in the energy of the initial state. This uncertainty, according to the present proposal, lies essentially in the uncertainty of the gravitational self-energy of the superposed state. Such a gravitational self-energy involves the nebulous non-local field energy that causes so much difficulty in general relativity, and which is not provided by the adding up of local energy density contributions. It also involves the essential uncertainty of identifying the points in two superposed different space–time geometries referred to in §6.10. If we indeed regard this gravitational contribution as representing an *essential* 'uncertainty' in the energy of the superposed state, then we get agreement with the lifetime for that state that is being proposed here. Thus, the present scheme seems to provide a clear link of consistency between the two energy problems, and is at least highly suggestive of the possibility that a fully coherent theory might eventually be found along these lines.

Finally, there are two important questions that are of particular relevance for us here. First: what possible roles might such considerations have for *brain* action? Second: are there any reasons to expect, on purely physical grounds, that *non-computability* (of the appropriate kind) might be a feature of this gravitationally induced reduction process? We shall find, in the next chapter, that there are indeed some intriguing possibilities.

Notes and references

1. There is a certain sense in which the 'bosonic' property of photons referred to in §5.16 may be regarded as an example of quantum entanglement in which case the observations of Hanbury Brown and Twiss (1954, 1956) indeed provide a confirmation over large distances (cf. footnote on p. 290).
2. Everett (1957), Wheeler (1957), DeWitt and Graham (1973), Geroch (1984).
3. Squires (1990, 1992*b*).
4. Bell (1992).
5. For a different argument supporting the objective reality of the wavefunction, see Aharonov, Anandan, and Vaidman (1993).
6. See, for example, d'Espagnat (1989).
7. See d'Espagnat (1989), Zurek (1991, 1993), Paz, Habib, and Zurek (1993).
8. This seems to be the conclusion of the SETI programme of F. Drake.
9. My own suggestions, although being firmly in the 'gravitational' camp, have not been very specific until recently, cf. Penrose (1993*a*, 1994*b*). This proposal shares with the original Ghirardi–Rimini–Weber proposal the idea that the reduction should be a sudden discontinuous process. Much of the current activity, however, is concerned with a *continuous* (stochastic) state reduction process, like the original one of Pearle (1976). See Diósi (1992), Ghirardi *et al.* (1990*b*), Percival (1994). For work of this nature concerned with making the scheme consistent with relativity, see Ghirardi *et al.* (1992), Gisin (1989), Gisin and Percival (1993).
10. Schrödinger (1935*a*); cf. also ENM, pp. 290–6.
11. See also Diósi (1989), Ghirardi *et al.* (1990*a*), Penrose (1993*a*).
12. Zeilinger *et al.* (1988).
13. Weber (1960), Braginski (1977).
14. However, the general motivations given in ENM, Chapter 7, would seem to support the proposal being promoted here (and as suggested in Penrose 1993*a*)

rather more clearly than they support the 'one-graviton criterion' as given in ENM. Further research is needed in order to make the connections more specific.

15. See Penrose (1991*a*); also ENM pp. 220–1.

7

Quantum theory and the brain

7.1 Large-scale quantum action in brain function?

Brain action, according to the conventional viewpoint, is to be understood in terms of essentially classical physics—or so it would seem. Nerve signals are normally taken to be 'on or off' phenomena, just as are the currents in the electronic circuits of a computer, which *either* take place *or* do not take place—with none of the mysterious *superpositions* of alternatives that are characteristic of quantum actions. Whilst it would be admitted that, at *underlying* levels, quantum effects must have their roles to play, biologists seem to be generally of the opinion that there is no necessity to be forced out of a classical framework when discussing the large-scale implications of those primitive quantum ingredients. The chemical forces that control the interactions of atoms and molecules are indeed quantum mechanical in origin, and it is largely chemical action that governs the behaviour of the *neurotransmitter* substances that transfer signals from one neuron to another—across tiny gaps that are called *synaptic clefts*. Likewise, the action potentials that physically control nerve-signal transmission itself have an admittedly quantum-mechanical origin. Yet it seems to be generally assumed that it is quite adequate to model the behaviour of neurons themselves, and their relationships with one another, in a completely classical way. It is widely believed, accordingly, that it should be entirely appropriate to model the physical functioning of the brain as a whole, as a *classical* system, where the more subtle and mysterious features of quantum physics do not significantly enter the description.

This would have the implication that any possible significant activity that might take place in a brain is indeed to be taken as *either* 'occurring' *or* 'not occurring'. The strange *superpositions* of quantum theory, that would allow *simultaneous* 'occurring' *and* 'not occurring'—with complex-number weighting factors—would, accordingly, be considered to play no significant role. Whilst it might be accepted that at some submicroscopic level of activity such quantum superpositions do 'really' take place, it would be felt that the interference effects that are characteristic of such quantum phenomena would have no role at the relevant larger scales. Thus, it would be considered adequate to treat any such superpositions as though they were statistical mixtures, and the classical modelling of brain activity would be perfectly satisfactory FAPP.

There are certain dissenting opinions from this, however. In particular, the renowned neurophysiologist John Eccles has argued for the importance of quantum effects in synaptic action (see, in particular, Beck and Eccles (1992), Eccles (1994)). He points to the presynaptic vesicular grid—a paracrystalline hexagonal lattice in the brain's pyramidal cells—as being an appropriate quantum site. Also, some people (even including myself; cf. ENM, pp. 400–401, and Penrose 1987) have tried to extrapolate from the fact that light-sensitive cells in the retina (which is technically part of the brain) can respond to a small number of photons (Hecht *et al.* 1941)—sensitive even to a *single* photon (Baylor *et al.* 1979), in appropriate circumstances—and to speculate that there might be neurons in the brain, proper, that are also essentially quantum detection devices.

With the possibility that quantum effects might indeed trigger much larger activities within the brain, some people have expressed the hope that, in such circumstances, *quantum indeterminacy* might be what provides an opening for the *mind* to influence the physical brain. Here, a *dualistic* viewpoint would be likely to be adopted, either explicitly or implicitly. Perhaps the 'free will' of an 'external mind' might be able to influence the quantum choices that actually result from such non-deterministic processes. On this view, it is presumably through the action of quantum theory's **R**-process that the dualist's 'mind-stuff' would have its influence on the behaviour of the brain.

The status of such suggestions is unclear to me, especially since, in standard quantum theory, quantum indeterminacy does *not* occur at quantum-level scales, since it is the deterministic **U**-evolution that always holds at this level. It is only in the magnification process from the quantum to the classical levels that the indeterminacy of **R** is deemed to occur. On the standard FAPP viewpoint, this indeterminacy is something that 'takes place' only when sufficient amounts of the environment become entangled with the quantum event. In fact, as we have seen in §6.6, on the standard view it is not even clear what 'taking place' actually means. It would be hard, on conventional quantum-physical grounds, to maintain that the theory does actually allow an indeterminacy to occur just at the level where a single quantum particle, such as a photon, atom, or small molecule, is critically involved. When (for example) a photon's wavefunction encounters a photon-sensitive cell, it sets in train a sequence of events that remains deterministic (action of **U**) so long as the system can be considered to stay 'at the quantum level'. Eventually, significant amounts of the environment become disturbed and, on the conventional view, one considers that **R** has occurred FAPP. One would have to contend that the 'mind-stuff' somehow influences the system only at this indeterminate stage.

According to the viewpoint on state reduction that I have myself been promoting in this book (cf. §6.12), to find the level at which the **R**-process *actually* becomes operative, we must look to the quite large scales that become relevant when considerable amounts of material (microns to millimetres in diameter—or perhaps a good deal more, if no significant mass movement is involved) become entangled in the quantum state. (I shall henceforth denote this fairly specific but putative procedure by **OR**, which stands for *objective reduction**. In any case, if we try to adhere to the above dualist viewpoint, where we are looking for somewhere where an external 'mind' might have an influence on physical behaviour—presumably by replacing the pure randomness of quantum theory by something more subtle—then we must indeed find how the 'mind's' influence could enter at a much larger scale than single quantum particles. We must look to wherever the cross-over point occurs between the quantum and classical levels. As we have seen in the previous chapter, there is no general agreement about what, whether, or where such a cross-over point might be.

In my own opinion, it is not very helpful, from the scientific point of view, to think of a dualistic 'mind' that is (logically) *external* to the body, somehow influencing the choices that seem to arise in the action of **R**. If the 'will' could somehow influence Nature's choice of alternative that occurs with **R**, then why is an experimenter not able, by the action of 'will power', to influence the result of a quantum experiment? If this were possible, then violations of the quantum probabilities would surely be rife! For myself, I cannot believe that such a picture can be close to the truth. To have an external 'mind-stuff' that is not itself subject to physical laws is taking us outside anything that could be reasonably called a scientific explanation, and is resorting to the viewpoint $\mathscr{D}$ (cf. §1.3).

It is hard, however, to argue against such a viewpoint in a rigorous way, since by its very nature it is devoid of the clear rules that would allow it to be subject to scientific argument. Those readers who, for whatever reason, retain a conviction (viewpoint $\mathscr{D}$) that science must remain forever incompetent to address issues of the mind, I ask merely that they continue to bear with me to see what room there might eventually be found within a science which will undoubtedly become extended far beyond the limited scope that it admits today. If the 'mind' is something quite external to the physical body, it is hard to see why so many of its attributes can be very closely associated with properties of a physical brain. My own viewpoint is that we must search more

*In ENM I used the description 'correct quantum gravity'—abbreviated CQG—for this kind of thing. Here the emphasis is a little different. I do not wish to stress the connection of this procedure with the profound problem of finding a fully coherent theory of quantum gravity. The emphasis is more on a procedure that would be in line with the specific suggestions put forward in §6.12, but together with some fundamental missing non-computational ingredient. The use of the acronym **OR** has the additional point that, in an objective reduction procedure, the physical result is indeed one thing *or* the other, in place of the combined superposition that had occurred before.

deeply within the actual physical 'material' structures that constitute brains—and also more deeply into the very question of what a 'material' structure, at the quantum level of things, actually *is*! In my view, there is ultimately no escape from having to probe more profoundly into the truths that actually lie at the roots of Nature.

Be that as it may, at least one thing seems to be clear. We must not look simply to the quantum effects of single particles, atoms, or even small molecules, but to the effects of quantum systems that retain their manifest quantum nature at a much larger scale. If there is no large-scale quantum coherence involved, then there would be no chance of any of the subtle quantum-level effects, such as non-locality, quantum parallelism (several superposed actions being carried out simultaneously), or effects of counterfactuality, having any significance when the classical level of brain activity is reached. Without adequate 'shielding' of the quantum state from its environment, such effects would become immediately lost in the randomness inherent in that environment—i.e. in the random motions of those biological materials and fluids that constitute the bulk of the brain.

What is *quantum coherence*? This phenomenon refers to circumstances when large numbers of particles can collectively cooperate in a single quantum state which remains essentially unentangled with its environment. (The word 'coherence' refers, generally, to the fact that oscillations at different places beat time with one another. Here, with *quantum* coherence, we are concerned with the oscillatory nature of the wavefunction, and the coherence refers to the fact that we are dealing with a single quantum state.) Such states occur most dramatically in the phenomena of superconductivity (where electrical resistance drops to zero) and superfluidity (where fluid friction, or viscosity, drops to zero). The characteristic ingredient of such phenomena is the presence of an *energy gap* that has to be breached by the environment if it is to disturb this quantum state. If the temperature in that environment is too high, so that the energy of many of the ambient particles is great enough for them to breach this gap and entangle with the state, then the quantum coherence is destroyed. Consequently, phenomena of the nature of superconductivity and superfluidity have been found normally to occur only at very low temperatures, just a few degrees above absolute zero. For reasons such as this, there had been a general scepticism about the possibility of quantum coherence effects having any relevance to such a 'hot' object as the human brain—or, indeed, any other biological system.

In recent years, however, some remarkable experimental findings have shown that, with suitable substances, superconductivity can occur at very much higher temperatures, even up to 115 K (cf. Sheng *et al.* 1988). This is still very cold, from the biological point of view, being about −158°C or −212°F, only a little warmer than liquid nitrogen. But even more remarkable still are the observations of Laguës *et al.* (1993) which seem to indicate the presence of superconductivity at the merely 'Siberian' temperatures −23°C or −10°F.

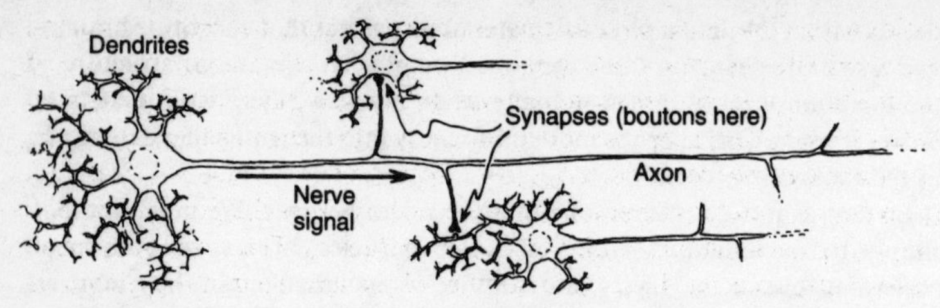

Fig. 7.1. A sketch of a neuron, connected to some others via synapses.

Though still somewhat on the 'cold' side, in biological terms, such *high-temperature superconductivity* gives some strong support to the conjecture that there could be quantum-coherent effects that are indeed relevant to biological systems.

In fact, long before the phenomenon of high-temperature superconductivity was observed, the distinguished physicist Herbert Fröhlich (who, in the 1930s had made one of the fundamental breakthroughs in the understanding of 'ordinary' low-temperature superconductivity) suggested a possible role for collective quantum effects in biological systems. This work was stimulated by a puzzling phenomenon that had been observed in biological membranes as far back as 1938, and Fröhlich was led to propose, in 1968 (employing a concept due to my brother Oliver Penrose and Lars Onsager (1956)—as I learnt to my surprise when looking into these matters), that there should be vibrational effects within active cells, which would resonate with microwave electromagnetic radiation, at 10^{11} Hz, as a result of a biological quantum coherence phenomenon. Instead of needing a low temperature, the effects arise from the existence of a large energy of metabolic drive. There is now some respectable observational evidence, in many biological systems, for precisely the kind of effect that Fröhlich had predicted in 1968. We shall try to see later (§7.5) what relevance this might have to brain action.

7.2 Neurons, synapses, and computers

So far, although it is encouraging to find a distinct possibility of quantum coherence having a genuinely significant role to play in biological systems, there is, as yet, no clear link between this and what might be directly relevant in brain activity. Much of our understanding of the brain, though still very rudimentary, had led us to a classical picture (essentially one put forward by McCullogh and Pitts back in 1943) in which neurons and their connecting synapses seem to play a role essentially similar to those of transistors and wires (printed circuits) in the electronic computers of today. In more detail, the biological picture is of classical nerve signals travelling out from the central bulb (soma) of the neuron, along the very long fibre called an *axon*, this axon bifurcating into separate strands at various places (Fig. 7.1). Each strand finally terminates at a *synapse*—the junction at which the signal is transferred, usually to a subsequent neuron, across a synaptic cleft. It is at this stage that the neurotransmitter chemicals carry the message that the previous neuron has fired, by moving from one cell (neuron) to the next. This synaptic junction would often occur at the treelike *dendrite* of the next neuron, or else on its soma. Some synapses are excitatory in nature, with neurotransmitters that tend to enhance the firing of the next neuron, whereas some are inhibitory, and their (different) neurotransmitter chemicals tend to inhibit a neuron's firing. The effects of the different synaptic actions on the next neuron essentially add up ('plus' for excitatory and 'minus' for inhibitory), and when a certain threshold is reached, the next neuron will fire.* More correctly there would be a strong *probability* that it would fire. In all such processes, there would be certain chance factors involved also.

There is, at least so far, no question but that this picture is one that could in principle be effectively simulated computationally, assuming that the synaptic connections and their individual strengths are kept fixed. (The random ingredients would, of course, pose no computational problem, cf. §1.9). Indeed, it is not hard to see that the neuron–synapse picture that is presented here (with fixed synapses and fixed strengths) is essentially *equivalent* to that of a computer (cf. ENM, pp. 392–6). However, because of a phenomenon known as *brain plasticity*, the strengths of at least some of these connections may change from time to time, perhaps even in a timescale of less than a second, as may the very connections themselves. An important question is: what procedures govern these synaptic changes?

In the connectionist models (as are adopted for artificial neural networks), there is a *computational rule* of some kind governing the synaptic changes. This rule would be specified in such a way that the system can improve on its past performance on the basis of certain preassigned criteria in relation to its external inputs. As far back as 1949, Donald Hebb suggested a simple rule of this kind. Modern connectionist models[1] have considerably modified the original Hebb procedure in various ways. In general models of this kind, there clearly has to be *some* clear-cut computational rule—because the models are always things that can be enacted by an ordinary computer; cf. §1.5. But the

*At least, this has been the conventional picture. There is now some evidence that this simple 'additive' description may be a considerable over-simplification, and some 'information processing' is likely to be taking place within the dendrites of individual neurons. This possibility has been emphasized by Karl Pribram and others (cf. Pribram 1991). Some early suggestions along these general lines were made by Alwyn Scott (1973, 1977; and for the possibility of 'intelligence' within individual cells cf. Albrecht–Buehler 1985, for example). That there may be complex 'dendritic processing' taking place within single neurons is consistent with the discussion of §7.4.

thrust of the arguments that I have put forward in Part I is that no such computational procedure could be adequate to explain all the operational manifestations of human conscious understanding. Thus, we must look for something different, as the appropriate type of controlling 'mechanism'—at least in the case of synaptic changes that might have some relevance to actual *conscious* activity.

Certain other ideas have been suggested, such as those of Gerald Edelman in his recent book *Bright Air, Brilliant Fire* (1992) (and his earlier trilogy Edelman 1987, 1988, 1989), in which it is proposed that rather than having rules of a Hebbian type, a form of 'Darwinian' principle operates within the brain, enabling it to improve its performance continually by means of a kind of natural selection principle that governs these connections—there being, in this model, significant associations with the way the immune system develops its facility to 'recognize' substances. Importance is placed on the complicated role of the neurotransmitter and other chemicals that are involved in communicating between neurons. However, these processes, as they are presently conceived, are still treated in a classical and computational way. Indeed, Edelman and his colleagues have constructed a series of computationally controlled devices (called DARWIN I, II, III, IV, etc.) that are intended to simulate, in increasing orders of complexity, the very kinds of procedure that he is proposing lie at the basis of mental action. By the very fact that an ordinary general-purpose computer carries out the controlling action, it follows that this particular scheme is still a computational one—with some particular 'bottom-up' system of rules. It does not matter how different in detail such a scheme might be from other computational procedures. It still comes under the heading of those included under the discussion of Part I—cf. particularly §1.5, §3.9, and the arguments that are summarized in the fantasy dialogue of §3.23. Those arguments alone render it exceedingly improbable that anything that is solely of this nature can provide an actual model of the conscious mind.

In order to escape from the computational strait-jacket, some other means of controlling synaptic connections is needed—and whatever it is, it must presumably involve some physical process in which some form of quantum coherence has a significant role to play. If that process is similar in an essential way to the action of the immune system, then the immune system itself must be dependent on quantum effects. Perhaps, indeed, there is something in the particular way in which the immune system's recognition mechanism operates that has an essentially quantum character—as has been argued, in particular, by Michael Conrad (1990, 1992, 1993). This would not surprise me; but such possible roles for quantum action in the operation of the immune system do not form any key part of Edelman's model of the brain.

Even if synaptic connections are controlled in some way by coherent quantum-mechanical effects, it is difficult to see that there can be anything essentially quantum-mechanical about actual nerve-signal activity. That is to say, it is hard to see how one could usefully consider a quantum superposition consisting of one neuron *firing*, and simultaneously *not firing*. Nerve signals would appear to be quite sufficiently macroscopic to make it hard to believe in such a picture, despite the fact that nerve transmission is indeed rather well insulated by the presence of the nerve's surrounding fatty sheath of myelin. On the view (**OR**) that I have been promoting in §6.12, we must expect that objective state reduction would take place rapidly when a neuron fires, not because there is much large-scale movement of mass (there is not nearly enough, by the standards required), but because the electric field propagating along the nerve—caused by the nerve signal—would be likely to be detectable in the surrounding environment of material in the brain. This field would disturb, in random ways, reasonably large amounts of that material—quite enough, it would seem, that the criterion of §6.12 for the enaction of **OR** should be satisfied almost as soon as the signal is initiated. Thus, the maintaining of quantum superpositions of neuron firings and neuron not-firings seems an implausible possibility.

7.3 Quantum computation

This environment-disturbing property of neuron firing is the feature that I had always found to be most uncomfortable for the rough kind of proposal that I had previously argued for in ENM, in which quantum superposition of the simultaneous firing and non-firing of families of neurons seems indeed to be needed. With the present **OR** criterion for state reduction, the process **R** would be effected with even less of an environmental disturbance than would have been the case before, and it is even harder to believe in the possibility that such superpositions could be significantly maintained. The idea had been that if it were possible to perform many superposed separate 'calculations' in different patterns of neuron firing simultaneously, then something of the nature of a *quantum computation* would be achieved by the brain, rather than mere Turing computation. Despite the apparent implausibility of quantum computation being operative at this level of brain activity, it will be helpful for us to examine certain aspects of what is involved in this notion.

Quantum computation is a theoretical concept that had been put forward in its essentials by David Deutsch (1985) and Richard Feynman (1985, 1986) cf. also Benioff (1982), Albert (1983), and is now being actively explored by a number of people. The idea is that the classical notion of a Turing machine is extended to a corresponding quantum one. Accordingly, all the various operations that this extended 'machine' undertakes are subject to the quantum laws—with superpositions allowed—that apply to a quantum-level system. Thus, for the most part, it is the action of **U** that governs the evolution of the device, with the preservation of such superpositions being an essential part of its action. The **R**-procedure would become relevant mainly only at the *end* of the operation, when the system is 'measured' in order to ascertain the result of the computation. In fact (although this is not always recognized) the action of **R** must also be invoked from time to time in a more minor way during the course of the computation, in order to ascertain whether or not the computation has yet terminated.

It is found that although a quantum computer cannot achieve anything beyond what could already be done *in principle* by conventional Turing computation, there are certain classes of problem for which quantum computation is able to outperform Turing computation in the sense of *complexity theory* (cf. Deutsch 1985). That is to say for these classes of problem, the quantum computer is in principle much *faster*—but *merely* faster—than the conventional computer. See, particularly, Deutsch and Jozsa (1992) for a class of interesting (yet somewhat artificial) problems in which the quantum computer excels. Moreover, the important problem of factorizing large integers can now be solved (in polynomial time) by quantum computation, according to a recent argument by Peter Shor.

In 'standard' quantum computation, the usual rules of quantum theory are adopted, in which the system carries on according to the procedure **U** for essentially the entire operation, but where **R** comes in at certain specified places. There is nothing 'non-computational' in such a procedure, in the *ordinary* sense of 'computational', since **U** is a computable operation and **R** is a purely probabilistic procedure. What can be achieved in principle by a quantum computer could also be achieved, in principle, by a suitable Turing-machine-with-randomizer. Thus, even a quantum computer would not be able to perform the operations required for human conscious understanding, according to the arguments of Part I. The hope would have to be that the subtleties of what is *really* going on when the state vector 'appears' to get reduced, rather than just the stop-gap random procedure **R**, would lead us to something *genuinely* non-computable. Thus, the complete theory of the putative **OR** process would have to be an *essentially non-computable* scheme.

The idea in ENM had been that superposed Turing computations could be carried on for a while, but these would be interspersed with some non-computable action that could only be understood in terms of whatever new physics would come in (e.g. **OR**) to replace **R**. But if such superpositions of neuron computations are forbidden to us, because too much of the environment gets disturbed by each neuron signal, then it is hard to see how it could be possible even to make use of the idea of standard quantum computation, let alone any modification of this procedure that takes advantage of some putative non-computational **R**-replacement such as **OR**. However, we shall be seeing in a moment that there is another much more promising possibility. In order to understand how this might be the case, we shall have to look more deeply into the biological nature of brain cells.

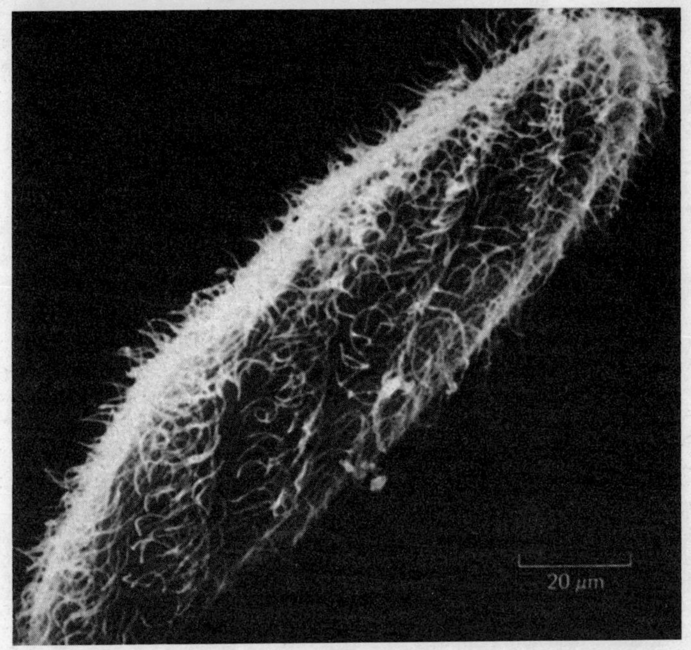

Fig. 7.2. A *paramecium*. Note the hair-like cilia that are used for swimming. These form the external extremities of the paramecium's *cytoskeleton*.

7.4 Cytoskeletons and microtubules

If we are to believe that neurons are the only things that control the sophisticated actions of animals, then the humble paramecium presents us with a profound problem. For she swims about her pond with her numerous tiny hairlike legs—the *cilia*—darting in the direction of bacterial food which she senses using a variety of mechanisms, or retreating at the prospect of danger, ready to swim off in another direction. She can also negotiate obstructions by swimming around them. Moreover, she can apparently even *learn* from her past experiences[2]—though this most remarkable of her apparent faculties has been disputed by some[3]. How is all this achieved by an animal without a single neuron or synapse? Indeed, being but a single cell, and not being a neuron herself, she has no place to accommodate such accessories (see Fig. 7.2).

Yet there must indeed be a complicated control system governing the behaviour of a paramecium—or indeed other one-celled animals like amoebas—but it is not a nervous system. The structure responsible is apparently part of what is referred to as the *cytoskeleton*. As its name suggests, the cytoskeleton provides the framework that holds the cell in shape, but it does much more. The

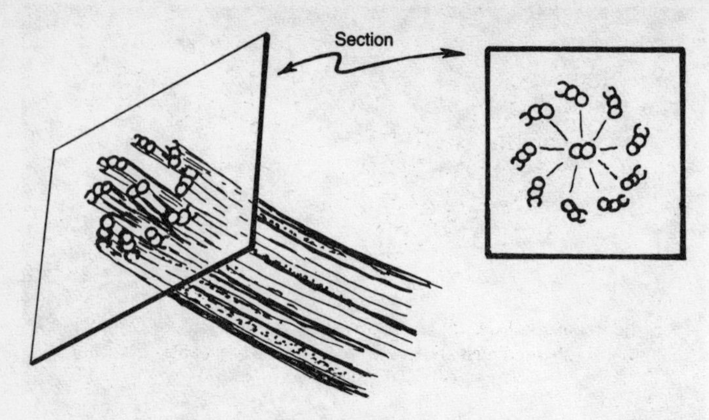

Fig. 7.3. Important parts of the cytoskeleton consist of bundles of tiny tubes (microtubules) organized in a structure with a fan-like cross-section. The paramecium's cilia are bundles of this nature.

cilia themselves are endings of the cytoskeletal fibres, but the cytoskeleton seems also to contain the control system for the cell, in addition to providing 'conveyor belts' for the transporting of various molecules from one place to another. In short, the cytoskeleton appears to play a role for the single cell rather like a combination of skeleton, muscle system, legs, blood circulatory system, and nervous system all rolled into one!

It is the cytoskeleton's role as the cell's 'nervous system' that will have the main importance for us here. For our own neurons are themselves single cells, and each neuron has its *own* cytoskeleton! Does this mean that there is a sense in which each individual neuron might itself have something akin to its own 'personal nervous system'? This is an intriguing issue, and a number of scientists have been coming round to the view that something of this general nature might actually be true. (See Stuart Hameroff's pioneering 1987 book *Ultimate Computing: Biomolecular Consciousness and NanoTechnology*; also Hameroff and Watt (1982) and numerous articles in the new journal *Nanobiology*.)

In order to address such issues, we should first glimpse the basic organization of the cytoskeleton. It consists of protein-like molecules arranged in various types of structure: actin, microtubules, and intermediate filaments. It is the *microtubules* that will be our main concern here. They consist of hollow cylindrical tubes, some 25 nm in diameter on the outside and 14 nm on the inside (where 'nm' = 'nanometre', which is 10^{-9} m), sometimes organized into larger tubelike fibres that consist of nine doublets, triplets, or partial triplets, of microtubules, organized in an arrangement with a fanlike cross-section, as indicated in Fig. 7.3, with sometimes a pair of microtubules running down the centre. The paramecium's cilia are structures of this kind. Each microtubule is itself a protein polymer consisting of subunits referred to as *tubulin*. Each

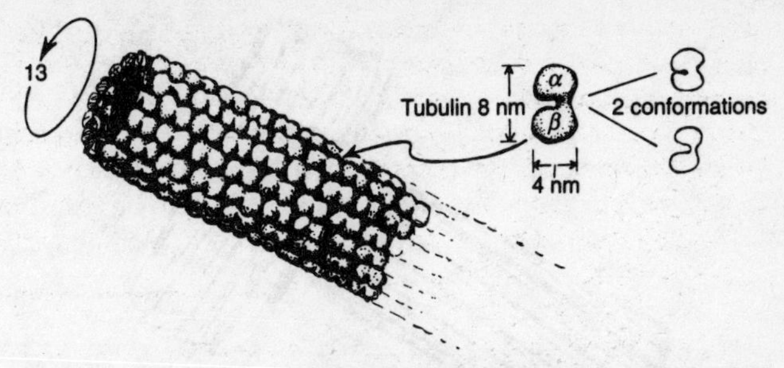

Fig. 7.4. A *microtubule*. It is a hollow tube, normally consisting of 13 columns of tubulin dimers. Each tubulin molecule is capable of (at least) two conformations.

tubulin subunit is a 'dimer', i.e. it consists of two essentially separate parts called α-tubulin and β-tubulin each being composed of about 450 amino acids. It is a globular protein pair, somewhat 'peanut shaped' and organized in a slightly skew hexagonal lattice along the entirety of the tube, as indicated in Fig. 7.4. There are generally 13 columns of tubulin dimers to each microtubule. Each dimer is about 8 nm $\times$ 4 nm $\times$ 4 nm and its atomic number is about 11×10^4 (which means that it has about that many nucleons in it, so its mass, in absolute units, is about 10^{-14}).

Each tubulin dimer, as a whole, can exist in (at least) two different geometrical configurations—called different *conformations*. In one of these, they bend to about 30° to the direction of the microtubule. There is evidence that these two conformations correspond to two different states of the dimer's electric polarization, where these come about because an electron, centrally placed at the α-tubulin/β-tubulin juncture, can shift from one position to another.

The 'control centre' of the cytoskeleton (if indeed this is really an appropriate term) is a structure known as the *centriole*. This seems to consists essentially of two cylinders of nine triplets of microtubules, where the cylinders form a kind of separated 'T' (Fig. 7.5). (The cylinders are similar, in a general way, to those that occur in cilia, as illustrated in Fig. 7.3.) The centriole forms the critical part of a structure called the *microtubules organizing centre* or *centrosome*. Whatever the role of the centriole might be during the normal course of an ordinary cell's existence, it has at least one fundamentally important task. At a critical stage, each of the two cylinders in the centriole grows another, so as to make *two* centriole 'T's that then *separate* from each other, each apparently dragging a bundle of microtubules with it—although it would be more accurate to say that each becomes a focal point around which microtubules assemble. These microtubule fibres somehow connect the centriole to the separate DNA strands in the nucleus (at central points, known as their centromeres) and the DNA

Fig. 7.5. The *centriole* (which appears to be the 'control centre' of the cytoskeleton—if such exists) consists essentially of a separated 'T' built from two bundles of microtubules very like that illustrated in Fig. 7.3.

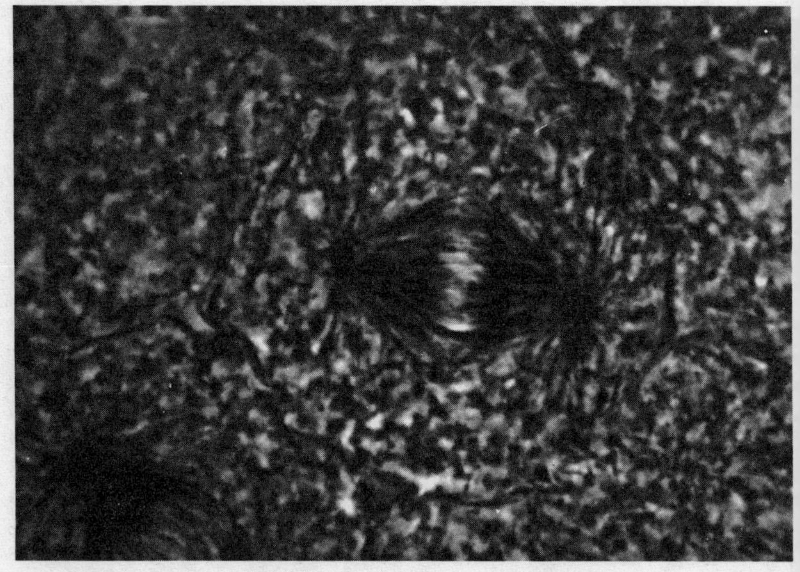

Fig. 7.6. In mitosis (cell division) the chromosomes separate, being pulled apart by microtubules.

strands separate—initiating the extraordinary process technically known as *mitosis*, which simply means *cell division* (see Fig. 7.6).

It may seem odd that there should be two quite different 'headquarters' in a single cell. On the one hand there is the *nucleus*, where the fundamental genetic material of the cell resides, which controls the cell's heredity and its own particular identity, and governs the production of the protein materials of which the cell itself is composed. On the other hand, there is the *centrosome* with its chief component the *centriole*, which seems to be the focal point of the cytoskeleton, a structure that apparently controls the cell's movements and its detailed organization. The presence of these two different structures in eukaryotic cells (the cells of all animals and almost all plants on this planet—but excluding bacteria, blue-green algae, and viruses) is believed to be the result of an ancient 'infection' that took place some thousands of millions of years ago. The cells that previously inhabited the earth were the prokaryotic cells that still exist today as bacteria and blue-green algae, and which possess no cytoskeletons. One suggestion (Sagan 1976) is that some early prokaryotes became entangled with—or, perhaps, 'infected by'—some kind of spirochete, an organism that swam with a whiplike tail composed of cytoskeletal proteins. These mutually alien organisms subsequently grew to live permanently together in a symbiotic relationship as single *eukaryotic* cells. Thus, these 'spirochetes' ultimately became the cells' cytoskeletons—with all the implications for the future evolution that thereby made *us* possible!

The organization of mammalian microtubules is interesting from a mathematical point of view. The number 13 might seem to have no particular mathematical significance, but this is not entirely so. It is one of the famous *Fibonacci numbers*:

$$0, 1, 1, 2, 3, 5, 8, 13, 21, 34, 55, 89, 144, \ldots$$

where each successive number is obtained as a sum of the previous two. This might be fortuitous, but Fibonacci numbers are well known to occur frequently (at a much larger scale) in biological systems. For example, in fir cones, sunflower heads, and palm tree trunks, one finds spiral or helical arrangements involving the interpenetration of right-handed and left-handed twists, where the number of rows for one handedness and the number for the other handedness are two successive Fibonacci numbers (see Fig. 7.7). (As one examines the structures from one end to the other, one may find that a 'shunt' takes place, and the numbers then shift to an adjacent pair of successive Fibonacci numbers.) Curiously, the skew hexagonal pattern of microtubules exhibits a very similar feature—generally of an even more precise organization—and it is apparently found (at least normally) that this pattern is made up of 5 right-handed and 8 left-handed helical arrangements, as depicted in Fig. 7.8. In Fig. 7.9, I have tried to indicate how this structure might appear as actually 'viewed' from within a microtubule. The number 13 features here in its role as the sum: $5+8$. It is curious, also, that the double microtubules that frequently occur seem normally to have a total of 21 columns of tubulin dimers forming the outside boundary of the composite tube—the next Fibonacci number! (However, one should not get carried away with such considerations; for example, the '9' that occurs in the bundles of microtubules in cilia and centrioles is *not* a Fibonacci number.)

Fig. 7.7. A sunflower head. As with many other plants, Fibonacci numbers feature strongly. In the outer regions, there are 89 clockwise and 55 anti-clockwise spirals. Nearer the centre we can find other Fibonacci numbers.

Why do Fibonacci numbers arise in microtubule structure? In the case of fir cones and sunflower heads, etc., there are various plausible theories—and Alan Turing himself was someone who thought seriously about the subject (Hodges 1983, p. 437). But it may well be that these theories are not appropriate for microtubules, and different ideas are probably relevant at this level. Koruga (1974) has suggested that these Fibonacci numbers may provide advantages for the microtubule in its capacity as an 'information processor'. Indeed, Hameroff and his colleagues have argued, for more than a decade[4], that microtubules may play roles as *cellular automata*, where complicated signals could be transmitted and processed along the tubes as waves of differing electric polarization states of the tubulins. Recall that tubulin dimers can exist in (at least) two different conformational states that can switch from one to the other, apparently because of alternative possibilities for their electric polarizations. The state of each dimer would be influenced by the polarization states of each of its six neighbours (because of van der Waals interactions between them) giving rise to certain specific rules governing the conformation of each dimer in terms of the conformations of its neighbours. This would allow all kinds of messages to be propagated and processed along the length of each microtubule. These propagating signals appear to be relevant to the way that microtubules

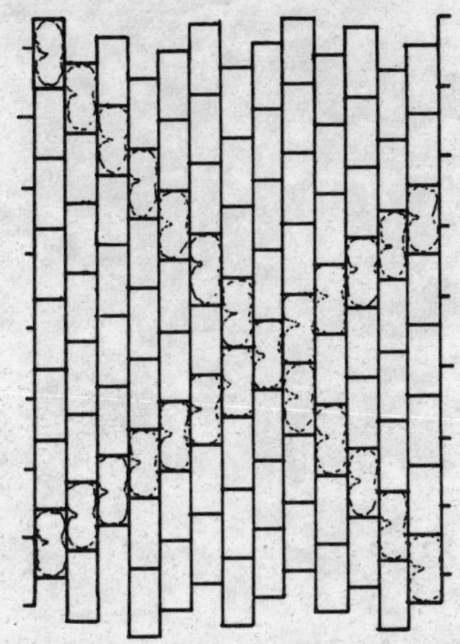

Fig. 7.8. Imagine a microtubule slit along is length, and then opened out flat into a strip. We find that the tubulins are ordered in sloping lines which rejoin at the opposite edge 5 or 8 places displaced (depending upon whether the lines slope to the right or to the left).

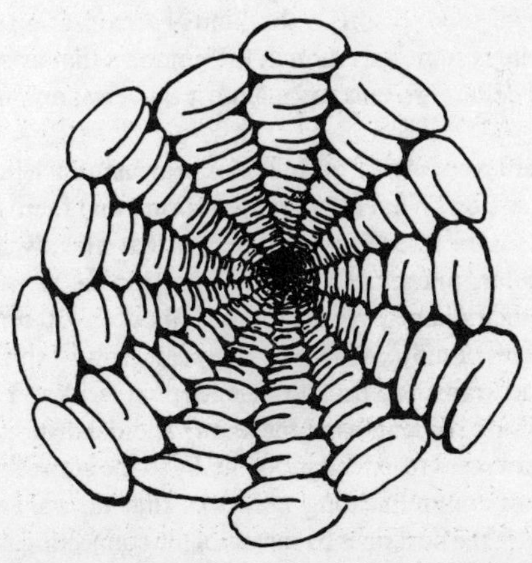

Fig. 7.9. View down a microtubule! The 5+8 spiral arrangement of the tubulins in this microtubule can be seen.

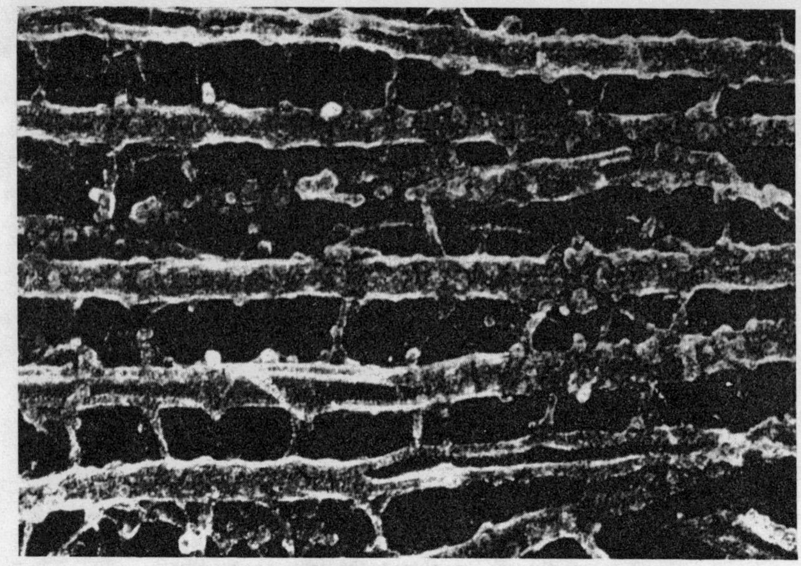

Fig. 7.10. Microtubules tend to be interconnected with neighbouring ones by bridges of *microtubule associated proteins* (MAPs).

transport various molecules alongside them, and to the various interconnections between neighbouring microtubules—in the form of bridge-like connecting pro-teins referred to as MAPs (microtubule associated proteins). See Fig. 7.10. Koruga argues for a special efficiency in the case of a Fibonacci-number-related structure of the kind that is actually observed for microtubules. There must indeed be some good reason for this kind of organization in microtubules, since although there is some variation in the numbers that apply to eukaryotic cells generally, 13 columns seems to be almost universal amongst mammalian microtubules.

What is the significance of microtubules for neurons? Each individual neuron has its own cytoskeleton. What is its role? I am sure that there is a great deal to be uncovered by future research, but it seems that already a fair amount is known. In particular, microtubules in neurons can be very long indeed, in comparison with their diameter (which is only about 25–30 nm) and can reach lengths of millimetres or more. Moreover, they can grow or shrink, according to circumstances, and transport neurotransmitter molecules. There are microtubules running along the lengths of the axons and dendrites. Although single microtubules do not seem to extend individually to the entire length of an axon, they certainly form communicating networks that do so, each microtubule communicating with the next ones by means of the connecting MAPs referred to above. Microtubules seem to be responsible for maintaining the strengths of synapses and, no doubt, for effecting alterations of these strengths when the need

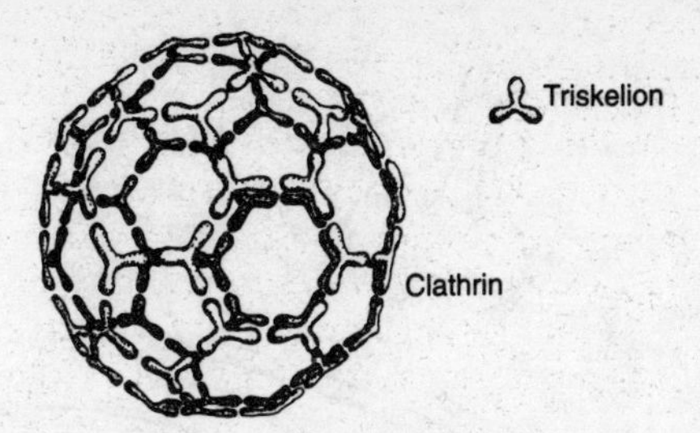

Fig. 7.11. A clathrin molecule (similar in overall structure to a fullerene, but made of more complicated substructures—triskelion proteins rather than carbon atoms). The clathrin depicted resembles an ordinary soccer ball in structure.

arises. Moreover, they seem to organize the growth of new nerve endings, guiding them towards their connections with other nerve cells.

Since neurons do not divide after the brain is fully formed, there is not a role of this particular kind for a centriole in a neuron. Indeed, centrioles seem to be absent in the neuron's centrosome—which is found close to the neuron's nucleus. Microtubules extend from there right up to the vicinity of the presynaptic endings of the axon, and also, in the other direction, into the dendrites and, via contractile actin, into dendritic spines, which frequently form the postsynaptic end of a synaptic cleft (Fig. 7.12). These spines are subject to growth and degeneration, a process which seems to form an important part of brain plasticity, whereby the overall interconnections in the brain are undergoing continual subtle changes. There would seem to be significant evidence that microtubules are indeed importantly involved in the control of brain plasticity.

As an apparent curiosity, it may also be mentioned that in the presynaptic endings of axons there are certain substances associated with microtubules which are fascinating from the geometrical point of view, and which are important in connection with the release of neurotransmitter chemicals. These substances—called *clathrins*—are built from protein trimers known as clathrin triskelions, which form three-pronged (polypeptide) structures. The clathrin triskelions fit together to make beautiful mathematical configurations that are identical in general organization to the carbon molecules known as 'fullerenes' (or 'bucky balls') owing to their similarity with the famous geodesic domes constructed by the American architect Buckminster Fuller.[5] Clathrins are much larger than fullerene molecules, however, since an entire clathrin triskelion, a structure involving several amino acids, takes the place of the fullerene's single carbon atom. The particular clathrins that are concernedwith the release of neurotransmitter chemicals at synapses seem mainly to have the structure of a *truncated icosahedron*—which is familiar as the polyhedron demonstrated in the modern soccer ball (see Fig. 7.11 and 7.12)!

In the previous section, the important question was raised: what *is* it that

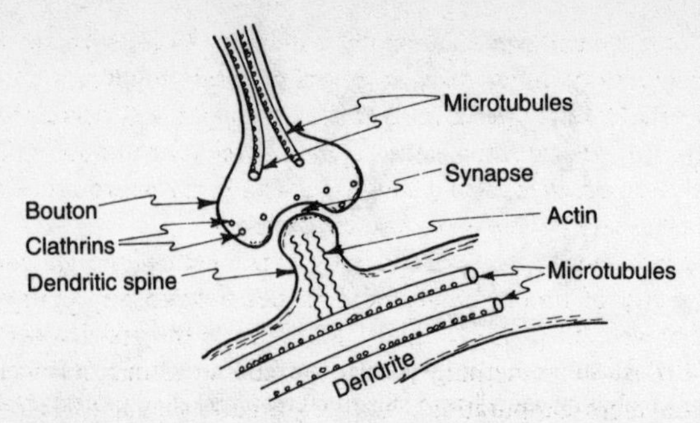

Fig. 7.12. Clathrins, like those of Fig. 7.11 (and microtubule endings) inhabit the axon's synaptic boutons and seem to be involved in controlling the strength of the synapse; this strength could also be influenced by the contractile actin filaments in dendritic spines, which are controlled by microtubules.

governs the variation in the strengths of synapses and organizes the places where functioning synaptic connections are to be made? We have been guided to a clear belief that it is the *cytoskeleton* that must play a central role in this process. How does this help us in our quest for a non-computational role for the mind? So far, all we seem to have gained is an enormous potential increase in computing power over and above what could have been achieved if the units were simply the neurons alone.

Indeed, if tubulin dimers are the basic computational units, then we must envisage the possibility of a potential computing power in the brain that vastly exceeds that which has been contemplated in the AI literature. Hans Moravec, in his book *Mind Children* (1988), assumed, on the basis of a 'neuron alone' model, that the human brain might in principle conceivably achieve some 10^{14} basic operations per second, but no more, where we consider that there might be some 10^{11} operational neurons, each capable of sending about 10^3 signals per second (cf. §1.2). If, on the other hand, we consider the tubulin dimer as the basic computational unit, then we must bear in mind that there are some 10^7 dimers per neuron, the elementary operations now being performed some 10^6 times faster, giving us a total of around 10^{27} operations per second. Whereas present-day computers may be beginning to close in on the first figure of 10^{14} operations per second, as Moravec and others would strongly argue, there is no prospect of the 10^{27} figure being achieved in the foreseeable future.

Of course it could reasonably be claimed that the brain is operating nowhere remotely close to the 100% microtubular efficiency that these figures assume. Nevertheless, it is clear that the possibility of 'microtubular computing' (cf. Hameroff 1987) puts a completely different perspective on some of the arguments for imminent human-level artificial intelligence. Can we even trust suggestions[6] that the mental faculties of a nematode worm have already been computationally achieved merely because its neural organization appears to have been mapped and computationally simulated? As remarked in §1.15, the actual capabilities of an ant seem to outstrip by far, anything that has been achieved by the standard procedures of AI. One might well wonder how much an ant gains from its enormous array of nano-level 'microtubular information processors', as opposed to what it could do if it had only 'neuron-type switches'. As for a paramecium, there is no case to answer.

Yet the arguments of Part 1 are making a stronger claim. I am contending that the faculty of human understanding lies beyond any computational scheme whatever. If it is microtubules that control the activity of the brain, then there must be something within the action of microtubules that is different from mere computation. I have argued that such non-computational action must be the result of some reasonably large-scale quantum-coherent phenomenon, coupled in some subtle way to macroscopic behaviour, so that the system is able to take advantage of whatever new physical processes must replace the stop-gap **R**-procedure of present-day physics. As a first step, we must look for a genuine role for *quantum coherence* in cytoskeletal activity.

7.5 Quantum coherence within microtubules?

Is there any evidence for this? Let us recall Fröhlich's (1975) ideas for the possibility of quantum-coherent phenomena in biological systems, referred to in the discussion of §7.1. He argued that so long as the energy of metabolic drive is large enough, and the dielectric properties of the materials concerned are sufficiently extreme, then there is the possibility of large-scale quantum coherence similar to that which occurs in the phenomena of superconductivity and superfluidity—sometimes referred to as *Bose–Einstein* condensation—even at the relatively high temperatures that are present in biological systems. It turns out that not only is the metabolic energy indeed high enough and the dielectric properties unusually extreme (a striking observational fact of the 1930s that started Fröhlich on this whole line of thinking), but there is now also some direct evidence for the 10^{11} Hz oscillations within cells that Fröhlich had predicted (Grundler and Keilmann 1983).

In a Bose–Einstein condensate (which occurs also in the action of a laser), large numbers of particles participate collectively in a single quantum state. There is a wavefunction, for this state, of the kind that would be appropriate for a single particle—but now it applies all at once to the entire collection of particles that are participating in the state. We recall the very counter-intuitive nature of the (spread-out) quantum state of a single quantum particle (§5.6, §5.11). With a Bose–Einstein condensate, it is as though the entire system containing a large number of particles behaves as a whole very much as the quantum state of a single particle would, except that everything is scaled up appropriately. There is a coherence on a large scale, where many of the strange features of quantum wavefunctions hold at a macroscopic level.

Fröhlich's original ideas seem to have been that such large-scale quantum states would be likely to occur in cell membranes*, but now the additional—and perhaps more plausible—possibility presents itself that it is in the *microtubules* that we should seek quantum behaviour of this kind. There appears to be some evidence that this may be the case.[7] Even as early as 1974, Hameroff (1974) had proposed that microtubules might act as 'dielectric waveguides'. It is indeed tempting to believe that Nature has chosen hollow tubes in her cytoskeletal structures for some good purpose. Perhaps the tubes themselves serve to provide the effective insulation that would enable the quantum state in the interior of the tube to remain unentangled with its environment for an appreciable time. It is interesting to note, in this connection, that Emilio del Giudice, and his colleagues at the University of Milan (del Giudice *et al.* 1983), have argued that a quantum self-focusing effect of electromagnetic waves within the cytoplasmic material in cells causes signals to be confined to a size that is just the internal diameter of microtubules. This could give substance to the waveguide theory, but the effect could also be instrumental in the very formation of microtubules.

There is another matter of interest here, and this concerns the very nature of *water*. The tubes themselves appear to be empty—a curious and possibly significant fact in itself, if we are looking to these tubes to provide for us the controlled conditions favourable to some kind of collective quantum oscillations. 'Empty', here, means that they essentially contain just water (without even dissolved ions). We might think that 'water', with its randomly moving molecules, is not a sufficiently organized kind of structure for quantum-coherent oscillations to be likely to occur. However, the water that is found in cells is not at all like the ordinary water that is found in the oceans—disordered, with molecules moving about in an incoherent random way. Some of it—and it is a controversial matter how much—exists in an *ordered* state (sometimes referred to as 'vicinal' water; cf. Hameroff (1987), p. 172). Such an ordered state of water may extend some 3 nm or more outwards from cytoskeletal surfaces. It seems not unreasonable to suppose that the water inside the microtubules is also of an ordered nature, and this would strongly favour the possibility of quantum-coherent oscillations within, or in relation to, these tubes. (See, in particular, Jibu *et al.* 1994.)

Whatever the ultimate status of these intriguing ideas, one thing seems clear to me: there is little chance that an entirely classical discussion of the cytoskeleton could properly explain its behaviour. This is quite unlike the situation for neurons themselves, where discussions in entirely classical terms

*A strong proponent of the idea that Bose–Einstein condensation may provide the 'unitary sense of self' that seems to be characteristic of consciousness, in relation to Fröhlich's ideas is Ian Marshall (1989), cf. also Zohar (1990), Zohar and Marshall (1994), and Lockwood (1989). A strong early proponent of global (essentially quantum) large-scale coherent 'hologram' activity in the brain was Karl Pribram (1966, 1975, 1991).

do seem to be largely appropriate. Indeed, an examination of the current literature on cytoskeletal action reveals the fact that quantum-mechanical concepts are continually being appealed to, and I have little doubt that this will be increasingly the case in the future.

However, it is also clear that there will be many who remain unconvinced that there are likely to be any significant quantum effects of relevance to cytoskeletal or brain action. Even if there are indeed important effects of a quantum nature which are essential to microtubules' functioning and to conscious brain action, it may not be easy to demonstrate their presence by definitive experiment. If we are lucky, some of the standard procedures that already serve to show the presence of Bose–Einstein condensates in physical systems—such as with high-temperature superconductivity—may also be found to be applicable in microtubules. On the other hand we may not be so lucky, and something quite new is needed. One intriguing possibility might be to demonstrate that microtubule excitations exhibit the kind of non-locality that occurs with EPR phenomena (Bell inequalities, etc.; cf. §5.3, §5.4, §5.17), since there is no (local) classical explanation of effects of this kind. One might, for example, imagine measurements being made at two points on a microtubule—or on separate microtubules—where the results of the measurements cannot be explained in terms of classical independent actions taking place at these two points.

Whatever the status of such suggestions, microtubule research is still in its relative infancy. There is no doubt in my own mind that there are significant surprises in store for us.

7.6 Microtubules and consciousness

Is there any direct evidence that the phenomenon of *consciousness* is related to the action of the cytoskeleton, and to microtubules in particular? Indeed, there *is* such evidence. Let us try to examine the nature of this evidence—which addresses the issue of consciousness by considering what causes it to be *absent*!

An important avenue towards answering questions concerning the physical basis of consciousness comes from an examination of precisely what it is that very specifically turns consciousness off. *General anaesthetics* have precisely this property—completely reversibly, if the concentrations are not too high—and it is a remarkable fact that general anaesthesia can be induced by a large number of completely different substances that seem to have no chemical relationship with one another whatever. Included in the list of general anaesthetics are such chemically different substances as nitrous oxide (N_2O), ether ($CH_3CH_2OCH_2CH_3$), chloroform ($CHCl_3$), halothane ($CF_3CHClBr$), isofluorane ($CHF_2OCHClCF_3$) and even the chemically inert gas xenon!

If it is not chemistry that is responsible for general anaesthesia, then what can it be that *is* responsible? There are other types of interaction that can take place between molecules, which are much weaker than chemical forces. One of these is referred to as the *van der Waals* force. The van der Waals force is a weak attraction between molecules which have *electric dipole moments* (the 'electric' equivalent of the magnetic dipole moments that measure the strength of ordinary magnets). Recall that tubulin dimers are capable of two different conformations. These appear to arise because there is an electron, centrally placed in a water-free region in each dimer, which can occupy one of two separate positions. The overall shape of the dimer is affected by this positioning, as is its electric dipole moment. The ability of the dimer's capability of 'switching' from one conformation to the other is influenced by the van der Waals force exerted by neighbouring substances. Accordingly, it has been suggested (Hameroff and Watt 1983) that general anaesthetics may act through the agency of their van der Waals interactions (in 'hydrophobic' regions—where water has been expelled—see Franks and Lieb (1982)), which interfere with the normal switching actions of tubulin. As anaesthetic gases diffuse into individual nerve cells, their electric dipole properties (which need have little directly to do with their ordinary chemical properties) can thereby interrupt the actions of microtubules. This is certainly a plausible way in which general anaesthetics might operate. Although there seems to be no generally accepted detailed picture of the actions of anaesthetics, one coherent view seems to be that it is indeed the van der Waals interactions of these substances with the comformational dynamics of brain proteins that are responsible. It is a strong possibility that the relevant proteins are the tubulin dimers in neuronal microtubules—and that it is the consequent interruption of the functioning of microtubules that results in the loss of consciousness.

As support for suggestions that it is the *cytoskeleton* that is directly affected by general anaesthetics, it may be remarked that it is not only the 'higher animals' such as mammals or birds that are rendered immobile by these substances. A paramecium, an amoeba, or even green slime mould (as was noticed by Claude Bernard as early as 1875) is similarly affected by anaesthetics at about the same kind of concentration. Whether it is on the paramecium's cilia or its centriole that the anaesthetic exerts its immobilizing effect, it seems that it must be on *some* part of its cytoskeleton. If we may take it that the controlling system of such a one-celled animal is indeed its cytoskeleton, then a consistent picture is obtained if we take it that it is on the cytoskeleton that general anaesthetics indeed act.

This is not to say that such one-celled animals need be considered to be conscious. That is a separate matter altogether. For there may well be a great deal, in *addition* to properly functioning cytoskeletons, that is needed to evoke a conscious state. What does seem to be strongly indicated, however, on the basis of such arguments as have been indicated here, is that our state (or states) of consciousness *require* such a functioning cytoskeleton. Without a properly operating system of cytoskeletons, consciousness is removed, being instantly knocked out as soon as the functioning of the cytoskeletons is inhibited—and it instantly returns as soon as this functioning is restored, so long as there has been no other damage incurred in the meanwhile. The question is significantly raised, of course, as to whether a paramecium—or, indeed, an individual human liver cell—might actually possess some rudimentary form of consciousness, but this question is not answered by such considerations. On any account, it must also be the case that the detailed neural organization of the brain is fundamentally involved in governing what *form* that consciousness must take. Moreover, if that organization were not important, then our livers would evoke as much consciousness as do our brains. Nevertheless, what the preceding arguments strongly suggest is that it is not *just* the neuronal organization of our brains that is important. The cytoskeletal underpinnings of those very neurons seem to be essential for consciousness to be present.

Presumably, for consciousness to arise generally, it is not a cytoskeleton *as such* that is relevant, but some *essential physical action* that biology has so cleverly contrived to incorporate into the activity of its microtubules. What is this essential physical action? The thrust of the arguments of Part I of this book has been that we need something that is beyond computational simulation if we are to find a physical basis for conscious acts. The arguments of Part II, in the chapters preceding this one, have told us to look to the borderline between the quantum and classical levels, where present-day physics tells us to use the stop-gap procedure **R**, but where I am claiming a *new* physical theory of **OR** is needed. In the present chapter, we tried to pinpoint the place in the brain where quantum action might be important to classical behaviour, and have apparently been driven to consider that it is through the *cytoskeletal control of synaptic connections* that this quantum/classical interface exerts its fundamental influence on the brain's behaviour. Let us try to explore this picture a little more fully.

7.7 A model for a mind?

As was remarked in §7.1, it seems appropriate to accept that nerve signals themselves are things that can be treated in a completely classical way—in view of the probable fact that such signals disturb their environment to an extent that quantum coherence cannot be maintained at that stage. If the synaptic connections and their strengths are kept fixed, then the way in which each neuron's firing affects the next one will again be something that can be treated classically, apart from a random ingredient that comes in at this stage. The action of the brain in such circumstances would be entirely computational, in the sense that a computational simulation would in principle be possible. By this, I do not mean that such a simulation would imitate precisely the actions of a *specific* brain that was wired up in this way—because of these random ingredients—but that it would provide a simulation of a *typical* action of that brain and therefore of a typical behaviour of some individual controlled by such a brain (cf. §1.7). Moreover, this is very much an *in principle* statement. There is no suggestion that with present-day technology such a simulation could actually be carried out. I am also assuming, in this, that the random ingredients are *genuinely* random. The possibility of a dualistic external 'mind' coming in to influence these probabilities is not under consideration here (cf. §7.1).

Thus we are accepting (provisionally, at least) that, with *fixed* synaptic connections, the brain is indeed acting as some kind of *computer*—albeit a computer with built-in random ingredients. As we have seen from the arguments of Part I, it is exceedingly improbable that such a scheme could ever provide a model for human conscious understanding. On the other hand, if the specific synaptic connections that define the particular neural computer under consideration are subject to continual change, where the control of those changes is governed by some *non*-computational action, then it remains possible that such an extended model could indeed simulate the behaviour of a conscious brain.

What non-computational action could this be? In this connection, we should bear in mind the *global* nature of consciousness. If it were merely the case that some 10^{11} individual cytoskeletons were each separately supplying some non-computational input, it is hard to see that this would be of much use to us. According to the arguments of Part I, the non-computational behaviour is indeed linked to the action of consciousness—at least to the extent that *some* conscious actions, specifically the quality of *understanding*, are argued to be non-computational. But this is not something relevant to individual cytoskeletons or to individual microtubules within a cytoskeleton. There can be no suggestion that any particular cytoskeleton or microtubule 'understands' any part of the Gödel argument! The understanding is something that operates at a much more global scale; and if cytoskeletons are involved, then it must be some collective phenomenon which concerns very large numbers of cytoskeletons all at once.

Recall Fröhlich's idea that large-scale collective quantum phenomenon—perhaps of the nature of a Bose–Einstein condensate—is a definite biological possibility, even within the 'hot' brain (cf. also Marshall 1989). Here we envisage that not only must single microtubules be involved in a relatively large-scale quantum-coherent state, but that such a state must extend from one microtubule to the next. Thus, not only must this quantum coherence stretch to the length of an entire microtubule (and we recall that microtubules can extend to considerable length), but a good many of the different microtubules in the cytoskeleton within a neuron, if not all of them, must together take part in this same quantum-coherent state. Not only this, but the quantum coherence must leap the synaptic barrier between neuron and neuron. It is not much of a globality if it involves only individual cells! The unity of a single mind can arise, in such a description, only if there is some form of quantum coherence extending across at least an appreciable part of the entire brain.

Such a feat would be a remarkable one—almost an incredible one—for Nature to achieve by biological means. Yet I believe that the indications must be that she has done so, the main evidence coming from the fact of our own mentality. There is much to be understood about biological systems and how they achieve their magic. There is much, in biology, that surpasses by far what can be done with present-day direct physical techniques. (Think, for example, of a tiny millimetre-sized spider delicately weaving her elaborate web.) We recall, moreover, that certain quantum-coherent effects over a distance of several metres—the EPR entanglements involved in pairs of photons—have already been observed (by *physical* means) in the experiments of Aspect and others (cf. §5.4). Despite the technical difficulty of performing experiments which can detect such large-distance quantum effects, we should not rule out the possibility of Nature having found biological ways of doing a great deal more. The 'ingenuity' that is to be found in biology should never be underestimated.

However, the arguments I have been presenting require more than just quantum coherence on a large scale. They require that the biological systems that are our brains have somehow contrived to harness the details of a physics that is yet unknown to human physicists! This physics is the missing **OR** theory that straddles the quantum and classical levels and, as I am arguing, replaces the stop-gap **R**-procedure by a highly subtle non-computational (but undoubtedly still mathematical) physical scheme.

That human physicists are, as yet, largely ignorant of this missing theory is, of course, no argument against Nature having made use of it in biology. She took advantage of the principles of Newtonian dynamics long before Newton, of electromagnetic phenomena long before Maxwell, and of quantum mechanics long before Planck, Einstein, Bohr, Heisenberg, Schrödinger, and Dirac—by some thousands of millions of years! It is only the arrogance of our present age that leads so many to believe that we now know all the basic principles that can underlie all the subtleties of biological action. When some organism is blessed with the fortune of stumbling upon such subtle action, it may reap the benefits that this physical process confers upon it. Then Nature smiles upon that organism and upon its descendants, and allows that subtle physical action to be preserved from generation to generation in increasing numbers—through her powerful process of natural selection.

When the first eukaryotic cell-creatures emerged, they must have found that they obtained great benefit from the presence of the primitive microtubules within them. Some kind of organizational influence arose, according to the picture that I am presenting here, which perhaps enabled them to behave in some rudimentary kind of purposeful way, and this helped them to survive better than their competitors. It would, no doubt, be quite inappropriate to refer to such an influence as a 'mind'; yet it arose, I suggest, by virtue of some subtle interplay between quantum-level and classical-level processes. The subtle nature of this interplay owed its very existence to the sophisticated physical **OR** action—still in detail unknown to us—that, in less subtly organized circumstances, appears as the crude quantum-mechanical **R**-process we now adopt. The distant descendants of these cell-creatures—today's parameciums and amoebas, and also our ants, trees, frogs, buttercups, and human beings—have kept the benefits that this sophisticated action had conferred upon those ancient cell-creatures and have twisted them to serve many completely different-looking purposes. Only when incorporated into a highly developed nervous system has this action finally been able to realize a good measure of its tremendous potential—and given rise to what we actually refer to as 'mind'.

Let us then accept the possibility that the totality of microtubules in the cytoskeletons of a large family of the neurons in our brains may well take part in global quantum coherence—or at least that there is sufficient quantum entanglement between the states of different microtubules across the brain—so that an overall *classical* description of the collective actions of these microtubules is *not* appropriate. We might envisage complicated 'quantum oscillations' within microtubules, where the isolation that the tubes themselves provide is sufficient to ensure that not all quantum coherence is lost. It is tempting to suppose that the cellular-automata-like computations that are envisaged as taking place *along* the tubes by Hameroff and his colleagues can be coupled to the presumed quantum oscillations (e.g. those of del Giudice *et al.* 1983 or Jibu *et al.* 1994) taking place *within* the tubes.

In this connection, it may be remarked that the kind of frequency that Fröhlich had envisaged for his collective quantum oscillations, as supported by the observations of Grundler and Keilmann (1983)—namely in the frequency region of 5×10^{10} Hz (that is, 5×10^{10} oscillations per second)—is the same kind of frequency that is envisaged, by Hameroff and colleagues, as the 'switching time' for the tubulin dimers in their microtubular cellular automata. Thus, if Fröhlich's mechanism is indeed what is operating within microtubules, then some kind of coupling between the two kinds of activity is indeed indicated.*

If the coupling between the two were too strong, however, then it would not be possible to maintain a quantum nature for the internal oscillations without the computations along the tubes themselves having to be treated quantum-mechanically. If this were the case, then it would have to be some kind of *quantum computation* that is taking place along microtubules (cf. §7.3)! We must ask whether this is a serious possibility.

*It is much less clear, however, whether there can be any direct connection between such comparatively high frequencies and the more familiar 'brain-wave' activity (such as the 8–12 Hz α-rhythm). It is just conceivable that such lower frequencies might arise as 'beat frequencies', but no link has been established. Of particular note, in this connection, are recent observations of 35–75 Hz oscillations that seemingly appear in association with regions of the brain involved with conscious attention. These seem to have some puzzling non-local properties. (See Eckhorn *et al.* 1988, Gray and Singer 1989, Crick and Koch 1990, 1992, Crick 1994).

The difficulty is that this would seem to require that the changes in the dimer conformations do not significantly disturb the outside ambient material. In this connection, it should be pointed out that there appears to be a region surrounding a microtubule of *ordered* water and where other materials are excluded (cf. Hameroff 1987, p. 172), which might provide some measure of quantum shielding. On the other hand, there are connecting MAPs (cf. §7.4) reaching outwards from microtubules, some concerned with their role as transporters of other materials, and which appear to be influenced by the movement of the signals along the tubes (cf. Hameroff, p. 122). This latter fact seems to be telling us that the 'computations' that the tube indulges in may indeed disturb the environment to an extent that they must be treated classically. The amount of disturbance remains fairly small in terms of mass movement, according to the **OR** criterion put forward in §6.12, but for the entire system to remain at the quantum level, it would be necessary that these disturbances do not for long extend outwards into the cell and then outwards beyond the cell's boundaries. To my own mind, there remains sufficient uncertainty, both in relation to the actual physical situation and with regard to how the **OR** criterion of §6.12 is to be applied, that one cannot be sure whether or not an entirely classical picture is appropriate at this stage.

For the purposes of argument, however, let us assume that the microtubules' computations must indeed be treated as essentially classical—in the sense that we do not consider quantum superpositions of different computations to be playing a significant role. On the other hand, let us also envisage that there are genuinely quantum oscillations of some kind taking place *within* the tubes, with some kind of delicate coupling between the quantum internal and classical external aspects of each tube. According to this picture, it would be in this delicate coupling that the *details* of the needed new **OR** theory would most importantly come into play. There would have to be some influence of the quantum interior 'oscillations' on the exterior computations that take place, but that is not unreasonable—in view of the mechanisms that are envisaged as being responsible for the microtubule's cellular-automata-like behaviour, namely weak van der Waals' type of influences between neighbouring tubulin dimers.

Our picture, then, is of some kind of global quantum state which coherently couples the activities taking place within the tubes, concerning microtubules collectively right across large areas of the brain. There is some influence that this state (which may not be simply a 'quantum state', in the conventional sense of the standard quantum formalism) exerts on the computations taking place along the microtubules—an influence which takes delicate and precise account of the putative, missing, non-computational **OR** physics that I have been strongly arguing for. The 'computational' activity of conformational changes in the tubulins controls the way that the tubes transport materials along their outsides (see Fig. 7.13), and ultimately influences the synapse strengths at pre- and postsynaptic endings. In this way, a little of this coherent

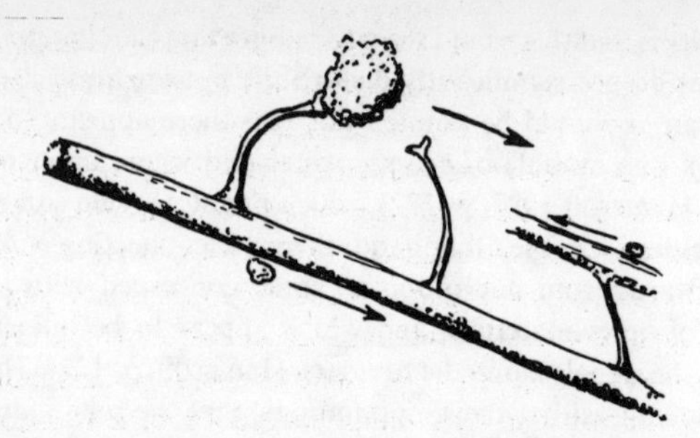

Fig. 7.13. MAPs also transport large molecules, while other molecules move directly along microtubules.

quantum organization *within* the microtubules is 'tapped off' to influence changes in the synaptic connections of the neural computer of the moment.

One might speculate in various ways, in relation to such a picture. There is, for example, a possible role for the puzzling non-locality of the EPR-type effects of quantum entanglement. The strange quantum roles of counterfactuals may be playing their part also. Perhaps the neural computer is poised to enact some computation which it does not actually perform, but (as with the bomb-testing problem) the mere fact that it *might* have performed the computation causes an effect that is different from that which would be the case if it could not perform it. In this way, the classical 'wiring' of the neural computer at any one moment could have an influence on the internal cytoskeletal state, even though the neuron firings that would activate that particular 'wired-up' computer might not actually take place. One may contemplate possible analogies with this sort of thing, in many of the familiar mental activities that we continually indulge in—but I feel that it is better not to pursue matters of this nature further here!

On the view that I am tentatively putting forward, consciousness would be some manifestation of this quantum-entangled internal cytoskeletal state and of its involvement in the interplay (**OR**) between quantum and classical levels of activity. The computer-like classically interconnected system of neurons would be continually influenced by this cytoskeletal activity, as the manifestation of whatever it is that we refer to as 'free will'. The role of neurons, in this picture, is perhaps more like a *magnifying device* in which the smaller-scale cytoskeletal action is transferred to something which can influence other organs of the body—such as muscles. Accordingly, the neuron level of description that provides the currently fashionable picture of the brain and mind is a mere *shadow* of the deeper level of cytoskeletal action—and it is at this deeper level where we must seek the physical basis of *mind*!

There is admittedly speculation involved in this picture, but it is not out of line with our current scientific understanding. We have seen in the last chapter that there are powerful reasons, coming from considerations within present-day physics itself, for expecting that these present physical ideas must be modified—giving new effects at just such a level that could well be relevant to microtubules, and perhaps to the cytoskeleton/neuron interface. By the arguments of Part I, we need an opening for a non-computational physical action if we are to find a physical home for consciousness, and I have argued in Part II that the only plausible place for such action is in a cogent replacement (**OR**) for the quantum-state-reduction process that I have denoted by **R**. We must now address the question of whether there are any purely *physical* grounds for believing that **OR** might indeed be of a non-computational nature. We shall find that, in keeping with the suggestions that I have put forward in §6.12, there are indeed some such grounds.

7.8 Non-computability in quantum gravity: 1

A key requirement in the preceding discussion is that some kind of non-computability should be a feature of whatever new physics should come in to replace the probabilistic **R**-procedure that is used in ordinary quantum theory. I have argued in §6.10 that this new physics, **OR**, should combine the principles of quantum theory with those of Einstein's general relativity—i.e. it should be a quantum-*gravitational* phenomenon. Is there any evidence that non-computability might be an essential feature of whatever theory eventually emerges to unify correctly (and modify appropriately) both quantum theory and general relativity?

In a particular approach to quantum gravity, Robert Geroch and James Hartle (1986) found themselves confronted with a computationally unsolvable problem, namely the *topological equivalence problem for 4-manifolds*. Basically, their approach involved the question of deciding when two four-dimensional spaces are 'the same', from the topological point of view (i.e. when it is possible to deform one of them continuously until it coincides with the other, where the deformation does not allow tearing or gluing the spaces in any way). In Fig. 7.14, this is illustrated in the two-dimensional case, where we see that the surface of a teacup is topologically the same as the surface of a ring, but the surface of a ball is different. In two dimensions, the topological equivalence problem is computationally solvable, but it was shown by A. A. Markov in 1958 that there is no algorithm for solving this problem in the *four*-dimensional case. In fact, what is shown effectively demonstrates that if there were such an algorithm, then one could convert that algorithm into another algorithm which could solve the *halting problem*, i.e. it could decide whether or not a Turing-machine action will stop. Since, as we have seen in §2.5, there is no such algorithm, it follows that there cannot be any algorithm for solving the equivalence problem for 4-manifolds either.

There are many other classes of mathematical problem which are

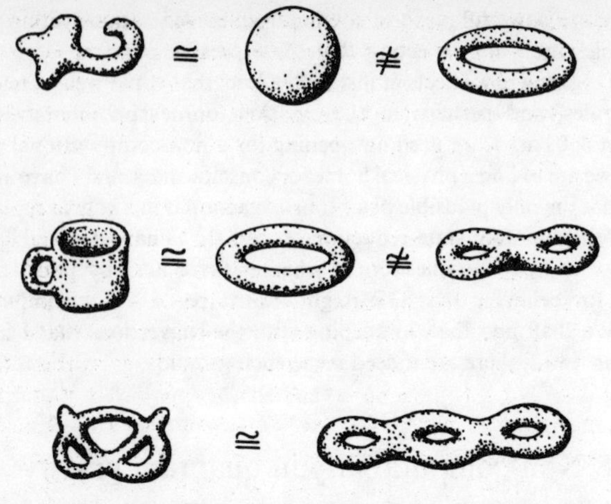

Fig. 7.14. Two-dimensional closed surfaces can be computably classified (roughly by counting the number of 'handles'). On the other hand, four-dimensional closed 'surfaces' *cannot* be computably classified.

computationally unsolvable. Two of these, Hilbert's tenth problem and the tiling problem, were discussed in §1.9. For another example, the word problem (for semigroups), see ENM, pp. 130–2.

It should be made clear that 'computationally unsolvable' does not imply that there are any problems in the class that are individually insoluble in principle. It merely asserts that there is no systematic (algorithmic) means of solving all the problems in the class. In any individual case, it might turn out to be possible to arrive at a solution by means of human ingenuity and insight, perhaps aided by computation, It *might* also happen to be the case that there are members of the class which *are* humanly (or machine-aided humanly) inaccessible. Nothing really seems to be known about this, so one can form one's own opinion on the matter. However, what the Gödel–Turing-type argument as given in §2.5, together with the arguments of Chapter 3, *does* effectively show is that the problems of such a class that *are* accessible by human understanding and insight (aided by computation if you like) form a class that is itself computationally inaccessible. (In the case of the halting problem, for example, §2.5 shows that the class of computations that can be humanly ascertained not to halt cannot be encapsulated by any knowably sound algorithm *A*—and the arguments of Chapter 3 take over from there.)

In the Geroch–Hartle approach to quantum gravity, the equivalence problem for 4-manifolds enters their analysis because, according to the *standard* rules of quantum theory, the quantum-gravitational state would involve superpositions of all possible geometries—here *space–time* geometries, which are four-dimensional things—with complex weighting factors. In order to understand how to specify such superpositions in some kind of unique way (i.e. without 'overcounting'), it is necessary to know when two of these space–times are to be taken to be different and when they are taken to be the same. The topological equivalence problem thus arises as part of that decision.

One may ask: if something of the nature of the Geroch–Hartle approach to quantum gravity turned out to be physically correct, would that mean that there would be something essentially non-computable in the evolution of a physical system? I do not think that the answer to this question is at all clear. It is not even clear to me that the computational unsolvability of the topological equivalence need make the more complete *geometrical* equivalence problem also unsolvable. It is also unclear how (and if) their approach might relate to the **OR** ideas that I have been arguing for here, where an actual change in the structure of quantum theory is expected at the very stage when gravitational effects become involved. Nevertheless, the Geroch–Hartle work does indicate the clear possibility that non-computability may have a genuine role in whatever quantum gravity theory finally emerges as being physically correct.

7.9 Oracle machines and physical laws

All this notwithstanding, one may ask a separate question: suppose that the emergent quantum gravity theory indeed turns out to be a non-computational theory, in the specific sense that it would enable a physical device to be constructed which could solve the halting problem. Would this be sufficient to resolve all the problems that arose from our considerations of the Gödel–Turing argument in Part I? Surprisingly the answer to this question is *no*!

Let us try to see why an ability to solve the halting problem will not help. In 1939, Turing introduced an important concept of relevance to this issue, which he referred to as an *oracle*. The idea of an oracle was that it would be something (presumably a fictional thing, in his mind, that need not be physically constructable) which could indeed solve the halting problem. Thus, if we present the oracle with a pair of natural numbers q, n, then it would, after some finite time, give us the answer **YES** or **NO**, depending upon whether the computation $C_q(n)$ ultimately stops or does not stop (see §2.5). The arguments of §2.5 provide us with a proof of Turing's result that no such oracle can be constructed which works in an entirely computational way, but it does not tell us that an oracle could not be physically constructed. For that conclusion to follow, we would need to know that the physical laws are computational in nature—which issue, after all, is what the discussion of Part II is all about. It should be pointed out, also, that the physical possibility of constructing an oracle is not, as far as I can tell, an implication of the viewpoint that I am promoting. As mentioned above, there is no requirement that all halting problems are accessible to human understanding and insight, so one need not conclude that a constructable device could do this either.

Turing, in his discussion of these matters, considered a modification of computability in which such an oracle could be called in at any stage where desired. Thus, an *oracle machine* (which enacts an *oracle algorithm*) would be like an ordinary Turing machine, except that adjoined to its ordinary computational operations would be another operation: 'Call in the oracle and ask it whether $C_q(n)$ stops; when the answer comes back, continue calculating, making use of that answer.' The oracle can be called in again and again if needed. Note that an oracle machine is just as much of a *deterministic* thing as is an ordinary Turing machine. This illustrates the fact that computability is not at all the same thing as determinism. It would be just as possible, in principle, to have a universe that runs deterministically as an oracle machine as it would to have one that runs deterministically as a Turing machine. (The 'toy universes' that were described in §1.9 and on p. 170 of ENM, would, in effect, be oracle-machine universes.)

Might it be the case that *our* universe actually *does* run as an oracle machine? Curiously, the arguments of Part I of this book can be applied equally well against an oracle-machine model of mathematical understanding as they were against the Turing-machine model, almost without change. All we need do, in the discussion of §2.5, is to read '$C_q(n)$' as now standing for the 'qth oracle machine applied to the natural number n'. Let us rewrite this as, say, $C'_q(n)$. Oracle machines can be (computably) listed just as well as can ordinary Turing machines. As far as their specification is concerned, the only additional feature is that one must take note of the stages at which the oracle is to be brought into the operation, and this presents no new problem. We now replace the *algorithm* $A(q,n)$ of §2.5 with an *oracle algorithm* $A'(q,n)$, which we try to consider represents the totality of means, available to human understanding and insight, for deciding for sure that the oracle operation $C'_q(n)$ does not stop. Following through the arguments, exactly as before, we conclude:

> $\mathscr{G}'$ Human mathematicians are not using a knowably sound oracle algorithm in order to ascertain mathematical truth.

From this, we conclude that a physics which works like an oracle machine will not solve our problems either.

In fact, the whole process can be repeated again, and applied to 'second-order oracle machines' which are allowed to call in, when required, a *second-order oracle*—which can tell us whether or not an ordinary oracle machine ever comes to a halt. Just as above, we conclude:

> $\mathscr{G}''$ Human mathematicians are not using a knowably sound second-order oracle algorithm in order to ascertain mathematical truth.

It should be clear that this process can be repeated again and again, rather in the manner of repeated Gödelization, as discussed in relation to **Q19**. For every recursive (computable) ordinal α, we have a concept of α-order oracle machine, and we seem to conclude:

> $\mathscr{G}^{\alpha}$ Human mathematicians are not using a knowably sound α-order oracle machine in order to ascertain mathematical truth, for any computable ordinal α.

The final conclusion of all this is rather alarming. For it suggests that we must seek a non-computable physical theory that reaches beyond every computable level of oracle machines (and perhaps beyond).

No doubt there are readers who believe that the last vestige of credibility of my argument has disappeared at this stage! I certainly should not blame any reader for feeling this way. But this gives no excuse for not coming to terms with all the arguments that I have given in detail. In particular, the arguments of Chapters 2 and 3 must all be retraced, with α-order oracle machines replacing the Turing machines of that discussion. I do not think that the arguments are affected in any significant way, but I have to confess to boggling, rather, at the prospect of presenting everything all over again in these terms. There is another point that should be made, however, and this is that it need not be the case that human mathematical understanding is in principle as powerful as *any* oracle machine at all. As noted above, the conclusion $\mathscr{G}$ does *not* necessarily imply that human insight is powerful enough, in principle, to solve each instance of the halting problem. Thus, we need not necessarily conclude that the physical laws that we seek reach, in principle, beyond every computable level of oracle machine (or even reach the first order). We need only seek something that is not equivalent to *any* specific oracle machine (including also the zero*th*-order machines, which are Turing machines). Physical laws could perhaps lead to something that is just *different*.

7.10 Non-computability in quantum gravity: 2

Let us return to the question of quantum gravity. It should be stressed that there is no accepted theory at present—there is not even an acceptable candidate. There are, however many different and fascinating proposals.[8] The particular idea that I wish to refer to now has, in common with the Geroch–Hartle approach, the requirement that quantum superpositions of different *space–times* are to be considered. (Many approaches require superpositions only of three-dimensional spatial geometries, which is a little different.) The suggestion, due to David Deutsch[9], is that one must superpose, alongside the 'reasonable' space–time geometries in which *time* behaves fairly sensibly, 'unreasonable' space–times in which there are *closed timelike lines*. Such a space–time is depicted in Fig. 7.15. A *timelike line* describes the possible

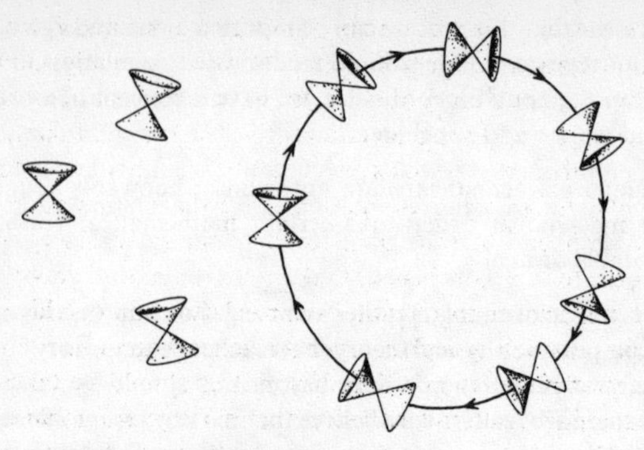

Fig. 7.15. With severe enough light-cone tilting in a space–time, closed timelike lines can occur.

history of a (classical) particle, 'timelike' referring to the fact that the line is always directed within the local light cone at each of its points, so that the local absolute speed is not exceeded—as is required by relativity theory (see §4.4). The significance of a *closed* timelike line is that one could contemplate an 'observer' who actually has that line as his* own world-line, i.e. as the line that describes, within the space–time, the history of his own body. Such an observer would, after a finite passage of his own perceived time, find himself back in his past (time travel!). The possibility seems to be open to him to do something to himself (assuming that he has some kind of 'free will') that he never actually experienced, this leading to a contradiction. (Usually such discussions have him kill his own grandfather before he was born—or something else equally alarming.)

Arguments of this sort provide sufficient reason for not taking seriously space–times with closed timelike lines as possible models of the actual classical universe. (Curiously it was Kurt Gödel who, in 1949, first proposed space–time models with close timelike lines. Gödel did not regard the paradoxical aspects of such space–times as adequate reasons for ruling them out as cosmological models. For various reasons, we would normally take a stronger position on this issue these days, but see Thorne (1994). It would have been interesting to have Gödel's reaction to the use that such space–times will be put to in a moment!) Though it indeed seems reasonable to rule out space–time geometries with closed timelike lines as descriptions of the *classical* universe, a case can be made that they should not be ruled out as potential occurrences that could be involved in a *quantum superposition*. This, indeed, is Deutsch's point. Although the contributions of such geometries to the total state vector may well be utterly minute, their potential presence has (according

*See Notes to the reader, p. xvi.

to Deutsch) a startling effect. If we now consider what it means to perform a quantum computation in such a situation, we apparently come to the conclusion that *non-computable* operations can be performed! This arises from the fact that in the space–time geometries with closed timelike lines, a Turing-machine operation can feed on to its own output, running round indefinitely, if necessary, so that the answer to the question 'does that computation ever stop' has an actual influence on the final result of the quantum computation. Deutsch comes to the conclusion that in his quantum gravity scheme, quantum oracle machines are possible. As far as I can make out, his arguments would apply just as well to higher-order oracle machines also.

Of course, many readers may feel that all this should be taken with an appropriate amount of salt. Indeed, there is no real suggestion that the scheme provides us with a consistent (or even plausible) theory of quantum gravity. Nonetheless, the ideas are logical within their own framework and are suggestively interesting—and it seems quite reasonable to me that when the appropriate scheme for quantum gravity *is* eventually found, then some important vestiges of Deutsch's proposal will indeed survive. In my own view, as was stressed particularly in §6.10 and §6.12, the very laws of quantum theory must become modified (in accordance with **OR**) when the correct union between quantum theory and general relativity is found. But I take it as considerable support for the possibility of a resulting non-computational action that, in Deutsch's approach, non-computability—even to the degree that seems to be required for $\mathscr{G}^{\alpha}$—is a feature of his quantum gravity ideas.

As a final point, it should be remarked that it is precisely the potential tilting of the light cones in Einstein's general relativity (cf. §4.4) that gives us the non-computable effects that Deutsch points out. Once the light cones are allowed to tilt *at all*, even by the minute amounts that occur with Einstein's theory in ordinary circumstances, then there is the *potential* possibility for them to tilt to such a degree that closed timelike lines will be the result. This potential possibility need only play a role as a counterfactual, according to quantum theory, for it to have an *actual* effect!

7.11 Time and conscious perceptions

Let us return to the issue of consciousness. It was the particular role that consciousness played in the perception of mathematical truth, after all, that has led us along the road to the strange territory in which we now find ourselves. But clearly there is far more to consciousness than the perception of mathematics. We followed this particular road only because it seemed possible to move somewhere along it. No doubt many readers will not much like the 'somewhere' at which we have more or less arrived. However, if we look back from our new vantage point, we may find that some of our old problems appear in a new light.

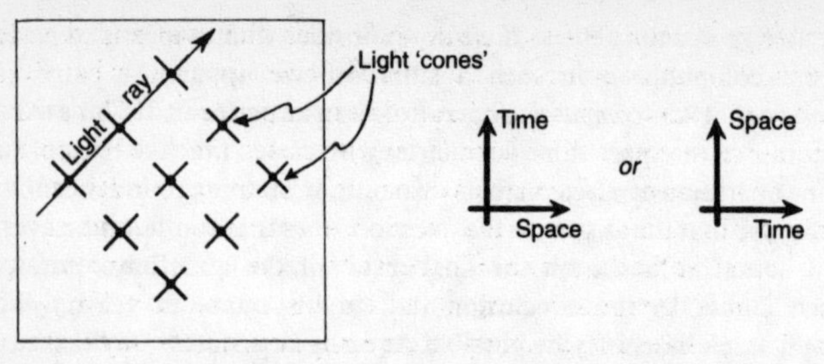

Fig. 7.16. In a two-dimensional space–time there is nothing to choose between space and time—yet no-one would claim that space should 'flow'!

One of the most striking and immediate features of conscious perception is the *passage of time*. It is something so familiar to us that it comes as a shock to learn that our wonderfully precise theories of the behaviour of the physical world have had, up to this point, virtually nothing to say about it. Worse than this, what our best physical theories *do* say is almost in flat contradiction with what our perceptions seem to tell us about time.

According to general relativity, 'time' is merely a particular choice of coordinate in the description of the location of a space–time event. There is nothing in the physicists' space–time descriptions that singles out 'time' as something that 'flows'. Indeed, physicists quite often consider model space–times in which there is only *one* space dimension in addition to the single time dimension; and in such two-dimensional space–times there is nothing to say which is space and which is time (see Fig. 7.16). Yet, no one would consider *space* to 'flow'! It is true that time evolutions are often considered in physical problems, where one may be concerned with computing the future from the present state of the system (cf. §4.2). But this is not at all a necessary procedure, and calculations are normally carried out this way *because* one is concerned with modelling, mathematically, our experiences of the world in terms of the 'flowing' time that we seem to experience—and because we are interested in predicting the future.[10] It is our apparent experiences that tempt us to bias our computational models of the world in terms of time evolutions (frequently, but not invariably) whilst the physical laws themselves do not contain such a compelling inbuilt bias.

In fact it is *only* the phenomenon of consciousness that requires us to think in terms of a 'flowing' time at all. According to relativity, one has just a 'static' four-dimensional space–time, with no 'flowing' about it. The space–time is just *there* and time 'flows' no more than does space. It is only consciousness that seems to need time to flow, so we should not be surprised if the relationship between consciousness and time is strange in other ways too.

Indeed, it would be unwise to make too strong an identification between the phenomenon of conscious awareness, with its seeming 'flowing' of time, and the physicists' use of a real-number parameter t to denote what would be referred to as a 'time coordinate'. In the first place, relativity tells us that there is no uniqueness about the choice of the parameter t, if it is to apply to the space–time as a whole. Many different mutually incompatible alternatives are possible, with nothing particular to choose between one and any other. Second, it is clear that the precise concept of a 'real number' is not completely relevant to our conscious perception of the passage of time, if only for the reason that we have no sensibility of very tiny timescales—say timescales of even just one hundredth of a second, for example—whereas the physicists' time scales hold good down to some 10^{-25} seconds (as is demonstrated by the accuracy of quantum electrodynamics, the quantum theory of electromagnetic fields interacting with electrons and other charged particles), or perhaps even down to the Planck time of 10^{-43} seconds. Moreover, the mathematicians' concept of time as a real number would require that there be *no* limit of smallness whatsoever, below which the concept meaningfully applies, whether or not this concept remains physically relevant at all scales.

Is it possible to be more specific about the relationship between conscious experience and the parameter t that physicists use as the 'time' in their physical descriptions? Can there really be any experimental way to test 'when' a subjective experience 'actually' takes place, in relation to this physical parameter? Does it even *mean* anything, in an objective sense, to say that a conscious event takes place at any particular time? In fact certain experiments of definite relevance to this issue have indeed been carried out, but it turns out that the results are distinctly puzzling, and have almost paradoxical implications. A description of some of these experiments was given in ENM, pp. 439–44, but it will be appropriate to re-examine them here.

In the mid 1970s, H. H. Kornhuber and his associates (cf. Deeke *et al.* 1976) used electroencephalograms (EEGs) in order to record electrical signals at a number of points on the heads of several human volunteers, in order to try to time any brain activity that might be associated with an act of *free will* (the *active* aspect of consciousness). They were asked to flex an index finger, at various times, but to do this suddenly at moments that were *entirely of their own choosing* in the hope that the brain activity involved with 'willing' this finger movement could be timed. Significant signals from the EEG traces could only be obtained by averaging over a number of different runs. What was found was the surprising result that the recorded electric potential appeared to build up gradually for something like a second to a second and one-half, *before* the actual finger-flexing. Does this mean that a conscious act of will takes a second or more in order to act? As far as the subject was actually aware, the decision to flex the finger would have occurred only momentarily before the finger was flexed, and certainly not as long as a second or so before. (It should be borne in mind that a 'preprogrammed' reaction time in response to an external signal is much shorter than this—about one-fifth of a second.)

One appears to conclude from these experiments that: (i) the conscious act of 'free will' is a pure illusion, having been, in some sense, already preprogrammed in the preceding unconscious activity of the brain; *or* (ii) there is a possible 'last-minute' role for the will, so that it can sometimes (but not usually) reverse the decision that had been unconsciously building up for a second or so before; *or* (iii) the subject actually consciously wills the finger-flexing at the earlier time of a second or so before the flexing takes place, but mistakenly perceives, in a consistent way, that the conscious act occurs at the much later time, just before the finger is indeed flexed.

In more recent experiments, Benjamin Libet and his co-workers have repeated these experiments, but with added refinements designed to time more directly the actual act of willing the finger-flexing, by asking the subject to note the positioning of a clock hand at the moment the decision was made (see Libet 1990, 1992). The conclusions seem to confirm the earlier results but also to tell against (iii), and Libet himself appears to favour (ii).

In another class of experiments, the timing of *sensory* (or *passive*) aspects of consciousness was investigated, in 1979, by Libet and Feinstein. They tested subjects who had consented to having electrodes placed in a part of the brain concerned with receiving sensory signals from certain points on the skin. In conjunction with this direct stimulation, there would be occasions when the corresponding point on the skin would be stimulated. The general conclusion of these experiments was that it would take something like half a second of neuronal activity (but with some variation, depending on the circumstances) before the subjects could become consciously aware of any sensation at all, yet in the case of the direct skin stimulation, they would have the impression that they had already been aware of the stimulus at the earlier time when the skin was actually stimulated.

Each of these experiments, by itself, would not be paradoxical, although perhaps somewhat disturbing. Perhaps one's apparent conscious decisions are really made *unconsciously* at some earlier time, at least a second before. Perhaps one's conscious sensations *do* need something like half a second of brain activity before they can be actually evoked. But if we take these two findings together, then we seem to be driven to the conclusion that in any action in which an external stimulus leads to a consciously controlled response, a time delay of some one and one-half seconds would seem to be needed before that response can occur. For awareness would not even take place until half a second has passed; and if that awareness is to be put to use, then the apparently sluggish machinery of free will would then have to be brought into play, with perhaps another second's delay.

Are our conscious responses really that slow? In ordinary conversation, for example, this does not seem to be the case. Accepting (ii) would lead one to conclude that most acts of response are entirely unconscious, whilst from time to time one might be able to override such a response, in about a second, with a conscious one. But if the response is usually *un*conscious, then unless it is as slow as a conscious one, there is no chance for consciousness to override it—otherwise, when the conscious act comes into play, the unconscious response has already been made and it is now too late for consciousness to affect it! Thus, unless conscious acts can *sometimes* be swift, the unconscious response would *itself* have to take about a second. In this connection, we recall that a 'preprogrammed' unconscious reaction can occur much more quickly—in about one-fifth of a second.

Of course we could still have a rapid unconscious response (of, say, one-fifth of a second) together with the possibility (i), which could go along with the total ignoring, by the unconscious response system, of any conscious (sensory) activity that might come along later. In this case (and the situation with (iii) is even worse) the only role for our consciousness in reasonably rapid conversation would be that of a spectator, being aware only of an 'action replay' of the entire drama.

There is no actual contradiction here. It is possible that natural selection has produced consciousness just for its role in deliberate thinking, whilst in any reasonably rapid activity consciousness is just a passenger. The entire discussion of Part I, after all, was in relation to the kind of conscious contemplation (mathematical understanding) that is indeed notoriously slow. Perhaps the faculty of consciousness *has* evolved only for the purpose of such slow and contemplative mental activity, while the more rapid response times are entirely unconscious in action—yet accompanied by a delayed conscious perception of them which plays no active role.

It is certainly true that consciousness comes into its own when it is allowed a long time to work. But I have to confess to a disbelief in the possibility that there can be *no* role for consciousness in such reasonably rapid activities like ordinary conversation—or in table tennis, squash, or motor racing, for that matter. It seems to me that there is at least one profound loophole in the above discussion, and that is in the assumption that the precise timing of conscious events actually makes sense. Is there *really* an 'actual time' at which a conscious experience does take place, where that particular 'time of experience' must precede the time of any effect of a 'free-willed response' to that experience? In view of the anomalous relation that consciousness has to the very physical notion of time, as was described at the beginning of this section, it seems to me to be at least possible that there is *no* such clear-cut 'time' at which a conscious event must occur.[11]

The mildest possibility in accordance with this would be a non-local spread in time, so that there would just be an inherent fuzziness about the relationship between conscious experience and physical time. But my own guess would be that there is something much more subtle and puzzling at work. If consciousness is a phenomenon which cannot be understood in physical terms without an essential input from quantum theory, then it might well be the case that the **Z**-mysteries of that theory are interfering with our seemingly watertight conclusions about the causality, non-locality, and counterfactuality properties that might actually exist between awareness and free will. For example, perhaps some kind of role is played by the type of counterfactuality that occurs with the bomb-testing problem of §5.2 and §5.9: the mere fact that some act or thought *might* be going to take place, even though it actually does not, can affect behaviour. (This might invalidate seemingly logical deductions as in, say, ruling out possibility (ii) as above.)

In a general way, one must be very wary about coming to seemingly logical conclusions concerning temporal ordering of events when quantum effects are involved (as the EPR considerations of the following section will serve to emphasize). The converse of this is that *if*, in some manifestation of consciousness, classical reasoning about the temporal ordering of events leads us to a contradictory conclusion, then this is a strong indication that quantum actions are indeed at work!

7.12 EPR and time: need for a new world-view

There are reasons for being suspicious of our physical notions of time, not just in relation to consciousness, but in relation to physics itself, when quantum non-locality and counterfactuality are involved. If we take a strongly 'realistic' view of the state vector $|\psi\rangle$ in EPR situations—and I have argued strongly in §6.3 and §6.5 about the difficulties in *not* doing so—then we are presented with a profound puzzle. Such puzzles lead to genuine difficulties for any detailed theory of the nature of the GRW theory, described in §6.9, and potentially also with the **OR** type of scheme of §6.12, which I am promoting myself.

Recall the magic dodecahedra of §5.3 and their explanation in §5.18. We ask: which of the following two possibilities represents the 'reality' of things? Is it my *colleague's* button-pressing that instantaneously reduces (and disentangles) the original entangled total state—so that the state of the atom in my own dodecahedron is instantaneously created, disentangled, by *his* button-pressing, and it is *that* reduced state which defines the possibilities that can result from my own subsequent button-pressing? Or does my *own* button-pressing come first, acting on the original entangled state, to reduce the state instantaneously on the atom in my colleague's dodecahedron, so that it is my colleague who encounters the reduced disentangled state? It does not matter, as far as the results are concerned, which way we treat the problem, as was remarked upon in the discussion of §6.5. It is just as well that it does not matter, for if it did, then this would violate the principles of Einstein's relativity in which the notion of 'simultaneity' for distant (spacelike-separated) events can have no physically observable effects. However, if we believe that $|\psi\rangle$ represents *reality*, then this reality is indeed different in the two pictures. Some would regard this as reason enough not to take such a realistic view of $|\psi\rangle$. Others would accept the other strong reasons in favour of the realistic view (cf.

§6.3)—and would be quite prepared to jettison the Einsteinian picture of the world.

My own inclinations are to try to hang on to both—quantum realism *and* the spirit of the relativistic space–time view. But to do so will require a fundamental change in our present way of representing physical reality. Rather than insisting that the way in which we describe a quantum state (or even space–time itself) must follow the descriptions that are familiar to us now, we should seek, instead, something that looks very different, though (initially at least) it would be mathematically equivalent to the familiar descriptions.

In fact there is a good precedent for this kind of thing. Before Einstein discovered general relativity, we had become thoroughly used to Newton's wonderfully accurate theory of gravity in which particles, moving about in a flat space, attracting each other according to the inverse square law of gravitational force. One would have thought that introducing any fundamental change into that picture would be bound to destroy the remarkable accuracy of Newton's scheme. Yet, such a fundamental change is just what Einstein did introduce. In his alternative view of gravitational dynamics, the picture is completely transformed. Space is no longer flat (and it is not even 'space', it is 'space–time'); there is no gravitational force, this being replaced by the tidal effects of space–time curvature. And particles do not even move, being represented as 'static' curves drawn on space–time. Was the remarkable accuracy of Newton's theory destroyed? Not at all; it was even improved upon, to an extraordinary degree! (See §4.5.)

Might we expect that something of a similar kind could occur with quantum theory? I think that it is extremely likely. It will need a *profound* change of viewpoint, which makes it hard to speculate on the specific nature of the change. Moreover, it will undoubtedly look crazy!

To end this section, I shall mention two crazy-looking ideas, neither of which is crazy enough, but each of which has its merits. The first is due to Yakir Aharonov and Lev Vaidman (1990) and to Costa de Beauregard (1989) and Paul Werbos (1989). According to this idea, quantum reality is described by *two* state vectors, one of which propagates forwards in time from the last occurrence of **R**, in the normal way, and the other propagates *backwards in time*, from the next occurrence of **R** in the future. This second state vector* behaves 'teleologically' in the sense that it is governed by what is going to happen to it in the future, rather than what happened to it in the past, a feature that some might feel would be unacceptable. But the implications of the theory are precisely the same as in standard quantum theory, so it cannot be ruled out on grounds of this nature. Its *advantage* over standard quantum theory is that

*There is some mathematical significance in assigning the backwards-evolving state vector a 'bra-vector' $\langle\phi|$, while the normal forward-evolving one is assigned a standard 'ket-vector' $|\psi\rangle$. The pair of state vectors might be represented as a product $|\psi\rangle\langle\phi|$. This is in keeping with the density-matrix notation of §6.4.

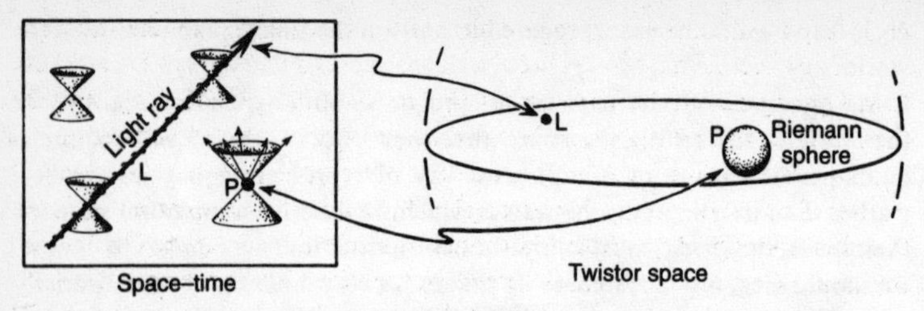

Fig. 7.17. *Twistor theory* provides an alternative physical picture to that of space–time, whereby entire light rays are represented as points, and events by entire Riemann spheres.

it enables one to have a completely objective description of the state in EPR situations which can be represented in space–time terms consistently with the spirit of Einstein's relativity. Thus it provides a (kind of) solution to the puzzle referred to at the beginning of this section—but at the expense of having to have a teleologically behaving quantum state, which many may find worrying. (For myself, these teleological aspects of descriptions are perfectly acceptable, so long as they do not lead to problems with actual physical behaviour.) For details, I refer the reader to the literature.

The other idea that I wish to mention is that of *twistor theory* (see Fig. 7.17). This theory was strongly motivated by the same EPR puzzles, but it does not (as such) *so far* provide a solution to them. Its strengths lie elsewhere, in providing unexpected and elegant mathematical descriptions of certain fundamental physical notions (such as Maxwell's electromagnetic equations, cf. §4.4 and ENM, pp. 184–7, for which it yields an attractive mathematical formulation). It provides a non-local description of space–time, where entire light rays are represented as single points. It is this space–time non-locality that relates it to the quantum non-locality of EPR situations. It is also fundamentally based on *complex numbers* and their related geometry, so that an intimate relationship between the complex numbers of **U**-quantum theory and space–time structure is obtained. In particular, the Riemann sphere of §5.10 plays a fundamental role in relation to the light cone of a space–time point (and to the 'celestial sphere' of an observer at that point). (See David Peat's (1988) account, for a non-technical description of the relevant ideas, or the book by Stephen Huggett and Paul Tod (1985) for a relatively brief but technical one.[12])

It would be inappropriate for me to dwell on these matters further here. I mention them only to indicate that there are various possibilities for changing our already extraordinarily accurate picture of the physical world into something that looks quite different from the pictures that we hold to today.

As an important consistency requirement for such a change, we must be able to use the new description to reproduce all the successful results of **U**-quantum theory (and general relativity as well). But we should also be able to go further than this, and incorporate the physically appropriate modification of quantum theory in which the **R** procedure is replaced by some real physical process. At least that is my own strong belief; and it is also my present opinion that this 'appropriate modification' must be something along the **OR** lines of the ideas described in §6.12. It should be mentioned that theories in which relativity is combined with a 'realistic' state reduction, such as the GRW theory, have so far encountered very severe problems (particularly with regard to energy conservation. This tends to reinforce my own belief that a fundamental change in the way that we view the world is needed, before we can make any profound progress in these central physical issues.

It is also my own belief that any genuine progress in the physical understanding of the phenomenon of *consciousness* will also need—as a prerequisite—that same fundamental change in our physical world-view.

Notes and references

1. See, for example, Lisboa (1992).
2. French (1940), Gelber (1958), Applewhite (1979), Fukui and Asai (1976).
3. Dryl (1974).
4. Hameroff and Watt (1982), Hameroff (1987), Hameroff *et al.* (1988).
5. See Koruga *et al.* (1993), for an accessible reference on clathrins, and Curl and Smalley (1991), for a popular account of fullerenes.
6. See Stretton *et al.* (1987).
7. For example, Hameroff's switching time for the tubulin dimers seems to agree with Fröhlich's frequency of about 5×10^{10} Hz.
8. See, for example, Isham (1989, 1994), Smolin (1993, 1994).
9. This idea was to be found in an early draft of the paper by Deutsch (1991), but it did not appear in the published paper. David Deutsch has assured me that the reason that he removed this portion from the final version was not that the idea is 'wrong' but that it is not relevant to the particular purpose of that paper. In any case, for my own purposes, the value of the idea is not for it to be 'correct' within any existing framework of quantum gravity—since there is no such consistent framework at the present time—but that it should be suggestive as to future developments, as indeed it is!

   For an alternative approach to non-computability in 'quantum computing' see Castagnoli *et al.* (1992).
10. In any case, our normal physical representations of time do not distinguish 'flowing' into the future from 'flowing' into the past. (However, because of the second law of thermodynamics, 'retrodicting the past' is not something that can be effectively achieved by means of the time-evolution of dynamical equations.)

11. See also Dennett (1991).

    Some people who have seen the film *A Brief History of Time*, concerning Stephen Hawking and his work, may have obtained a very curious view of my own opinions concerning the relation of consciousness to the flow of time. I wish to take this opportunity to point out that this was due to some highly misleading and inappropriate cutting in the filmed sequence.
12. For further information about twistors, see also Penrose and Rindler (1986), Ward and Wells (1990), Bailey and Baston (1990).

8

Implications?

8.1 Intelligent artificial 'devices'

What do we conclude, from the discussions above, is the ultimate potential of artificial intelligence? The arguments of Part I strongly made the case that the technology of electronic computer-controlled robots will *not* provide a way to the artificial construction of an *actually* intelligent machine—in the sense of a machine that understands what it is doing and can act upon that understanding. Electronic computers have their undoubted importance in clarifying many of the issues that relate to mental phenomena (perhaps, to a large extent, by teaching us what genuine mental phenomena are *not*), in addition to their being extremely powerful and valuable aids to scientific, technological, and social progress. Computers, we conclude, do something very different from what *we* are doing when we bring our awareness to bear on some problem.

However, it should be clear from the later discussions of Part II that I am by no means arguing that it would be necessarily impossible to build a genuinely intelligent *device*, so long as such a device were not a 'machine' in the specific sense of being computationally controlled. Instead it would have to incorporate the same kind of physical action that is responsible for evoking our own awareness. Since we do not yet have any physical theory of that action, it is certainly premature to speculate on when or whether such a putative device might be constructed. Nevertheless, its construction can still be contemplated within the viewpoint $\mathscr{C}$ that I am espousing (cf. §1.3), which allows that mentality can eventually be understood in scientific though non-computational terms.

I do not perceive any necessity that such a device be biological in nature. I perceive no essential dividing line between biology and physics (or between biology, chemistry, and physics). Biological systems indeed tend to have a subtlety of organization that far outstrips even the most sophisticated of our (often very sophisticated) physical creations. But, in a clear sense, these are still early days in the physical understanding of our universe—particularly in relation to mental phenomena. Thus, it must be expected that our physical constructions will increase greatly in their sophistication in the future. We may anticipate that this future sophistication may involve physical effects that can no more than dimly be perceived at present.

I see no reason to doubt that, in the more immediate future, some of the puzzling (**Z**-mystery) effects of quantum theory will find surprising applications in appropriate circumstances. Already there are ideas for employing quantum effects in cryptography to achieve things that no classical device can. In particular, there are theoretical proposals, depending essentially upon quantum effects (cf. C. Bennett *et al.* 1983), whereby secret information may be sent from one person to another in such a way that it is not possible for a third party to eavesdrop without being detected. Experimental devices have already been constructed based on these ideas, and it is by no means out of the question that they will find commercial applicability within a few years. Numerous other schemes have been put forward within the general area of cryptography which make use of quantum effects, and the embryonic subject of *quantum cryptography* is now developing rapidly. Moreover, someday it may perhaps become possible actually to build a *quantum computer*, but at the moment these theoretical constructs are very far from practical realization, and it is hard to predict when—or even whether—they might eventually be physically constructed (see Obermayer *et al.* 1988*a*,*b*).

It is even harder to predict the possibility (or the timescale) for building a device whose action depends on a physical theory that we do not even *know* at present. Such a theory would be needed, I am claiming, before we could understand the physics underlying a non-computably acting device—'non-computably', that is, in the Turing-machine-inaccessible sense that I have been using in this book. According to my own arguments, in order to build such a device we should first need to find the appropriate physical (**OR**) theory of quantum-state reduction—and it is very hard to know how far we are from such a theory—before we could begin to contemplate its construction. It is also possible that the specific nature of that **OR** theory might itself provide an unexpected complexion on the very task at hand.

At least, I *suppose* that we should need to find the theory first, if we are to construct such a non-computational device. But conceivably not: in actual practice, it has often been the case that surprising new physical effects have been discovered many years before their theoretical explanation. A good example was superconductivity, which was originally observed experimentally (by Heike Kammerlingh Onnes in 1911) nearly 50 years before the full quantum-theoretic explanation was eventually found, by Bardeen, Cooper, and Schrieffer in 1957. Moreover, high-temperature superconductivity was discovered in 1986, cf. Sheng *et al.* (1988), also without prior good reason to believe in it on purely theoretical grounds. (As of early 1994, there is still no adequate theoretical explanation for this phenomenon.) On the other hand, in the case of non-computable activity, it would be hard to see how one could even *tell* when a given insentient object is behaving non-computably. The whole concept of computability is very much bound up with *theory*, rather than being directly an observational matter. But within some non-computable theory, there could well be behaviour, characteristic of the non-computational features of that theory, which could be tested, and which an actual device might exhibit. My guess is that without first having the theory it would be very unlikely that non-computational behaviour could be observed or exhibited in a physically constructed object.

For the purpose of further argument, let us now try to imagine that we *have* the required physical theory—which, as I have argued, should be a non-computational **OR** theory of quantum-state reduction—and that we also have some experimental confirmation of that theory. How then would we proceed to construct an *intelligent* device? We *could not*—just on the basis of this. We would need yet another breakthrough in theory: that breakthrough which tells us how consciousness actually comes about as a result of some appropriate organization, in which non-computational **OR** effects are being harnessed appropriately. I, for one, have no idea what kind of a theoretical development this might be. As with the above examples of superconductivity, one might again imagine that a device of the required nature could be hit upon partly accidentally, without there being a proper theory of consciousness. It goes without saying that this seems very improbable—unless, of course, advantage is taken of some Darwinian evolution process, so that the intelligence might eventually arise simply through the direct benefits that this consciousness confers, without there being any understanding on our part as to how it was done (which, indeed, is how it happened with us!). This could hardly be other than an extremely lengthy business, especially when one considers how slowly it takes consciousness to manifest its advantages. The reader may well come to the conclusion that the much more satisfactory way of constructing intelligent devices is by adopting the haphazard but remarkably effective and compelling procedures that we have already been using for millennia!

Of course none of this will stop us from wanting to know what it is that is really going on in consciousness and intelligence. I want to know too. Basically the arguments of this book are making the point that what is *not* going on is solely a great deal of computational activity—as is commonly believed these days—and what *is* going on will have no chance of being properly understood until we have a much more profound appreciation of the very nature of matter, time, space, and the laws that govern them. We shall need also to have much better knowledge of the detailed physiology of brains, particularly at the very tiny levels that have received little attention until recent years. We shall need to know more about the circumstances under which consciousness arises or disappears, about the curious matter of its timing, of what it is used for, and what are the specific advantages of its possession—in addition to many other issues where objective testing is possible. It is a very broad field indeed, in which progress in many different directions is surely to be anticipated.

8.2 Things that computers do well—or badly

Even if it is accepted that the present concept of a computer will *not* achieve any actual intelligence or awareness, we are still presented with the extraordinary power that modern computers possess, and with the potential prospect for an absolutely enormous increase in this power in the future (cf. §1.2, §1.10 and Moravec 1988). Although these machines will not *understand* the things they are doing, they will do them almost incredibly fast and accurately. Will such activity—although still mindless—be able to achieve things that we use our minds for, perhaps more effectively than we can? Can we get any feeling for the kinds of things that computer systems will become very good at, or that minds may always remain better at?

Already, computers can play chess extraordinarily well—approaching the level of the very best human grandmasters. At draughts (which, to US readers, is the game of checkers), the computer Chinook has proved itself superior to all but the supreme champion Marion Tinsley. However, with the ancient oriental game of go, computers seem to have got almost nowhere. When such games are required to be played very fast, this is a factor in the computer's favour; whereas being allowed a lot of time is to the human player's advantage. Chess problems two or three moves deep can be solved almost instantly by a computer, no matter how hard a human may find the problem. On the other hand, a problem with a simple idea but requiring, say 50 or 100 moves, might defeat the computer completely, whereas the experienced human solver might not encounter too much difficulty; cf. also §1.15, Fig. 1.7.

These differences can be largely understood in terms of certain distinctions between what computers and humans are good at. The computer just performs calculations without any understanding of what it is doing—although it employs certain of the understandings of its *programmers*. It can contain a great deal of stored knowledge, whereas a human player can do this also. The computer can make extremely rapid and accurate repeated applications of the programmers' understandings in an entirely mindless way, but it does this to a degree which far outstrips the ability of any human being. The human player needs to keep reapplying judgements and forming meaningful plans, with an overall understanding of what the game is all about. These are qualities that are not available to the computer at all, but to a large extent, it can use its computational power to offset its lack of actual understanding.

Suppose that the number of possibilities per move that the computer need consider is p, on the average; then for a depth of m moves, there would be about p^m alternatives for it to consider. If the calculation of each alternative takes a time t on the average, say, then we have something like

$$T = t \times p^m$$

for the total time T needed to calculate to that depth. In draughts, the number

p is not very large, say about four, which enables the computer to calculate to a considerable depth in the available time, in fact to about 20 moves ($m=20$), whereas with the game of go we might have something like $p=200$, so a comparable computer system could probably manage no more than five ($m=5$) moves or so. The case of chess is somewhat intermediate. Now, we must bear in mind that the human being's judgements and understandings would be much slower things than those of the computer (large t for the human, small t for the computer) but that these judgements could cut down the *effective* number p very considerably (small effective p for the human, large p for the computer), because only a small number of the available alternatives would be judged by the human player to be worthy of consideration.

It follows, in a general way, that games for which p is large, but can effectively be cut down significantly by the use of understanding and judgement, are relatively to the advantage of the human player. For, given a reasonably large T, the human act of reducing the 'effective p' makes much more difference in achieving a large m in the formula $T=t\times p^m$ than does making the time t very small (which is what computers are good at). But for a *small* T, it is making t very small that can be more effective (since the m-values of relevance are likely to be small). These facts are simple consequences of the 'exponential' form of the expression $T=t\times p^m$.

This consideration is a little crude, but I believe that the essential point is reasonably clear-cut. (If you are a non-mathematical reader, and you wish to get a feeling for the behaviour of $t\times p^m$, just try a few examples of values t, p, and m.) It is not worth going into a great deal of further detail here, but one point of clarification may be helpful. It might be argued that 'large calculation depth' as measured by 'm' is not really what the human player strives for. But *in effect* it is. When the human player judges the value of a position at a level a few moves deep, and then considers that it is not helpful to calculate further, this is an *effective* calculation to a much greater depth, since the human judgement encapsulates the probable effect of continued moves. In any case, on the rough basis of considerations of this kind, it is possible to obtain some understanding of why it is much harder to make computers play good go than good draughts, why computers are good at short but not long chess problems, and why they have a relative advantage with short time limits for play.

These arguments are not particularly sophisticated, but the essential point is that the quality of human *judgement*, which is based on human *understanding*, is an essential thing that the computers lack, and this is generally supported by the above remarks—as it is supported also by the considerations of the chess position of Fig. 1.7 in §1.15. Conscious understanding is a comparatively slow process, but it can cut down considerably the number of alternatives that need to be seriously considered, and thereby greatly increase the *effective* depth of calculation. (The alternatives need not even be tried, beyond a certain point.) In fact, it seems to me that if one considers what computers might achieve in the future, one good guide to the answer would be obtained if one asks the question 'is actual understanding needed in the performance of the task?' Many things in our daily lives do not need much understanding in order to perform them, and it is quite possible that computer-controlled robots will get very good at such things. Already there are machines controlled by artificial neural networks that perform very creditably in tasks of this kind. For example, they can do reasonably well with recognizing faces, prospecting for minerals, recognizing faults in machinery by distinguishing different sounds, checking up on credit-card fraud, etc.[1] In a general way, where these methods are successful, the abilities of these machines approach, or sometimes might somewhat exceed, those of average human experts. But with such bottom-up programming, we do not see the kind of powerful machine 'expertise' that occurs with top-down systems, say with chess computers, or—even more impressively—with straightforward numerical calculation, where the best human calculators come nowhere close to what can be achieved by electronic computers. In the case of tasks that are effectively dealt with by (bottom-up) artificial neural network systems, it is probably fair to say that there is not a great deal of understanding involved in the way the *humans* perform these tasks any more than there is in the way that the computer does it, and some limited degree of success on the part of computers is to be expected. Where there is a good deal of specific top-down organization in the computer's programming, as there is with numerical calculation, chess computers, or computing for scientific purposes, then the computers can become very powerfully effective. In these cases, again the computer does not need any actual understanding on its own part, but now because the relevant understanding has been provided by the human programmers (see §1.21).

It should also be mentioned that, very frequently, computer errors occur in top-down systems—because the programmer has made a mistake. But that is a result of human error, which is another matter altogether. Automatic error-correcting systems can be introduced—and these have their value—but errors which are too subtle cannot be caught in this way.

The kind of situation in which it might be dangerous to put too much faith in an entirely computer-controlled system is when the system can carry on for a long time performing reasonably well, and perhaps even giving people the *impression* that it is understanding what it is doing. Then, unexpectedly, it might do something that appears completely crazy, revealing that it never *really* had any understanding at all (as was the case with Deep Thought's failure, with the chess position of Fig. 1.7). Thus, one should always be on one's guard. Armed with the realization that 'understanding' is simply not a computational quality, we can proceed in the knowledge that there is no possibility of a purely computer-controlled robot possessing any of that quality.

Of course, human beings themselves differ very much with regard to their own possession of the quality of understanding. And, as with computers, it is quite possible for a human to give the impression that understanding is present when it is not. There is something of a trade-off between genuine understanding on the one hand, and memory and calculational powers on the other. Computers are capable of the latter but not of the former. As is very familiar to teachers at all levels (but *not*, alas, always familiar to governments) it is the quality of *understanding* which is by far the more valuable. It is *this* quality, rather than the mere parroting of rules or information, that one wishes to encourage in one's pupils. Indeed, it is one of the skills in the construction of examination questions (as is particularly the case in mathematics) that they should test the candidate's understanding, as distinct merely from memory and calculational powers—though there is indeed value in those other qualities also.

8.3 Aesthetics, etc.

I have concentrated, in the above discussion, on the quality of 'understanding' as something essential that is missing from any purely computational system. That particular quality was, after all, what featured in the Gödel argument of §2.5—and whose absence, within the mindlessness of computational action, revealed the essential limitations of computation, thus spurring us on to try to find something better. Yet 'understanding' is but one of the qualities for which conscious awareness is of value to us. More generally, we conscious beings gain benefit from any circumstance where we can directly 'feel' things; and *this*, I am arguing, is just what a purely computational system can never achieve.

We may well ask: in what way is a computer-controlled robot *disadvantaged* by its inability to feel, so that it could not appreciate, say, the beauty of a starlit sky, or the magnificent splendour of the Taj Mahal in a still evening, or the magical complexities of a Bach fugue—or even the stark beauty of the Pythagorean theorem? We could simply say that it is the robot's loss that it cannot feel what we are capable of feeling when confronted by such manifestations of quality. Yet there is more to the matter than this. We might ask a different question. Accepting that the robot is not actually capable of *feeling* anything, might not a cleverly programmed computer be nevertheless capable of producing great works of art?

This is a delicate question, it seems to me. The short answer, I believe, is simply 'no'—if only because the computer cannot possess the sensual qualities that are necessary in order to judge the good from the bad, or the superb from the merely competent. But, we may ask: why is it *necessary* for the computer actually to 'feel', in order for it to develop its own 'aesthetic criteria' and to form its own judgements? One might imagine that such judgements could simply 'emerge' after a long period of (bottom-up) training. However, as with the quality of understanding, I feel that it is much more probable that the criteria would have to be part of the computer's deliberate input, these criteria having been carefully distilled from a detailed top-down analysis (very possibly computer aided) that has been carried out by aesthetically sensitive human beings. Indeed, schemes of this very kind have been put into action by a number of AI researchers. For example, Christopher Longuet-Higgins, in work performed at the University of Sussex, has implemented various computer systems that compose music according to criteria that he has provided. Even in the eighteenth century, Mozart and his contemporaries showed how to construct 'musical dice' that could be used to combine known aesthetically pleasing ingredients with random elements in order to produce vaguely creditable compositions. Similar devices have been adopted in the visual arts, such as the 'AARON' system programmed by Harold Cohen, which can produce numerous 'original' line drawings by invoking random elements to combine fixed-input ingredients according to certain rules. (See Margaret Boden's (1990) book *The Creative Computer* for many examples of this kind of 'computer creativity'; also Michie and Johnston (1984).)

I think that it would be generally accepted that the product of this sort of activity has not, as yet, been anything that could stand comparison with what can be achieved by a moderately competent human artist. I feel that it would not be inappropriate to say that what is lacking, when the computer's input reaches any significant level, is any 'soul' in the resulting work! That is to say, the work *expresses* nothing because the computer itself *feels* nothing.

Of course, from time to time, such a randomly generated computer-produced work might, simply by chance, have genuine artistic merit. (This is related to the old matter of generating the play *Hamlet* by typing letters entirely at random.) Indeed, it must be admitted, in this context, that Nature herself is capable of producing many works of art by random means, as in the beauty of rock formations or the stars in the sky. But without the ability to *feel* that beauty, there is no means of distinguishing what is beautiful from what is ugly. It is in this *selection* process that an entirely computational system would show its fundamental limitations.

Again, one could envisage that computational criteria could be fed in to the computer by a human being, and these might work fairly well, so long as it is just a matter of generating large numbers of examples of the same sort of thing (as one might imagine could be done with run-of-the-mill popular art)—until the products of this activity become boring, and something new is needed. At that point, some *genuine* aesthetic judgements would be required, in order to perceive which 'new idea' has artistic merit and which has not.

Thus, in addition to the quality of *understanding*, there are other qualities that will always be lacking in any entirely computational system, such as *aesthetic* qualities. To these must be added, it seems to me, other kinds of things that require our awareness, such as *moral* judgements. We have seen in Part I that the judgement of what is or is not *true* cannot be reduced to pure computation. The same (perhaps more obviously) applies to the *beautiful*, or to the *good*. These are matters which require awareness and are thus inaccessible to entirely computer-controlled robots. There must always be a continuing controlling input from a sensitive, outside, conscious—presumably human—presence.

Irrespective of their non-computational nature we may ask: are the qualities of 'beauty' and 'goodness' *absolute* ones, in the Platonic sense in which the term 'absolute' is applied to truth—especially to mathematical truth? Plato himself argued in favour of such a standpoint. Might it be the case that our awareness is somehow able to make contact with such absolutes, and it is *this* that gives consciousness its essential strength? Perhaps there might be some clue, here, as to what our consciousness actually 'is' and what it is 'for'. Does awareness play some kind of role as a 'bridge' to a world of Platonic absolutes? These issues will be touched upon again in the final section of this book.

The question of the absolute nature of morality is relevant to the legal issues of §1.11. There is relevance, also, to the question of 'free will', as was raised at the end of §1.11: might there be something that is beyond our inheritance, beyond environmental factors, and beyond chance influences—a separate 'self' that has a profound role in controlling our actions? I believe that we are very far from an answer to this question. As far as the arguments of this book go, all that I could claim with any confidence would be that whatever is indeed involved must lie in principle beyond the capabilities of those devices that we presently call 'computers'.

8.4 Some dangers inherent in computer technology

With any broad-ranging technology, there are likely to be dangers as well as benefits. Thus, in addition to the evident advantages that computers provide, there are also many potential threats to our society that are inherent in the rapid advance of this particular technology. One of the main problems would seem to be the extraordinary interconnected complication that computers present us with, so that there is no chance that any individual human being can comprehend their implications in totality. It is not just a matter of computer technology, but also of the almost instantaneous global communication that links computers to each other almost completely around the planet. We see something of the problems that can arise in the unstable way that the stock market behaves, where transactions are performed virtually instantaneously on the basis of world-wide computer predictions. Here, perhaps, the problem may be not so much the individual's lack of understanding of the interconnected system as a whole, but an instability (not to mention unfairness) inherent in a system which is geared to enabling individuals to make instantaneous fortunes simply by outcomputing or outguessing their competitors. But it is very likely that other instabilities and potential dangers may well arise simply because of the sheer complication of the interconnected system as a whole.

I suspect that some people might believe that it may *not* be so serious a problem if, in the future, the interconnected system gets so complicated that it is beyond human comprehension. Such people might have faith in the prospect that eventually the computers *themselves* would acquire the necessary understanding of the system. But we have seen that understanding is not a quality that computers are even *capable* of, so there can be no chance of genuine relief from that quarter.

There are additional problems of a different kind which result merely from the fact that advances in technology are very rapid, so that a computer system can become 'out of date' soon after it has appeared on the market. The resulting requirements of continual updating, and of having to use systems that are often not adequately tested because of competition pressures, must surely get even worse in the future.

The profound problems that we are just beginning to face in the new technology of the computer-aided world, and the rapidity of the pace of change, are all too numerous, and it would be foolish for me to attempt to summarize them here. Issues such as personal privacy, industrial espionage, and computer sabotage are among the things that come to mind. Another alarming possibility for the future is the ability to 'forge' an image of a person, so that this image could be presented on a television screen expressing views that the actual individual would have no wish to express.[2] There are also social issues that are not so specifically computer matters but which are computer related, such as the fact that, because of the wonderfully accurate powers of reproduction of musical sound, or visual images, the expertise of a very small number of popular performers can be propagated world-wide, perhaps to the disadvantage of those who are not so favoured. With 'expert systems' where the expertise and understanding of a small number of individuals—in, say, the legal or medical profession—can be put into a computer software package, perhaps to the detriment of the local lawyers or doctors, we encounter something similar. My guess, however, is that the local *understanding* that a personal involvement provides will mean that such computer-controlled expert systems will remain aids rather than replacements to local expertise.

Of course there is an 'upside' for the rest of us in all these developments, when they can be carried out successfully. For expertise becomes much more freely available and can be appreciated by a much larger public. Likewise, on the matter of personal privacy, there are now 'public key' systems (see Gardner 1989), which can in principle be used by individuals or by small concerns—just as effectively as by large ones—and which *appear* to provide complete security from eavesdropping. These depend, by their very nature, on the availability of very fast powerful computers—though their effectiveness depends upon the computational difficulty of factorizing large numbers—something now threatened by advances in quantum computation (see §7.3; also Obermayer *et al.* (1988*a*,*b*) for ideas pointing to a future feasibility of quantum computing). As mentioned in §8.1, there is the possibility of using quantum cryptography as security against eavesdropping—which would also depend for its effectiveness on significant amounts of computing. Clearly it is no straightforward matter to assess the benefits and dangers of any new technology, be it directly computer related or not.

As a final comment on such computer–social issues, I should like to provide a little piece of fiction, but which expresses a worry that I have had in relation to a whole new area of potential problems. I have not heard anyone express this worry before, but it seems to me to present a new class of possible computer-related dangers.

8.5 The puzzling election

The date of a long-awaited election approaches. Numerous opinion polls are held over a period of several weeks. To a very consistent degree, the ruling party trails by three or four percentage points. Expectedly, there are fluctuations and deviations one way or another from this figure—expectedly because the opinion poll figures are based on relatively small samples of perhaps a few hundred voters at a time, whereas the total population (of many tens of millions) has considerable variations of opinion from place to place. Indeed, the margin of error for each of these opinion polls may itself run to that same 3 or 4%, so no one of them can really be trusted. Yet the totality of evidence is more impressive. The polls taken together have a much smaller margin of error, and the agreement between them seems to have just the right kind of slight variation that would be anticipated on statistical grounds. The averaged results can perhaps now be trusted to an error of less than 2%. Some might argue that there is a slight swing to the ruling party noticeable in these poll figures on the eve of the actual polling day; and that on the day itself a small proportion of previously undecided (or even fairly committed) people may well be finally persuaded to change their votes to the ruling party. Even so, a swing away from the poll figures towards the ruling party would not do it much good unless their resulting majority in the votes were perhaps some 8% over their nearest rivals, since only then would they achieve the overall majority that they need in order to stop a coalition between their opponents. But opinion polls are merely guesses of a sort, are they not? Only the *true* vote will express the actual voice of the people, and this will be obtained from the actual voting figures on polling day.

Polling day arrives, and passes. The votes are counted, and the result is a complete surprise to almost everyone—especially to the opinion poll organizations who had devoted so much of their energies and expertise, not to mention their reputations, on their previous findings. The ruling party is back with a comfortable majority, having achieved their target of 8% over their nearest rivals. A large number of voters are stunned—even horrified. Others, even though completely surprised, are delighted. Yet the result is false. The vote-rigging has been achieved by a highly subtle means that has eluded everyone. There have been no stuffed ballot boxes; and none has been lost, substituted, or duplicated. The people counting the votes have done their work conscientiously, and for the most part accurately. Yet the result is horribly wrong. How has this been achieved, and who is responsible?

It might be that the entire cabinet of the ruling party is completely ignorant of what has happened. They need not be the ones directly responsible, though they are the beneficiaries. There are others behind the scenes who had feared for their very existence if the ruling party were defeated. They are part of an organization more trusted (with good reason!) by the ruling party than by their opponents—an organization the strict secrecy of whose undercover activities the ruling party has been careful to preserve and even extend. Though the organization is legal, much of their actual activity is not, and they are not above indulging in illegal acts of political chicanery. Perhaps the members of the organization have a genuine (but misguided) fear that the opponents of the ruling party will destroy the country, or even 'sell it out' to the ideals of foreign powers. Amongst the members of the organization are experts—experts of extraordinary skill—at the construction of computer viruses!

Recall what a computer virus can do. The ones that are most familiar to people are those that on some appointed day may destroy all the records on any computer that has become infected with the virus. Perhaps the operator watches horrified as the very letters on the computer screen drop from their positions to the bottom and then disappear. Perhaps some obscene message appears on the screen. In any case, all the data may have become irreparably lost. Moreover, any disk that has been inserted in the machine and opened will itself become infected, and will have transferred this infection to the next machine. Anti-viral programs could in principle be used to destroy such an infection if spotted, but only if the nature of the virus is known beforehand. After the virus has struck, nothing can be done.

Such viruses are normally constructed by amateur hackers, often disgruntled computer programmers who wish to cause mischief, sometimes for understandable reasons, sometimes not. But these members of our organization are not amateurs; they are highly paid, and very expert professionals. Perhaps many of their activities are 'genuine' ones, entirely in the interests of their country; yet they also act, under the direction of their immediate superiors, in less morally excusable ways. Their viruses cannot be spotted by the standard anti-viral programs, and are preprogrammed to strike on just the one appointed day—the election day known, for sure, to the leader of the ruling party and those in the leader's confidence. After the job is done—a job much more subtle than just to destroy data—the viruses self-destruct, leaving no record whatever, bar the evil deed itself, to indicate their previous existence.

For such a virus to be effective in an election, it is necessary that there should be some stage in the vote-counting that is unchecked by humans, either by hand or by hand calculator. (A virus can infect only a generally programmable computer.) Perhaps the contents of the individual ballot boxes are correctly counted; but the results of these counts must be added. How much more efficient, accurate, and up to date it is to do this adding on a computer—adding perhaps 100 individual such numbers—than by hand or by hand calculator! Surely there is no scope for error. For exactly the same result is obtained no matter whose computer is used to perform the sum. The members of the ruling party get just the same results as do those of their principal opponents, or of any of the other interested parties, or of any neutral observer. Perhaps they all use different makes or models of computer systems, but this is of no real consequence. The experts in our organization will know these different systems, and have designed a separate virus for each. Though the construction of each of these different viruses will differ slightly, so that each is specific to each separate system, their results will be identical, and an agreement from machine to machine convinces even the most suspicious of sceptics.

Though agreement between the machines is exact, the figures are uniformly wrong. They have been cleverly concocted according to some precise formula, depending to some extent on the actual votes cast—whence the agreement between the different machines and vague plausibility of the result—so as to give the ruling party precisely the majority they need; and though credulity may be a little strained, the result seemingly must be accepted. It *seems* that, at the last minute, a significant number of voters took fright and voted for the ruling party.

In the hypothetical situation that I have described in this tale, they did not and the result was false. Although the inspiration for the story actually arose from our recent (1992) British election, I must strongly emphasize that the official system of vote-counting that is adopted in Britain does *not allow* for this kind of fraud. All counting stages are done by hand. Whilst it may seem that this is an outdated and inefficient method, it is important to retain it—or at least to retain some system where there are clear safeguards against even the mere suspicion of frauds of this kind.

In fact, on the positive side, modern computers offer wonderful opportunities for the employment of voting systems in which the opinion of the electorate can be much more fairly represented than it is now. This is not the place to enter into such matters, but the essential point is that it is possible for each voter to convey much more information than the casting of just a single vote for a single individual. With a computer-controlled system, this information could be instantly analysed, so that the result could be known immediately after all the polls are closed. However, as the above tale shows, one would have to be extremely wary of such a system unless there are thorough and manifest checks that would convincingly guard against any kind of fraud of the general nature of that described above.

It is not necessarily just in elections that one should be wary; the sabotaging of a rival's company's accounts, for example, would be another possibility where a 'computer-virus' technique might be employed. There are many other ways in which one could envisage that carefully constructed insidious computer viruses could be put to devastating use. I hope that my tale brings home the continued necessity of the manifest human overriding of the apparent, and apparently reliable, authority of computers. It is not just that computers do not understand anything, but they are extremely prone to manipulation by those few who do understand the detailed ways in which they are specifically programmed.

8.6. The physical phenomenon of consciousness?

The purpose of Part II of this book has been to search, within scientific explanation, for some place where subjective experience might find a physical home. I have argued that this will need an extension of our present-day scientific understanding. There is not much doubt in my own mind that we must look to the phenomenon of quantum-state reduction to see where our present picture of physical reality must indeed be fundamentally changed. For physics to be able to accommodate something that is as foreign to our current physical picture as is the phenomenon of consciousness, we must expect a profound change—one that alters the very underpinnings of our philosophical viewpoint as to the nature of reality. I shall have some comments to make on this shortly—in the final section of this book. For now, let us try to ask a somewhat simpler-sounding question: where might one expect, on the basis of the arguments that I have been putting forward in these pages, that consciousness is to be found in the known world?

I must make clear from the very start that the arguments I have been presenting have very little to say on the positive side. They say that present-day computers are not conscious, but they do not have much to say about when an object *would* be expected to be conscious. Our experience would tend to suggest, so far at least, that it is in biological structures that we are likely to find this phenomenon. At one end of the scale, we have human beings, and the case is surely clear that whatever consciousness is, it is a phenomenon that we must assume is normally present in association with wakeful (and probably also dreaming) human brains.

What about the other end of the scale? I have been making the case that it is to the microtubules in the cytoskeleton, rather than to neurons, that we must look for the place where collective (coherent) quantum effects are most likely to be found—and that without such quantum coherence we shall not find a sufficient role for the new **OR** physics that must provide the non-computational prerequisite for the encompassing of the phenomenon of consciousness within scientific terms. Yet cytoskeletons are ubiquitous amongst eukaryiotic cells—the kind of cell that constitutes plants and animals, and also single-celled animals like parameciums and amoebas, but not bacteria. Must we expect that some vestige of consciousness is present in a paramecium? Does a paramecium, in any sense of the word, 'know' what it is doing? What about *individual* human cells, perhaps in the brain or perhaps in the liver? I have no idea whether we shall be forced into accepting such apparent absurdities when our understanding of the physical nature of awareness becomes adequate for us to be able to answer such questions. But there is one thing that I *do* believe in relation to this problem, and that is that it is a *scientific question* that eventually should become answerable, no matter how far from being able to answer it we may be at present.

It is sometimes claimed, on general philosophical grounds, that there can be no way of knowing whether *any* entity other than oneself might be in possession of the quality of conscious awareness, let alone whether a paramecium could possess any vestige of it. To my way of thinking, this is much too narrow and pessimistic a line to take. One is never concerned, after all, with matters of *absolute certainty*, when establishing the presence of some physical quality in an object. I see no reason why there should not come a stage at which we can answer questions concerning the possession of conscious awareness with the same kind of certainty that astronomers have when they make assertions about celestial bodies many light years away. It was not so long ago when people would argue that the material composition of the sun and stars would never be known, nor would the features of the far side of the moon. However, the entire moon's surface is now well mapped (surveys from space) and the composition of the sun is now understood in great detail (observations of spectral lines in the sun's light and thoroughly detailed modelling of the physics of its interior). The detailed composition of many distant stars is also known, to a very good degree. And even the overall composition of the entire universe in its initial stages is, in a number of respects, very well understood (see end of §4.5).

But in the absence of the needed theoretical ideas, judgements with regard to the possession of consciousness remain, as yet, largely matters of guesswork. To express my own guesses on this issue, I find myself believing strongly that, on this planet, consciousness is *not* restricted to human beings. In one of the most deeply moving of David Attenborough's television programmes[3] was an episode that leaves us hard pressed not to believe that elephants, for example, not only have strong feelings but that these feelings are not far removed from those that instil religious belief in human beings. The leader of a herd—a female, whose sister had died some five years earlier—took the herd on a long detour to the place of her sister's death, and when they came upon her bones, the leader picked up her skull with great tenderness, whence the elephants passed it from one to another, caressing it with their trunks. That elephants also possess understanding is convincingly, if horrifically, displayed in another television programme.[4] Films from a helicopter, engaged in what was generously termed a 'culling' operation, showed clearly the elephants' terror as full knowledge of the impending slaughter of their entire herd was convincingly portrayed in their frighteningly agonized awful cries.

There is good evidence, also, for awareness (and self-awareness) in apes, and I would myself have little doubt that the phenomenon of consciousness is still a feature of animal life considerably 'lower' than that. For example, in yet another television programme[5]—concerning the extraordinary agility, determination, and resourcefulness of (some) squirrels—I was particularly struck by a sequence in which a squirrel realized that by biting through the wire along which it was crawling it could release the container of nuts suspended at some horizontal distance away. It is hard to see how this insight could have been instinctive or any part of the squirrel's previous experience. To appreciate this positive consequence of its action, the squirrel must have had some rudimentary understanding of the *topology* that is involved (compare §1.19). It seems to me that this was an act of genuine *imagination* on the part of the squirrel—which surely requires consciousness!

There would seem to be little doubt that consciousness can be a matter of degree, and it is not just a matter of it being 'there' or 'not there'. Even in my own experience, at different times I feel that it is present to a greater or lesser degree (such as, in a dreaming state, there appears to be a good deal less of it than in a fully wakeful one).

How far down, then, are we to go? There is enormous scope for differing opinions. As for myself, I have always had difficulties in believing that insects have much or any of this quality after watching another documentary film, where an insect was portrayed ravenously consuming another insect, apparently oblivious of the fact that it was itself being eaten by a third. Nevertheless, as mentioned in §1.15, the behaviour pattern of an ant is enormously complex and subtle. Need we believe that their wonderfully effective control systems are unaided by whatever principle it is that give us our own qualities of understanding? Their controlling neuronal cells have their own cytoskeletons, and if these cytoskeletons contain microtubules that are capable of sustaining the quantum-coherent states that I am suggesting are, at root, necessary for our own awareness, then might they not also be beneficiaries of this elusive quality? If the microtubules in our brains do possess the enormous sophistication needed for the maintaining of collective quantum-coherent activity, then it is difficult to see how natural selection could have evolved this facility just for us and (some of) our multicellular cousins. These quantum-coherent states must also have been valuable structures to the early eukaryotic one-celled animals, although it is quite possible that the value to them may have been very different from what it is to us.

Large-scale quantum coherence does *not*, in itself, imply consciousness of course—otherwise superconductors would be conscious! Yet it is quite possible that such coherence could be *part* of what is needed for consciousness.

In our own brains there is enormous organization, and since consciousness appears to be a very *global* feature of our thinking, it seems that we must look to some kind of coherence on a much larger scale than the level of single microtubules or even single cytoskeletons. There must be significant quantum entanglements between the states in the separate cytoskeletons of large numbers of different neurons, so that large areas of the brain would be involved in some kind of collective quantum state. But much more would be required than just that. In order that some kind of useful *non-computable* action can be involved—which I am taking to be an essential part of consciousness—it would be necessary that the system can make specific use of the genuinely *non-random* (non-computable) aspects of **OR**. The particular proposal that I have been promoting in §6.12 gives us at least some idea as to the *scales* involved, at which a precise, mathematical non-computable **OR** action could begin to have importance.

Thus, on the basis of considerations that I have been putting forward in this book, one might anticipate some way of at least *guessing* a level at which conscious awareness could begin to be present. Processes that can be adequately described according to computable (or random) physics would not, according to my viewpoint, involve consciousness. On the other hand, even the essential involvement of a precise non-computable **OR** action would not, in itself, necessarily *imply* the presence of consciousness—although, on my view, it would be a *prerequisite* for consciousness. Certainly, this is not a very clear-cut criterion, but it is the best that I can do for the moment. Let us see how far we can get with it.

I shall try to develop the picture that arises, based upon the suggestions of §6.12 as to where the quantum/classical boundary should occur—and also on the biological speculations of §7.5–§7.7, according to which we should find this boundary having relevance at the internal/external interface of the system of microtubules in a cell, or system of cells. An essential additional idea is that if state-vector reduction occurs merely because too much of the environment becomes entangled with the system under consideration, the **OR** occurs effectively as a *random* process for which the standard FAPP arguments (outlined in §6.6) are adequate, and **OR** behaves just like **R**. What is needed is that this reduction takes place just at the point when the (unknown) non-computational *details* of our putative **OR** theory come into play. Although the details of this theory are not known, at least we can in principle obtain some idea of the level at which that theory should begin to become relevant. Thus, in order for these non-computable aspects of **OR** to play an important role, it would be necessary that some kind of quantum coherence is maintained until the coupling only *just* brings in enough movement of material so that **OR** acts *before* the random environment becomes significantly involved.

The picture that I am proposing for microtubules is that there are 'quantum-coherent oscillations' taking place *within* the tubes, and these are weakly coupled to 'computational cellular-automata-like' activity taking place in the conformational switchings of tubulin dimers *on* the tubes. So long as the quantum oscillations remain isolated, the level would be too low for **OR** to take place. However, the coupling would entail that the tubulins also get involved in the state, and at a certain point **OR** would be effected. What we need is that **OR** comes in *before* the environment of the microtubules becomes entangled with the state, because as soon as that happens the non-computable aspects of **OR** are lost, and the action is just the random **R**-process.

Thus, we can ask whether in a single cell (such as a paramecium, say, or a human liver cell) the amount of tubulin conformational activity can involve enough mass movement that the criterion of §6.12 is satisfied, so **OR** comes into effect at that stage—as required—or whether there is not enough, and the **OR** action is delayed until the environment *is* disturbed—and the (non-computational) game is lost. On the face of it, there would seem to be much too little mass movement in tubulin conformational activity, and the game would indeed seem to be lost at this level. But with large collections of cells, the situation would appear to be much more promising.

Perhaps this picture, as it stands so far, indeed favours a viewpoint in which the non-computational prerequisite for consciousness can only occur with large collections of cells, as in the case of a reasonable-sized brain.[6] One should obviously be very cautious, at this stage, about coming to any clear-cut conclusions of this kind. Both the physical and the biological sides of the picture are too crudely formulated for any clear conclusions to be drawn, just yet, as to the implications of the viewpoint that I am presenting. It is clear that even with the specific proposals that I have been arguing for here, a great deal of further research on both the physical and biological sides would be needed before a clearly reasoned guess can be made as to the place at which consciousness could enter.

There are also some other issues that should be considered. How much of the brain, we may ask, is actually involved in a conscious state? Very likely the entire brain is *not* involved. Much of the brain's action indeed appears not to be conscious. The cerebellum (cf. §1.14), most strikingly, seems to act entirely *un*consciously. It governs the delicate and precise control of our actions—at times when we are *not* consciously performing those actions (cf. ENM, pp. 379–81, for example). The cerebellum is indeed, frequently referred to as 'just a computer' because of its entirely unconscious activity. It would surely be instructive to know what essential differences there might be in the cellular or cytoskeletal organization of the cerebellum, as opposed to that of the cerebrum, since it is with the latter structure that consciousness seems to have its much closer relationship. It is interesting that simply on the basis of neuron count, there is not that much difference between the two, there being perhaps just twice the number of neurons in the cerebrum as in the cerebellum and generally many more synaptic connections between individual cells in the cerebellum (cf. §1.14, Fig. 1.6). There must indeed be something more subtle at work than mere neuron counting.*

Perhaps, also, there would be something instructive to be gained from a study of the way in which the unconscious cerebellar control is 'learnt' from a conscious cerebral one. There could well be a strong similarity, in the case of the cerebellum's learning procedures, with the way that artificial neural networks are trained, in accordance with the connectionist philosophy. But even if so, and even if it is *also* true that some actions in the *cerebrum* can be (partially) understood in this way—as implicit in the connectionist approach to understanding the visual cortex[7]—there is no reason to expect that the same need be true of those aspects of cerebral action that are involved in consciousness. Indeed, as I have strongly argued in Part I of this book, there must indeed be something very different from connectionism that is concerned with those higher cognitive functions where consciousness itself comes into play.

8.7 Three worlds and three mysteries

I shall try to bring the themes of this work together. The central issue that I have been attempting to address throughout this book is how the phenomenon of consciousness can relate to our scientific world-view. Admittedly, I have not had much to say about the issue of consciousness in general. Instead, I have concentrated, in Part I, on just one particular mental quality: *conscious understanding*, especially mathematical understanding. It is only with this mental quality that I have been able to make the necessary strong claim: that it is essentially *impossible* that such a quality can have arisen as a feature of mere computational activity nor can computation even properly simulate it—and I should emphasize that there is no suggestion here that there is anything special about *mathematical* as opposed to any other kind of understanding. The conclusion is that whatever brain activity is responsible for consciousness (at least in this particular manifestation) it must depend upon a physics that lies beyond computational simulation. Part II represents an attempt to find scope, within the framework of science, for a relevant

*As an outsider to the subject of neuroanatomy, I cannot but be struck by the fact that there is an (unexplained?) oddity about cerebral organization that does *not* seem to be shared by the cerebellum. Most of the sensory and motor nerves cross over, so that the left side of the cerebrum is mainly concerned with the right side of the body, and *vice versa*. Not only this, but the part of the cerebrum concerned with vision is right at the back, whereas the eyes are at the front; that part concerned with the feet is at the top, whereas the feet themselves are at the bottom; that part mainly concerned with each ear's hearing is diametricaly opposite to the ear concerned. This is not completely a universal feature of the cerebrum, but I cannot help feeling that it is not an accident. For the cerebellum is not organized in this way. Could it be that consciousness benefits in some way from the nerve signals having to take a long route?

physical action that might indeed take us beyond the limitations of mere computation. In order to encapsulate the profound issues that confront us, I shall phrase things in terms of three different worlds, and the three deep mysteries that relate each of these worlds to each of the others. The worlds are somewhat related to those of Popper (cf. Popper and Eccles 1977), but my emphasis will be very different.

The world that we know most directly is the *world of our conscious perceptions*, yet it is the world that we know least about in any kind of precise scientific terms. That world contains happiness and pain and the perception of colours. It contains our earliest childhood memories and our fear of death. It contains love, understanding, and the knowledge of numerous facts, as well as ignorance and revenge. It is a world containing mental images of chairs and tables, and where smells and sounds and sensations of all kinds intermingle with our thoughts and our decisions to act.

There are two other worlds that we are also cognisant of—less directly than the world of our perceptions—but which we now know quite a lot about. One of these worlds is the world we call the *physical world*. It contains actual chairs and tables, television sets and automobiles, human beings, human brains, and the actions of neurons. In this world are the sun and the moon and the stars. So also are clouds, hurricanes, rocks, flowers, and butterflies; and at a deeper level there are molecules and atoms, electrons and photons, and space-time. It also contains cytoskeletons and tubulin dimers and superconductors. It is not at all clear why the world of our perceptions should have anything to do with the physical world, but apparently it does.

There is also one other world, though many find difficulty in accepting its actual existence: it is the *Platonic world of mathematical forms*. There, we find the natural numbers 0, 1, 2, 3, . . ., and the algebra of complex numbers. We find Lagrange's theorem that every natural number is the sum of four squares. We find Euclidean geometry's Pythagorean theorem (about the squares on the sides of a right-angled triangle). There is the statement that for any pair of natural numbers, $a \times b = b \times a$. In this same Platonic world is also the fact that this last result no longer holds for certain other types of 'number' (such as with the Grassmann product referred to in §5.15). This same Platonic world contains geometries other than the Euclidean one, in which the Pythagorean theorem fails to hold. It contains infinite numbers and non-computable numbers and recursive and non-recursive ordinals. There are Turing-machine actions that never come to a halt as well as oracle machines. In it are many classes of mathematical problem which are computationally unsolvable, such as the polyomino tiling problem. Also in this world are Maxwell's electromagnetic equations as well as Einstein's gravitational equations and innumerable theoretical space-times that satisfy them—whether physically realistic or not. There are mathematical simulations of chairs and tables, as would be made use of in 'virtual reality', and simulations also of black holes and hurricanes.

What right do we have to say that the Platonic world is actually a 'world', that can 'exist' in the same kind of sense in which the other two worlds exist? It may well seem to the reader to be just a rag-bag of abstract concepts that mathematicians have come up with from time to time. Yet its existence rests on the profound, timeless, and universal nature of these concepts, and on the fact that their laws are independent of those who discover them. The rag-bag—if indeed that is what it is—was not of our creation. The natural numbers were there before there were human beings, or indeed any other creature here on earth, and they will remain after all life has perished. It has always been true that each natural number is the sum of four squares, and it did not have to wait for Lagrange to conjure this fact into existence. Natural numbers that are so large that they are beyond the reach of any conceivable computer are still the sums of four squares, even though there may be no chance of ever finding what these particular square numbers might be. It will always remain the case that there is no general computational procedure for deciding whether a Turing-machine action ever halts, and it always was the case long before Turing came across his notion of computability.

Nevertheless, many might still argue that the absolute nature of mathematical truth is no argument for assigning an 'existence' to mathematical concepts and mathematical truths. (I have sometimes heard it said that mathematical Platonism is 'out of date'. It is true that Plato himself died some 2340 years ago, but that is hardly a reason! A more serious objection is the difficulty that philosophers sometimes have with an entirely abstract world having any influence on the physical world. This profound issue is indeed part of one of the mysteries that we shall be addressing in a moment.) In fact the reality of mathematical concepts is a much more natural idea for mathematicians than it is for those who have not had the fortune to spend time exploring the wonders and mysteriousness of that world. However, for the moment, it will not be necessary for the reader to accept that mathematical concepts truly form a 'world' with an actual reality comparable with that of the physical world and the mental world. How one really chooses to view mathematical concepts will not be too important for us for now. Take 'the Platonic world of mathematical forms' as merely a figure of speech, if you will, but it will be a useful phrase for our descriptions here. When we come to consider the three mysteries that relate these three 'worlds' we shall perhaps begin to see something of the significance of this phraseology.

What, then, are the mysteries? These are illustrated in Fig. 8.1. There is the mystery of why such precise and profoundly mathematical laws play such an important role in the behaviour of the physical world. Somehow the very world of physical reality seems almost mysteriously to emerge out of the Platonic world of mathematics. This is represented by the arrow down on the right, from the Platonic world to the physical world. Then there is the second mystery of how it is that perceiving beings can arise from out of the physical world. How is it that subtly organized material objects can mysteriously

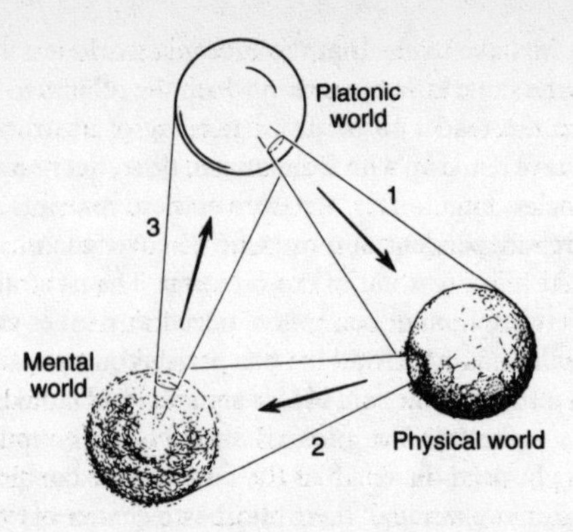

Fig. 8.1 In some way, each of the three worlds, the Platonic mathematical, the physical, and the mental, seems mysteriously to 'emerge' from—or at least to be intimately related to—a small part of its predecessor (the worlds being taken cyclically).

conjure up mental entities from out of its material substance? This is represented in Fig. 8.1 by the arrow pointing at the bottom, from the physical to the mental world. Finally, there is the mystery of how it is that mentality is able seemingly to 'create' mathematical concepts out of some kind of mental model. These apparently vague, unreliable, and often inappropriate mental tools, with which our mental world seems to come equipped, appear nevertheless mysteriously able (at least when they are at their best) to conjure up abstract mathematical forms, and thereby enable our minds to gain entry, by understanding, into the Platonic mathematical realm. This is indicated by the arrow that points upwards on the left, from the mental to the Platonic world.

Plato himself had been very much concerned with the first of these arrows (and also, in his own way, with the third), and he was careful to distinguish between the perfect mathematical form and its imperfect 'shadow' in the physical world. Thus a mathematical triangle (or a Euclidean one, as we would have to be careful to specify today) would have its angles sum to exactly two right angles, whereas a physical one made of wood, say, to as much precision as could be mustered, might have angles that summed very closely indeed to the required amount but would not be perfectly precise. Plato described such ideas in terms of a parable. He imagined some citizens confined to a cave, chained so that they could not see the perfect shapes behind them that were casting shadows in the light of a fire on the part of the cave that they faced. All that they could directly see would be the imperfect shadows of these shapes, somewhat distorted by the flickering of the fire. These perfect shapes represented the mathematical forms, and the shadows, the world of 'physical reality'.

Since Plato's day, the underlying role of mathematics in the perceived structure and in the actual behaviour of our physical world has increased enormously. The esteemed physicist Eugene Wigner delivered a famous lecture, in 1960, with the title 'The unreasonable effectiveness of mathematics in the physical sciences'. In this, he expressed something of the amazing precision and subtle applicability of sophisticated mathematics that physicists continually and increasingly find in their descriptions of reality.

To me, the most impressive example of all is Einstein's general relativity. One not infrequently hears the viewpoint expressed that physicists are merely noticing patterns, from time to time, where mathematical concepts may happen to apply quite well to physical behaviour. It would be claimed, accordingly, that physicists tend to bias their interests in the directions of those areas where their mathematical descriptions work well, so there is no real mystery that mathematics is found to work in the descriptions that physicists use. It seems to me, however, that such a viewpoint is extraordinarily wide of its mark. It simply provides no explanation of the deep underlying unity that Einstein's theory, in particular, shows that there is between mathematics and the workings of the world. When Einstein's theory was first put forward, there was really no need for it on observational grounds. Newton's gravitational theory had stood for some 250 years, and had achieved an extraordinary accuracy, of something like one part in ten million (already an impressive enough justification for taking seriously a deep mathematical underpinning to physical reality). An anomaly had been observed in Mercury's motion, but this certainly did not provide cause to abandon Newton's scheme. Yet Einstein perceived, from deep physical grounds, that one could do better, if one changed the very framework of gravitational theory. In the early years after Einstein's theory was put forward, there were only a few effects that supported it, and the increase in precision over Newton's scheme was marginal. However, now, nearly 80 years after the theory was first produced, its overall precision has grown to something like ten million times greater. Einstein was not just 'noticing patterns' in the behaviour of physical objects. He was uncovering a profound mathematical substructure that was already hidden in the very workings of the world. Moreover, he was not just searching around for whatever physical phenomenon might best fit a good theory. He found this precise mathematical relationship in the very structure of space and time—the most fundamental of physical notions.

In our other successful theories of basic physical processes, there has always been an underlying mathematical structure that has not only proved extraordinarily accurate, but also mathematically sophisticated. (And lest the reader think that the 'overthrowing' of earlier ideas of physics, such as Newton's theory, invalidates the appropriateness of those earlier ideas, I should make clear that this is *not so*. The older ideas, when they are good enough, such as those of Galileo and Newton, still survive, and have their place within the newer scheme.) Moreover, mathematics itself gains much inspiration of subtle and unexpected kinds from input from the detailed behaviour of Nature. Quantum theory—whose close relation with subtle mathematics (e.g. complex numbers) is, I hope, apparent even from the glimpse into the subject that we have had in these pages—as well as general relativity and Maxwell's electromagnetic equations, have all provided enormous stimulation for the progress of mathematics. But this is true not only for relatively recent theories like these. It was at least as true for the more ancient ones like Newtonian mechanics (yielding the calculus) and the Greek analysis of the structure of space (giving us the very notion of geometry). The extraordinary accuracy of mathematics within physical behaviour (such as the 11- or 12-figure accuracy of quantum electrodynamics) has often been stressed. But there is much more to the mystery than just this. There is a very remarkable depth, subtlety and *mathematical fruitfulness* in the concepts that lie latent within physical processes. This is something not so familiar to people—unless they are directly concerned with the mathematics involved.

It should be made clear that this mathematical fruitfulness, providing a valuable stimulus to the actual activities of mathematicians, is not just a matter of mathematical fashion (though fashion is not without its role to play also). Ideas that were developed for the sole purpose of deepening our understanding of the workings of the physical world have very frequently provided profound and unexpected insights into mathematical problems that had *already* been objects of considerable interest for quite separate reasons. One of the most striking recent examples of this was the use made of Yang–Mills-type theories (which had been developed by physicists in their mathematical explanations of the interactions that hold between sub-atomic particles), by Oxford's Simon Donaldson, to obtain totally unexpected properties of four-dimensional manifolds[8]—properties that had eluded understanding for many years previously. Moreover, such mathematical properties, though often not at all anticipated by human beings before the appropriate insights came to light, have lain timelessly within the Platonic world, as unchangeable truths waiting to be discovered—in accordance with the skills and insights of those who strive to discover them.

I hope that I have persuaded the reader of the close and genuine relationship—still deeply mysterious—between the Platonic mathematical world and the world of physical objects. I hope, also, that the very presence of this extraordinary relationship will help the Platonic sceptics to take that world a little more seriously *as* a 'world' than they may have been prepared to do previously. Indeed, some might well go further than I have been prepared to go in this discussion. Perhaps a Platonic reality should be assigned to other abstract concepts, not just mathematical ones. Plato himself would have insisted that the ideal concept of 'the good' or 'the beautiful' must also be attributed a reality (cf. §8.3), just as mathematical concepts must. Personally, I am not averse to such a possibility, but it has yet played no important part in my deliberations here. Issues of ethics, morality, and aesthetics have had no significant role in my present discussions, but this is no reason to dismiss them as being not, at root, as 'real' as the ones I have been addressing. Clearly there are important separate issues to be considered here, but they have not been my particular concern in this book.[9]

Nor, in this book, have I been so *very* much concerned with the particular mystery (first arrow, down to the right in Fig. 8.1) of the puzzling precise underlying role that the Platonic mathematical world has in the physical world—but with the other two, which are even less well understood. In Part I, I have been addressing questions raised mainly by the third arrow: the mystery of our very perceptions of mathematical truth; that is, with the apparent way in which, through mental contemplation, we seem to be able to 'conjure up' those very Platonic mathematical forms. It is as though the perfect forms might be merely shadows of our imperfect thoughts. To view the Platonic world in this way—as only a product of our own mentality—would be very much at variance with Plato's own conceptions. To Plato, the world of perfect forms is primary, being timeless and independent of ourselves. In the Platonic view, my third arrow in Fig. 8.1 should perhaps be thought of as pointing downwards instead of upwards: from the world of perfect forms to the world of our mentality. To think of the mathematical world as being a product of our modes of thinking would be to take a *Kantian* view rather than the Platonic one that I am espousing here.

Likewise, some might argue for a reversal of the directions of some of my other arrows. Perhaps Bishop Berkeley would have preferred my *second* arrow to point from the mental world to the physical one, 'physical reality' being, on this view, a mere shadow of our mental existence. There are some others (the 'nominalists') who would argue for a reversal of my *first* arrow, the world of mathematics being a mere reflection of aspects of the world of physical reality. My own sympathies, as should be evident from this book, would be strongly against a reversal of these first two arrows, though it may be equally evident that I am somewhat uncomfortable about directing the *third* arrow in the seemingly 'Kantian' orientation that is depicted in Fig. 8.1! To me the world of perfect forms is primary (as was Plato's own belief)—its existence being almost a logical necessity—and *both* the other two worlds are its shadows.

Owing to such differing viewpoints about which of the worlds of Fig. 8.1 might be regarded as primary and which as secondary, I shall recommend regarding the arrows in a different light. The essential point about the arrows in Fig. 8.1 is not so much their direction but the fact that in each case they represent a correspondence in which a *small* region of one world encompasses the *entire* next world. With regard to my first arrow, it has often been remarked to me that by far the greatest part of the world of mathematics (as judged in terms of the activity of mathematicians) seems to have little if any relation to actual physical behaviour. Thus, it is but a tiny part of the Platonic world that can underlie the structure of our physical universe. Likewise, the second of my arrows expresses the fact that our mental existence emerges from but a minute portion of the physical world—a portion where conditions are organized in the very precise way needed for consciousness to arise, as in human brains. In the same way, my third arrow refers to but a tiny part of our mental activity, namely that which is concerned with absolute and timeless issues—most particularly, with mathematical truth. For the most part, our mental lives are concerned with quite other matters!

There is a seemingly paradoxical aspect to these correspondences, where each world seems to 'emerge' from but a *tiny* part of the one which precedes it. I have drawn Fig. 8.1 so as to emphasize this paradox. However, by regarding the arrows merely as expressing the various correspondences, rather than asserting any actual 'emergence', I am trying not to prejudge the question as to which, if any, of the worlds are to be regarded as primary, secondary, or tertiary.

Yet, even so, Fig. 8.1 reflects another aspect of my opinions or prejudices. I have depicted things as though it is to be assumed that each *entire* world is indeed reflected within a (small) portion of its predecessor. Perhaps my prejudices are wrong. Perhaps there are aspects of the behaviour of the physical world that *cannot* be described in precise mathematical terms; perhaps there is mental life that is *not* rooted in physical structures (such as brains); Perhaps there are mathematical truths that remain, *in principle, inaccessible* to human reason and insight. To encompass any of these alternative possibilities, Fig. 8.1 would have to be redrawn, so as to allow for some or all of these worlds to extend beyond the compass of its preceding arrow.

In Part I, I have been much concerned with some of the implications of Gödel's famous incompleteness theorem. Some readers might have been of the opinion that Gödel's theorem indeed tells us that there are parts of the world of Platonic mathematical truths that lie in principle beyond human understanding and insight. I hope that my arguments have made it clear that this is *not* the case.[10] The specific mathematical propositions that Gödel's ingenious argument provides are ones which are humanly accessible—provided that they are constructed from mathematical (formal) systems that have been already accepted as valid means of assessing mathematical truth. Gödel's argument does not argue in favour of there being inaccessible mathematical truths. What it *does* argue for, on the other hand, is that human insight lies beyond formal argument and beyond computable procedures. Moreover, it argues powerfully for the very existence of the Platonic mathematical world. Mathematical truth is not determined arbitrarily by the rules of some 'man-made' formal system, but has an absolute nature, and lies beyond any such system of specifiable rules. Support for the Platonic viewpoint (as opposed to the formalist one) was an important part of Gödel's initial motivations. On the other hand, the arguments from Gödel's theorem serve to illustrate the deeply mysterious nature of our mathematical perceptions. We do not just 'calculate', in order to form these perceptions, but something else is profoundly involved—something that would be impossible without the very conscious awareness that is, after all, what the world of perceptions is all about.

Part II has been concerned mainly with questions to do with the second arrow (though these cannot be properly addressed without some reference to the first)—whereby the concrete physical world can somehow conjure up the *shadowy* phenomenon we refer to as consciousness. How is it that consciousness can arise from such seemingly unpromising ingredients as matter, space, and time? We have not come to an answer, but I hope that at least the reader may be able to appreciate that matter *itself* is mysterious, as is the space–time within whose framework physical theories now operate. We simply do not know the nature of matter and the laws that govern it, to an extent that we shall need in order to understand what kind of organization it is, in the physical world, which gives rise to conscious beings. Moreover, the more deeply we examine the nature of matter, the more elusive, mysterious, and mathematical, matter itself appears to be. We might well ask: what *is* matter, according to the best theories that science has been able to provide? The answer comes back in the form of mathematics, not so much as a system of equations (though equations are important too) but as subtle mathematical concepts that take a long time to grasp properly.

If Einstein's general relativity has shown how our very notions of the nature of space and time have had to shift, and become more mysterious and mathematical, then it is quantum mechanics that has shown, to an even greater extent, how our concept of *matter* has suffered a similar fate. Not just matter, but our very notions of actuality have become profoundly disturbed. How is it that the mere counterfactual *possibility* of something happening—a thing which does *not* actually happen—can have a decisive influence upon what actually *does* happen? There is something in the mystery of the way that quantum mechanics operates that at least *seems* much closer than is classical physics, to the kind of mystery needed to accommodate mentality within the world of physical reality. I have no doubt myself that when deeper theories are at hand, then the place of mind in relation to physical theory will not seem so incongruous as it does today.

In §7.7 and §8.6, I have tried to come to grips with the issue of what physical circumstances might be appropriate for the phenomenon of consciousness. Yet, it should be made clear that I do *not* regard consciousness as merely a matter of the right amount of coherent mass movement according to some **OR** theory of the quantum/classical boundary. As I hope I have made clear, such things would provide merely the appropriate opening for a non-computable action within the confines of our present-day physical picture. Genuine consciousness involves an awareness of an endless variety of qualitatively different things—such as of the green colour of a leaf, the smell of a rose, the song of a blackbird, or the soft touch of a cat's fur; also of the passage of time, of emotional states, of worry, of wonder, and of the appreciation of an idea. It involves hopes, ideals, and intentions, and the actual willing of innumerable different bodily movements in order that such intentions may be realized. The study of neuroanatomy, of neurological disorders, psychiatry, and psychology has told us much about the detailed relationship between the physical nature of the brain and our mental conditions. There is no question of our being able to understand such matters merely in terms of the physics of critical amounts of coherent mass movement. Yet without such an opening into a new physics, we shall be stuck within the strait-jacket of an entirely computational physics, or of a computational cum random physics. Within that strait-jacket, there can be no scientific role for intentionality and subjective experience. By breaking loose from it, we have at least the potentiality of such a role.

Many who might agree with this would argue that there can be no role for such things within *any* scientific picture. To those who argue this way, I can ask only that they be patient, and that they wait to see how science moves in the future. I believe that there is already an indication, within the mysterious developments of quantum mechanics, that the concepts of mentality are a little closer to our understandings of the physical universe than they had been before—although only a *little* closer. I would argue that when the necessary *new* physical developments come to light, these indications should become a good deal clearer than that. Science has a long way to develop yet; of *that*, I am certain!

Moreover, the very possibility of a human understanding of such matters tells us something about the abilities that consciousness confers upon us. Admittedly there are some, such as Newton or Einstein, or Archimedes, Galileo, Maxwell, or Dirac—or Darwin, Leonardo da Vinci, Rembrandt, Picasso, Bach, Mozart, or Plato, or those great minds who could conceive the *Iliad* or *Hamlet*—who seem to have more of this faculty of being able to 'smell' out truth or beauty than is given to the rest us. But a unity with the workings of Nature is potentially present within all of us, and is revealed in our very faculties of conscious comprehension and sensitivity, at whatever level they may be operating. Every one of our conscious brains is woven from subtle physical ingredients that somehow enable us to take advantage of the profound organization of our mathematically underpinned universe—so that we, in turn, are capable of some kind of direct access, through that Platonic quality of 'understanding', to the very ways in which our universe behaves at many different levels.

These are deep issues, and we are yet very far from explanations. I would argue that no clear answers will come forward unless the interrelating features of *all* these worlds are seen to come into play. No one of these issues will be resolved in isolation from the others. I have referred to three worlds and the mysteries that relate them one to another. No doubt there are not really three worlds but *one*, the true nature of which we do not even glimpse at present.

Notes and references

1. See, for example, Lisboa (1992).
2. This idea was described to me by Joel de Rosnay.
3. *Echo of the elephants* (BBC, January 1993).
4. *If the rains don't come* (BBC, September 1992).
5. *Daylight robbery* (BBC, August 1993)
6. One might speculate on the absence of centrioles in neurons (cf. p. 365). The cytoskeletons of other types of individual cells seem to require centrioles as their 'control centres'; yet the cytoskeletons of neurons perhaps defer to a more global authority!
7. Marr (1982) and, for example, Brady (1993).
8. Donaldson (1983); cf. Devlin (1988), Chapter 10, for a non-technical account.
9. Popper's 'World 3' contains mental constructs with some similarity to those that would reside in this extended Platonic world; see Popper and Eccles (1977). However, his World 3 is not viewed as having a timeless existence independent of ourselves, nor as a world underlying the very structure of physical reality. Accordingly, its status is very different from the 'Platonic world' under consideration here.
10. Mostowski (1957) makes it clear, in the introduction to his book, that arguments such as those of Gödel have no bearing on the question of whether there might exist *absolutely* undecidable mathematical questions. The issue must be regarded as completely open, as of now, as far as what can be proved or disproved. This question remains, like the other two, purely a matter of faith!

EPILOGUE

Jessica and her father emerged from the cave. It was now getting quite dark and still, and some stars were clearly visible. Jessica said to her father:

'Daddy, you know, when I look up into the sky, I still find it hard to believe that the earth really *does* move—spinning around at all those thousands of kilometres per hour—even though I think I really know it *must* be true.'

She paused, and stood looking up at the sky for some while.

'Daddy, tell me about the stars . . .'

BIBLIOGRAPHY

Aharonov, Y. and Albert, D. Z. (1981). Can we make sense out of the measurement process in relativistic quantum mechanics? *Phys. Rev.*, **D24**, 359–70.

Aharonov, Y. and Vaidman, L. (1990). Properties of a quantum system during the time interval between two measurements. *Phys. Rev.*, **A41**, 11.

Aharonov, Y., Anandan, J., and Vaidman, L. (1993). Meaning of the wave function. *Phys. Rev.*, **A47**, 4616–26.

Aharonov, Y., Bergmann, P. G., and Liebowitz, J. L. (1964). Time symmetry in the quantum process of measurement. In *Quantum theory and measurement* (ed. J. A. Wheeler and W. H. Zurek). Princeton University Press, 1983; originally in *Phys. Rev.*, **B134**, 1410–16.

Aharonov, Y., Albert, D. Z., and Vaidman, L. (1986). Measurement process in relativistic quantum theory. *Phys. Rev.*, **D34**, 1805–13.

Albert, D. Z. (1983). On quantum-mechanical automata. *Phys. Lett.*, **98A(5, 6)**, 249–52.

Albrecht-Buehler, G. (1985). Is the cytoplasm intelligent too? *Cell and Muscle Motility*, **6**, 1–21.

Anthony, M. and Biggs, N. (1992). *Computational learning theory, an introduction.* Cambridge University Press.

Applewhite, P. B. (1979). Learning in protozoa. In *Biochemistry and physiology of protozoa*, Vol. 1 (ed. M. Levandowsky and S. H. Hunter), pp. 341–55. Academic Press, New York.

Arhem, P. and Lindahl, B. I. B. (ed.) (1993). Neuroscience and the problem of consciousness: theoretical and empirical approaches. In *Theoretical medicine*, **14**, Number 2. Kluwer Academic Publishers.

Aspect, A. and Grangier, P. (1986). Experiments on Einstein–Podolsky–Rosen-type correlations with pairs of visible photons. In *Quantum concepts in space and time* (ed. R. Penrose and C. J. Isham). Oxford University Press.

Aspect, A., Grangier, P., and Roger, G. (1982). Experimental realization of Einstein–Podolsky–Rosen–Bohm *Gedankenexperiment*: a new violation of Bell's inequalities. *Phys. Rev. Lett.*, **48**, 91–4.

Baars, B. J. (1988). *A cognitive theory of consciousness*. Cambridge University Press.

Bailey, T. N. and Baston, R. J. (ed.) (1990). *Twistors in mathematics and physics*. London Mathematical Society Lecture Notes Series, 156. Cambridge University Press.

Baylor, D. A., Lamb, T. D., and Yau, K.-W. (1979). Responses of retinal rods to single photons. *J. Physiol.*, **288**, 613–34.

Beck, F. and Eccles, J. C. (1992). Quantum aspects of consciousness and the role of consciousness. *Proc. Nat. Acad. Sci.*, **89**, 11357–61.

Becks, K.-H. and Hemker, A. (1992). An artificial intelligence approach to data analysis. In *Proceedings of 1991 CERN School of Computing* (ed. C. Verkerk). CERN, Switzerland.

Bell, J. S. (1964). On the Einstein Podolsky Rosen paradox. *Physics*, **1**, 195–200.

Bell, J. S. (1966). On the problem of hidden variables in quantum theory. *Revs. Mod. Phys.*, **38**, 447–52.

Bell, J. S. (1987). *Speakable and unspeakable in quantum mechanics*. Cambridge University Press.

Bell, J. S. (1990). Against measurement. *Physics World*, **3**, 33–40.

Benacerraf, P. (1967). God, the Devil and Gödel. *The Monist*, **51**, 9–32.

Benioff, P. (1982). Quantum mechanical Hamiltonian models of Turing Machines. *J. Stat. Phys.*, **29**, 515–46.

Bennett, C. H., Brassard, G., Breidbart, S., and Wiesner, S. (1983). Quantum cryptography, or unforgetable subway tokens. In *Advances in cryptography*. Plenum, New York.

Bernard, C. (1875). *Leçons sur les anesthésiques et sur l'asphyxie*. J. B. Bailliere, Paris.

Blakemore, C. and Greenfield, S. (ed.) (1987). *Mindwaves: thoughts on intelligence, identity and consciousness*. Blackwell, Oxford.

Blum, L., Shub, M., and Smale, S. (1989). On a theory of computation and complexity over the real numbers: NP completeness, recursive functions and universal machines. *Bull Amer. Math. Soc.*, **21**, 1–46.

Bock, G. R. and Marsh, J. (1993). *Experimental and theoretical studies of consciousness*. Wiley.

Boden, M. (1977). *Artificial intelligence and natural man*. The Harvester Press, Hassocks.

Boden, M. A. (1990). *The creative mind: myths and mechanisms*. Wiedenfeld and Nicolson, London.

Bohm, D. (1952). A suggested interpretation of the quantum theory in terms of 'hidden' variables, I and II. In *Quantum theory and measurement* (ed. J. A.

Wheeler and W. H. Zurek). Princeton University Press 1983. Originally in *Phys. Rev.*, **85**, 166–93.

Bohm, D. and Hiley, B. (1994). *The undivided universe*. Routledge, London.

Boole, G. (1854). *An investigation of the laws of thought*. 1958, Dover, New York.

Boolos, G. (1990). On seeing the truth of the Gödel sentence. *Behavioural and Brain Sciences*, **13(4)**, 655.

Bowie, G. L. (1982). Lucas' number is finally up. *J. of Philosophical Logic*, **11**, 279–85.

Brady, M. (1993). Computational vision. In *The simulation of human intelligence* (ed. D. Broadbent). Blackwell, Oxford.

Braginsky, V. B. (1977). The detection of gravitational waves and quantum non-disturbtive measurements. In *Topics in theoretical and experimental gravitation physics* (ed. V. de Sabbata and J. Weber), p. 105. Plenum, London.

Broadbent, D. (1993). Comparison with human experiments. In *The simulation of human intelligence* (ed. D. Broadbent). Blackwell, Oxford.

Brown, H. R. (1993). Bell's other theorem and its connection with nonlocality. Part I. In *Bell's Theorem and the foundations of physics* (ed. A. Van der Merwe and F. Selleri). World Scientific, Singapore.

Butterfield, J. (1990). Lucas revived? An undefended flank. *Behavioural and Brain Sciences*, **13(4)**, 658.

Castagnoli, G., Rasetti, M., and Vincenti, A. (1992). Steady, simultaneous quantum computation: a paradigm for the investigation of nondeterministic and non-recursive computation. *Int. J. Mod. Phys. C*, **3**, 661–89.

Caudill, M. (1992). *In our own image. Building an artificial person*. Oxford University Press.

Chaitin, G. J. (1975). Randomness and mathematical proof. *Scientific American*, (May 1975), 47.

Chalmers, D.J. (1990). Computing the thinkable. *Behavioural and Brain Sciences*, **13(4)**, 658.

Chandrasekhar, S. (1987). *Truth and beauty. Aesthetics and motivations in science*. The University of Chicago Press.

Chang, C.-L. and Lee, R. C.-T. (1987). Symbolic logic and mechanical theorem proving, 2nd edn (1st edn 1973). Academic Press, New York.

Chou, S.-C. (1988). Mechanical geometry theorem proving. Ridel.

Christian, J. J. (1994). On definite events in a generally covariant quantum world. Unpublished preprint.

Church, A. (1936). An unsolvable problem of elementary number theory. *Am. Jour. of Math.*, **58**, 345–63.

Church, A. (1941). *The calculi of lambda-conversion*. Annals of Mathematics Studies, No. 6. Princeton University Press.

Churchland, P. M. (1984). *Matter and consciousness*. Bradford Books, MIT Press, Cambridge, Massachusetts.

Clauser, J. F. and Horne, M. A. (1974). Experimental consequences of objective local theories. *Phys. Rev.*, **D10**, 526–35.

Clauser, J. F., Horne, M. A., and Shimony, A. (1978). Bell's theorem: experimental tests and implications. *Rpts. on Prog. in Phys.*, **41**, 1881–927.

Cohen, P. C. (1966). *Set theory and the continuum hypothesis*. Benjamin, Menlo Park, CA.

Conrad, M. (1990). Molecular computing. In *Advances in computers* (ed. M. C. Yovits), Vol. 31. Academic Press, London.

Conrad, M. (1992). Molecular computing: the lock–key paradigm. *Computer* (November 1992), 11–20.

Conrad, M. (1993). The fluctuon model of Force, Life, and computation: a constructive analysis. *Appl. Math. and Comp.*, **56**, 203–59.

Costa de Beauregard, O. (1989). In *Bell's theorem, quantum theory, and conceptions of the universe* (ed. M. Kafatos). Kluwer, Dordrecht.

Craik, K. (1943). *The nature of explanation*. Cambridge University Press.

Crick, F. (1994). *The astonishing hypothesis. The scientific search for the soul*. Charles Scribner's Sons, New York, and Maxwell Macmillan International.

Crick, F. and Koch, C. (1990). Towards a neurobiological theory of consciousness. *Seminars in the Neurosciences*, **2**, 263–75.

Crick, F. and Koch, C. (1992). The problem of consciousness. *Sci. Amer.*, **267**, 110.

Curl, R. F. and Smalley, R. E. (1991). Fullerenes. *Scientific American*, **265**, No. 4, pp. 32–41.

Cutland N. J. (1980). *Computability. An introduction to recursive function theory*. Cambridge University Press.

Davenport, H. (1952). *The higher arithmetic*. Hutchinson's University Library.

Davies, P. C. W. (1974). *The physics of time asymmetry*. Surrey University Press, Belfast.

Davies, P. C. W. (1984). *Quantum mechanics*. Routledge, London.

Davis, M. (ed.) (1965). *The undecidable—basic papers on undecidable propositions, unsolvable problems and computable functions*. Raven Press, Hewlett, New York.

Davis, M. (1978). What is a computation? In *Mathematics today; twelve informal essays* (ed. L. A. Steen). Springer-Verlag, New York.

Davis M. (1990). Is mathematical insight algorithmic? *Behavioural and Brain Sciences*, **13(4)**, 659.

Davis, M. (1993). How subtle is Gödel's theorem? *Behavioural and Brain Sciences*, **16**, 611–12.

Davis, M. and Hersch, R. (1975). Hilbert's tenth problem. *Scientific American* (Nov 1973), 84

Davis, P. J. and Hersch, R. (1982). *The mathematical experience*. Harvester Press.

de Broglie, L. (1956). *Tentative d'intérpretation causale et nonlinéaire de la mécanique ondulatoire*. Gauthier–Villars, Paris.

Deeke, L., Grötzinger, B., and Kornhuber, H. H. (1976). Voluntary finger movements in man: cerebral potentials and theory. *Biol. Cybernetics*, **23**, 99.

del Giudice, E., Doglia, S., and Milani, M. (1983). Self-focusing and ponderomotive forces of coherent electric waves—a mechanism for cytoskeleton formation and dynamics. In *Coherent excitations in biological systems* (ed. H. Fröhlich and F. Kremer). Springer-Verlag, Berlin.

Dennett, D. (1990). Betting your life on an algorithm. *Behavioural and Brain Sciences*, **13(4)**, 660.

Dennett, D. C. (1991). *Consciousness explained*. Little, Brown and Company.

d'Espagnat, B. (1989). *Conceptual foundations of quantum mechanics*, 2nd edn. Addison-Wesley, Reading, Massachusetts.

Deutsch, D. (1985). Quantum theory, the Church–Turing principle and the universal quantum computer. *Proc. Roy. Soc.* (*Lond.*), **A400**, 97–117.

Deutsch, D. (1989). Quantum computational networks. *Proc. Roy. Soc. Lond.*, **A425**, 73–90.

Deutsch, D. (1991). Quantum mechanics near closed time-like lines. *Phys. Rev.*, **D44**, 3197–217.

Deutsch, D. (1992). Quantum computation. *Phys. World*, **5**, 57–61.

Deutsch, D. and Ekert, A. (1993). Quantum communication moves into the unknown. *Phys. World*, **6**, 22–3.

Deutsch, D. and Jozsa, R. (1992). Rapid solution of problems by quantum computation. *Proc. R. Soc. Lond.*, **A439**, 553–8.

Devlin, K. (1988). *Mathematics: the New Golden Age*. Penguin Books, London.

DeWitt, B. S. and Graham, R. D. (ed.) (1973). *The many-worlds interpretation of quantum mechanics*. Princeton University Press.

Dicke, R. H. (1981). Interaction-free quantum measurements: a paradox? *Am. J. Phys.*, **49**, 925–30.

Diósi, L. (1989). Models for universal reduction of macroscopic quantum fluctuations. *Phys. Rev.*, **A40**, 1165–74.

Diósi, L. (1992). Quantum measurement and gravity for each other. In *Quantum chaos, quantum measurement*; NATO AS1 Series C. Math. Phys. Sci 357 (ed. P. Cvitanovic, I. C. Percival, A. Wirzba). Kluwer, Dordrecht.

Dirac, P. A. M. (1947). *The principles of quantum mechanics*, 3rd edn. Oxford University Press.

Dodd, A. (1991). Gödel, Penrose, and the possibility of AI. *Artificial Intelligence Review*, **5**.

Donaldson, S. K. (1983). An application of gauge theory to four dimensional topology. *J. Diff. Geom.*, **18**, 279–315.

Doyle, J. (1990). Perceptive questions about computation and cognition. *Behavioural and Brain Sciences*, **13(4)**, 661.

Dreyfus, H. L. (1972). *What computers can't do*. Harper and Row, New York.

Dummett, M. (1973). *Frege: philosophy of language*. Duckworth, London.

Dustin, P. (1984). *Microtubules*, 2nd revised edn. Springer-Verlag, Berlin.

Dryl, S. (1974). Behaviour and motor responses in paramecium. In *Paramecium—a current survey* (ed. W. J. Van Wagtendonk), pp. 165–218. Elsevier, Amsterdam.

Eccles, J.C. (1973). *The understanding of the brain*. McGraw-Hill, New York.

Eccles, J. C. (1989). *Evolution of the brain: creation of the self*. Routledge, London.

Eccles, J. C. (1992). Evolution of consciousness. *Proc. Natl. Acad. Sci.*, **89**, 7320–4.

Eccles, J. C. (1994). *How the self controls its brain*. Springer-Verlag, Berlin.

Eckert, R., Randall, D., and Augustine, G. (1988). *Animal physiology. Mechanisms and adaptations*, Chapter 11. Freeman, New York.

Eckhorn, R., Bauer, R., Jordan, W., Brosch, M., Kruse, W., Munk, M., and Reitboeck, H. J. (1988). Coherent oscillations: a mechanism of feature linking in the visual cortex? *Biol. Cybern.*, **60**, 121–30.

Edelman, G. M. (1976). Surface modulation and cell recognition on cell growth. *Science*, **192**, 218–26.

Edelman, G. M. (1987). *Neural Darwinism, the theory of neuronal group selection*. Basic Books, New York.

Edelman, G. M. (1988). *Topobiology, an introduction to molecular embryology*. Basic Books, New York.

Edelman, G. M. (1989). *The remembered present. A biological theory of consciousness*. Basic Books, New York.

Edelman, G. M. (1992). *Bright air, brilliant fire: on the matter of the mind*. Allen Lane, The Penguin Press, London.

Einstein, A., Podolsky, P., and Rosen, N. (1935). Can quantum-mechanical description of physical reality be considered complete? In *Quantum theory and measurement* (ed. J. A. Wheeler and W. H. Zurek). Princeton University Press, 1983. Originally in *Phys. Rev.*, **47**, 777–80.

Elitzur, A. C. and Vaidman, L. (1993). Quantum-mechanical interaction-free measurements. *Found. of Phys.*, **23**, 987–97.

Elkies, Noam G. (1988). On $A^4+B^4+C^4=D^4$. *Maths. of Computation*, **51**, (No. 184) 825–35.

Everett, H. (1957). 'Relative State' formulation of quantum mechanics. In *Quantum theory and measurement* (ed. J. A. Wheeler and W. H. Zurek).

Princeton University Press 1983; originally in *Rev. of Modern Physics*, **29**, 454–62.

Feferman, S. (1988). Turing in the Land of O(z). In *The universal Turing machine: a half-century survey* (ed. R. Herken). Kammerer and Unverzagt, Hamburg.

Feynman, R. P. (1948). Space–time approach to non-relativistic quantum mechanics. *Revs. Mod. Phys.*, **20**, 367–87.

Feynman, R. P. (1982). Simulating physics with computers. *Int. J. Theor. Phys.*, **21(6/7)**, 467–88.

Feynman, R. P. (1985). Quantum mechanical computers. *Optics News*, Feb, 11–20.

Feynman, R. P. (1986). Quantum mechanical computers. *Foundations of Physics*, **16(6)**, 507–31.

Fodor, J. A. (1983). *The modularity of mind*. MIT Press, Cambridge, Massachusetts.

Franks, N. P. and Lieb, W. R. (1982). Molecular mechanics of general anaesthesia. *Nature*, **300**, 487–93.

Freedman, D. H. (1994). *Brainmakers*. Simon and Schuster, New York.

Freedman, S. J. and Clauser, J. F. (1972). Experimental test of local hidden-variable theories. In *Quantum theory and measurement* (ed. J. A. Wheeler and W. H. Zurek). Princeton University Press, 1983; originally *Phys. Rev. Lett.*, **28**, 938–41.

Frege, G. (1893). *Grundgesetze der Arithmetik, begriffsschriftlich abelgeleitet*, Vol 1. H. Pohle, Jena.

Frege, G. (1964). *The basic laws of arithmetic*, translated and edited with an introduction by Montgomery Firth. University of California Press, Berkeley.

French, J. W. (1940). Trial and error learning in paramecium. *J. Exp. Psychol.*, **26**, 609–13.

Fröhlich, H. (1968). Long-range coherence and energy storage in biological systems. *Int. Jour. of Quantum. Chem.*, **II**, 641–9.

Fröhlich, H. (1970). Long range coherence and the actions of enzymes. *Nature*, **228**, 1093.

Fröhlich, H. (1975). The extraordinary dielectric properties of biological materials and the action of enzymes, *Proc. Natl. Acad. Sci*, **72(11)**, 4211–15.

Fröhlich, H. (1984). General theory of coherent excitations on biological systems. In *Nonlinear electrodynamics in biological systems* (ed. W. R. Adey and A. F. Lawrence). Plenum Press, New York.

Fröhlich, H. (1986). Coherent excitations in active biological systems. In *Modern bioelectrochemistry* (ed. F. Gutmann and H. Keyzer). Plenum Press, New York.

Fukui, K. and Asai, H. (1976). Spiral motion of paramecium caudatum in small capillary glass tube. *J. Protozool.*, **23**, 559–63.

Gandy, R. (1988). The confluence of ideas in 1936. In *The universal Turing machine: a half-century survey* (ed. R. Herken). Kammerer and Unverzagt, Hamburg.

Gardner, M. (1965). *Mathematical magic show*. Alfred Knopf, New York and Random House, Toronto.

Gardner, M. (1970). Mathematical games: the fantastic combinations of John Conway's new solitaire game 'Life'. *Scientific American*, **223**, 120–3.

Gardner, M. (1989). *Penrose tiles to trapdoor ciphers*. Freeman, New York.

Gelber, B. (1958). Retention in paramecium aurelia. *J. Comp. Physiol. Psych.*, **51**, 110–15.

Gelernter, D. (1994). *The muse in the machine*. The Free Press, Macmillan Inc., New York and Collier Macmillan, London.

Gell-Mann, M. and Hartle, J. B. (1993). Classical equations for quantum systems. *Phys. Rev.*, **D47**, 3345–82.

Gernoth, K. A., Clark, J. W., Prater, J. S., and Bohr, H. (1993). Neural network models of nuclear systematics. *Phys. Lett.*, **B300**, 1–7.

Geroch, R. (1984). The Everett interpretation. *Nous*, **4** (special issue on the foundations of quantum mechanics), 617–33.

Geroch, R. and Hartle, J. B. (1986). Computability and physical theories. *Found. Phys.*, **16**, 533.

Ghirardi, G. C., Rimini, A., and Weber, T. (1980). A general argument against superluminal transmission through the quantum mechanical measurement process. *Lett. Nuovo. Chim.*, **27**, 293–8.

Ghirardi, G. C., Rimini, A., and Weber, T. (1986). Unified dynamics for microscopic and macroscopic systems. *Phys. Rev.*, **D34**, 470.

Ghirardi, G. C., Grassi, R., and Rimini, A. (1990*a*). Continuous-spontaneous-reduction model involving gravity. *Phys. Rev.*, **A42**, 1057–64.

Ghirardi, G. C., Grassi, R., and Pearle, P. (1990*b*). Relativistic dynamical reduction models: general framework and examples. *Foundations of Physics*, **20**, 1271–316.

Ghirardi, G. C., Grassi, R., and Pearle, P. (1992). Comment on 'Explicit collapse and superluminal signals'. *Phys. Lett.*, **A166**, 435–8.

Ghirardi, G. C., Grassi, R., and Pearle, P. (1993). Negotiating the tricky border between quantum and classical. *Physics Today*, **46**, 13.

Gisin, N. (1989). Stochastic quantum dynamics and relativity. *Helv. Phys. Acta.*, **62**, 363–71.

Gisin, N. and Percival, I. C. (1993). Stochastic wave equations versus parallel world components. *Phys. Lett.*, **A175**, 144–5.

Gleick, J. (1987). *Chaos. Making a new science*. Penguin Books.

Glymour, C. and Kelly, K. (1990). Why you'll never know whether Roger Penrose is a computer. *Behavioural and Brain Sciences*, **13(4)**, 666.

Gödel, K. (1931). Über formal unentscheidbare Sätze per Principia Mathematica und verwandter Systeme I. *Monatshefte für Mathematik und Physik*, **38**, 173–98.

Gödel, K. (1940). *The consistency of the axiom of choice and of the generalized continuum-hypothesis with the axioms of set theory*. Princeton University Press and Oxford University Press.

Gödel, K. (1949). An example of a new type of cosmological solution of Einstein's field equations of gravitation. *Rev. of Mod. Phy.*, **21**, 447.

Gödel, K. (1986). *Kurt Gödel, collected works*, Vol. I (publications 1929–1936) (ed. by S. Feferman *et al.*). Oxford University Press.

Gödel, K. (1990). *Kurt Gödel, collected works*, Vol. II (publications 1938–1974) (ed. S. Feferman *et al.*). Oxford University Press.

Gödel, K. (1995). *Kurt Gödel, collected works*, Vol. III (ed. S. Feferman *et al.*). Oxford University Press.

Golomb, S. W. (1965). *Polyominoes*. Scribner and Sons.

Good, I. J. (1965). Speculations concerning the first ultraintelligent machine. *Advances in Computers*, **6**, 31–88.

Good, I. J. (1967). Human and machine logic. *Brit, J. Philos. Sci.*, **18**, 144–7.

Good, I. J. (1969). Gödel's theorem is a red herring. *Brit. J. Philos. Sci.*, **18**, 359–73.

Graham, R. L. and Rothschild, B. L. (1971). Ramsey's theorem for *n*-parameter sets. *Trans. Am. Math. Soc.*, **59**, 290.

Grant, P. M. (1994). Another December revolution? *Nature*, **367**, 16.

Gray, C. M. and Singer, W. (1989). Stimulus-specific neuronal oscillations in orientation columns of cat visual cortex. *Proc. Natl. Acad. Sci. USA*, **86**, 1689–1702.

Grangier, P., Roger, G., and Aspect, A. (1986). Experimental evidence for a photon anticorrelation effect on a beam splitter: a new light on single-photon interferences. *Europhysics Letters*, **1**, 173–9.

Green, D. G. and Bossomaier, T. (ed.) (1993). *Complex systems: from biology to computation*. IOS Press.

Greenberger, D. M., Horne, M. A., and Zeilinger, A. (1989). Going beyond Bell's theorem. In *Bell's theorem, quantum theory, and conceptions of the universe* (ed. M. Kafatos) pp. 73–76. Kluwer Academic, Dordrecht, The Netherlands.

Greenberger, D. M., Horne, M. A., Shimony, A., and Zeilinger, A. (1990). Bell's theorem without inequalities. *Am. J. Phys.*, **58**, 1131–43.

Gregory, R. L. (1981). *Mind in science; a history of explanations in psychology and physics*. Weidenfeld and Nicholson Ltd. (also Penguin, 1984).

Grey Walter, W. (1953). *The living brain*. Gerald Duckworth and Co. Ltd.

Griffiths, R. (1984). Consistent histories and the interpretation of quantum mechanics. *J. Stat. Phys.*, **36**, 219.

Grossberg, S. (ed.) (1987). *The adaptive brain I: Cognition, learning, reinforcement and rhythm* and *The adaptive brain II: Vision, speech, language and motor control.* North-Holland, Amsterdam.

Grünbaum, B. and Shephard, G. C. (1987). *Tilings and Patterns.* Freeman, New York.

Grundler, W. and Keilmann, F. (1983). Sharp resonances in yeast growth proved nonthermal sensitivity to microwaves. *Phys. Rev. Letts.*, **51**, 1214–16.

Guccione, S. (1993). Mind the truth: Penrose's new step in the Gödelian argument. *Behavioural and Brain Sciences*, **16**, 612–13.

Haag, R. (1992). *Local quantum physics: fields, particles, algebras.* Springer-Verlag, Berlin.

Hadamard, J. (1945). *The psychology of invention in the mathematical field.* Princeton University Press.

Hallett, M. (1984). *Cantorian set theory and limitation of size.* Clarendon Press, Oxford.

Hameroff, S. R. (1974). Chi: a neural hologram? *Am. J. Clin. Med.*, **2(2)**, 163–70.

Hameroff, S. R. (1987). *Ultimate computing. Biomolecular consciousness and nano-technology.* North-Holland, Amsterdam.

Hameroff, S. R. and Watt, R. C. (1982). Information in processing in microtubules. *J. Theor. Biol.*, **98**, 549–40.

Hameroff, S. R. and Watt, R. C. (1983). Do anesthetics act by altering electron mobility? *Anesth. Analg.*, **62**, 936–40.

Hameroff, S. R., Rasmussen, S., and Mansson, B. (1988). Molecular automata in microtubles: basic computational logic of the living state? In *Artificial Life, SFI studies in the sciences of complexity* (ed. C. Langton). Addison-Wesley, New York.

Hanbury Brown, R. and Twiss, R. Q. (1954). A new type of interferometer for use in radio astronomy. *Phil. Mag.*, **45**, 663–82.

Hanbury Brown, R. and Twiss, R. Q. (1956). The question of correlation between photons in coherent beams of light. *Nature*, **177**, 27–9.

Harel, D. (1987). *Algorithmics. The spirit of computing.* Addison-Wesley, New York.

Hawking, S. W. (1975). Particle creation by Black Holes. *Commun. Math. Phys.*, **43**, 199–220.

Hawking, S .W. (1982). Unpredictability of quantum gravity. *Comm. Math. Phys.*, **87**, 395–415.

Hawking, S. W. and Israel, W. (ed.) (1987). *300 years of gravitation.* Cambridge University Press.

Hebb, D. O. (1949). *The organization of behaviour.* Wiley, New York.

Hecht, S., Shlaer, S., and Pirenne, M. H. (1941). Energy, quanta and vision. *Journal of General Physiology*, **25**, 891–40.

Herbert, N. (1993). *Elemental mind. Human consciousness and the new physics.* Dutton Books, Penguin Publishing.

Heyting, A. (1956). *Intuitionism: an introduction.* North-Holland, Amsterdam.

Heywood, P. and Redhead, M. L. G. (1983). Nonlocality and the Kochen–Specker Paradox. *Found. Phys.*, **13**, 481–99.

Hodges, A. P. (1983). *Alan Turing: the enigma.* Burnett Books and Hutchinson, London; Simon and Schuster, New York.

Hodgkin, D. and Houston, A. I. (1990). Selecting for the con in consciousness. *Behavioural and Brain Sciences*, **13(4)**, 668.

Hodgson, D. (1991). *Mind matters: consciousness and choice in a quantum world.* Clarendon Press, Oxford.

Hofstadter, D. R. (1979). *Gödel, Escher, Bach: an eternal golden braid.* Harvester Press, Hassocks, Essex.

Hofstadter, D. R. (1981). A conversation with Einstein's brain. In *The mind's I* (ed. D. R. Hofstadter and D. Dennett). Basic Books; Penguin, Harmondsworth, Middlesex.

Hofstadter, D. R. and Dennett, D. C. (ed.) (1981). *The mind's I.* Basic Books; Penguin, Harmondsworth, Middlesex.

Home, D. (1994). A proposed new test of collapse-induced quantum nonlocality. Preprint.

Home, D. and Nair, R. (1994). Wave function collapse as a nonlocal quantum effect. *Phys. Lett.*, **A187**, 224–6.

Home, D. and Selleri, F. (1991). Bell's Theorem and the EPR Paradox. *Rivista del Nuovo Cimento*, **14**, N.9.

Hopfield, J. J. (1982). Neural networks and physical systems with emergent collective computational abilities. *Proc. Natl. Acad. Sci.*, **79**, 2554–8.

Hsu, F.-H., Anantharaman, T., Campbell, M., and Nowatzyk, A. (1990). A grandmaster chess machine. *Scientific American*, **263**.

Huggett, S. A. and Tod, K. P. (1985). *An introduction to twistor theory.* London Math. Soc. student texts. Cambridge University Press.

Hughston, L. P., Jozsa, R., and Wootters, W. K. (1993). A complete classification of quantum ensembles having a given density matrix. *Phys. Letters*, **A183**, 14–18.

Isham, C. J. (1989). Quantum gravity. In *The new physics* (ed. P. C. W. Davies), pp. 70–93. Cambridge University Press.

Isham, C. J. (1994). Prima facie questions in quantum gravity. In *Canonical relativity: classical and quantum* (ed. J. Ehlers and H. Friedrich). Springer-Verlag, Berlin.

Jibu, M., Hagan, S., Pribram, K., Hameroff, S. R., and Yasue, K. (1994). Quantum optical coherence in cytoskeletal microtubules: implications for brain function. *Bio. Systems* (in press).

Johnson-Laird, P. N. (1983). *Mental models.* Cambridge University Press.

Johnson-Laird, P. (1987). How could consciousness arise from the computations of the brain? In *Mindwaves: thoughts on intelligence, identity and consciousness* (ed. C. Blakemore and S. Greenfield). Blackwell, Oxford.

Károlyházy, F. (1966). Gravitation and quantum mechanics of macroscopic bodies. *Nuo. Cim. A*, **42**, 390–402.

Károlyházy, F. (1974). Gravitation and quantum mechanics of macroscopic bodies. *Magyar Fizikai Polyoirat*, **12**, 24.

Károlyházy, F., Frenkel, A., and Lukács, B. (1986). On the possible role of gravity on the reduction of the wave function. In *Quantum concepts in space and time* (ed. R. Penrose and C.J. Isham). Oxford University Press.

Kasumov, A. Y., Kislov, N. A., and Khodos, I. I. (1993). Can the observed vibration of a cantilever of supersmall mass be explained by quantum theory? *Microsc. Microanal. Microstruct.*, **4**, 401–6.

Kentridge, R. W. (1990). Parallelism and patterns of thought. *Behavioural and Brain Sciences.*, **13(4)**, 670.

Khalfa, J. (ed.) (1994). *What is intelligence? The Darwin College lectures.* Cambridge University Press.

Klarner, D. A. (1981). My life among the Polyominoes. In *The mathematical gardner* (ed. D. A. Klarner). Prindle, Weber and Schmidt, Boston MA and Wadsworth Int., Belmont CA.

Kleene, S. C. (1952). *Introduction to metamathematics.* North-Holland, Amsterdam and van Nostrand, New York.

Klein, M. V. and Furtak, T. E. (1986). *Optics*, 2nd edn. Wiley, New York.

Kochen, S. and Specker, E. P. (1967). The problem of hidden variables in quantum mechanics. *J. Math. Mech.*, **17**, 59–88.

Kohonen, T. (1984). *Self-organisation and associative memory.* Springer-Verlag, New York.

Komar, A. B. (1969). Qualitative features of quantized graviation *Int. J. Theor. Phys.*, **2**, 157–60.

Koruga, D. (1974). Microtubule screw symmetry: packing of spheres as a latent bioinformation code. *Ann. NY Acad. Sci.*, **466**, 953–5.

Koruga, D., Hameroff, S., Withers, J., Loutfy, R., and Sundareshan, M. (1993). *Fullerene C_{60}. History, physics, nanobiology, nanotechnology.* North-Holland, Amsterdam.

Kosko, B. (1994). *Fuzzy thinking: the new science of fuzzy logic.* Harper Collins, London.

Kreisel, G. (1960). Ordinal logics and the characterization of informal concepts of proof. *Proc. of the Internat. Cong. of Mathematics, Aug. 1958.* Cambridge University Press.

Kreisel, G. (1967). Informal rigour and completeness proofs. In *Problems in the philosophy of mathematics* (ed. I. Lakatos), pp. 138–86. North-Holland, Amsterdam.

Laguës, M., Xiao Ming Xie, Tebbji, H., Xiang Zhen Xu, Mairet, V., Hatterer,

C., *et al.* (1993). Evidence suggesting superconductivity at 250 K in a sequentially deposited cuprate film. *Science*, **262**, 1850–1.

Lander, L. J. and Parkin, T. R. (1966). Counterexample to Euler's conjecture on sums of like powers. *Bull. Amer. Math. Soc.*, **72**, 1079.

Leggett, A. J. (1984). Schrödinger's cat and her laboratory cousins. *Contemp. Phys.*, **25(6)**, 583.

Lewis, D. (1969). Lucas against mechanism. *Philosophy*, **44**, 231–3.

Lewis, D. (1989). Lucas against mechanism II. *Can. J. Philos.*, **9**, 373–6.

Libet, B. (1990). Cerebral processes that distinguish conscious experience from unconscious mental functions. In *The principles of design and operation of the brain* (ed. J. C. Eccles and O. D. Creutzfeldt), Experimental Brain research series 21, pp. 185–205. Springer-Verlag, Berlin.

Libet, B. (1992). The neural time-factor in perception, volition and free will. *Revue de Métaphysique et de Morale*, **2**, 255–72.

Libet, B., Wright, E. W., Jr., Feinstein, B., and Pearl, D. K. (1979). Subjective referral of the timing for a conscious sensory experience. *Brain*, **102**, 193–224.

Linden, E. (1993). Can animals think? *Time Magazine* (March) 13.

Lisboa, P. G. J. (ed.) (1992). *Neural networks: current applications*. Chapman Hall, London.

Lockwood, M. (1989). *Mind, brain and the quantum*. Blackwell, Oxford.

Longair, M. S. (1993). Modern cosmology—a critical assessment. *Q. J. R. Astr. Soc.*, **34**, 157–99.

Longuet-Higgins, H. C. (1987). Mental processes: studies in cognitive science, Part II. MIT Press, Cambridge, Massachusetts.

Lucas, J. R. (1961). Minds, machines and Gödel. *Philosophy*, **36**, 120–4; reprinted in Alan Ross Anderson (ed.) (1964) *Minds and Machines*. Englewood Cliffs.

Lucas, J. R. (1970). *The freedom of the will*. Oxford University Press.

McCarthy, J. (1979). Ascribing mental qualities to machines. In *Philosophical perspectives in artificial intelligence* (ed. M. Ringle). Humanities Press, New York.

McCulloch, W. S. and Pitts, W. H. (1943). A logical calculus of the idea immanent in nervous activity. *Bull. Math. Biophys.*, **5**, 115–33. (Reprinted in McCulloch, W.S., *Embodiments of mind*, MIT Press, 1965.)

McDermott, D. (1990). Computation and consciousness. *Behavioural and Brain Sciences*, **13(4)**, 676.

MacLennan, B. (1990). The discomforts of dualism. *Behavioural and Brain Sciences*, **13(4)**, 673.

Majorana, E. (1932). Atomi orientati in campo magnetico variabile. *Nuovo Cimento*, **9**, 43–50.

Manaster-Ramer, A., Savitch, W. J., and Zadrozny, W. (1990). Gödel redux. *Behavioural and Brain Sciences*, **13(4)**, 675.

Margulis, L. (1975). *Origins of eukaryotic cells*. Yale University Press, New Haven, CT.

Markov, A. A. (1958). The insolubility of the problem of homeomorphy. *Dokl. Akad. Nauk. SSSR*, **121**, 218–20.

Marr, D. E. (1982). *Vision: a computational investigation into the human representation and processing of visual information*. Freeman, San Francisco.

Marshall, I. N. (1989). Consciousness and Bose–Einstein condensates. *New Ideas in Psychology*, **7**.

Mermin, D. (1985). Is the moon there when nobody looks? Reality and the quantum theory. *Physics Today*, **38**, 38–47.

Mermin, D. (1990). Simple unified form of the major no-hidden-variables theorems. *Phys. Rev. Lett.*, **65**, 3373–6.

Michie, D. and Johnston, R. (1984). *The creative computer. Machine intelligence and human knowledge*. Viking Penguin.

Minsky, M. (1968). Matter, mind and models. In *Semantic information processing* (ed. M. Minsky). MIT Press, Cambridge, Massachusetts.

Minsky, M. (1986). *The society of mind*. Simon and Schuster, New York.

Minsky, M., and Papert, S. (1972). *Perceptrons: an introduction to computational geometry*. MIT Press, Cambridge, Massachusetts.

Misner, C. W., Thorne, K. S., and Wheeler, J. A. (1973). *Gravitation*. Freeman, New York.

Moore, A. W. (1990). *The infinite*. Routledge, London.

Moravec, H. (1988). *Mind children: the future of robot and human intelligence*. Harvard University Press, Cambridge, Massachusetts.

Moravec, H. (1994). *The Age of Mind: transcending the human condition through robots*. In press.

Mortensen, C. (1990). The powers of machines and minds. *Behavioural and Brain Sciences*, **13(4)**, 678.

Mostowski, A. (1957). *Sentences undecidable in formalized arithmetic: an exposition of the theory of Kurt Gödel*. North-Holland, Amsterdam.

Nagel, E. and Newman, J. R. (1958). *Gödel's proof*. Routledge and Kegan Paul.

Newell, A. and Simon, H. A. (1976). Computer science as empirical enquiry: symbols and search. *Communications of the ACM*, **19**, 113–26.

Newell, A., Young, R., and Polk, T. (1993). The approach through symbols. In *The simulation of human intelligence* (ed. D. Broadbent). Blackwell, Oxford.

Newton, I. (1687). *Philosophiae Naturalis Principia Mathematica*. Reprint: Cambridge University Press.

Newton, I. (1730). *Opticks*. 1952, Dover, New York.

Oakley, D. A. (ed.) (1985). *Brain and mind*. Methuen, London.

Obermayer, K., Teich, W. G., and Mahler, G. (1988*a*). Strutural basis of multistationary quantum systems. I. Effective single-particle dynamics. *Phys. Rev.*, **B37**, 8096–110.

Obermayer, K., Teich, W. G., and Mahler, G. (1988*b*). Structural basis of multistationary quantum systems. II. Effetive few-particle dynamics. *Phys. Rev.*, **B37**, 8111–121.

Omnès, R. (1992). Consistent interpretations of quantum mechanics *Rev. Mod. Phys.*, **64**, 339–82.

Pais, A. (1991). *Neils Bohr's times*. Clarendon Press, Oxford.

Pauling L. (1964). The hydrate microcrystal theory of general anesthesia. *Anesth. Analg.*, **43**, 1.

Paz, J. P. and Zurek, W. H. (1993). Environment induced-decoherence, classicality and consistency of quantum histories. *Phys. Rev.*, **D48(6)**, 2728–38.

Paz, J. P., Habib, S., and Zurek, W. H. (1993). Reduction of the wave packet: preferred observable and decoherence time scale. *Phys. Rev.*, **D47(2)**, 3rd Series, 488–501.

Pearle, P. (1976). Reduction of the state-vector by a nonlinear Schrödinger equation. *Phys. Rev.*, **D13**, 857–68.

Pearle, P. (1989). Combining stochastic dynamical state-vector reduction with spontaneous localization. *Phys. Rev.*, **A39**, 2277–89.

Pearle, P. (1992). Relativistic model statevector reduction. In *Quantum chaos—quantum measurement*, NATO Adv. Sci. Inst. Ser. C. Math. Phys. Sci. 358 (Copenhagen 1991). Kluwer, Dordrecht.

Peat, F. D. (1988). *Superstrings and the search for the theory of everything*. Contemporary Books, Chicago.

Penrose, O. (1970). *Foundations of statistical mechanics: a deductive treatment*. Pergamon, Oxford.

Penrose, O. and Onsager, L. (1956). Bose–Einstein condensation and liquid helium. *Phys. Rev.*, **104**, 576–84.

Penrose, R. (1980). On Schwarzschild causality—a problem for 'Lorentz covariant' general relativity. In *Essays in general relativity* (A. Taub Festschrift) (ed. F. J. Tipler), pp. 1–12. Academic Press, New York.

Penrose, R. (1987). Newton, quantum theory and reality. In *300 Years of gravity* (ed. S.W. Hawking and W. Israel). Cambridge University Press.

Penrose, R. (1990). Author's response, *Behavioural and Brain Sciences*, **13(4)**, 692.

Penrose, R. (1991*a*). The mass of the classical vacuum. In *The philosophy of vacuum* (ed. S. Saunders and H. R. Brown). Clarendon Press, Oxford.

Penrose, R. (1991*b*). Response to Tony Dodd's 'Gödel, Penrose, and the possibility of AI'. *Artificial Intelligence Review*, **5**, 235.

Penrose, R. (1993*a*). Gravity and quantum mechanics. In *General relativity* *and gravitation 1992. Proceedings of the Thirteenth International Conference on General Relativity and Gravitation held at Cordoba, Argentina 28 June–4 July 1992. Part 1: Plenary lectures.* (ed. R. J. Gleiser, C. N. Kozameh, and O. M. Moreschi). Institute of Physics Publications, Bristol.

Penrose, R. (1993*b*). Quantum non-locality and complex reality. In *The Renaissance of general relativity* (in honour of D. W. Sciama) (ed. G. Ellis, A. Lanza, and J. Miller). Cambridge University Press.

Penrose, R. (1993*c*). Setting the scene: the claim and the issues. In *The simulation of human intelligence* (ed. D. Broadbent). Blackwell, Oxford.

Penrose, R. (1993*d*). An emperor still without mind, *Behavioural and Brain Sciences*, **16**, 616–22.

Penrose, R. (1994*a*). On Bell non-locality without probabilities: some curious geometry. In *Quantum reflections* (in honour of J. S. Bell) (ed. J. Ellis and A. Amati). Cambridge University Press.

Penrose, R. (1994*b*). Non-locality in and objectivity in quantum state reduction. In *Fundamental aspects of quantum theory* (ed. J. Anandan and J. L. Safko). World Scientific, Singapore.

Penrose, R. and Rindler, W. (1984). *Spinors and space–time*, Vol. 1: *Two-spinor calculus and relative fields.* Cambridge University Press.

Penrose, R. and Rindler, W. (1986). *Spinors and space–time*, Vol. 2: *Spinor and twistor methods in space–time geometry.* Cambridge University Press.

Percival, I. C. (1994). Primary state diffusion. *Proc. R. Soc. Lond.*, **A**, submitted.

Peres, A. (1985). Reversible logic and quantum computers. *Phys. Rev.*, **A32(6)**, 3266–76.

Peres, A. (1990). Incompatible results of quantum measurements. *Phys. Lett.*, **A151**, 107–8.

Peres, A. (1991). Two simple proofs of the Kochen–Specker theorem. *J. Phys. A: Math. Gen*, **24**, L175–L178.

Perlis, D. (1990). The emperor's old hat. *Behavioural and Brain Sciences*, **13(4)**, 680.

Planck, M. (1906). *The theory of heat radiation* (trans M. Masius, based on lectures delivered in Berlin, in 1906/7). 1959, Dover, New York.

Popper, K. R. and Eccles, J. R. (1977). *The self and its brain.* Springer International.

Post, E.L. (1936). Finite combinatory processes-formulation I, *Jour. Symbolic Logic*, **1**, 103–5.

Poundstone, W. (1985). *The recursive universe: cosmic complexity and the limits of scientific knowledge.* Oxford University Press.

Pour-El, M. B. (1974). Abstract computability and its relation to the general purpose analog computer. (Some connections between logic, differential equations and analog computers.) *Trans. Amer. Math. Soc.*, **119**, 1–28.

Pour-El, M. B. and Richards, I. (1979). A computable ordinary differential equation which possesses no computable solution, *Ann. Math. Logic*, **17**, 61–90.

Pour-El, M. B. and Richards, I. (1981). The wave equation with computable initial data such that its unique solution is not computable. *Adv. in Math.*, **39**, 215–39.

Pour-El, M. B. and Richards, I. (1982). Noncomputability in models of physical phenomena. *Int. J. Theor. Phys.*, **21**, 553–5.

Pour-El, M. B. and Richards, J. I. (1989). *Computability in analysis and physics*. Springer-Verlag, Berlin.

Pribram, K. H. (1966). Some dimensions of remembering: steps toward a neuropsychological model of memory. In *Macromolecules and behaviour* (ed. J. Gaito), pp. 165–87. Academic Press, New York.

Pribram, K. H. (1975). Toward a holonomic theory of perception. In *Gestalttheorie in der modern psychologie* (ed. S. Ertel), pp. 161–84. Erich Wengenroth, Köln.

Pribram K. H. (1991). *Brain and perception: holonomy and structure in figural processing*. Lawrence Erlbaum Assoc., New Jersey.

Putnam, H. (1960). Minds and machines. In *Dimensions of mind* (ed. S. Hook), New York symposium. Reprinted in *Minds and machines* (ed. A. R. Anderson), pp. 43–59, Prentice-Hall, 1964; also reprinted in *Dimensions of mind: a symposium* (*Proceedings of the third annual NYU Institute of Philosophy*), pp. 148–79, NYU Press, 1964.

Ramon y Cajal, S. (1955). *Studies on the cerebral cortex* (trans. L. M. Kroft). Lloyd-Luke, London.

Redhead, M. L. G. (1987). *Incompleteness, nonlocality, and realism*. Clarendon Press, Oxford.

Rosenblatt, F. (1962). *Principles of neurodynamics*. Spartan Books, New York.

Roskies, A. (1990). Seeing truth or just seeming true?, *Behavioural and Brain Sciences*, **13(4)**, 682.

Rosser, J. B. (1936). Extensions of some theorems of Gödel and Church *Jour. Symbolic Logic*, **1**, 87–91.

Rubel, L. A. (1985). The brain as an analog computer. *J. Theoret. Neurobiol.*, **4**, 73–81.

Rubel, L. A. (1988). Some mathematical limitations of the general-purpose analog computer. *Adv. in Appl. Math.*, **9**, 22–34.

Rubel, L. A. (1989). Digital simulation of analog computation and Church's thesis. *Jour. Symb. Logic.*, **54(3)**, 1011–17.

Rucker, R. (1984). *Infinity and the mind: the science and philosophy of the infinite*. Paladin Books, Granada Publishing Ltd., London. (First published by Harvester Press Ltd., 1982.)

Sacks, O. (1973). *Awakenings*. Duckworth, London.

Sacks, O. (1985). *The man who mistook his wife for a hat*. Duckworth, London.

Sagan, L. (1967). On the origin of mitosing cells. *J. Theor. Biol.*, **14**, 225–74.

Sakharov, A. D. (1967). Vacuum quantum fluctuations in curved space and the theory of gravitation. *Doklady Akad, Nauk SSSR*, **177**, 70–1. English translation: *Sov. Phys. Doklady*, **12**, 1040–1 (1968).

Schrödinger, E. (1935*a*). 'Die gegenwartige Situation in der Quantenmechanik'. *Naturwissenschaftenp*, **23**, 807–12, 823–8, 844–9. (Translation by J. T. Trimmer (1980) in *Proc. Amer. Phil. Soc.*, **124**, 323–38.) In *Quantum theory and measurement* (ed. J. A. Wheeler and W. H. Zurek). Princeton University Press, 1983.

Schrödinger, E. (1935*b*). Probability relations between separated systems. *Proc. Camb. Phil. Soc.*, **31**, 555–63.

Schrödinger, E. (1967). '*What is Life?*' and '*Mind and Matter*' Cambridge University Press.

Schroeder, M. (1991). *Fractals, chaos, power laws. Minutes from an infinite paradise*. Freeman, New York.

Scott, A. C. (1973). Information processing in dendritic trees. *Math. Bio. Sci*, **18**, 153–60.

Scott, A. C. (1977). *Neurophysics*. Wiley Interscience, New York.

Searle, J. R. (1980). Minds, brains and programs. In *The behavioral and brain sciences*, Vol. 3. Cambridge University Press. (Reprinted in *The mind's I* (ed. D. R. Hofstadter and D. C. Dennett). Basic Books, Inc.; Penguin Books Ltd., Harmondsworth, Middlesex, 1981.)

Searle, J. R. (1992). *The rediscovery of the mind*. MIT Press, Cambridge, Massachusetts.

Seymore, J. and Norwood, D. (1993). A game for life. *New Scientist*, **139**, No. 1889, 23–6.

Sheng, D., Yang, J., Gong, C., and Holz, A. (1988). A new mechanism of high Tc superconductivity. *Phys. Lett.*, **A133**, 193–6.

Sloman, A. (1992). The emperor's real mind: review of Roger Penrose's The Emperor's New Mind. *Artificial Intelligence*, **56**, 355–96.

Smart, J. J. C. (1961). Gödel's theorem, Church's theorem and mechanism. *Synthèse*, **13**, 105–10.

Smith, R. J. O. and Stephenson, J. (1975). *Computer simulation of continuous systems*. Cambridge University Press.

Smith, S., Watt, R. C., and Hameroff, S. R. (1984). Cellular automata in cytoskeletal lattice proteins. *Physica D*, **10**, 168–74.

Smolin, L. (1993). What have we learned from non-pertubative quantum gravity? In *General relativity and gravitation 1992. Proceedings of the thirteenth international conference on GRG, Cordoba Argentina* (ed. R. J. Gleiser, C. N. Kozameh, and O. M. Moreschi). Institute of Physics Publications, Bristol.

Smolin, L. (1994). Time, structure and evolution in cosmology. In *Temponelle scienziae filosofia* (ed. E. Agazzi). Word Scientific, Singapore.

Smorynski, C. (1975). *Handbook of mathematical logic*. North-Holland, Amsterdam.

Smorynski, C. (1983). 'Big' news from Archimedes to Friedman. *Notices Amer. Math. Soc.*, **30**, 251–6.

Smullyan, R. (1961). *Theory of Formal Systems*. Princeton University Press.

Smullyan, R. (1992). *Gödel's incompleteness theorem*, Oxford Logic Guide No. 19. Oxford University Press.

Squires, E. J. (1986). *The mystery of the quantum world*. Adam Hilger Ltd., Bristol.

Squires, E. J. (1990). On an alleged proof of the quantum probability law. *Phys. Lett.*, **A145**, 67–8.

Squires, E. J. (1992*a*). Explicit collapse and superluminal signals. *Phys. Lett.*, **A163**, 356–8.

Squires, E. J. (1992*b*). History and many-worlds quantum theory. *Found. Phys. Lett.*, **5**, 279–90.

Stairs, A. (1983). Quantum logic, realism and value-definiteness. *Phil. Sci.*, **50(4)**, 578–602.

Stapp, H. P. (1979). Whiteheadian approach to quantum theory and the generalized Bell's theorem. *Found. Phys.*, **9**, 1–25.

Stapp, H. P. (1993). *Mind, matter, and quantum mechanics*. Springer-Verlag, Berlin.

Steen, L. A. (ed.) (1978). *Mathematics today: twelve informal essays*. Springer-Verlag, Berlin.

Stoney, G. J. (1881). On the physical units of nature. *Phil. Mag.*, (Series 5) **11**, 381.

Stretton, A. O. W., Davis, R. E., Angstadt, J. D., Donmoyer, J. E., Johnson, C. D., and Meade, J. A. (1987). Nematode neurobiology using Ascaris as a model system. *J. Cellular Biochem.*, **511A**, 144.

Thorne, K. S. (1994). *Black holes & time warps: Einstein's outrageous legacy*. W.W. Norton and Company, New York.

Torrence, J. (1992). *The concept of nature. The Herbert Spencer lectures*. Clarendon Press, Oxford.

Tsotsos, J. K. (1990). Exactly which emperor is Penrose talking about? *Behavioural and Brain Sciences*, **13(4)**, 686.

Turing, A. M. (1937). On computable numbers, with an application to the Entscheidungsproblem. *Proc. Lond. Math. Soc.* (*ser. 2*), **42**, 230–65; a correction **43**, 544–6.

Turing, A. M. (1939). Systems of logic based on ordinals. *P. Lond. Math. Soc.*, **45**, 161–228.

Turing, A. M. (1950). Computing machinery and intelligence. *Mind*, **59** No. 236; reprinted in *The mind's I* (ed. D. R. Hofstadter and D. C. Dennett), Basic Books; Penguin, Harmondsworth, Middlesex, 1981.

Turing, A. M. (1986). Lecture to the London Mathematical Society on 20

February 1947. In *A. M. Turing's ACE report of 1946 and other papers* (ed. B. E. Carpenter and R.W. Doran). The Charles Babbage Institute, vol. 10, MIT Press, Cambridge, Massachusetts.

von Neumann, J. (1932). *Mathematische Grundlagen der Quantenmechanik*, Springer-Verlag, Berlin; Engl. trans: *Mathematical foundations of quantum mechanics*. Princeton University Press, 1955.

von Neumann, J. and Morgenstern, O. (1944). *Theory of games and economic behaviour*. Princeton University Press.

Waltz, D. L. (1982). Artificial intelligence. *Scientific American*, **247**(4), 101–22.

Wang, Hao (1974). *From mathematics to philosophy*. Routledge, London.

Wang, Hao (1987). *Reflections on Kurt Gödel*. MIT Press, Cambridge, Massachusetts.

Wang, Hao (1993). On physicalism and algorithmism: can machines think? *Philosophia mathematica* (Ser. III), pp. 97–138.

Ward, R. S. and Wells, R. O. Jr (1990). *Twistor geometry and field theory*. Cambridge University Press.

Weber, J. (1960). Detection and generation of gravitational waves. *Phys. Rev.*, **117**, 306.

Weinberg, S. (1977). *The first three minutes: a modern view of the origin of the universe*. Andre Deutsch, London.

Werbos, P. (1989). Bells' theorem; the forgotten loophole and how to exploit it. In *Bell's theorem, quantum theory, and conceptions of the universe* (ed. M. Kafatos). Kluwer, Dordrecht.

Wheeler, J. A. (1957). Assessment of Everett's 'relative state' formulation of quantum theory. *Revs. Mod. Phys.*, **29**, 463–5.

Wheeler, J. A. (1975). On the nature of quantum geometrodynamics. *Annals of Phys.*, **2**, 604–14.

Wigner, E. P. (1960). The unreasonable effectiveness of mathematics. *Commun. Pure Appl. Math.*, **13**, 1–14.

Wigner, E. P. (1961). Remarks on the mind–body question. In *The scientist speculates* (ed. I. J. Good). Heinemann, London. (Reprinted in E. Wigner (1967), *Symmetries and reflections*. Indiana University Press, Bloomington; and in *Quantum theory and measurement* (ed. J.A. Wheeler and W.H. Zurek) Princeton University Press, 1983.)

Wilensky, R. (1990). Computability, consciousness and algorithms, *Behavioural and Brain Sciences*, **13(4)**, 690.

Will, C. (1988). *Was Einstein right? Putting general relativity to the test*. Oxford University Press.

Wolpert, L. (1992). *The unnatural nature of science*. Faber and Faber, London.

Woolley, B. (1992). *Virtual worlds*. Blackwell, Oxford.

Wykes, A. (1969). *Doctor Cardano. Physician extraordinary*. Frederick Muller.

Young, A. M. (1990). *Mathematics, physics and reality*. Robert Briggs Associates, Portland, Oregon.

Zeilinger, A., Gaehler, R., Shull, C. G., and Mampe, W. (1988). Single and double slit diffraction of neutrons. *Revs. Mod. Phys.*, **60**, 1067.

Zeilinger, A., Horne, M. A., and Greenberger, D. M. (1992). Higher-order quantum entanglement. In *Squeezed states and quantum uncertainty* (ed. D. Han, Y. S. Kim, and W. W. Zachary), NASA Conf. Publ. 3135. NASA, Washington, DC.

Zeilinger, A., Zukowski, M., Horne, M. A., Bernstein, H. J., and Greenberger, D. M. (1994). Einstein–Podolsky–Rosen correlations in higher dimensions. In *Fundamental aspects of quantum theory* (ed. J. Anandan and J. L. Safko). World Scientific, Singapore.

Zimba, J. (1993). Finitary proofs of contextuality and nonlocality using Majorana representation of spin-3/2 states, M.Sc. thesis, Oxford.

Zimba, J. and Penrose, R. (1993). On Bell non-locality without probabilities: more curious geometry. *Stud. Hist. Phil. Sci.*, **24(5)**, 697–720.

Zohar, D. (1990). *The quantum self. Human nature and consciousness defined by the New Physics*. William Morrow and Company, Inc., New York.

Zohar, D. and Marshall, I. (1994). *The quantum society. Mind, physics and a new social vision*. Bloomsbury, London.

Zurek, W. H. (1991). Decoherence and the transition from quantum to classical. *Physics Today*, **44** (No. 10), 36–44.

Zurek, W. H. (1993). Preferred states, predictability, classicality and the environment-induced decoherence. *Prog. of Theo. Phys.*, **89(2)**, 281–302.

Zurek, W. H., Habib, S., and Paz, J. P. (1993). Coherent states via decoherence. *Phys. Rev. Lett.*, **70(9)**, 1187–90.

INDEX

Also available in Vintage

Roger Penrose

THE EMPEROR'S NEW MIND

Winner of the 1990 Science Book Prize

Arguing against artificial intelligence, and exploring the mystery of the mind and consciousness, Roger Penrose takes the reader on the most engaging and creative tour of modern physics, cosmology, mathematics and philosophy that has ever been written.

'In *The Emperor's New Mind*, a bold, brilliant, groundbreaking work, he argues that we lack a fundamentally important insight into physics, without which we will never be able to comprehend the mind. Moreover, he suggests, this insight may be the same one that will be required before we can write a unified theory of everything. This is an astonishing claim...'
New York Times Book Review

'The reader might feel privileged indeed to accompany Penrose on his magical mystery tour'
Sunday Times

Also available in Vintage

Timothy Ferris

COMING OF AGE IN THE MILKY WAY

Timothy Ferris takes us from Aristotle to the Big Bang, and tells the story of man's attempts to understand the universe in the best book on the subject since Carl Sagan's *Cosmos*.

'Timothy Ferris is a very good expositor of science... *Coming of Age in the Milky Way* is a very readable account of the remarkable progress we have made towards understanding the universe'
Stephen Hawking

'This spell-binding popular history of scientific discovery...quite literally expands the reader's horizons and fills one with awe at the power of scientifically creative human intelligence'
New York Times

'The prose...sings and shines'
New Scientist

Also available in Vintage

Lawrence Krauss

THE FIFTH ESSENCE

The Search for Dark Matter in the Universe

In this superbly written book, Professor Krauss not only sets out in diamond-clear prose arguments for the existence of dark matter, he also explains what methods are being devised by scientists to find it. A professor of physics and astronomy at Yale, Krauss himself is intimately involved in the search.

'Particle Astrophysics is the science of the 90s and "dark matter" its holy grail, to which *The Fifth Essence* provides a lucid, comprehensive and eminently readable introduction'

Joseph Silk, author of *The Big Bang*

'This book is about the mystery of the missing mass. Either we have failed to see 99% of the universe, or we are wrong about how the universe began'

Stephen Hawking